# International Conference on Measurement Instrumentation and Electronics 2016 (ICMIE 2016)



MATEC Web of Conferences Volume 75 (2016)

Munich, Germany
6 - 8 June 2016

**Editors:**

**Axel Sikora**
**Ben Choi**
**Wang Shuhui**

ISBN: 978-1-5108-2928-2

**Printed from e-media with permission by:**

Curran Associates, Inc.
57 Morehouse Lane
Red Hook, NY 12571

**Some format issues inherent in the e-media version may also appear in this print version.**



Printed by Curran Associates, Inc. (2016)

For additional information, please contact EDP Sciences – Web of Conferences
at the address below.

EDP Sciences – Web of Conferences
17, Avenue du Hoggar
Parc d'Activité de Courtabœuf
BP 112
F-91944 Les Ulis Cedex A
France

Phone: +33 (0) 1 69 18 75 75
Fax: +33 (0) 1 69 28 84 91

contact-edps@webofconferences.org

**Additional copies of this publication are available from:**

Curran Associates, Inc.
57 Morehouse Lane
Red Hook, NY 12571 USA
Phone: 845-758-0400
Fax: 845-758-2633
Email: curran@proceedings.com
Web: www.proceedings.com

# TABLE OF CONTENTS

ISSN (electronic edition): 2261-236X

# Conference Committees

**Conference Chairs**

Prof. Axel Sikora, Offenburg University of Applied Sciences, Germany

Prof. Ben Choi, Computer Science Louisiana Tech University, USA

**Program Committee Chairs**

Dr. Hadj Hamma Tadjine, Technical University of Clausthal Germany, Germany

Prof. Wang Shuhui, Tongji University, China

**Technical Committee**

Assoc. Prof. WojciechGiernacki, Poznań University of Technology, Poland

Assoc. Prof. Lamia Youb, University of Batna, Algeria

Asst. Prof. Paulo Gil, University of Coimbra, Portugal

Assoc. Prof. Vladimir Shurenkov, National Research Nuclear University MEPhI, Russia

Prof. Dr. vimala juliet, SRM University, India

Prof. Yong Yue, University of Bedfordshire, UK

Prof. Anithamary Johnson, Karunya University, India

Ing. Petr Doležel, Slovak University of Technology, Slovak

Dr. Tahmina Ajmal, University of Bedfordshire, UK

Dr. Masood Ur Rehman, University of London, UK

Prof. Song Qiang, Beijing Institute of Technology, China

Prof. Takahiro Numai, Ritsumeikan University, Japan

Prof. Koichi Takiguchi, Ritsumeikan University, Japan

Prof. Jia-Yush Yen, National Taiwan University, Taiwan

Dr. Kang Li, National Taiwan University, Taiwan

Prof. Kuan-Kung Chung, Changhua Taiwan Normal University, Taiwan

Assis. Prof. Veysel Gokhan Bocekci, Marmara University, Turkey

Assis. Prof. Kazim Yildiz, Marmara University, Turkey

Assis. Prof. Zaib Hussain, University of the Punjab, Pakistan

Dr. Farzin Shama, Islamic Azad University, Iran

Prof. Khaled Nagaty, British University in Egypt

Prof. Mounia Zaim, Mohammed V University in Rabat Morocco, Morocco

Prof. Mohamed Sbihi, Mohammed V University in Rabat Morocco, Morocco

# Preface

2016 International Conference on Measurement Instrumentation and Electronics provides a forum for accessing to the most up-to-date and authoritative knowledge from both industrial and academic worlds, sharing best practice in the field of measurement instrumentation and electronic.

The objective of the conference is to address and deliberate on the latest technical status and recent trends in the research and applications of measurement instrumentation and electronics. This conference has been designed with the view to provide an opportunity for researchers and engineers from both industrial and academic to interact and exchange their new ideas and research outcomes for future collaboration.

The first ICMIE was held in Munich, Germany during June 6-8, 2016. ICMIE 2016 was held with presentations delivered by researchers and scholars from the international community, including keynote speeches and highly selective lectures.

The present volume contains selected papers which were submitted to this conference and provides most comprehensive and worldwide state-of-the art knowledge in the field of measurement instrumentation and electronics. All papers included in this collection had undergone the careful peer-review by the conference technical committee members and international reviewers before they're selected for publication. Studies presented in this volume cover the following 10 research areas: electronic instruments and measurement technology, material properties analysis and measurement, signal processing and pattern recognition, mechanical automation, theory of measurement error, network technology, artificial intelligence, computer application technology, system modeling and analysis, the technology of electronic circuit.

We would like to take this opportunity to thank many people. First and foremost we want to express my deep appreciation to keynote speakers, session chairs, as well as all the reviewers for their efforts and kind help in this congress. Final thanks go to all authors and participants at ICMIE 2016 for contributing to the conference and make it a successful event.

Prof. Axel Sikora, Offenburg University of Applied Sciences, Germany
Prof. Ben Choi, Computer Science Louisiana Tech University, USA

June 2016.

edp sciences

MATEC Web of Conferences

# Statement of Peer review

In submitting conference proceedings to MATEC Web of Conferences, I certify to the Publisher that I adhere to the **Policy on Publishing Integrity** of the journal in order to safeguard good scientific practice in publishing.

1. All articles have been subjected to peer review administered by the proceedings editors.
2. Reviews have been conducted by expert referees, who have been requested to provide unbiased and constructive comments aimed, whenever possible, at improving the work.
3. Proceedings editors have taken all reasonable steps to ensure the quality of the materials they publish and their decision to accept or reject a paper for publication has been based only on the merits of the work and the relevance to the journal.

Title, date and place of the conference

2016 International Conference on Measurement Instrumentation and Electronics

Munich, German

June 6-8, 2016

Proceedings editor(s):

Prof. Dr.-Ing. Axel Sikora
Offenburg University of Applied Sciences, Germany
Hahn-Schickard, Germany

Dr. Ben Choi
Computer Science Louisiana Tech University, United States

Dr. Wang Shuhui, Tongji University, China

Date and editor's signature

# Hardware design of a submerged buoy system based on electromagnetic inductive coupling

Dalei Song [1], Shan Gao [1], Ming Xu [2], Xiangdong Wang [1] and Yutao Wang [2]

[1]*Department of Automation, Engineering College, Ocean University of China, Qingdao 266100, China*

[2]*College of Information Science and Technology, Ocean University of China, Qingdao 266100, China*

**Abstract.** This paper mainly introduces the hardware design of a new type of ocean buoy for multi-scale marine dynamic process. The buoy system can collect a number of real-time marine environment data and then transmit all the data back to the landing site through wireless module. The authors mainly designed the hardware circuit of the buoy system, including data collection system, data communication system, data storage system. Due to the buoy system will complete the marine observation work continuously for at least a month, so we add the low power consumption function which can realize the intermittent work for the data collection system. This paper also introduces the electromagnetic induction coupling technology of underwater sensors, the sea surface communication network technology, etc. The system can also extends to the ecological regional anomaly monitoring and the early warning of disaster weather.

## 1 Introduction

As an important support of marine science, marine observation technology is very important to study of marine internal mechanism. In recent years, marine instruments such as ocean turbulence, Argo develops rapidly. Because the marine environment is relatively complex and long-term observation is difficult, some marine observation instruments, such as Argo is chosen the self-contained design [1]. In comparison, the ocean buoy is a very effective method for long-term real-time monitoring of the marine environment. However, most of the observed buoy select one perspective of unilateral physics to observe part of the sea, what's more, most of the message is transmitted based on satellite communication system which has poor real-time performance [2]. This will bring a certain delay for early warning of real-time marine disaster.

This paper introduces a hardware design of a buoy system for multi-scale marine process observation. The buoy system can collect data such as the temperature and velocity from different depth and different sampling point, which uses stable and reliable electromagnetic induction coupling. The communication function of the observation buoy system depends on the point-to-point wireless communication module, and it realizes network construction in the sea using the wireless module in accordance with the self-developed protocol. This paper mainly introduces the hardware design of the buoy system, including real-time data acquisition module for various sensors, multiple storage module, real-time data transmission module etc. Moreover, this paper also puts forwards the low-power mode design of microcontroller, which plays a vital role in the buoy system for long-term submersible deployment.

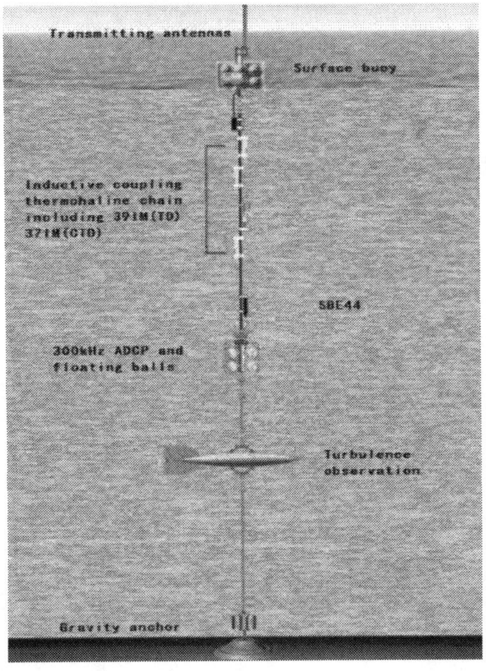


**Figure 1.** Diagram of the buoy system deployment.

 DOI: 10.1051/matecconf/20167501001

## 2 Design of buoy system

In order to realize the multi-scale ocean dynamic process observation, the buoy system will carry a variety of sensors, and according to different sampling points, changes in the physical parameters of real-time observations in various regions can be read on shore. The buoy system is mainly composed of the wireless communication device, floating body, a set of Acoustic Doppler Current Profiler, 4 sets of the CTD sensor, 16 sets of the TD sensor, a turbulence instrument, a GPS system and mooring recycling system. Figure 1 shows the deployment of all the sensors of the system.

The buoy system mainly collects real-time TD, CTD, ADCP and GPS data. Among them, TD and CTD sensors all work by electromagnetic induction coupling. Due to the ADCP has no function of inductive coupling, we use SBE44 to achieve transparent transmission. Electromagnetic induction coupling module also includes IMM module, power supply module. In the laboratory test, we use the loop wire to go through IMM coupler. Of course, in order to ensure the reliability of future sea trial, we will use the plastic coated steel pipe instead of loop wire to connect all the sensors. The turbulence observation instrument of the buoy system is designed and tested alone by Ocean University of China [3], which can realize micro-scale marine environment observation. Because of the large amount of marine data, the turbulence observation instrument is self-contained designed and does not participate in the real-time communication. The instrument can simultaneously capture turbulence, three direction acceleration and angular velocity and depth data synchronous sampling. The turbulence observation instrument circuit design is based on ARM7 processor as the core control chip, and the turbulent signal is obtained via a process of charge amplifier, differential, filtering, voltage amplifier and filtering conversion by the shear probe. All the sensors and instruments will be hung in order across the plastic coated steel pipe and then launched in the sea with the mooring recycling system.

The buoy system uses network construction in the sea to communicate with the wireless module on shore. We choose FGR2 wireless communication module to construct network, which mainly contains FGR2 transmission unit, protocol controlled module and antenna module. It depends on its long distance transmission function, strong anti-interference ability and high transmission rate to achieve network construction in the sea. The main principle of the communication method is to use a FGR2 as the master node and the other FGR2 as slave node fixed to each buoy system. The master node will send handshake request packet continuously for 10 times to node network by polling each buoy sequentially. If the slave nodes receive handshake request package, they will respond handshake response package to the master node immediately. As long as the master node received the response more than 5 times, it grants the slave nodes a network access. After adding slave nodes in the network, the master node interact with slave nodes data in accordance with our self-developed network protocol and the specified data format. The slave nodes will send the collected real-time data to the master node, and the master node will parse data at the same time, then the slave node will wait for a next data request from the master node.

## 3 Hardware design

For the hardware design of the buoy system, the most important section consists of two parts. On the one hand, it needs to acquire real-time data of the underwater sensors, including the TD, CTD, ADCP and GPS signals; on the other hand, it needs to transmit collected data to the wireless communication module according to the network construction protocol in accordance with the requirements for a certain period of time. The hardware block diagram of the whole system is shown in figure 2.

### 3.1. Data collection system

The entire hardware circuit of the data collection control system mainly includes: Stm32l151c8t6 microcontroller, MAX3223 serial transceiver module, external flash memory, battery, etc [4]. The control system is mainly composed of MCU Stm32l151c8t6 and peripheral circuit, which is the control center of the module. In addition, the system adopts the external flash memory to store collected sensor data. The data acquisition system can exchange data with various sensors through the communication interface.

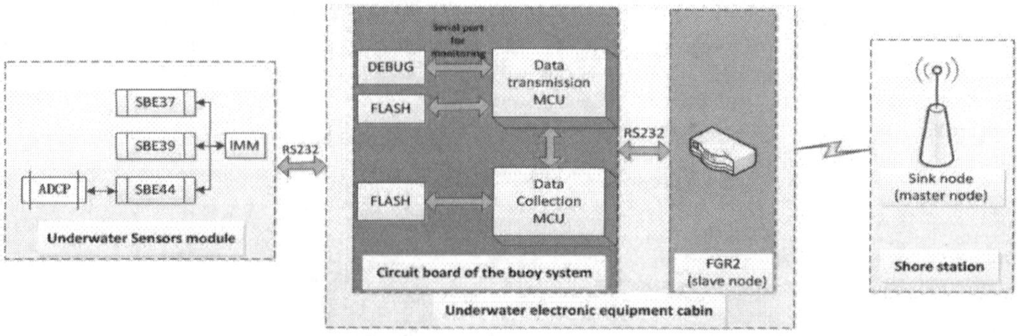


**Figure 2.** Hardware block diagram of the buoy system.

The design of data acquisition circuit is completed based on Stm32 series MCU, which model is Stm32l151c8t6. In order to satisfy the requirement of the system for long-term submersible deployment, we select the type of microcontroller because it has low power mode. Moreover, the standby current of the microcontroller is only 0.3uA. In addition, the Stm32l series microcontroller is a typical ultra low power and high performance microprocessor which is compatible with Stm32. It works on 32MHz and its power supply is 1.65V to 3.6V. It has analog and digital peripherals including watchdog, 12-bit ADC, 12-bit DAC, timer, power supply manager, internal low frequency oscillator, etc. It provides a variety of bus interface, including a USB, three UART, two 16 Mbits/s SPI, two I2C. And additionally, it has 2 ultra-low-power-comparators and supports serial wire debug, JTAG and trace. It uses the flash ROM technology and has 128KB flash ROM, 16KB RAM, up to 83 fast I/Os (73 I/Os 5V tolerant), up to 4KB of true EEPROM with ECC. It is a system which can work independently.

According to the communication protocol between master node and slave node, the master node request data every 10 minutes and during the 10 minutes, each sensor needs collect data 5 times at each sampling point. All data should be uploaded to the communication module microcontroller after collecting, but before uploading, all data must be written to the external flash memory to guarantee against data loss and compare with the received data from master node conveniently in the future. Due to buoy system needs to work continuously for at least a month, and all data collected by the sensors every 10 minutes is a total of 2190 bytes, according to the results of calculation, we select a 16M flash memory W25Q128. The flash memory supports SPI work mode, it can connect with the microcontroller through interface to realize read and write real-time data.

Electromagnetic induction coupling has the advantages of high reliability and strong stability, therefore, we select the sensors which can work in this way. Among them, TD and CTD can apply to the work mode directly, while the ADCP needs transparent transmission via SBE-44 to use the work mode. When the underwater instruments work, they need insulated wire looped through modem coupling core to connect to the IMM module. The IMM module is a low power microprocessor peripheral that interfaces to inductive mooring by serial RS-232 or RS-485. We select MAX3232 serial transceiver module as the communication mode. We set aside two serial ports when we design the circuit, one is connected with the IMM module and the other is connected with GPS.

In the design of data acquisition circuit, we added the low power consumption mode. According to the demands of the protocol, underwater sensors take two minutes to acquire data for once. Actually, it takes less than two minutes to collect all the data once. So after the acquisition of the data, the data collection system will send commands to make all sensors enter hibernation immediately and the microcontroller enter standby mode at the same time. The control board only needs to provide power for the clock module to reduce the ineffective power dissipation which saves the electricity consumption.

### 3.2 Data transmission system

The hardware circuit of the data transmission control system mainly includes: MAX3232 serial transceiver module, Stm32l151c8t6 MCU, external flash memory, voltage conversion module etc. The design of the data transmission control system is mainly to receive data from the data acquisition control system [5]. It also needs to control the slave node when it receives network access request command from the master node. Then the data collected in recent 10 minutes will be uploaded to the master node.

The data transmission module also has a piece of external flash storage which purpose is to store all the data sent to the master node after finishing packing. In this way, the collected data not only can be compared with data which the data collection system gets, also can be compared with received data through the wireless transmission module so as to ensure the accuracy of the data.

The data transmission module is mainly for data exchange with the FGR2 wireless communication module. FGR2 communication module supports RS-232 communication interface, so we set aside a serial port on the circuit board with FGR2. The data transmission control system will send real-time data to FGR2 slave node every 10 minutes via a serial port. If FGR2 master node send data request package to FGR2 slave node, FGR2 slave node will send the last ten minutes data to the FGR2 master node. In addition, we also set aside a serial port for monitoring when we design the circuit board to facilitate the debugging of the system. The serial port can send the data collected, the data transferred and the data stored in two pieces of flash to the computer which realizes the data real-time monitoring and verification. What's more, it also provides great convenience for debugging programs.

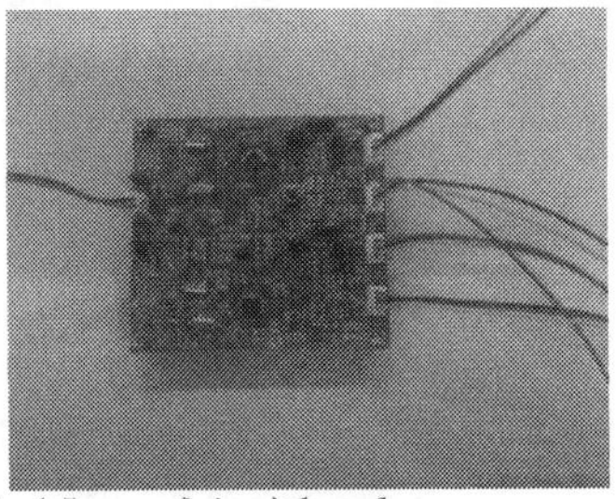

**Figure 3.** Real figure of circuit board.

Because of limitations of the submergence depth and the limited space of the electronic cabin, the data acquisition circuit and transmission circuit are integrated on a circuit board. Communication between the two pieces of MCU are realized through an IO port. Both IMM module of data gathering and FGR2 module for data transmission adopt RS-232 communication with microcontroller. In order to facilitate the system for long distance transmission in the future, the RS-485 communication module is also reserved in this dual-core control circuit board. The real figure of the circuit board is shown in figure 3.

### 3.3 Power system

The hardware circuit of the submersible buoy system needs a total of 4 modules for external power supply. The design voltage of data acquisition and transmission control circuit is 3.6V-6.5V, the power supply voltage of wireless communication module FGR2 is 12V, and electromagnetic induction coupling IMM module voltage is between 6 and 30V, in addition, GPS module needs additional power supply with the voltage of 5V. Therefore, in order to ensure that multi-scale ocean dynamic process observation system is able to complete at least one month's long-term stability deployment, we choose to add DC-DC converter module and unify with the 12V alkaline battery to supply power which reduce the risk on a variety of batteries according to the calculation.

Except the hardware circuit system, the underwater sensors TD, CTD and ADCP all bring supply battery and can work continuously more than 6 months. The turbulence instrument also comes with batteries without the need for additional battery power supply.

## 4 Laboratory tests

The function of data acquisition and communication control circuit board was tested in the laboratory. The data collection module successfully collected data of 16 TD sensors, 4 CTD sensors, 1 ADCP sensor and GPS, and the data collected was stored into the flash while also uploaded to the data communication module at the same time. Moreover, the collected data from each sensor can also be monitored by the serial port in real time, the transfer rate of the serial port is 115200 bps.

At the same time, we tested the communication effect between the data communication module and FGR2 module, the network construction effect of FGR2 master node and slave nodes in the laboratory. At the beginning of the experiments we found that the network access effect for the slave node is not very good, what's more, our slave node was often kicked off by the master node because of the slow data transmission. After updating the firmware and modification, it can enter the network smoothly. We also test the packet transmission peak rate, channel rate and so on to prove its stability and the communication effect of the FGR2 module with antenna and without antenna in respectively to simulate poor sea conditions.

The test scene photo was shown in figure 4. In order to facilitate the test of the whole system, we put the data collection and communication control circuit board, FGR2 communication module and GPS module in a test box and unified with a 12V lithium battery to supply power. In laboratory tests, we took a 5m long wire instead of the plastic coated steel pipe to complete induction coupling test. Although the data collection and communication control circuit board is unstable at first, some problems may occur like system crashed. We redesigned several versions and the last version has ran for a month stability after improvement. Acquisition data of temperature by 16 TD sensors in ten minutes was shown in figure 5, there was no electrical conductivity and flow velocity data due to the test was operated in the lab.

**Figure 4.** Test scene of system combined adjusting.

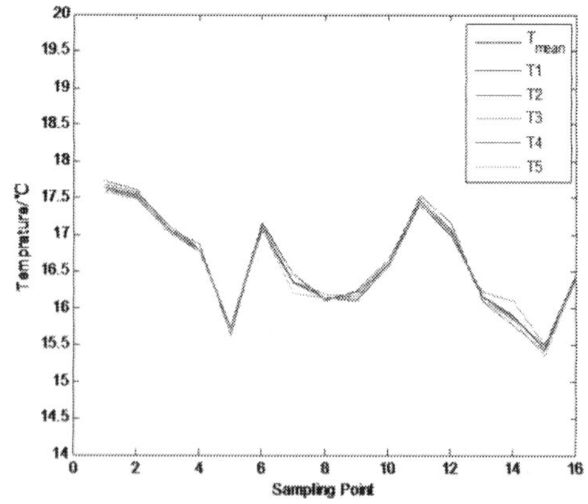


**Figure 5.** Lab temperature collected by TD sensors.

## 5 Conclusion and future work

This article aims to introduce the hardware design of a buoy system for marine dynamic process observation and the system of the control circuit board has worked accumulated for a month through the test which verified the stability of the hardware system and achieved the design function requirement, it also provides for the buoy system strong support which can maintain a long time deployment in the future. All of the sensors data we collected has carried on the comparison with the real environment in the month, we found that the data are normal and discrete degree is small. The stored data in the circuit board has been compared with the data received in the landing site to ensure the consistency of the data. And the hardware system is not only used in this buoy system, it also can be applied to other similar underwater sensor data acquisition system after further development, including long distance communication mode like satellite communication.

Now we are testing the software part of the buoy system. Especially, we test it to communicate with the landing site together with other buoy system and upload real-time collected data to several servers to parse the data. In addition to complete the function indexes, we will test communication status in bad communication condition to simulate poor sea condition as far as possible in order to ensure that the sea trials would be able to complete in the future.

## References

1. Liu, Chun-hu, and B. Fu. Development of a Compact Marine Environmental Monitoring Module Based on SMT32[C]. Engineering and Technology (S-CET), 2012 Spring Congress on IEEE, 2012:1-4.
2. Zhang H, Liu J, Yu H. Design of New Marine Environmental Data Acquisition and Transmission System[C]// Applied Informatics and Communication - International Conference, Icaic 2011, Xi'an, China, August 20-21, 2011, Proceedings. 2011:422-429.
3. Song Da-lei, Sun Jing-jing, Xue Bing, et al. Mooring system of ocean turbulence observation based on submerged buoy[J]. China Ocean Engineering, vol.27, pp 369-378,2013.
4. Liu S H, Gong D J, Yong-Ping X U, et al. The control circuit of the marine environment monitoring system based on microcontroller[J]. Marine Sciences, 2009.
5. Zhang Y, Cai W. The Marine Environment Monitoring Sensor Network Research Based on the Coupling Communication Chain[C]// International Conference on Control Engineering and Communication Technology. 2012:358-361.

# Non-Contact Measurement of Cereal Quality by Image Sensing and Numerical Regression Techniques

W. B. Hussein[1], A. A. Moaty[2], K. A. Nagaty[1] and M. A. Hussein[2]

[1]*Faculty of Informatics and Computer Science, The British University in Egypt, Cairo, Egypt*
[2]*Department of Life Science Engineering, Technische Universität München, Freising, Germany*

**Abstract.** In this paper, digital image processing techniques are applied to measure some of the quality parameters of the durum wheat semolina. One of these parameters is the semolina colour value in the lab colour space L*a*b*, which is the commonly employed colour space in food field. Several numerical methods are developed and analysed for mapping the RGB digital images to L*a*b*. These methods are direct, polynomial regression, and neural network methods. The accuracy of each method is obtained with respect to the measured L*a*b* values captured with a Chroma-Meter instrument. The numerical models outcomes showed lowest colour deviations of 0.72. The results also demonstrated a significant effect of the training data set on the numerical L*a*b* outputs. Moreover, a partial least-squares regression model was developed to numerically predict the β–carotene content in semolina, as another important quality parameter. The model proved a correlation coefficient of 0.94 between numerical predictions and experimental measurements according to the ICC standard method 152 for extracting the durum carotenoids, thus bears a high potential for facilitating carotene detection in durum.

## 1 Introduction

Food colour is the primary quality parameter checked by consumers, who use it as a tool to accept or reject the product. In addition, the colour has been widely demonstrated that it correlates with physical, chemical and sensorial indicators of product quality [1]. Therefore, many researches indicated that the observation of colours drives a potential stream for detection of several defects that the food items may have [2]. For example, the colour in durum is directly related to its trade value; at which high colour durum preserve higher value.

The analysis of the surface colour of a food product is normally performed by obtaining colour distribution and average values for the tested samples [3]. The colour measuring instruments (colorimeters) are commonly used since they are measuring the colours values mapped on a device-independent colour space with the coordinate L* for the lightness and a* and b* the colour-opponent dimensions. Some of these instruments require measuring samples that have extremely smooth and flat surfaces to avoid the dissipation of light in/out the measured area under the device prop. Within the food products, the flatness and smoothness of the tested samples is not guaranteed as well the accuracy of the colour measurements.

As an alternative solution, the RGB digital cameras can offer a reasonable solution, which can obtain information about the colour of the whole surface of the food product in terms of pixels.

It is always a challenge to drive computational models capable to convert the RGB digital images to L*a*b colour space with reasonable accuracy [4], [5]. As a result, a room has been made for researchers to develop and improve several conversion models, and to implement them as quality measurement criteria [6].

In this work, a novel quantitative analysis of the (Triticum durum) semolina colour was performed. According to the expectations of consumers to have uniform, amber-yellow coloured pasta without shades of grey or red, the yellowness of the semolina was used to evaluate the quality of the semolina [7]. Therefore, the b* value was used as an indicator of the yellowness since it represents the colours variation from blue to yellow [8]. Currently, the b* value is determined by gathering some measurement using the colorimeter devices. Due to the previously illustrated drawbacks of this technique, the RGB digital images were selected to be used as an alternative solution to get the b* value of semolina in a pixels wise approach. Three different numerical colour calibration methods are proposed to obtain accurate information about the b* value. The first method is the direct transformation from RGB to L*a*b*, by mapping the RGB on the XYZ colour space then transform the XYZ values to L*a*b* [9]. The second method is based on polynomial regression functions [10]. $1^{st}$, $2^{nd}$ and $3^{rd}$ order polynomials are investigated in order to find the proper function that accurately relates the RGB values to the L*a*b* space. The third method is using artificial neural network to design a nonlinear transformation model. Finally, since studies indicate that the b* value is correlated with the carotene content [11], a numerical model implementing the partial least-squares regression method (PLS) is constructed to estimate the β–carotene content (yellow pigments) in semolina by having the colour (b* value) and the percentage humidity as input variables.

## 2 Material and methods

Sixty-four semolina samples of 0.5mm particle size were randomly selected from different locations. Each sample was placed in measuring cup with 5cm diameter to measure the L*a*b* colour components using a Minolta CR-200 Chroma-Meter instrument. For each sample, five enclosed regions of 1 $cm^2$ area were marked as shown in Figure 1.

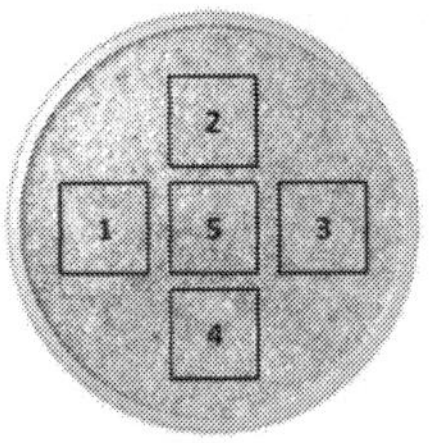


**Figure 1.** One semolina sample showing the selected five locations for measuring the colour values.

The mean RGB values with their corresponding L*a*b* measurements were calculated for each enclosed region. A data set out of 320 mean RGB values with their corresponding L*a*b* measurements were constructed to calibrate and validate the numerical colour calibration models proposed in this paper. The β–carotene content in these samples were extracted according to the ICC-152 standard method for yellow pigment extraction. The β–carotene content was calculated using equation (1).

$$\beta - carotene\ (mg) = \left(\frac{a * 5 * 10}{1000}\right)\left(\frac{100}{100 - b}\right) \quad (1)$$

where $a$ is the mg of β-Carotene in 10 ml of extract and $b$ is the percentage relative humidity. After constructing the acquisition system, a set of RGB-L*a*b* data was prepared using several colour charts. The best calibration process of the acquisition setup was achieved when 17 colour charts were employed, as in Figure 2-a.

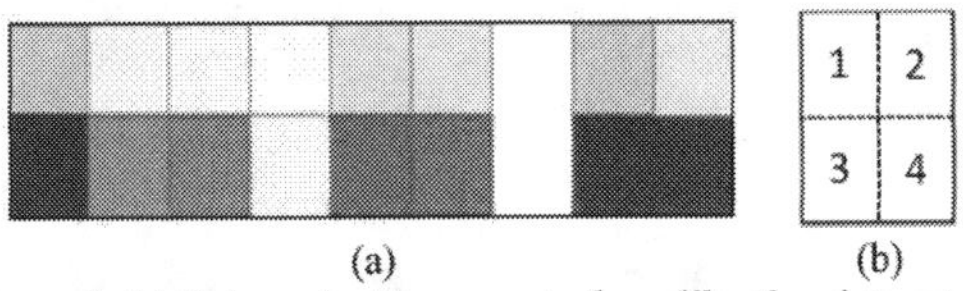


**Figure 2.** (a) Colour chart to generate the calibration data set. (b) The dividing method applied on each coloured area in order to increase the number of points.

Each colour was divided into 4 regions as in Figure 2-b. In each region, the L*a*b* values were measured using Minolta CR-200 Chroma-meter instrument. The mean of the colour values in each region was measured using equation (2).

$$R,G,B_{mean} = \frac{\sum_{i=1}^{imax}\sum_{j=1}^{jmax} P_{R,G,B}}{Np} \quad (2)$$

where $P_{R,G,B}$ is the pixel $R$, $G$ and $B$ values; the $N_p$ is the number of pixels enclosed in each region; $i, j$ is the row and column index respectively

### 2.1 Image acquisition

An industrial colour camera type Basler Pilot-piA2400 CCD camera with a resolution of 5 Mega Pixels (2456 x 2058) and pixel size (3.45 x 3.45) μm. The illumination was achieved by using 2 Philips day light fluorescent lamps with colour temperature of 6500 K approximately.

The RGB-L*a*b* transformation consists of two phases: first the background is eliminated using histogram analysis techniques as shown in Figure 3-(b).

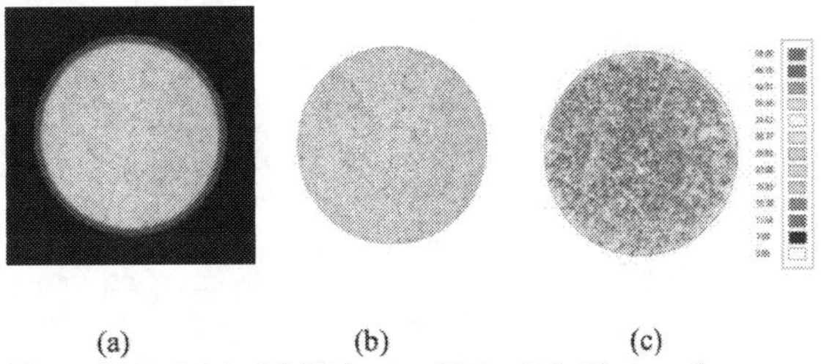


**Figure 3.** (a) Original RGB image, (b) the digital image after eliminating the background noise, (c) colour contoured image for representing the numerically calculated b* value

The RGB-L*a*b* colour conversion took place using the thresholded image with different numerical colour calibration models. Finally, the numerically calculated L*a*b* values were arranged over the pixels location and exported in a proper format such as colour contoured images as shown in Figure 3-(c).

Three types of colour calibration models were used to perform the RGB-L*a*b* transformation: the direct model, polynomial regression model and neural network model. The b* value in the colour space (L*a*b*) which represents the chromaticity coordinate for yellow-blue (+b* = yellow direction; −b* = blue direction) is used as an indicator of the yellowness of the tested sample since it represents the colours variation from blue to yellow while ignoring the L* and a* values. A special colour difference formula was used to evaluate the accuracy of the different models. This colour difference formula is based only on the Δb* as in equation (3).

$$\Delta E_{b*} = \sqrt{\Delta b^{*2}} \quad (3)$$

The accuracy of each model will be evaluated with the mean and the maximum value of the $\Delta E_{b*}$ difference calculated for all the tested samples.

### 2.2 Direct model

This calibration model performs the RGB-L*a*b* by firstly using equation (4).

$$\begin{bmatrix} X \\ Y \\ Z \end{bmatrix} = [M]_{3*3} \begin{bmatrix} R \\ G \\ B \end{bmatrix} \tag{4}$$

where matrix *M* depends on the chromaticity coordinates of the used RGB system and its reference white point. Convert the obtained XYZ values to L*a*b* colour space according to equation (5).

$$L^* = \begin{cases} 116\left(\sqrt[3]{\frac{Y}{Y_r}}\right) - 16 & if \frac{Y}{Y_r} > \varepsilon \\ 903.3\left(\frac{Y}{Y_r}\right) & if \frac{Y}{Y_r} \le \varepsilon \end{cases}$$
$$a^* = 500\left[\left(\sqrt[3]{\frac{X}{X_r}}\right) - \left(\sqrt[3]{\frac{Y}{Y_r}}\right)\right] \tag{5}$$
$$b^* = 200\left[\left(\sqrt[3]{\frac{Y}{Y_r}}\right) - \left(\sqrt[3]{\frac{Z}{Z_r}}\right)\right]$$

Secondly convert the obtained XYZ values to the L*a*b* colour space as shown in equation (5). The equation for mapping relative luminance (Y/Yr) into lightness L* is composed of two separate functions, f(Y/Yr) and g(Y/Yr). The two functions are grafted together at junction point $\varepsilon$ =0.008856 according to the CIE standard. Xr, Yr, Zr are the XYZ tri-stimulus values of the reference white point which equal to (0.95047, 1.0, 1.08883) according to the CIE standard illumination D65 (colour temperature of 6500 K approximately).

## 2.3 Polynomial regression model (PRM)

Polynomial functions of different orders (i.e., from order 1 to order 3) were investigated. For each polynomial function, the relation between the RBG and the L*a*b* colour components were assumed to follow a polynomial function of the RGB colour values and some other calibration parameters *M*, equation (6).

$$\begin{bmatrix} L^* \\ a^* \\ b^* \end{bmatrix} = \begin{bmatrix} f_L(M_L, R, G, B) \\ f_a(M_a, R, G, B) \\ f_b(M_b, R, G, B) \end{bmatrix} \tag{6}$$

where $M_{L,a,b} = \begin{bmatrix} M_1 & M_2 & \dots\dots & M_n \end{bmatrix}^T_{L,a,b}$ are 3 vectors of the model calibration parameters for L*, a* and b*; respectively. Using the PRM, the RBG-L*a*b* transformation process is performed by firstly, using a training data to calculate the model calibration parameters *M* based on the minimization of the mean absolute error between the estimated L*a*b* variables (model output) and the L*a*b* variables measured by the colorimeter device [2]. Secondly, after calculating the *M* parameters a test data is used to evaluate the accuracy of the PRM model. The numerical L*a*b* values are compared with the measured L*a*b* by calculating the colour differences using equation (3) for each tested point. The model calibration parameters (*M*) are calculated in equation (7), as a function of some RGB and L*a*b* inputs.

$$M = [X^T X]^{-1} X^T Y \tag{7}$$

where $f_L$, $f_a$, $f_b$ are the constructed regression functions of 1st, 2nd, or 3rd order. *X* and *Y* are the input matrices which carry (*n*) measurements of (RGB) and (L*a*b*) inputs respectively. The structure of the *X* input matrix is formed according to the order and number of terms in the polynomial functions. For a first order polynomial regression model, the relation between the RGB and the L*a*b* is shown in equation (8).

$$\begin{bmatrix} L^* = a_3R + a_2G + a_1B + a_0 \\ a^* = b_3R + b_2G + b_1B + b_0 \\ b^* = c_3R + c_2G + c_1B + c_0 \end{bmatrix} \tag{8}$$

Therefore, by setting the RGB inputs in a matrix form, the *X* matrix for *n* input points is given in equation (9).

$$X = \begin{bmatrix} R_1 & G_1 & B_1 & 1 \\ R_2 & G_2 & B_2 & 1 \\ \vdots & \vdots & \vdots & \vdots \\ R_n & G_n & B_n & 1 \end{bmatrix} \tag{9}$$

For the second order polynomial shown in equation (10), the *X* matrix of $10 * n$ elements as shown in equation (11) will be used to calculate the calibration parameters *M*.

$$2^{nd}PRM = a_9R + a_8G + a_7B + a_6RG + a_5RB + a_4GB + a_3R^2 + a_2G^2 + a_1B^2 + a_0 \tag{10}$$

$$X = \begin{bmatrix} R_1 & G_1 & B_1 & RG_1 & RB_1 & GB_1 & R^2{}_1 & G^2{}_1 & B^2{}_1 & 1 \\ R_2 & G_2 & B_2 & RG_2 & RB_2 & GB_2 & R^2{}_2 & G^2{}_2 & B^2{}_2 & 1 \\ \vdots & \vdots & \vdots & \vdots & \vdots & \vdots & \vdots & \vdots & \vdots & \vdots \\ R_n & G_n & B_n & RG_n & RB_n & GB_n & R^2{}_n & G^2{}_n & B^2{}_n & 1 \end{bmatrix} \tag{11}$$

The second order polynomial in equation (11) will be enriched with one additional RGB terms in order to investigate the influence of the combined third order term. In this case, the *X* matrix structure will have a structure of 11*n terms as seen in equation (12).

$$X = \begin{bmatrix} R_1 & G_1 & B_1 & RG_1 & RB_1 & GB_1 & R^2{}_1 & G^2{}_1 & B^2{}_1 & RGB_1 \\ R_2 & G_2 & B_2 & RG_2 & RB_2 & GB_2 & R^2{}_2 & G^2{}_2 & B^2{}_2 & RGB_2 \\ \vdots & \vdots & \vdots & \vdots & \vdots & \vdots & \vdots & \vdots & \vdots & \vdots \\ R_n & G_n & B_n & RG_n & RB_n & GB_n & R^2{}_n & G^2{}_n & B^2{}_n & RGB_n \end{bmatrix} \tag{12}$$

The same procedure is used for the third order polynomial, equation (13). The *X* matrix will follow the same structure like above but with 20 different terms in each row.

$$3^{rd}PRM = a_{19}R + a_{18}G + a_{17}B + a_{16}RG + a_{15}RB + a_{14}GB + a_{13}R^2 + a_{12}G^2 + a_{11}B^2 + a_{10}RGB + a_9R^2G + a_8R^2B + a_7G^2R + a_6G^2B + a_5B^2R + a_4B^2G + a_3R^3 + a_2G^3 + a_1B^3 + a_0 \tag{13}$$

DOI: 10.1051/matecconf/20167501002

For the L*a*b* input data, the $Y$ matrix of $3*n$ elements (the L*, a* and b* components for different $n$ points), equation (14), will be used for all the polynomial regression models.

$$Y = \begin{bmatrix} L_1 & a_1 & b_1 \\ L_2 & a_2 & b_2 \\ \vdots & \vdots & \vdots \\ L_n & a_n & b_n \end{bmatrix} \quad (14)$$

Finally, the test data set (RGB input and L*a*b* measurements) is combined with the previously calculated ($M$) parameters and supplied to model. The model will transform the inputs RGB points to L*a*b* then calculate the colour differences between the numerical outcomes and the L*a*b* input measurements. The RGB-L*a*b* transformation is performed through equation (15).

$$[L\ A\ B] = [M]^T[R\ G\ B] \quad (15)$$

where $[L\ A\ B]$ is a vector consists of three elements L*, a*and b* and The $[RGB]$ is a vector of R, G and B terms having the same structure and order of the implemented polynomial function.

### 2.4 Neural network model (NNM)

A feed forward back propagation multi-layer perceptron network was designed. The input layer contains three neurons for the three colour values (R, G, and B), 8 hidden layer neurons and the output layer has three neurons for the developed values of (L*, a*, and b*).

### 2.5 Partial least squares regression model

It is based on principal component scores of both the independent $X$ variables (b* value and the percentage of humidity of a sample) and dependent $Y$ variable (β-carotene content in semolina) to develop a regression model. Using singular value decomposition (SVD), matrix $X$ is decomposed into score matrix $T$ and loading matrix $P$ plus error matrix $E$, eq. (16). And matrix $Y$ is decomposed into $U$ and $Q$ plus error term $F$, eq. (17).

$$X = T.P^T + E \quad (16)$$

$$Y = U.Q^T + F \quad (17)$$

where $P$ and $Q$ are orthogonal matrices . The PLS algorithm minimizes $||F||$ while maintain correlation between $X$ and $Y$. The PLS algorithm computes the factor score matrix $T=XW$ for an appropriate weight matrix $W$. Once $Q$ matrix is computed, the regression model is equivalent to $Y=XB+E$, where $B=WQ$, which can be used as a predictive regression model. More details about the PLS algorithm can be found in [12].

## 3 Results and discussion

The different numerical colour calibration models were tested using 64 available samples. 80% of the samples (51 samples) were used for training purposes while (20% = 17 samples) were used for validation. The accuracy of the four polynomial models (1st order, 2nd order, 3rd order), direct model and the neural network model were investigated. By comparing the calculated L*, a* and b* with respect to the measured values at the same points, the results show that the 1st order model has higher fluctuations compared to other models. Moreover, the 3rd order model provided the highest regression accuracy, as presented in Table 1.

**Table 1.** The error in the L*a*b* Error calculations along 17 coloured samples using different polynomial regression models

| Model | Max. error |
|---|---|
| 1st order | 30.29 |
| 2nd Order + RGB Term | 7.11 |
| 3rd Order | 2.90 |

The deviations between the measured L*a*b* data and the numerically calculated values obtained from the 3rd order PRM and the NNM is displayed in Figure 4.

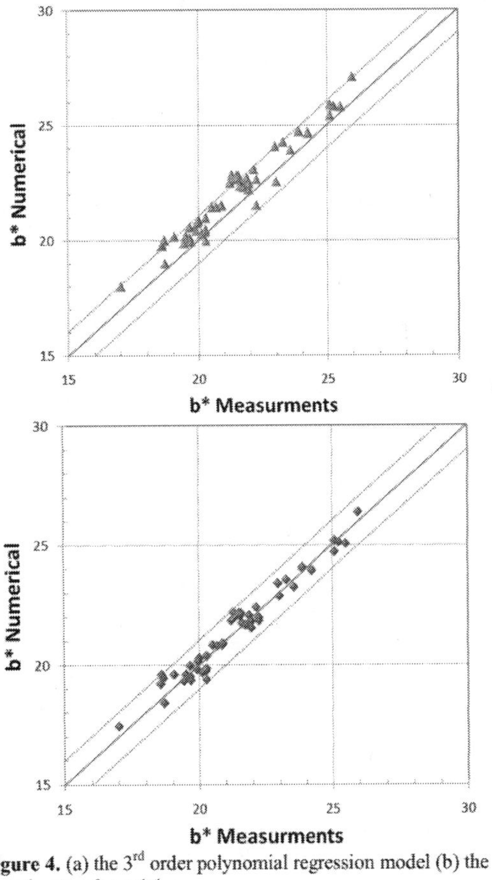

**Figure 4.** (a) the 3rd order polynomial regression model (b) the neural network model

Concerning the neural network model, a network of three neurons in both input and output layers was designed (3 input neurons for RGB values and 3 output neurons for the L*a*b* values). A parametric study was performed to determine the number of neurons in the hidden layer. The best performance was achieved with 10 hidden layer neurons that had the lower mean square error (MSE) of $2.32x10^{-5}$.

A data set of 50 semolina samples was divided into 80% (40 samples) for training the regression model and 20% (10 samples) for validating the model outputs. Figure 5 shows the comparison between the measured and predicted β-carotene content in 10 samples. The obtained results show that the predicted β-carotene values are highly correlated with the measured values with a correlation coefficient (r=0.94).

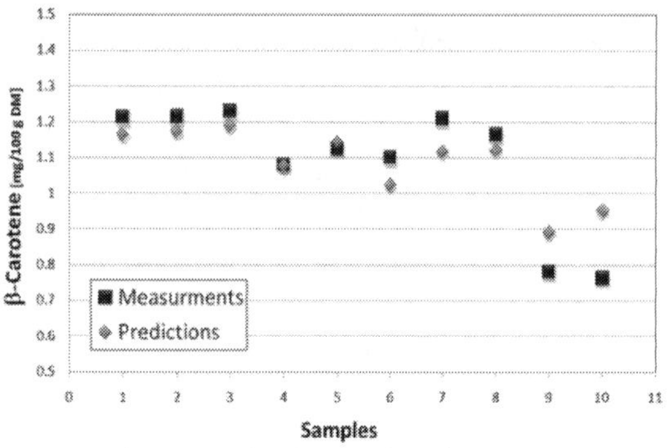


**Figure 5.** The measured and the predicted β-carotene content

## 4 Conclusion

The L*a*b* colour space was selected in this study as the standard device-independent colour space. The direct, polynomial regression and neural network models were developed to perform the RGB-L*a*b* transformation. Overall, the obtained results show that the lower error is achieved by the 3rd order polynomial model and the neural network models. Meanwhile, the PLS regression model for predicting the β-carotene shows a 0.94 correlation coefficient. The results indicate that the developed models are promising to present a consistent and reliable schemes for measuring L*a*b* and β-carotene as main quality measures of the semolina.

By increasing the order of the polynomial regression models enhanced the accuracy of the output results but also increased the computational time (form 2nd to 3rd order nearly one order of magnitude time increases).

Improving the imaging environment by using such a LED lamps will produces more stable environment without flickering as in the florescent lamps. The optimization of the lamps position will produce more uniform illumination over the tested sample and reduces the unrealistic variations of the RGB values induced by light which tends to more accurate L*a*b* predictions. Therefore, further parametric study will be performed to simulate the light distribution inside the acquisition setup by solving numerically the differential equation of the light diffusional motion.

## References

1. F. Mendoza, P. Dejmek and M. Aguilera: Calibrated color measurements of agricultural foods using image analysis. Postharvest Biology and Technology Volume **41**, Issue 3, (2006), Pages 285-295
2. K. León, D. Mery, F. Pedreschi: Color measurement in L*a*b* units from RGB digital images. Food Research International **39** (10): 1084-1091 (2006)
3. L. Kit, E. Spyridon: A simple digital imaging method for measuring and analyzing color of food surfaces. Journal of Food Engineering. Volume **61**, Issue 1, (2004), Pages 137-142
4. S. Basvaraj, C. Burkpalliy: Color Based Identification and Classification of Boiled Food Grain Images. International Journal of Food Engineering, Volume **5**, Issue 5 (2009) Article 7
5. A. Danti, M. Madgi, B. Anami: A Neural Network Based Recognition and Classification of Commonly Used Indian Non Leafy Vegetables. International Journal of Image, Graphics and Signal Processing. Vol. **6**, No. 10, pp. 62-68, (2014)
6. A. Danti, M. Madgi, S. Anami: Mean and Range Color Features Based Identification of Common Indian Leafy Vegetables. International Journal of Signal processing, Image Processing and Pattern Recognition, **5**, No. 3, September, (2012), pp. 151-160
7. Landi A.: Durum Wheat, semolina and pasta quality characteristics for an Italian food company. CIHEAM-IAMZ, (1995). p. 33-42
8. Y. Rharrabti, C. Royo, D. Villegas, N. Aparicio and L. F. García del Moral: Durum wheat quality in Mediterranean environments: I. Quality expression under different zones, latitudes and water regimes across Spain. Field Crops Research, Volume **80**, Issue 2, 20, 123-131 (2003)
9. Hunt, R.W.G. (1991): Measuring Color. 2nd Edition. Ellis Horwood. New York
10. Alvin G. Wee, Delwin T. Lindsey, Shanglun Kuo, and William M. Johnston: Color accuracy of commercial digital cameras for use in dentistry. Dent Mater. (2006); **22**(6): 553–559.
11. V. Hentschel, K. Kranl, J. Hollmann, G. Meinolf: Spectrophotometric Determination of Yellow Pigment Content and Evaluation of Carotenoids by High-Performance Liquid Chromatography in Durum Wheat Grain. J. Agric. Food Chem. (2002), **50**, 6663-6668.
12. P. Geladi and B. R. Kowalski, "Partial Least Squares Regression: A Tutorial," Analytica Chimica Acta, vol. **185**, pp. 1-17, (1986)

# An Approach of Electronic Subdividing Method for Measuring Straightness and Displacement of a Precision Linear Stage Simultaneously

Hsiu-An Tsai [1,2] and Shao-Ching Chen[1]

[1]Department of Industrial Upgrading Service, Metal Industries Research & Development Centre, Kaohsiung, Taiwan
[2]Department of Mechanical Engineering, National Cheng Kung University, Tainan, Taiwan

**Abstract.** Optical encoders are commonly used in modern positioning systems. The accuracy and resolution of the optical encoders are always restricted by generated sinusoidal signals and the assembly technique. In this study, an electronic signal subdividing system is developed. This system is based on FPGA in combination with A/D and D/A converting circuits. Subdividing algorithm improves the segmenting signal amplitude method. Furthermore, we also construct a laser encoder for measuring straightness error and displacement of a linear stage simultaneously. The laser encoder consisting of the stainless steel bar and the sensor are developed for two-axis (X- and Z-axis) position measurement. The two dimensional sinusoidal array on the stainless steel bar are machined by ultrasonic elliptical vibration cutting system. The stainless steel bar has a three dimensional micro-structured surface, which is a superposition of periodic sinusoidal waves in the X- and Z-directions with spatial wavelengths of 350 μm and amplitudes of 0.5 μm. The laser-based two-axis position sensor is used to detect local slope profiles of the grid surface, and the displacement and straightness error could be extracted from the X- and Z-axis sensing signal. The sensing signal is processed by FPGA subdividing system. In addition, the proposed subdividing method is verified by the performances and effects of measuring results.

## 1 Introduction

Precision metrology is still playing an important role in precision on-machine measurement, positioning and manufacturing from the point of a wide range of measurements with high accuracy and resolution [1]-[8].

With the rapid progress in the field of nanotechnology, there is high demand for new displacement measurement technologies to be able to achieve sub-nanometer level resolution with long-term reliability. In accordance with the increasing market demand for ultra-precision technology, a high precision multi-degree-of-freedom displacement measurement technology has become important for industrial applications such as the field of manufacturing and inspection because physical quantities, such as linear and angular displacements, are key parameters for keeping and improving quality control of a production system [1], [2], [5]-[8].

Linear stages are commonly used in machine tools, semiconductor instruments, precision assembly processing and nanometer level positioning [1]. In addition, high accuracy stages have commonly been used in semiconductor processing, PCB drilling processing, micromachining processing, precision assembly processing, inspection processing, etc. Any positioning stage is considered to have six-degrees-of-freedom. There errors can be defined as occurring along the reference coordinate system.

It should be noticed that the term "straightness error" is generally used to refer to many aspects of engineering quality, such as work-piece straightness, motion straightness, etc. Optical measurements of straightness errors have been widely used in the field of engineering metrology. The straightness datum that can be measured by means of various optical accessories such as: autocollimators, alignment telescopes and optical theodolites are commonly adopted. Another method employs a laser beam to generate a reference line, which has the properties of small divergence and high intensity. Moreover, many other methods were developed with high alignment accuracy, such as the polarimetry method [9], the polarizing method [10], the optical compensation method, and the Zeeman laser interferometer method [11]. However, the mentioned measurement devices are not suitable to be installed on precise positioning machinery for on-line measurement as a result of heavy overall structure. Recently, a laser diode was adopted for straightness measurement since it is small in size and low in power [12]. Furthermore, optical encoders are commonly used in modern positioning systems. Moreover, the accuracy and resolution of the optical encoder is always restricted by generated sinusoidal signals and the assembly technique. Therefore, an electronic signal subdividing system is developed for measuring straightness error and displacement of a linear stage simultaneously.

## 2 Measurement principles and instrument configuration

Figure 1 shows a diagram of the measurement system. Assume that the sensor is kept stationary and the two-dimensional sinusoidal grating was placed in the XZ plane of the linear X-axis stage.

An incident beam from the laser diode is collimated by the collimating lens. The collimated laser beam incidents to the beam-splitting device. It should be noticed that the beam-splitting device which is disposed in the light path of the incident light and comprises a beam splitter and a polarizer plate. And the beam splitter is used for guiding the incident light to pass the focus lens on the two-dimensional sinusoidal grating. The reflection light beam is reflected from the two-dimensional sinusoidal grating and reflected by the reflective mirror to pass the focus lens on the quadrant photo diodes (QPD).

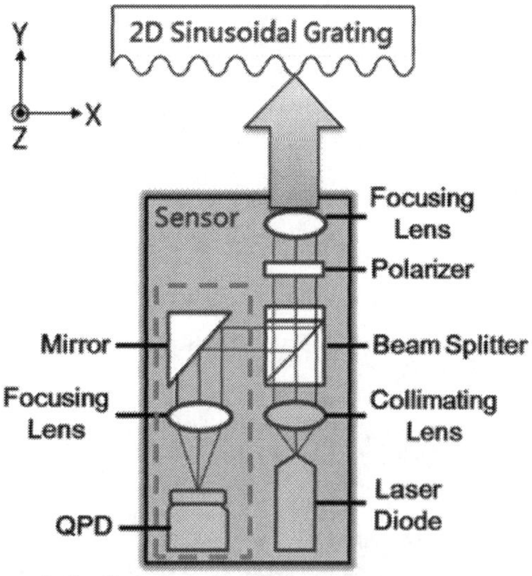


**Figure 1.** Configuration of the proposed measurement system.

The reflective-type two-dimensional sinusoidal grating, wherein a profile equation of the reflective-type two-dimensional sinusoidal grating is expressed as:

$$h(x,z) = -A_x\cos\left(\frac{2\pi x}{P_x}\right) - A_z\cos\left(\frac{2\pi z}{P_z}\right) \quad (1)$$

where $P_{x,z}$ are the wavelengths of the reflective-type two-dimensional sinusoidal grating in the x-axis direction and the z-axis direction. In addition, $A_{x,z}$ are the corresponding sinusoidal amplitudes. In this study, the $P_{x,z}$ and $A_{x,z}$ are 350μm and 0.5μm, respectively. As a result, separately performing partial differentiation to obtain the slope equations of the two-dimensional grating in the X-axis direction and the Z-axis direction are as shown as followed:

$$m_x(x,z) = \frac{\partial h(x,z)}{\partial x} = \frac{2\pi A_x}{P_x}\sin\left(\frac{2\pi x}{P_x}\right) \quad (2)$$

$$m_z(x,z) = \frac{\partial h(x,z)}{\partial z} = \frac{2\pi A_z}{P_z}\sin\left(\frac{2\pi z}{P_z}\right) \quad (3)$$

On the other hand, we can apply the auto-collimation principle to measure $m_x(x,z)$ and $m_z(x,z)$ from QPD:

$$u = (\tan^{-1} m_x(x,z)) \times 2f \quad (4)$$

$$v = (\tan^{-1} m_z(x,z)) \times 2f \quad (5)$$

where u and v are the displacement in the X-axis and Z-axis directions of the QPD, respectively. And the f is the focus distance from the lens to the QPD. With reference to the Eq. (4) and Eq. (5), the QPD receives the sensing signal which represents the position variation of the two-dimensional sinusoidal grating in the X- and Z- directions. Therefore, an axial motion displacement volume and a radial displacement volume of the linear stage are calculated by using a slope equation and a position signal equation to calculate the sensing signals of the QPD. The X- axis and Z-axis motion signal equations could be expressed as:

$$x = \frac{P_x}{2\pi}\sin^{-1}\left(\frac{P_x}{2\pi A_x} \times \tan\frac{u}{2f}\right) \quad (6)$$

$$z = \frac{P_z}{2\pi}\sin^{-1}\left(\frac{P_z}{2\pi A_z} \times \tan\frac{v}{2f}\right) \quad (7)$$

In other words, the X- axis and Z-axis motion signals are obtained as a function of the u and v values only. However, with reference to the Eq. (6) and Eq. (7), the signal discriminant analysis needs to be constructed.

In the first stage of the signal discriminant analysis procedure, the quadrant of the initial point signal ($S_i$) is defined before knowing the sign of $I_0$,$I_S$, and $I_C$ (i.e., positive or negative) are measured directly from the data acquisition products. It should be noticed that the corresponding signal output ($I_S$) from the QPD was converted to a square wave afterpassing through a comparator, as shown in Fig. 2. In addition, the corresponding signal output ($I_C$) was generated to a 90° phase shift square-wave signal of the $I_S$ signal.

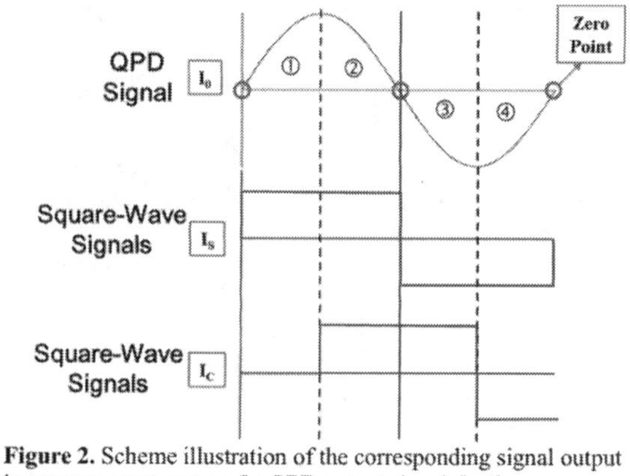


**Figure 2.** Scheme illustration of the corresponding signal output in measurement system. $I_0$: QPD output signal, $I_S$: the corresponding signal output from the QPD, and $I_C$: 90° phase shift signal output of the $I_S$ signal.

Moreover, if the value of $I_0>0$, $I_S>0$, and $I_C=0$, then the motion signal is located in the I-quadrant. Conversely, if the value of $I_0>0$, $I_S>0$, and $I_C>0$, then the motion signal is located in the II-quadrant. Meanwhile, if the value of $I_0<0$, $I_S<0$, and $I_C>0$, then the motion signal is located in the III-quadrant. Finally, if the value of $I_0<0$, $I_S<0$, and $I_C<0$, then the motion signal is located in the IV-quadrant. Having determined the quadrant of the initial point signal, the X- axis and Z-axis initial positions can be extracted from Eq. (6) and Eq. (7).

In the second stage of the signal discriminant analysis procedure, the quadrant of the terminal point signal is defined before knowing the quadrant within which the initial point lies and the number of the zero points of the sequence signal. Accordingly, we constructed the signal discriminant criterion as shown in Table 1. Having determined the quadrant of the terminal point signal, the X- axis and Z-axis terminal positions can be extracted from Eq. (6) and Eq. (7).

**Table 1.** Signal discriminant criterion for analyzing the quadrant of the terminal point signal. $S_i$ is the initial data point of $I_0$ and K is the whole numbers.

| Number of Zero Points / Quadrant of Initial point | 2K | | 2K+1 | |
|---|---|---|---|---|
| | $S_i>0$ | $S_i<0$ | $S_i>0$ | $S_i<0$ |
| I | I | IV | II | III |
| | $S_i>0$ | $S_i<0$ | $S_i>0$ | $S_i<0$ |
| II | II | III | I | IV |
| | $S_i>0$ | $S_i<0$ | $S_i>0$ | $S_i<0$ |
| III | II | III | I | IV |
| | $S_i>0$ | $S_i<0$ | $S_i>0$ | $S_i<0$ |
| IV | I | IV | II | III |

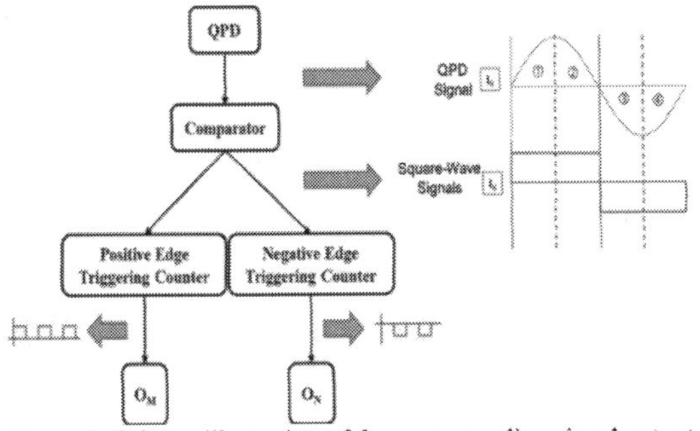


**Figure 3**. Scheme illustration of the corresponding signal output in measurement system. $O_M$ is the number of the positive edge triggering counting and $O_N$ is the number of the negative edge triggering counting.

In the last stage of the signal discriminant analysis procedure, we applied positive edge triggering counter and negative edge triggering counter as shown in Figure 3. Edge triggering is a trick to allow devices to create a very fine level trigger which is faster than all external feedback loops, allowing devices to accept inputs quickly, and then close off the entrance in time before their changing outputs will change the values of the inputs. The $O_M$ and $O_N$ are equal to the numbers of a half-wavelength long of the two-dimensional sinusoidal grating. Finally, having determined the $O_M$, $O_N$, initial positions, and terminal positions, the total displacements along the X- axis and Z-axis can be extracted as follows:

$$D_S = 0.5 \times P_{x,z} \times O_{M,N} + D_I + D_T \quad (8)$$

Figure 4 shows a prototype sensor designed and built for achieving the measurement principleof the straightness error and displacement of a linear stage simultaneously. The size of the slope sensor is 86(L) x 75(W) x 50(H)mm. As shown, the illumination collimated light (provided by a laser diode (Thorlab, Model: CPS532-C2)) is passed through a polarization beam-splitting device (Lambda, Model:OI-10.0BCQ-442)and enters a focus lens(Thorlab, Model: LB1757) on the two-dimensional sinusoidal grating. It should be noticed that the polarization beam-splitting device which is disposed in the light path of the incident light and comprises a beam splitter (PBS) and a polarizer. The light reflected from the sample is passed through a polarizer and reflected on the mirror to enter the lens and the QPD (Ontrak, Model: PSM2-4Q).The two-dimensional signal of the QPD is captured using a real-time controller system (National Instruments, Model:cRIO-9024) with 16-bit resolution. The real-time controller system which comprises the reconfigurable FPGA chip (Xilinx Virtex-5) and A&D (analog & digital signal) input module is used to efficiently develop the electronic signal subdividing method for data acquisition and analysis.

In this study, we need to make an incident light from a light source disposed at a moving stage illuminating a two-dimensional sinusoidal grating disposed at a guiding rail, wherein a reflection light of the two-dimensional grating forms a light spot on a quadrant photodiode therefore making the quadrant photodiode generating a plurality of sensing signals. The position differences of the light spot are variable according to the movement of the stage thus further altering the voltage of the sensing signals of the quadrant photodiode. In performing the experiments, the prototype sensor and two-dimensionalsinusoidal grating are set using precision linear air-bearing stage (Aerotech, Model: ABL1500) with a resolution of 1nm as shown in Fig. 5.

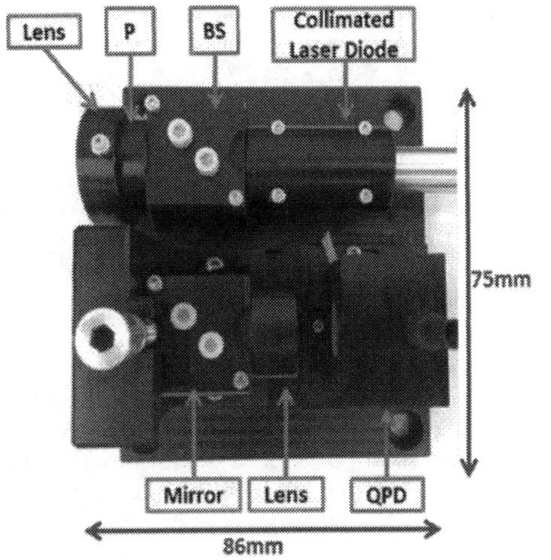


**Figure 4.** Photograph of the prototype sensor. P: polarizer, BS: beam splitter, and QPD: quadrant photo diodes.

DOI: 10.1051/matecconf/20167501003

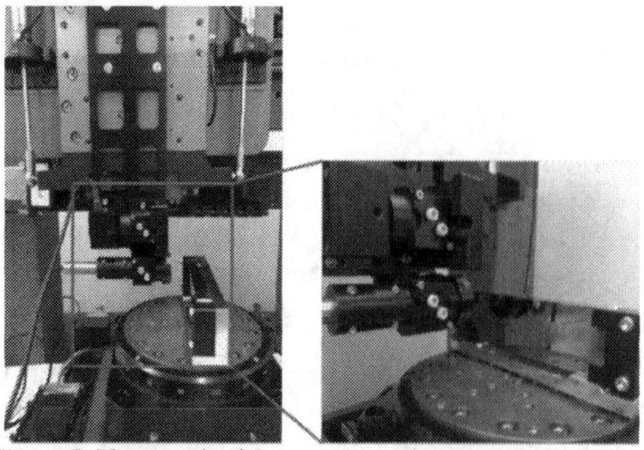

**Figure 5.** Photograph of the experimental setup.

On the other hand, the two dimensional sinusoidal structure on the stainless steel bar are machined by ultrasonic elliptical vibration cutting system. The two dimensional sinusoidal array on the stainless steel bar are machined by ultrasonic elliptical vibration cutting system. The stainless steel bar has a three dimensional micro-structured surface, which is a superposition of periodic sinusoidal waves in the X- and Z-directions with spatial wavelengths of 350μm and amplitudes of 0.5μm.

**Figure 6.** Photograph of the two dimensional sinusoidal grating.

## 3 Experimental results

The primary object of the study is to make an incident light from a light source disposed at a moving stage illuminating a two-dimensional sinusoidal grating disposed at a guiding rail, wherein a reflection light of the two-dimensional grating forms a light spot on a quadrant photodiode therefore making the quadrant photodiode generating a plurality of sensing signals. The position differences of the light spot are variable according to the movement of the stage thus further altering the intensity of the sensing signals of the quadrant photodiode.

Fig. 7 shows the three experimental results of sequence measurements in the X-axis corresponding to the travel ranges 3.5mm, is extracted from the QPD along with position variance of the moving stage. The two dimensional sinusoidal grating was moved by a precision linear air-bearing stage along the X-axis while the prototype sensor was kept stationary. As can be seen in the figure, the prototype sensor was able to measure the X- directional positions of the precision linear air-bearing stage by using the proposed phase-shift signal analysis method. The maximum repeatability error is 0.83 nm. The extracted signals could be inserted into Eq. (6), Eq.(7), and Eq. (8) to calculate the straightness and displacement. The maximum error corresponding to the travel ranges 100 mm in the measured displacement is 0.393 nm.

Fig. 8 shows the slope signal of sequency measurement corresponding to the travel ranges 3.5mm, is extracted from the QPD. It should be noticed that the blue datum lines of the figures are the X-axis direction along with position variance of the moving stage. And the red datum lines of the figures are the vertical direction of the linear stage. Therefore, the U and V axis are corresponding to the horizontal direction and vertical direction of the QPD.

The experimental results show good verification with the instantaneous measurement method of the precision linear air-bearing stage.

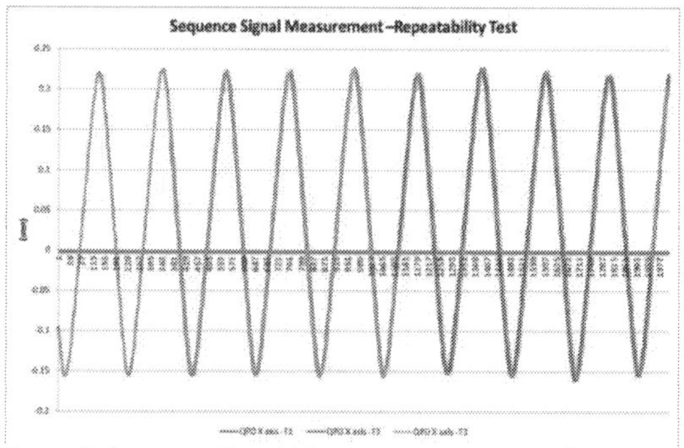


**Figure 7.** Sequence Signal Measurement –Repeatability Test of X axis.

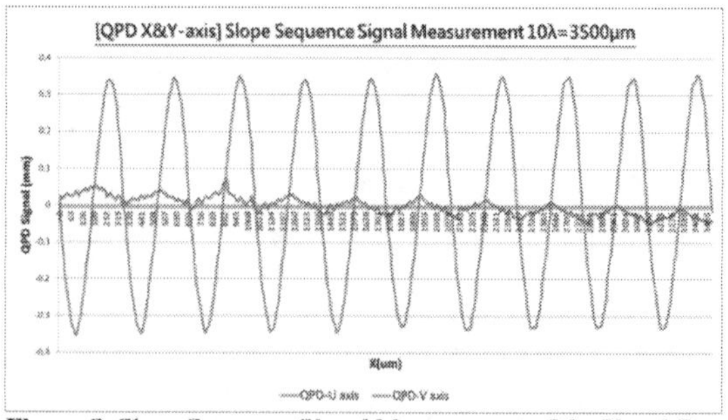


**Figure 8.** Slope Sequence Signal Measurement of the X and Z axis.

## 4 Conclusion

In this research, an electronic signal subdividing method is developed. This system is based on FPGA in combination with A/D and D/A converting circuits. Subdividing algorithm improves the segmenting signal amplitude method. Furthermore, we also construct a laser encoder for measuring straightness error and displacement of a linear stage simultaneously. The laser encoder consisting of the stainless steel bar and the sensor are developed for two-axis (X- and Z-axis) position measurement. Additionally, this methodology is especially suitable for linear stage of present invention is suitable to measure the displacement and the straightness error of the linear stage on line therefore instantaneously controlling the linear stage by information feedback. Moreover, the experimental results verified the instantaneous measurement method of the precision linear air-bearing stage.

## References

1. A. H. Slocum, *Precision machine design*(Englewood Cliffs, NJ: Prentice-Hall, (1992)
2. W. Gao, *Precision Nanometrology: Sensors and Measuring Systems for Nanomanufacturing,* Springer, (2010)
3. J. G. Webster, *The measurement, instrumentation, and sensors handbook*, CRC Press LLC, (1999)
4. O. Riemer, *Advances in ultraprecision manufacturing*, Proc. Jpn. Soc. Precis. Eng., (2011)
5. C. Lee, G. H. Kim, S. K. Lee, Meas. Sci. Technol. **22,** 105901 (2011)
6. C. Lee, G. H. Kim , S. K. Lee, Int. J. Prec. Manuf. Eng. **13,** 9 (2012)
7. S. R. Park, T. K. Hoang, S. H. Yang, J. Mech. Sci. Technol. **24,** 175 (2010)
8. C. B. Lee, S. K. Lee, J. Mech. Sci. Technol. **27,** 141 (2013)
9. R.J. King, K.W. Raine, Opt. Eng. **201**, 39 (1981)
10. C. Yin, W.Z. Chen, Opt. Eng. **304**, 480 (1991)
11. S. Shimizu, H.S. Lee, N. Imai, Int. J. Japan Soc. Prec. Eng. **28**, 273 (1994)
12. T.R. Armstrong, M.P. Fitzgerald, Measurement Sci., 1072 (1992)

# Testing Equation Method Modification for Demanding Energy Measurements Verification

Kochneva Elena[1], Pazderin Andrew[1] and Sukalo Alexandar[2]

[1]*Automated Electrical Systems Department, Ural Federal University named after the first President of Russia B.N. Yeltsin, Yekaterinburg, Russia*
[2]*Elektroperenos-Elektroprijenos BiH, Banja Luka, Bosnia i Hercegovina*

**Abstract.** The paper is devoted to the mathematical approaches of the measurements received from Automatic Meter Reading Systems verification. Reliability of metering data can be improved by application of the new issue named Energy Flow Problem. The paper considers demanding energy measurements verification method based on verification expressions groups analysis. Bad data detection and estimates accuracy calculation is presented using the Automatic Meter Reading system data from the Russian power system fragment.

## 1 Introduction

Electrical Energy (EE) measurements received from Automatic Meter Reading (AMR) systems are used to provide financial settlements on EE markets and also in calculations of various technical and economic parameters for solving power system development and planning. The current AMR systems development trend is directed to the intelligent digital devices using, as well as increasing the number of AMR complexes. Despite the considerable technical improvement of AMR systems the level of EE losses (primary the commercial losses) remains at a significant level. Rapid bad data detection is a difficult task that allows to identify AMR complexes with errors that exceeds the maximum allowable limits and to locate the potential sources of commercial EE losses.

Currently there are significant number of approaches for metering data validation [1]-[8]. Metrological control is the most legitimate method of EE measurements reliability assessing. Metrological inspections are carried out once in several years and are based on approved methodologies and procedures. If there are distortion of metrological characteristic during the calibration interval metering data will contain error up to the end of recalibration interval. More often than once in several years the reliability evaluation can be carried out by calculation and analysis of actual and allowable EE imbalances. One of the main drawbacks of the balance technique is the inability to determine the concrete source of bad data. Mathematical methods of EE measurements reliability increasing are the most promising and inexpensive. Research has shown that the verification methods developed in the frameworks of State Estimation (SE) theory [1]-[5] can successfully be used for EE measurements validation [9]-[11].

Energy Flow Problem (EFP) is the new issue concerning the electric condition parameters calculation. EFP allows to determine EE flows and technical losses on all elements of equivalent circuit using EE measurements and equivalent circuit parameters. EFP state equations system consists of the system of EE nodal balances, the system of EE balance in network branches and equations of technical losses calculation [9]. SE theory methods adopted to EE measurements in the frameworks of EFP allows to solve a variety of tasks such as determining the required amount and optimal location of AMR complexes for observability provision [12], determining the sources of EE commercial losses. This paper presents the method of Verifying Equations (VE) for demanding energy measurements verification

## 2 Theoretical background

Some EE measurements require more careful verification. These measurements include commercial ones, i.e. the measurements used for financial transactions, and important interconnections EE measurements. Additional approaches can be applied to such demanding measurements verification.

One of the promising approaches is testing equations (TE) technique. The core of the method and the process of TE system construction is discussed in details in [10]. TE are balance equations for any electrical network fragment that contain only actual EE measurements [8], [10]. TE imbalance occurs due to measurements errors presence. Any measurement error evaluation can be carried out using TE method by forming the group of linearly independent VE. Each measurement can be calculated through a linear combination of a number of other measurements in many ways using TE system. The number of such equations depends on the dimension of TE system. The previous papers demonstrate the number of TE is determined by metering system redundancy. The minimum number of measurements that provide observability of active (reactive) EE distribution is equal to the number of branches in network diagram [9]. If there is no duplicate measurements all the rest ones are redundant. The number of redundant measurements is equal to the number of TE in system. If there is no redundant measurements it is impossible to construct TE system and to provide mathematical measurements verification.

Construction of the VE is considered in five nodal network presented in Fig. 1. For clarity there is no measurement errors and no technical losses.

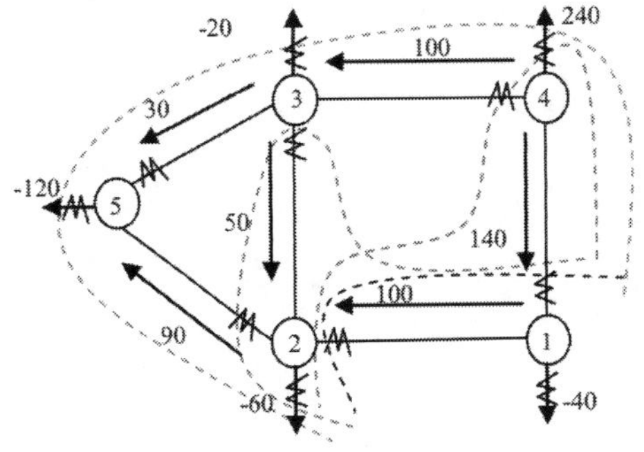


**Figure 1.** AMR complexes, measurements

There is 6 branches in the network under consideration. hence 6 EE measurements provide observability. There is 11 measurements in the scheme so 11-6=5 measurements are redundant and it is possible to draw up a system of five TE. Five VE can be received using TE system. Such VE are presented for nodal measurement $W_1$. Similarly to drawing up the system of TE, VE can be formed in two ways (algebraic and topological method). The algebraic method is more formalized. VE are formed by converting the TE system:

$$\begin{cases} -W_1 = -W_{1-2} + W_{1-4}, \\ -W_2 = W_{1-2} + W_{2-3} - W_{2-5}, \\ -W_3 = W_{3-4} - W_{2-3} - W_{3-5}, \\ -W_4 = -W_{1-4} - W_{3-4} \\ -W_5 = W_{3-5} + W_{2-5}. \end{cases} \quad (1)$$

The first VE is equation number 1 from (1). Here the measured EE values are substituted to the right side:

$$W^{(1)}_1 = -(W_{1-2} - W_{1-4}) = -(-100 + 140) = -40$$

This VE is the basis for the others. EE measurements along the branches 1-2 and 1-4 can be expressed through other measurements and substituted to the first VE.

$$W^{(2)}_1 = -(W_4 - W_{4-3} - W_{1-4}) = -(240 - 100 - 100) = -40,$$

here the forth equation from (1) is used. Measurement of branch 1-4 was expressed through $W_4$ and $W_{3-4}$.

The third VE can be obtained by expressing $W_{1-2}$ from the second equation of (1):

$$W^{(3)}_1 = -(W_{1-4} + (W_2 - W_{2-5} + W_{3-2}) = -(140 + (-60 - 90 + 50)) = -40.$$

The forth VE is recieved likewise:

$$W^{(4)}_1 = -(W_4 - W_{4-3} + (W_2 - W_{2-5} + W_{3-2}) =$$
$$= -(240 - 100 + (-60 - 90 + 50)) = -40.$$

Last VE is obtained by summing all equations from (1):

$$W^{(5)}_1 = -(W_2 + W_3 + W_4 + W_5) = -(-60 - 120 - 20 + 240) = -40.$$

There is no measurements errors in this hypothetical example, so all VE yield a result which coincides with the actual value $W_1$=-40.

VE can be constructed for any nodal or linear measurements. Topological technique of VE construction is more vivid. Topological interpretation of VE composed by algebraic technique is presented as dotted lines in Fig. 1.

Thus TE system application for VE groups construction allows to calculate any measurement through linear combination of the other measurements. The difference in the VE calculations corresponds to reliability of the measurement selected for VE. This approach allow to improve the reliability of demanding EE measurements.

## 3 Case study

Fragment of EE system is presented in Figure 2.The network consists of four-nodal ring and three radial sections. Locations of AMR complexes are marked by X. The number of nodes (without null) is 10, the number of branches is 10. There are 13 measurements in the network. 13 redundant measurements allow to get 13 TE and 13 VE.

Redundant measurements set allows to implement verification methods in particular TE technique. The source of metering data is AMR system installed in the affected substations. Calculations were performed for 1440 hour intervals.

EFP state equations system was performed for tje network presented in Fig. 2. Elimination of all unmeasured variables from EFP system resulted in TE system. $W_1$ was taken as demanding measurement as it presents ingoing EE flow from 500kV network.

Group of VE was obtained for $W_1$:

$$
\begin{aligned}
&W_1^1 = W_2 - \Delta W_{1\text{-}13};\\
&W_1^2 = -(W_4 + W_5 + W_{12} + W_{13} + W_{14} + W_{15} + W_{16} + W_3 +\\
&+ W_{10} + W_{11}) - \Delta W_{1\text{-}13};\\
&W_1^3 = -(W_4 + W_5 + W_{12} + W_{13} + W_{14} + W_{15} + W_{16} + W_3 +\\
&+ W_{10} + W_{34} - \Delta W_{1\text{-}6(II)}) - \Delta W_{1\text{-}13};\\
&\cdots\cdots\cdots\cdots\cdots\cdots\cdots\cdots\cdots\cdots\cdots\cdots\\
&W_1^{13} = W_6 + W_7 + W_8 + W_9 + W_{39} + W_{40} + W_{19} + W_{20} + W_{21} +\\
&+ W_{22} + W_{11} + W_{25} + W_{24} + W_{29} + W_{28} + W_{27} + W_{31} + W_{30} +\\
&+ W_{32} + W_{37} + W_{38} + W_{36} + \Delta W_{1\text{-}111\text{-}112} + \Delta W_{1\text{-}121\text{-}122} + \Delta W_{72\text{-}7111} +\\
&+ \Delta W_{1\text{-}7} + \Delta W_{1\text{-}2(I)} + \Delta W_{1\text{-}2(II)} + \Delta W_{2\text{-}212\text{-}2111} + \Delta W_{2\text{-}221\text{-}2221} +\\
&\Delta W_{1\text{-}3} + \Delta W_{3\text{-}312\text{-}3111} + \Delta W_{1\text{-}4} + \Delta W_{4\text{-}422\text{-}4211} + \Delta W_{5\text{-}4} +\\
&+ \Delta W_{5\text{-}6} + \Delta W_{6\text{-}61} + \Delta W_{6\text{-}62} + \Delta W_{1\text{-}6(I)} + \Delta W_{1\text{-}6(II)} + \Delta W_{1\text{-}13}.
\end{aligned}
\qquad (2)
$$

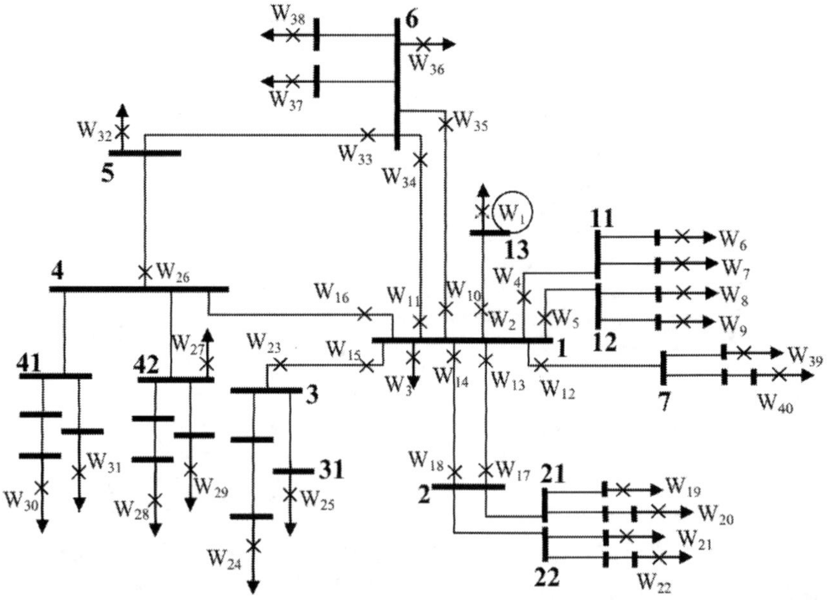


**Figure 2.** Case study. Network diagram, AMR complexes location

Imbalance error evaluation for each VE can be obtained on the basis of maximum permissible error of the constituent measurements:

$$\delta_{W_i} = \sqrt{\sum_{j=1}^{n} a_j^2 \cdot \delta_{W_j}^2} \qquad (3)$$

here $\delta_{W_j}$ is maximum permissible measuring error for the i-th AMR complex calculated according to fundamentals of metrology [13].

The value of maximum permissible measuring error can be seen as double value of standard deviation which is calculated on the basis of measuring instruments included in AMR complex accuracy classes.

Group (2) provides number of estimates calculation for the chosen demanding measurement on each measurement interval. So for the first hour actual measurement $W_1$ is 257600 kWh. Calculated estimates of $W_1$ according to each VE are presented in the second column of Table 1.

**Table 1.** Estimates and calculated errors of VE.

| VE | VE value, kWh | Difference, kWh | Difference, (r), % | VE error, % |
|---|---|---|---|---|
| $W_1^{calc(1)}$ | 253894 | 3706 | 1,438 | 1,15 |
| $W_1^{calc(2)}$ | 256446 | 1154 | 0,447 | 2,15 |
| $W_1^{calc(1)}$ | 256969 | 5516 | 0,244 | 2,15 |

 DOI: 10.1051/matecconf/20167501004

| | | | | |
|---|---|---|---|---|
| ......... | ......... | ......... | ......... | |
| $W_1^{calc(13)}$ | 280893 | -23294 | -9,042 | 4,09 |
| $W_1^{final\_calc}$ | **254623** | **2977** | **1,16** | **0,59** |

Third and fourth columns contain differences between actual measurement and estimates expressed in percentage and absolute value. Fifth column of Table 1 shows the evaluations of maximum permissible errors of VE.

Difference between actual measurement value and its calculated analogue allows to judge about this measurement reliability. The smaller the difference between actual and calculated measurement value the greater the confidence in the measurement. Analysis of Table 1 shows "steps" in VE errors. The more AMR complexes are involved into calculations the higher is the error of the separate VE. The first step is using the nodal EE balance. The second step (or belt) is formed by VE where each linear measurement from the first step VE is expresses through the measurement of the other branch end and technical losses value. Most often, the relative VE error of each "step" is about the same.

If the maximum permissible measuring errors of AMR complexes are the same, the error of each calculated analogues is exceed the error of the measurement. However the presence of a calculated analogues number obtained on the basis of VE allows to compute the final calculated estimate of the measurement by averaging all calculated analogues and taking the variances into account:

$$W_k^{final\_calc} = \frac{\sum_{i=1}^{N} \frac{1}{D_i} W_k^{calc(i)}}{\sum_{i=1}^{N} \frac{1}{D_i}} \qquad (4)$$

here $D_i$ is dispersion of $i$-th VE; $W_k^{final\text{-}calc}$ is final calculated estimate of the required measurement; $W_k^{calc(i)}$ is the value of $i$-th VE.

The final calculated value of $W_1$ obtained by averaging on the basis of (4) is 254623. The relative value of final calculated estimate error of $W_1$ is 0,59% which is almost two times less than the maximum permissible error of AMR complexes 1,1%. The maximum permissible error of AMR complex is determined on the basis of accuracy classes of AMR complex components [13]:

$$\delta_W = \pm 1{,}1\sqrt{\delta_U^2 + \delta_I^2 + \delta_C^2 + \delta_L^2} = \pm 1{,}1\sqrt{0{,}5^2 + 0{,}5^2 + 0{,}5^2 + 0{,}5^2} = 1{,}1,$$

here $\delta_I$, $\delta_U$, $\delta_C$ are accuracy ratings of a current transformer, voltage transformer and an energy meter correspondingly; $\delta_L$ is a voltage drop in a cable line.

The presence of numerous intervals for calculation allows collecting statistical information about estimates error for each AMR complex. Current AMR systems have sufficient depth of archives to provide the required number of measurements. Subsequent processing of such statistical series can rate the characteristics of the concrete AMR complexes. AMR complex is considered in good condition if the average value of errors series does not exceed the maximum permissible error for this AMR complex.

Fig. 3 shows the graphical processing of calculation errors for two AMR complexes 3a and 3b. The shaded area corresponds to the limits of permissible error for each AMR complex.

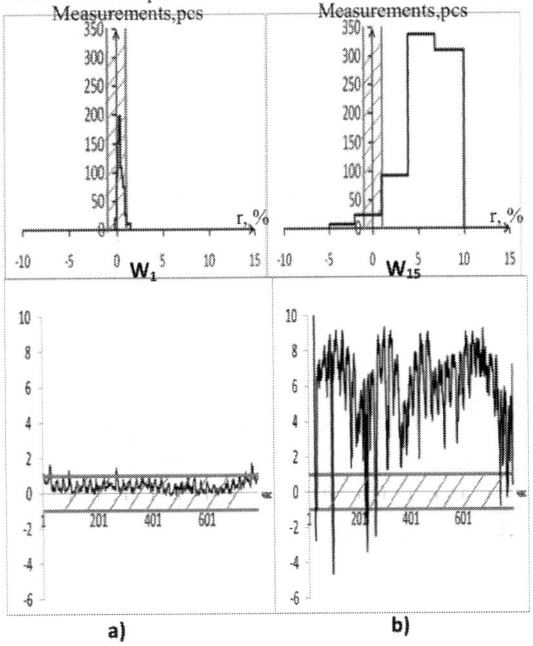


**Figure 3.** Graphical representation of the calculation results. Evaluation of statistical error components.

Fig. 3a shows the calculation results for AMR complex $W_1$. It functions within its accuracy class and do not need recalibration.

Fig. 3b) presents other AMR complex $W_{15}$. It has bias that exceeds maximum allowable limits defined by limits of maximum permissible error of AMR complex under consideration. Complex number 15 requires the attention of specialists to eliminate this error.

## 4 Conclusions

Application of verifying equations technique allows to evaluate the reliability of demanding electrical energy measurements. It also enables to calculate the estimate of energy flow. The error of this estimate is lower than the error of the actual measurement. Reduction is larger, the higher the information redundancy of AMR system.

Application of the proposed method allows to mathematically evaluate the statistical characteristics of the individual AMR complexes errors that identify the complexes operating out of the permissible errors bounds.

## Acknowledgement

The work was supported by Act 211 Government of the Russian Federation, contract № 02.A03.21.0006 and the Ministry of Education and Science of Russian

Federation (in the framework of state assignment, №13.1928.2014/K (project №1928)).

## References

1. A. Abur, A.G. Exposito Power System State Estimation. Markel Dekker inc., New York. 2004.
2. A. Monticelli. State Estimation in Electric Power Systems – a Generalized Approach. Norwell, M A: Kluwer, 1990.
3. F.C. Schweppe, J. Douglas, D. Rom. Power System State Estimation, Part I - Exact Model, IEEE Transactions on Power Apparatus and Systems, **89**, 1970, pp. 120-125.
4. F.C. Schweppe, J. Douglas, D. Rom. Power System State Estimation, Part II - Approximate Model, IEEE Transactions on Power Apparatus and Systems, **89**, 1970, pp. 125-130.
5. A. Monticelli. Electric Power Syaytem State Estimation. Proceedings of the IEEE, **88**, No.2 , pp. 262-282 (2000).
6. A. Monticelli, F.F. Wu, M. Yen. Multiple Bad Data Identification for State Estimation by Combinatorial Optimization. Proceesings of the PICA Conference, pp. 452-460 (1985).
7. X. Nian-De, W. Shi-Ying, Y. Er-Keng. A new approach for Detection and Identification of Multiple Bad Data in Power System State Estimation. IEEE Transactions on Power Systems, **101,** No. 2, pp.454-462 (1982).
8. A.Z. Gamm, I.N. Kolosok. Bad data detection in measurements in electric power system. Nauka, Novosibirsk, Sib. Enterpr. RAS (in Russian). (2000).
9. A.V. Pazderin, S.E. Kokin, A.O. Egorov, E.S. Kochneva. Solution of energy flow problem using state estimation technique, ECON Proceedings (Industrial Electronics Conference) 5414817, pp. 1736-1741. (2009).
10. A.O. Egorov, E.S. Kochneva, A.V. Pazderin. Improving of metering information validity on the basis of the control equation method. The first international conference on sustainable power generation and supply, pp.217-222, (2009).
11. A.V. Pazderin, V.O. Samoylenko. A Mathematical Method of Energy Resources Flows Data Validating Using the State Estimation Theory. Renewable Energy & Power Quality Journal. **10**. pp. 348-352. (2012).
12. A.V. Pazderin, A.O. Egorov, S.A. Eroshenko. The energy meters allocation in electric systems on the basis of observability theory, 9th Conference on Environment and Electrical Engineering, EEEIC 2010 5489950, pp. 167-170 (2010).
13. RD 34.09.101-94. Typical instruction for electricity metering in its generation, transmission and distribution. M. ORGRES, S. (in Russian) (1995).

MATEC Web of Conferences **75**, 01005 (2016) DOI: 10.1051/matecconf/20167501005
*ICMIE 2016*

# Study on the Detection of Transformer Oil Temperature

Kehui Zhou, Yuanliang Huang, Hongqi Li and Tengyu Sun

*Electrical Automation Institute of Jinan University, 519070, Zhuhai, China*

**Abstract.** Based on the influence of the ambient temperature, wind speed and other factors on the transformer monitoring system, the hot spot temperature model of the transformer outer wall is established by using thermoelectric analogy theory and lumped parameter model. Through the analysis of experimental data, mathematical model of influence of ambient temperature and wind speed on temperature of transformer is built, and the correction values on the temperature of the transformer are obtained under the different ambient temperature and wind speed. The correction values make the transformer hot-spot temperature more objective and more authentic, which has significant influence on the running of the transformer safely and efficiently.

## 1 Introduction

As the demand for electrical energy increasing, the security of corresponding devices has a higher demand. In the power system, the transformer is one of the most important and most expensive equipment, which is the key point whether the grid operate safely, efficiently and economically. Due to the nature of the transformer and long-term continuous operation in the power grid, the transformer inevitably occurs a variety of malfunctions and accidents. Therefore, when the power system is developing to EHV, power grid, high-capacity and automated direction, the reliability of the transformer to the safe operation of power system is also increasing [1].

Because the internal structure of the transformer is complex and the transformer hot spots are usually in the position of the high potential, it is difficult for the transformer to monitor constantly by sensor and we need to use temperature calculation method to estimate the temperature distribution within the transformer and forecast the maximum temperature. And it is inaccurate to restrict the load value of the transformer by the temperature rise of the transformer, because it is limited by the ambient temperature. By experience, the maximum temperature of the transformer will be different in different season. In the summer, because the temperature rises and the user load increases, the transformer increase the heat but the cooling conditions deteriorate, causing a marked increase in the internal temperature of the transformer. And after repeated experiments, we found that, under the same temperature, the faster wind speed is, the lower wall temperature of the transformer is. If not considering the impact of wind speed and using the previous formula, the estimated value is smaller than the actual value, which is not good for real-time monitoring of transformer normal operation. Therefore, in this paper, with the ambient temperature and wind speed as the input variable, the hot spot temperature model was established based on the transformer wall temperature.

## 2 Transformer surface temperature thermal model equivalent thermal circuit theory

When the oil-immersed power transformer runs, it has no-load loss and load loss, which transform the heat. Some of the heat increase the temperature of the transformer winding, the core and structure. The winding and the core dissipate the heat to the transformer oil, causing the temperature of the transformer oil rise, and the transformer oil dissipate the heat to the air by cooling device. When the temperature difference of each part achieve the balance between the heat generated and the heat dissipated, the parts reach a stable state and the temperature of the parts are no longer changed [3].

National standard GB1094.2-1996 stipulates that the transformer oil tank and structural surface temperature rising limit is 80K. If we want to discover the abnormal temperature of the transformer before all parts of the transformer temperature rising arrive the limit, we must be able to predict the normal temperature of transformer operation. In the literature [2] model, the transformer hot spot temperature model is modeling to the heat transfer process within the transformer wall, without considering the influence of the air and the transformer oil tank wall (outer) surface heat transfer resistance. However, in general, the transformer tank wall temperature is affected by the environmental temperature and wind speed. And if you ignore the impact, it will cause the calculated heat value becomes small, which is not good for real-time monitoring the operation condition of the transformer.

Therefore, the transformer tank wall temperature model in the paper considers the environmental factors.

According to heat transfer theory, the actual cooling structure of the transformer is refined, thinking that total thermal resistance is composed of four resistance in series. They are respectively thermal resistance $R_1$ of transformer oil, heat transfer resistance $R_2$ of the transformer oil and the tank wall (inside) surface, thermal resistance $R_3$ of the transformer tank wall and heat transfer resistance $R_4$ of air and transformer tank wall (outer) surface.

Temperature measuring point is located between $R_3$ and $R_4$, and the ratio of the temperature difference between measuring point temperature $\theta_{meas}$ and ambient temperature $\theta_{amb}$ to the temperature difference between top oil temperature $\theta_{oil}$ and ambient temperature $\theta_{amb}$ is similar to the relationship of the partial pressure in the circuit. And compared with the transformer oil heat capacity, the transformer tank wall heat capacity, the thermal boundary layer heat capacity between transformer oil and the tank wall (inside) and the thermal boundary layer heat capacity between air and the transformer tank wall (outer) are very small, so the three heat capacities are ignored. Heat of transformer is from core loss $q_{Fe}$ and winding copper loss $q_{Cu}$ and $C_{oil}$ represents the transformer oil heat capacity. So, the transformer surface temperature model equivalent circuit is shown in Figure 1.

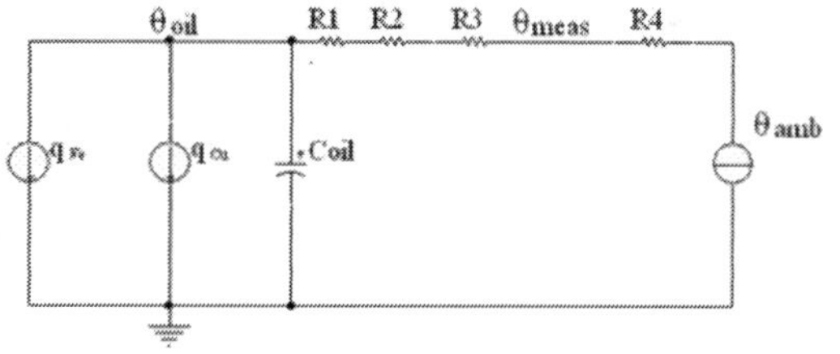


**Figure 1.** The transformer surface temperature model equivalent circuit

$$q_{Fe}+q_{Cu}=(1+\beta)C_{oil}\frac{d\theta_{meas}}{dt}-\beta C_{oil}\frac{d\theta_{amb}}{dt}+\frac{1}{R}[(1+\beta)(\theta_{meas}-\theta_{amb})]^{\frac{1}{n}} \tag{1}$$

wherein, $\theta_{meas}$ is transformer surface temperature measurements; $R$ is the total resistance, $R=R_1+R_2+R_3+R_4$. proportionality coefficient $\beta=(R_1+R_2+R_3)/R_4$.

Generally, the value of $n$ is between 0.8 and 1.0. To derive conveniently, $n$ is 1 in the paper.

Transformer core loss is proportional to the square of the voltage and taking into account the basically constant voltage in the operation of transformer, the core loss is constant, so $q_{Fe}=K_{Fe}$. Transformer winding copper loss is proportional to the square of the current, so $q_{Cu}=K_{Cu}I^2$.

Therefore, the transformer surface temperature prediction model is:

$$\theta_{meas}[k]=\frac{T_0}{T_0+\Delta t}\theta_{meas}[k-1]+\frac{\Delta t+\beta(T_0+\Delta t)}{(1+\beta)(T_0+\Delta t)}\theta_{amb}[k]-\frac{\beta T_0}{(1+\beta)(T_0+\Delta t)}\theta_{amb}[k-1]+\frac{\Delta t R K_{Cu}}{(1+\beta)(T_0+\Delta t)}I^2[k]+\frac{\Delta t R K_{Fe}}{(1+\beta)(T_0+\Delta t)} \tag{2}$$

wherein, $T_0$ is the time constant of the transformer top oil temperature ($T_0=R\cdot C_{oil}$), $\Delta t$ is the sampling interval, $\theta_{meas}^{[k]}$ is the tank wall surface temperature of the k-th sampling point. $\theta_{amb}^{[k]}$ is the environment temperature of the k-th sampling point. $I^{[k]}$ is the primary current RMS of k-th sampling point [4].

## 3 Experimental analysis and modeling of the inner and outer wall temperature of the transformer

To further study the correlation of the inner wall surface temperature and ambient temperature and wind speed, the paper carried out the following experiment.

When the ambient temperature is 23°C, 25°C or 29°C, we respectively measure the relationship between inner wall temperature and outer wall temperature in the no wind, first-class wind, second-class wind and third-class wind. Part of experimental data is shown in Table 1.

**Table 1**. Experimental data (part).

| wind speed | ambient temperature(°C) | hot spot temperature(°C) | 99 | 95 | 90 | 85 | 80 | 75 | 70 | 65 | 60 | 55 | 50 |
|---|---|---|---|---|---|---|---|---|---|---|---|---|---|
| 0 | 29°C | IT | 99 | 95 | 90 | 85 | 80 | 75 | 70 | 65 | 60 | 55 | 50 |
| | | OT | 94 | 90 | 85 | 81 | 76 | 72 | 67 | 62 | 58 | 53 | 49 |
| | 25°C | IT | 97 | 93 | 88 | 84 | 78 | 73 | 68 | 63 | 58 | 53 | 49 |
| | | OT | 87 | 83 | 79 | 75 | 71 | 68 | 63 | 58 | 54 | 49 | 46 |
| | 23°C | IT | 96 | 93 | 88 | 83 | 77 | 74 | 67 | 62 | 58 | 53 | 48 |
| | | OT | 85 | 81 | 77 | 73 | 70 | 66 | 61 | 57 | 53 | 48 | 45 |
| 1 | 29°C | IT | 99 | 94 | 89 | 84 | 78 | 73 | 68 | 63 | 59 | 53 | 48 |
| | | OT | 78 | 75 | 73 | 70 | 65 | 61 | 58 | 54 | 51 | 48 | 47 |
| | 25°C | IT | 95 | 90 | 86 | 81 | 76 | 72 | 67 | 62 | 57 | 52 | 47 |
| | | OT | 72 | 69 | 66 | 62 | 59 | 55 | 52 | 49 | 46 | 44 | 43 |
| | 23°C | IT | 96 | 93 | 89 | 84 | 79 | 74 | 69 | 64 | 58 | 54 | 48 |
| | | OT | 72 | 68 | 65 | 61 | 58 | 54 | 51 | 47 | 44 | 42 | 41 |
| 2 | 29°C | IT | 96 | 92 | 87 | 83 | 78 | 73 | 67 | 63 | 57 | 53 | 47 |
| | | OT | 80 | 77 | 74 | 71 | 67 | 64 | 60 | 56 | 53 | 49 | 48 |
| | 25°C | IT | 96 | 92 | 87 | 83 | 78 | 73 | 67 | 63 | 57 | 53 | 47 |
| | | OT | 70 | 68 | 64 | 62 | 57 | 55 | 51 | 49 | 45 | 44 | 43 |
| | 23°C | IT | 98 | 94 | 89 | 85 | 80 | 74 | 69 | 65 | 60 | 54 | 49 |
| | | OT | 69 | 65 | 61 | 57 | 54 | 50 | 47 | 45 | 43 | 41 | 40 |
| 3 | 29°C | IT | 99 | 95 | 89 | 84 | 79 | 73 | 69 | 62 | 55 | 50 | 46 |
| | | OT | 80 | 76 | 72 | 68 | 64 | 59 | 56 | 53 | 50 | 47 | 46 |
| | 25°C | IT | 95 | 91 | 85 | 81 | 77 | 71 | 66 | 62 | 57 | 52 | 47 |
| | | OT | 70 | 67 | 64 | 61 | 57 | 53 | 50 | 47 | 46 | 44 | 42 |
| | 23°C | IT | 98 | 94 | 88 | 84 | 79 | 74 | 70 | 64 | 58 | 54 | 49 |
| | | OT | 69 | 65 | 61 | 58 | 55 | 51 | 47 | 44 | 42 | 41 | 40 |
| Notes | IT: Inside temperature(°C), OT: Outside temperature(°C) | | | | | | | | | | | | |

After the experimental data for proper treatment, we do curves fitting by Matlab in different ambient temperatures $\theta_{amb}$ and wind speeds $v$. When the ambient temperature is 25°C, relationship between the inner and outer walls temperature of the transformer in different wind speed is shown in Figure 2.

In Figure 2, the fitting functions are:

Wind speed is 0:

$$y=1.18153x-5.64184$$

Wind speed is 1:

$$y=1.5384x-14.9263$$

Wind speed is 2:

$$y=1.6691x-19.8906$$

Wind speed is 3:

$$y=1.62x-17.2363$$

According to Figure 2 and the fitting functions, when the ambient temperature and the inner temperature of the transformer are constant values, different wind speed has different influence on the outer wall temperature of the transformer. The faster wind speed is, the greater the slope of the curve is. Therefore, it is better to estimate winding hot spot temperature when the heat transfer resistance $R_4$ of air and transformer tank wall (outer) surface are considered.

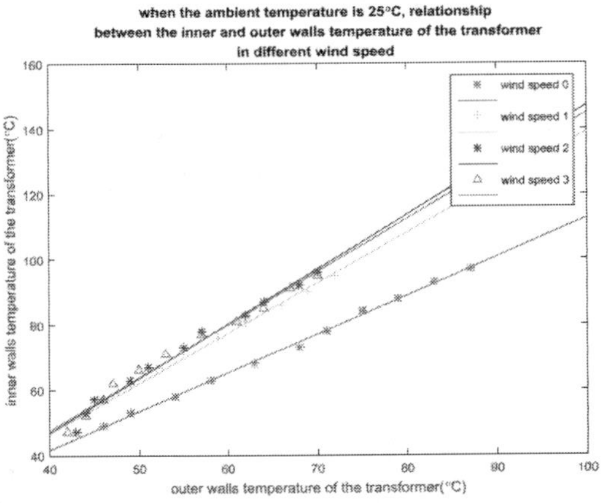


**Figure 2.** When the ambient temperature is 25°C, relationship between the inner and outer walls temperature of the transformer in different wind speed

When the ambient temperature is 29°C, relationship between the inner and outer walls temperature of the transformer in different wind speed is shown in Figure 3.

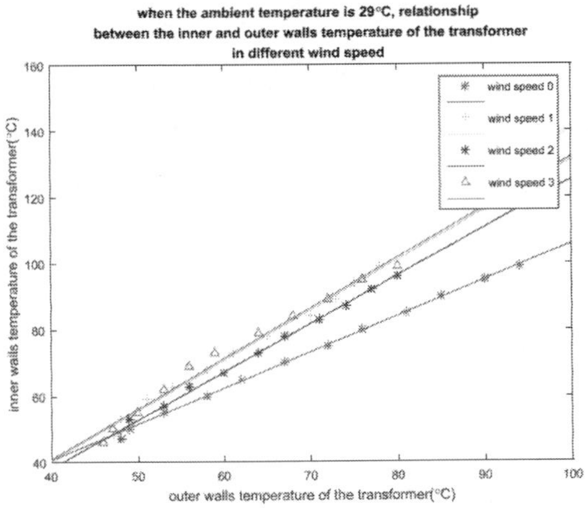


**Figure 3.** When the ambient temperature is 29°C, relationship between the inner and outer walls temperature of the transformer in different wind speed

In Figure 3, the fitting functions are:

Wind speed is 0:

$$y=1.08753x-2.89908$$

Wind speed is 1:

$$y=1.5123x-20.0422$$

Wind speed is 2:

$$y=1.4515x-19.8711$$

Wind speed is 3:

$$y=1.52x-19.9018$$

Compared with 25°C, when the ambient temperature is 29°C, the temperature difference values between inner wall and outer wall become smaller. So, when the inner wall temperature of the transformer is constant value, the higher ambient temperature is, the higher outer wall temperature of the transformer is. What's more, the wind has small influence on heat dissipation.

When the ambient temperature is 23°C, relationship between the inner and outer walls temperature of the transformer in different wind speed is shown in figure 4.

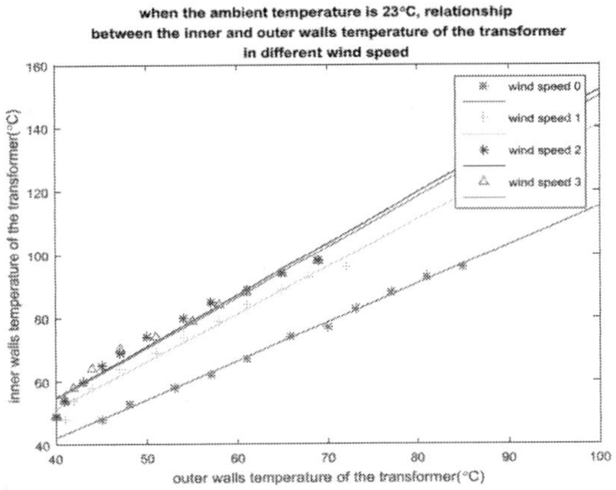


**Figure 4.** when the ambient temperature is 23°C , relationship between the inner and outer walls temperature of the transformer in different wind speed

In Figure 4, the fitting functions are:

Wind speed is 0:

$$y=1.21657x-6.55123$$

Wind speed is 1:

$$y=1.49184x-8.32535$$

Wind speed is 2:

$$y=1.61694x-9.80792$$

Wind speed is 3:

$$y=1.58742x-8.87181$$

The overall situation of 23°C is similar with 25°C, but when the wind speed is faster, the effect on heat dissipation is better in the situation.

## 4 Improved transformer hot spot temperature model

According to the experimental data, when the wind speed is constant value, if ambient temperatures are different, the inner and outer wall temperature difference will be different. It can tell us the change of ambient temperature has a great influence on the change of outer walls temperature. Similarly, wind speed also has obvious influence on the outer wall temperature. Therefore, based on the original model of literature [2], we need add the correction value of the influence on ambient temperature and wind speed, which is the temperature difference between the inner and outer wall.

When there is no wind, for example, according to the experimental data, the different temperatures of inner and outer wall are obtained in the different ambient temperatures, which is shown in Table 2.

According to Table 2, under condition of no wind, the fitting functions of the outer wall temperature and correction value are:(outer wall temperature: $x$, correction value: $\Delta T$)

Ambient temperature is 29°C:

$$\Delta T_1 = 0.08753x - 2.89908$$

Ambient temperature is 25°C:

$$\Delta T_2 = 0.18153x - 5.64184$$

Ambient temperature is 23°C:

$$\Delta T_3 = 0.21657x - 6.55123$$

**Table 2.** The different temperatures of inner and outer wall in the different ambient temperatures

| wind speed | ambient temperature | hot spot temperature(℃) | 99 | 95 | 90 | 85 | 80 | 75 | 70 | 65 | 60 | 55 | 50 |
|---|---|---|---|---|---|---|---|---|---|---|---|---|---|
| 0 | 29℃ | Inside temperature(℃) | 99 | 95 | 90 | 85 | 80 | 75 | 70 | 65 | 60 | 55 | 50 |
| | | Outside temperature(℃) | 94 | 90 | 85 | 81 | 76 | 72 | 67 | 62 | 58 | 53 | 49 |
| | | temperature difference (℃) | 5 | 5 | 5 | 4 | 4 | 3 | 3 | 3 | 2 | 2 | 1 |
| | 25℃ | Inside temperature(℃) | 97 | 93 | 88 | 84 | 78 | 73 | 68 | 63 | 58 | 53 | 49 |
| | | Outside temperature(℃) | 87 | 83 | 79 | 75 | 71 | 68 | 63 | 58 | 54 | 49 | 46 |
| | | temperature difference (℃) | 10 | 10 | 9 | 9 | 7 | 5 | 5 | 5 | 4 | 4 | 5 |
| | 23℃ | Inside temperature(℃) | 96 | 93 | 88 | 83 | 77 | 74 | 67 | 62 | 58 | 53 | 48 |
| | | Outside temperature(℃) | 85 | 81 | 77 | 73 | 70 | 66 | 61 | 57 | 53 | 48 | 45 |
| | | temperature difference (℃) | 11 | 12 | 11 | 10 | 7 | 8 | 6 | 5 | 5 | 5 | 3 |

**Table 3.** The different temperatures of inner and outer wall in the different wind speed

| wind speed | ambient temperature | hot spot temperature(℃) | 99 | 95 | 90 | 85 | 80 | 75 | 70 | 65 | 60 | 55 | 50 |
|---|---|---|---|---|---|---|---|---|---|---|---|---|---|
| 0 | 23℃ | Inside temperature(℃) | 96 | 93 | 88 | 83 | 77 | 74 | 67 | 62 | 58 | 53 | 48 |
| | | Outside temperature(℃) | 85 | 81 | 77 | 73 | 70 | 66 | 61 | 57 | 53 | 48 | 45 |
| | | temperature difference (℃) | 11 | 12 | 11 | 10 | 7 | 8 | 6 | 5 | 5 | 5 | 3 |
| 1 | 23℃ | Inside temperature(℃) | 96 | 93 | 89 | 84 | 79 | 74 | 69 | 64 | 58 | 54 | 48 |
| | | Outside temperature(℃) | 72 | 68 | 65 | 61 | 58 | 54 | 51 | 47 | 44 | 42 | 41 |
| | | temperature difference (℃) | 24 | 25 | 24 | 23 | 21 | 20 | 18 | 17 | 14 | 12 | 7 |
| 2 | 23℃ | Inside temperature(℃) | 98 | 94 | 89 | 85 | 80 | 74 | 69 | 65 | 60 | 54 | 49 |
| | | Outside temperature(℃) | 69 | 65 | 61 | 57 | 54 | 50 | 47 | 45 | 43 | 41 | 40 |
| | | temperature difference (℃) | 29 | 29 | 28 | 28 | 26 | 24 | 22 | 20 | 17 | 13 | 9 |
| 3 | 23℃ | Inside temperature(℃) | 98 | 94 | 88 | 84 | 79 | 74 | 70 | 64 | 58 | 54 | 49 |
| | | Outside temperature(℃) | 69 | 65 | 61 | 58 | 55 | 51 | 47 | 44 | 42 | 41 | 40 |
| | | temperature difference (℃) | 29 | 29 | 27 | 26 | 24 | 23 | 23 | 20 | 16 | 13 | 9 |

Similarly, when there is 23°C, for example, according to the experimental data, the different temperatures of inner and outer wall are obtained in the different wind speed, which is shown in Table 3.

According to Table 3, under condition of 23°C, the fitting functions of the outer wall temperature and correction value are:(outer wall temperature:$x$, correction value:$\Delta T$)

Wind speed is 0:

$$\Delta T_0 = 0.21657x - 6.55123$$

Wind speed is 1:

$$\Delta T_1 = 0.49184x - 8.32535$$

Wind speed is 2:

$$\Delta T_2 = 0.61694x - 9.80792$$

Wind speed is 3:

$$\Delta T_3 = 0.58742x - 8.87181$$

In the conditions of other ambient temperature and wind speed, we can derive the same way.

## 5 Summary

On the basis of referring to the classical model, through the analysis of the cooling process of the interior of transformer, thermoelectricity analogy principle and the influence of the environmental temperature and wind speed on the transformer, the paper improved the model of literature [2] and make a transformer winding hot spot temperature calculation model based on the transformer wall temperature. Under the condition of the constant hot-spot temperature of transformer, the major influencing factors of the transformer outer wall temperature are the ambient temperature and wind speed. Thereinto, the impact of the wind speed is significant.

By considering the influence of the external environment factors on the measurement of the transformer oil temperature, transformer hot-spot temperature can be monitored more objective and more authentic, which is of great significance to the operation of the transformer safely and efficiently.

## Acknowledgment

This paper is supported by the fund of Zhuhai collaborative and innovation plan project.

## References

1. S.Y. Deng. *Science and Technology Innovation Herald*, NO. 20 (2011)
2. X.P. Liu. *Master's degree thesis of Chongqing University* (2010)
3. Kurt Preis, etc. *IEEE transactions on magnetic*, **42**(4):999-1002. (2006)
4. H.Y. Wang, P.S. Su, X.H. Wang, etc. *Tsinghua Univ (Sci & Tech)*, Vol. **45**, No. 4. (2005)

MATEC Web of Conferences **75**, 01006 (2016) DOI: 10.1051/matecconf/20167501006
*ICMIE 2016*

# Research on the Optical Properties of Transformers Partial Discharge Based on Different Discharge Models

Wei Bengang[1], Huang Hua[1], Huang Chao[2] and Wang Liming[1]

[1]*Shanghai Municipal Electric Power Company Electric Power Research Institute, 200437 Shanghai, China*
[2]*Shanghai Dianji University, 201306 Shanghai, China*

**Abstract.** In this paper, the different types of discharge in transformer were simulated based on the real transformer fault model. The optical partial discharge detection system was established based on optical sensors which were capturing partial discharge accompanied by optical effects. In this research, surface discharge and suspended discharge defect model was pressurized to generate partial discharge signal. The results showed that: Partial discharge optical signals could effectively respond the production and development process of transformer partial discharge. It was able to assess discharge level also. When the discharge phenomenon stabilized, the phase of surface discharge mainly between 60°~150°and 240°~330°, the phase of suspended discharge mainly between 260°~320°. According to the phase characteristic of discharge pattern, the creeping discharge and suspended discharge phenomenon of transformer can be distinguished. It laid the foundation for the application of transformer optical partial discharge detection technology.

## 1 Introduction

Power transformers are the core equipment of power transmission and distribution system. Currently, the 220kV and above power transformer are almost oil-immersed power transformer. The internal structure of oil-immersed power transformer is complicated. Because of the internal partial discharge phenomenon, the transformer will be malfunction which might be caused by material defects, design defects and other factors. Therefore, test the transformer internal partial discharge phenomena is necessary [1].

Optical detection method is a non-contact direct of partial discharge detection method which is directly determine partial discharge failure by the discharge process of the optical signal [2]. The partial discharge of electrical equipment produce light radiation, use photoelectric detectors detect light radiation signal generated by the partial discharge, the optical radiation signal was turned into an electrical signal by the interception of the enlargement and processed to the monitoring system, the extent of the partial discharge is assessed by the characteristic of electrical signals [3]. It is a confined space inside transformer. Use photoelectric detectors detect light radiation signal generated by the internal partial discharge of transformer which is shield external light interference signal naturally. It has higher sensitivity measurement and anti-interference ability compared with other detection methods. Therefore, optical method has a unique advantage and gradually become a hot topic in the field [4].

In this paper, surface and suspended discharge in transformer was simulated based on the establishment of the real transformer fault model. In order to analysis of the different discharge types of optical discharge characteristics, the optical partial discharge detection system (OPDS) was established based on optical sensors in the shielded room which was capture partial discharge accompanied by optical effects.

## 2 The principle of optical measurement method to detect partial discharge

### 2.1 Partial discharge spectrum distribution

The research showed that spectral curve of discharge was distributed in this three regions: near ultraviolet, visible and near-infrared spectral region. The spectral curve of visible compose of gentle continuous spectrum. The spectral curve of near ultraviolet and near-infrared spectral superposition of banded spectrum and continuous spectrum superposition [5]. The spectra generated by discharge in the discharge process of high voltage electrical equipment mainly in the ultraviolet region, the visible region of the radiation intensity was weak. The ultraviolet region and infrared region radiation intensity was related to such conditions as the discharge voltage and electrode distance. As the voltage increases, the spectral radiation intensity of the ultraviolet region was also increased. And with the increase of the voltage, spectral amplitude also increases. There was a correspondence between the intensity of light radiation and discharge. So, we could judge the development of partial discharge through detecting spectral amplitude [6].

### 2.2 The spectral spectrum and intensity distribution of partial discharge

Suppose the defects electrons and ions in complex of insulation surrounding space was the reason for partial self-sustaining discharge emission. With electronic composite as an example, suppose space surrounding insulation defects was negative electric field. Put the electron mean free path: λ, then the probability of free path length was greater than or equal to $x_i$ to:

$$P_i = e^{-\frac{x_i}{\lambda}} \tag{1}$$

In addition, during the electrons leave the negative electrode and finally arrived at the $x_i$, the new number of electrons because of collisions should be:

$$n = n_0 e^{\alpha x_i} \qquad \alpha = \frac{1}{\lambda} e^{-\frac{x_i}{\lambda}} \tag{2}$$

α was impact ionization coefficient related to the electron mean free path. Then could be obtained :

$$n = n_0 e^{\frac{x_i}{\lambda} e^{-\frac{x_i}{\lambda}}} \tag{3}$$

By the formula for (4) could know in $x_i$=λ had great value. Visible, due to the impact intensified, the number of free electrons sharply reducing when the distance of electrons beyond the mean free path λ.

Obviously, due to electronic composite luminescence, assuming that all positive ions had same level, the spectrum should be consistent:

$$\lambda = \frac{Vh}{E_n - E_i} \tag{4}$$

In the above formula: V was the transmission speed of light in the media; h was Planck's constant; $E_n$ and $E_i$ were two neutral molecules of different level.

The magnitude of electron recombination luminescence was proportional to the number of electrons arriving at $x_i$ which. Obvious, partial discharge spectral intensity and discharge distance had a relationship with exponential function. This conclusion was also applicable in practical engineering [7].

### 2.3 The relationship between optical radiation power and discharge power

Optical radiation on the nature of the discharge process is a discharge of energy transfer and release, the optical radiation power was part of the discharge power. With the enhanced of discharge, the per unit volume density of ions and free electrons increase in ionizing area, the temperature of electronic increase, the corresponding optical radiation power increase. Assuming discharge power was P, and the corresponding optical radiation power was $P_l$, then the optical radiation power P and discharge power Pl approximation satisfy the following formula [8]:

$$P_1 = \lambda_1 P \tag{5}$$

In the above formula: $\lambda_1$ was the relationship coefficient between optical radiation power and electric power.

The spectral curve of discharge has a wide range: the near ultraviolet, visible and near-infrared spectral region. The spectra generated by discharge in the transformer mainly in the ultraviolet region. Suppose ultraviolet in the whole optical radiation power proportion coefficient was $\lambda_2$, between the ultraviolet radiation power $P_2$ and the discharge power could be expressed as:

$$P_2 = \lambda_1 \lambda_2 P \tag{6}$$

Optical signal intensity which detect by OPDS related to the power of discharge and spectral distribution of discharge. From a macro point of view it could be understood as an optical signal intensity related to the applied voltage level, the type of discharge and other factors. With the increase of discharge power, corresponding light signal radiation power will also increase [9].

## 3 Test platform and test plan

In this research, the transformer partial discharge fault simulation test platform was mainly composed of no partial discharge transformer, the discharge diagnosis typical defect model of partial discharge and OPDS. All the equipment was conducted in the shielding room in order to ensure the accuracy of the test.

### 3.1 No PD test transformer

As shown in Fig. 1, no partial discharge transformer formed by pressure system and the power frequency transformer tank model. The pressurized system (figure YD-25kVA/110kV) rated power frequency voltage was 110kV, rated capacity of 25kVA. Partial discharge under rated voltage 110 kV less than 10pc. Transformer tank simulation box size: 3000mm(L)× 2000mm(W)× 2000mm(W), for discharge within the oil output of light signal transmission provides a bigger space, tank shell grounding. Opening at the top of the box reserved for the installation simulation model of partial discharge defects. In order to provide partial discharge signal for the experimental, through the processing frequency voltage to simulated body in various discharge in transformer model to simulate the partial discharge in transformer oil.

DOI: 10.1051/matecconf/20167501006

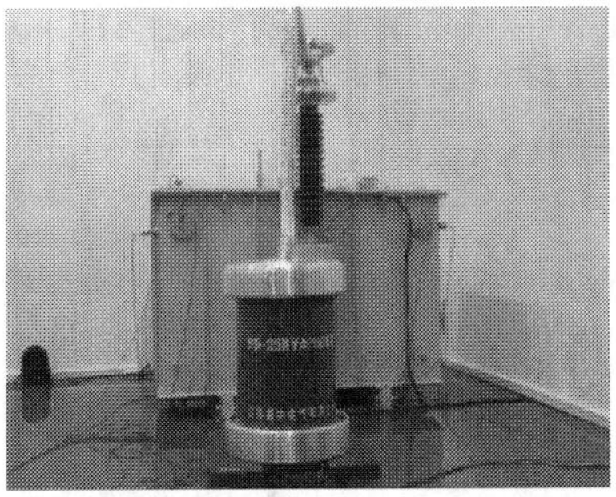

**Figure 1.** No PD test transformer.

### 3.2 Typical partial discharge defect model

Considering the type and location of partial discharge occurred in transformer insulation structure, two defect model was designed in this experiment to simulate partial discharge that common phenomenon of inside the transformer, namely creeping discharge and suspended discharge.

Surface discharge defect model was shown in Fig. 2(a). electrodes composed of upper and lower electrode which made of copper. The upper electrode diameter was 4cm. The side edge contacted insulating paper was a right angle. The lower electrode diameter was 4cm, the edge of electrode was arc-shaped. Insulation board was a 4mm thick circular piece of cardboard which diameter is 8cm.

Suspended discharge defect model was shown in Fig. 2(b). Tip electrode shape combination of cylindrical and conical, the diameter was 1.2cm, The height of cylindrical and conical are1cm. There was a cut wire fixed on the plastic sheet under the copper electrode.

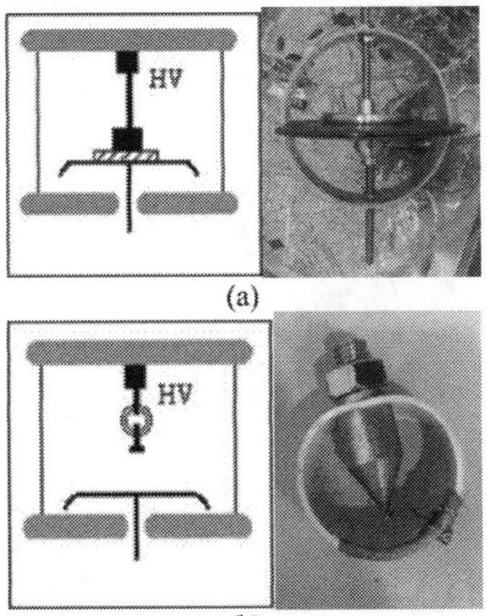


**Figure 2.** Defect model.

### 3.3 OPDS

OPDS structure was shown in Fig.3. OPDS was equivalent to a direct eye "view" technology. It could visual display partial discharge phenomenon that occurs inside the transformer.

The sensor for detecting the light generated by partial discharges functions according to the following principle: A highly sensitive photoelectric element captures the light generated by the partial discharge. The electrical voltage pulse resulting from the partial discharge passes through a signal processing chain at the end of which was displayed as a measuring point on the graphic user interface of the software. The sensors are connected via a connection line to the related interface box.

The spectral response of the sensor element with the wavelength range of 230~700nm. The most sensitive wavelength is 400nm which belongs to the ultraviolet wavelength range. X axis for phase axis 0~360 °, Y axis for amplitude which representative of the light intensity of partial discharge. The time for Signal acquisition was 4s. A atlas contains all discharge points within 200 frequency cycle.

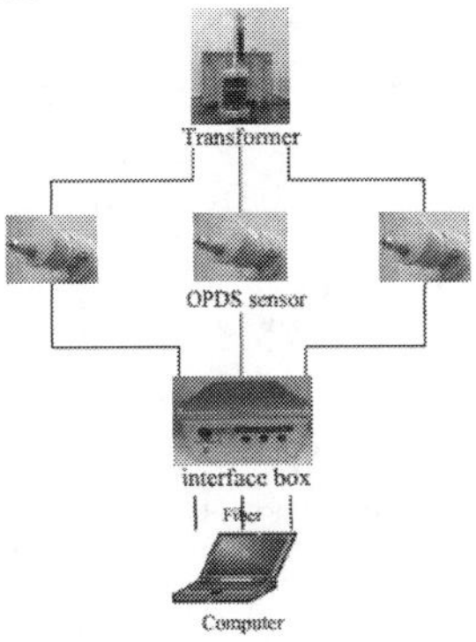


**Figure 3.** Optical partial discharge system.

### 3.4 Test program

The sensor was inserted into the transformer to direct detection of the optical signal generated by partial discharge. The sensor arrangement was shown in Fig. 4.

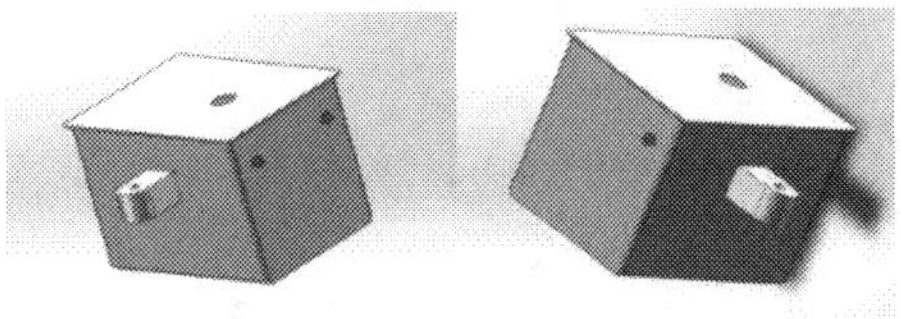

**Figure 4.** Sensor arrangement.

The test voltage discharge was weak as a starting discharge voltage model. As the applied voltage increases, the discharge in the development stage. Near the end of the breakdown voltage of the discharge, the discharge phenomenon was most serious. Thus, gradually increase the test voltage method was used to study the development process of partial discharge characteristics of the signal trends with creeping discharge and suspended discharge model. Between the initial discharge voltage and breakdown voltage, a few appropriate voltage levels was be select and each voltage level for a certain time to acquisition of partial discharge atlas.

## 4 Analysis of test results

### 4.1 Optical signals in the process of creeping discharge change rule

DOI: 10.1051/matecconf/20167501006

The creeping discharge atlas collected by system was shown in Fig. 5. The sample occurred creeping discharge when voltage was 16kv. We could found from Fig.5 (a) (b), in the initial discharge stage, the transformer discharge phenomenon was relatively weak. The discharge atlas collected by three sensor fewer and scattered. Gradually raising the voltage, creeping discharge tends to strong development. As shown in Fig.5(c), when the voltage was increased to 20KV, discharge atlas showed the number of discharge increased obviously which consistent with the development trend of the creeping discharge. Continue to raise the voltage to 26KV, internal transformer occurred strong partial discharge, the phase of surface discharge mainly between 60°~150°and 240°~330°.

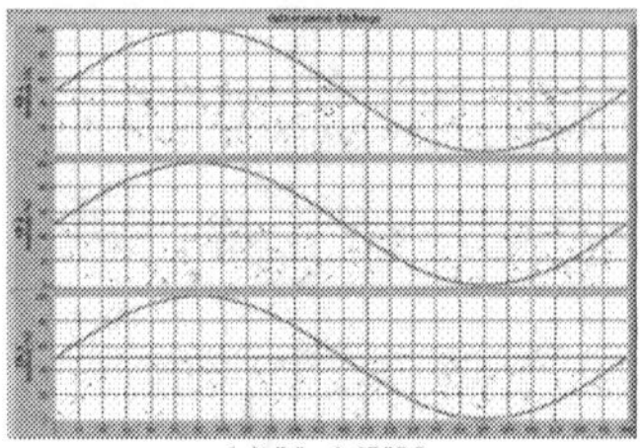

(a) U=16KV

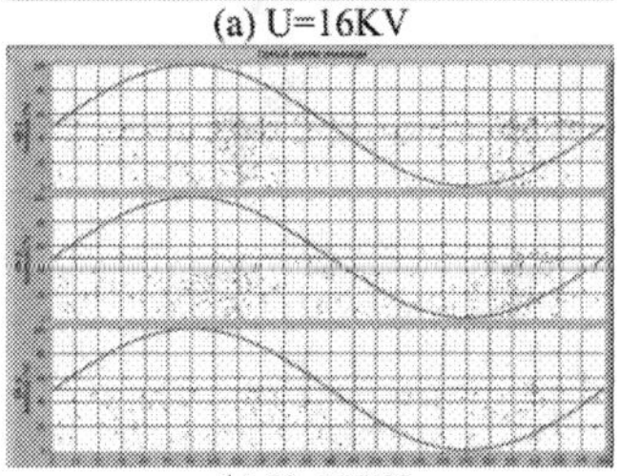

(b) U=18KV

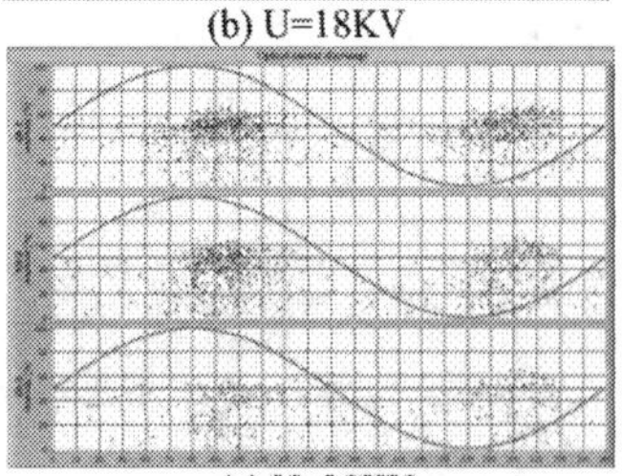

(c) U=20KV

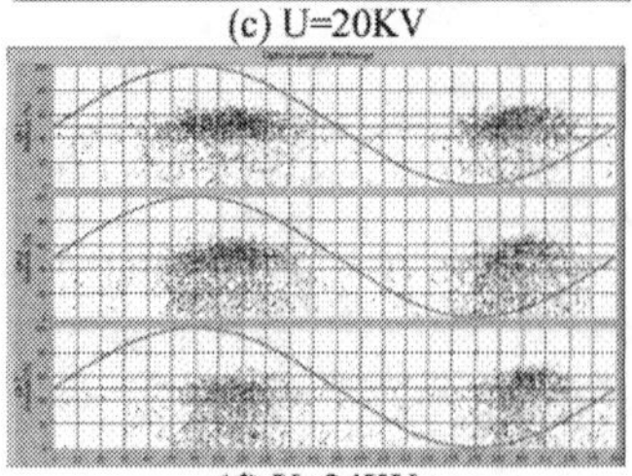

(d) U=24KV

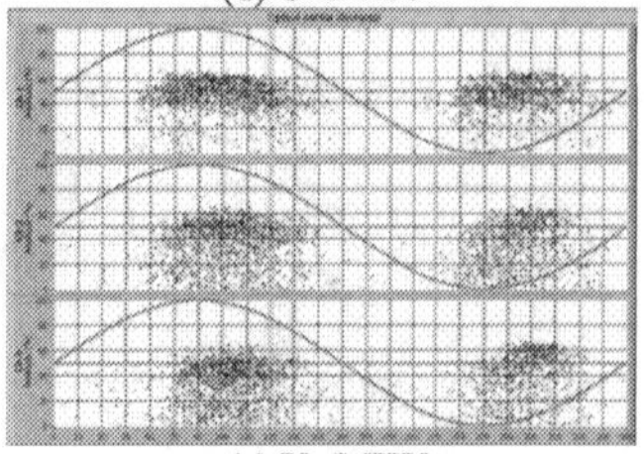

(e) U=26KV

**Figure 5.** Spectrum of surface discharge.

## 4.2 Optical signals in the process of suspended discharge change rule

The suspended discharge atlas collected by system was shown in Fig. 6. We could found from Fig. 6(a), when the voltage was increased to 44KV, the transformer discharge phenomenon was relatively weak. Gradually raising the voltage, suspended discharge tends to strong development. As shown in Fig. 6(c), when the voltage was increased to 70KV, internal transformer partial discharge times had a more significant increase when compared with 44KV. Continue to raise the voltage to 80KV, As shown in Fig. 6(d), internal transformer occurred strong partial discharge, the phase of surface discharge mainly in 260°~320°.

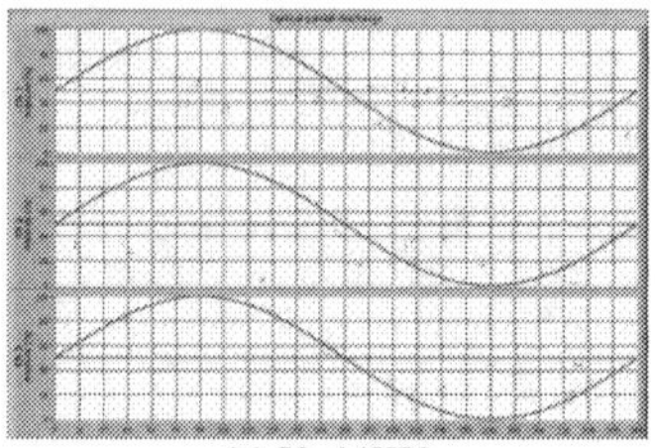

(a) U=44KV

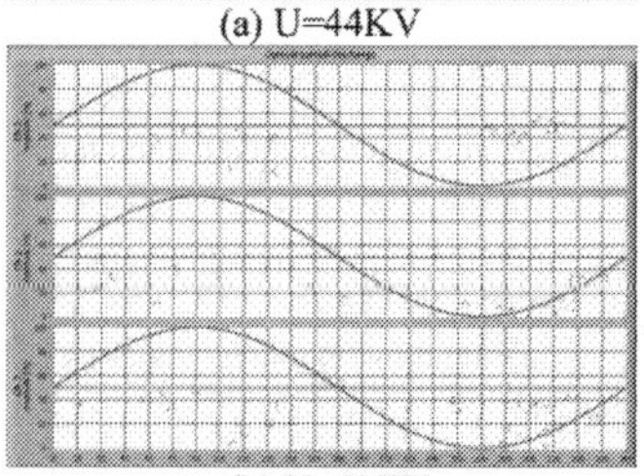

(b) U=60KV

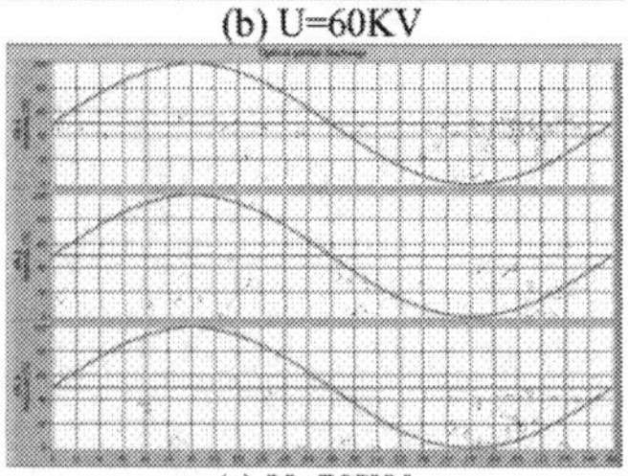

(c) U=70KV

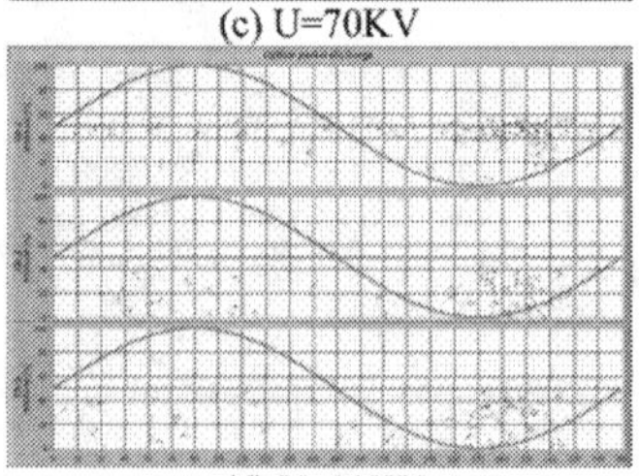

(d) U=80KV

**Figure 6.** Spectrum of suspended discharge.

# 5 Conclusion

In this paper, the surface discharge and suspended discharge in transformer was simulated based on the real transformer fault model. The OPDS which established based on optical sensors in the shielded room for capturing optical effects of partial discharge. The research findings conclude that:

1) OPDS could be used to detect the emergence and development of internal partial discharge of transformer. Discharge point number and energy acquired by the OPDS which could visually determine internal transformer partial discharge generation and discharge strength.

2) When the discharge phenomenon stabilized, the phase of surface discharge mainly between 60°~150°and 240°~330°, the phase of suspended discharge mainly between 260°~320°.

3) According to the phase characteristic of discharge pattern, the creeping discharge and suspended discharge phenomenon of transformer can be distinguished. It laid the foundation for the application of transformer optical partial discharge detection technology.

## References

1. Xie Yucheng, Manual of power transformers, Beijing: *China Machine Press* (2003)
2. Tang Ju, Ouyang Youpeng, Fan Min,System Using fluorescent fiber for partial discharge detection in transformer, *High Voltage Engineering*, **37**(5): 1129-1134 (2011)
3. Liu Yunpeng, Wang Jian, Zhao Tao, Optoelectronic partial discharge detection system under impulse voltage., *Electrical Measurement&Instrumentation*, **52**(12): 68-72 (2015)
4. Si Wenrong, Li Junhao, Yuan Peng, Current situation and development of optical detection for partial discharge, *High Voltage Apparatus*, **44**(3): 261-264 (2008)
5. Zhang Haifeng, Pang Qichang, Chen Xiuchun, The characteristics of high-voltage corona and its detection, *Electrical Measurement & Instrumentation*, **44**(3): 261-264 (2008)
6. Fu Zhong, Chen Shixiu, Chen Wei, Analysis of spectral characteristic in the corona discharge and experimental software design, *High Voltage Engineering*, **33**(7): 92-95 (2007)
7. Wang Yan, Liang Dakai, Zhao Guangxing, Detection of ultraviolet spectrum based on ICCD in the high voltage corona discharge, *Infrared and Laser Engineering*, **42**(9): 2431-2436 (2013)
8. Lin Qi, Ren Qinglei, The plasma optical emission characteristic test of atmospheric air glow discharge, *Journal of Xiamen University (Natural Science)*, **44**(5): 621-624 (2005)
9. Wang Shenghui, Detection and assessment of contaminated suspension insulator discharge based on ultraviolet imaging, *Baoding: North China Electric Power University*, (2011)

MATEC Web of Conferences **75**, 01007 (2016) DOI: 10.1051/mateconf/20167501007
*ICMIE 2016*

# Wall pressure signatures of turbulent flow over longitudinal

Hayder A. Abdulbari[1,2], Hassan D. Mahammed[1], Zulkefli B. Yaacob[1] and Wafaa K. Mahmood[2]

[1]*Faculty of Chemical and Natural Resources Engineering, Universiti Malaysia Pahang, Gambang 26300, Kuantan, Pahang, Malaysia*
[2]*Center of Excellence for Advanced Research in Fluid Flow, Universiti Malaysia Pahang, Gambang 26300, Kuantan, Pahang, Malaysia*

**Abstract.** Five triangular riblets longitudinal in the streamwise direction have been studied experimentally. The riblets have pick to pick spaced (s) equal to 1000 μm and with groove height to space ratio (h/s) 0.4, 0.6, 0.8 and 1. The tests were conducted in a full turbulence water channel on a flat plate for Reynolds numbers 13000 to 53000 based on channel hydraulic diameter. Pressure drop was measured using pressure transmitter gauge with pressure tap points of 12.7 mm in diameter were provided at the bottom of the channel. The main purpose of the present study is to investigate the response of turbulent flow to longitudinal grooves of triangular shaped riblets and compare the effect of the turbulence structure over smoothed and grooved surfaces with pressure drop measurements. 10.20 was the maximum drag reduction appear at h/s equal to (1).

## 1 Introduction

The effects of rib surfaces on the turbulent flow in rectangular channels and circular tubes have been extensively studied in literature [1-10]. Several challenges face surface textures and disabling this technology to be translated into industrial applications.

Choi [11] studied the wall pressure fluctuations in a wind tunnel to modified turbulent boundary layer with riblets. The tunnel had a test section of (4.8m x 2.4 m x 15 m) and 0.5 % flow speed controlled by a feedback control system. The results of this study were drawn comparing with smooth surface. The results revealed that surface modifications reduced RMS intensity up to 4 %. The reduction appeared as a result of decrease in turbulence intensity within a small volume near a riblet surface.

Reidy [12] and Reidy and Anderson[13], conducted experimental studies in 3M riblets in a high speed water tunnel over a flat plate in turbulent flow. The second experiment has been in 6-inch diameter pipe line system. 3M vinyl riblets have been used in both systems with height and spacing of V-groove h=s=76.2μm. the results shows a maximum drag reduction 8.1% Nonetheless, several challenges face surface textures and disabling this technology to be translated into industrial applications. Where there are no certain design (shape and dimensions) of riblets that can be used for all type of applications, due to turbulent regime where the turbulence coherent structure are completely random, as the viscous sublayer and the streamwise vortices are different from fluid to another due to properties of the fluid and the application circumstances . In the present work, triangular riblets will be designed, fabricated and tested in a closed loop liquid circulation system. the flow rate will be the controlled variable and the pressure drop across the rib surface will be measured to evaluate the drag reduction performance.

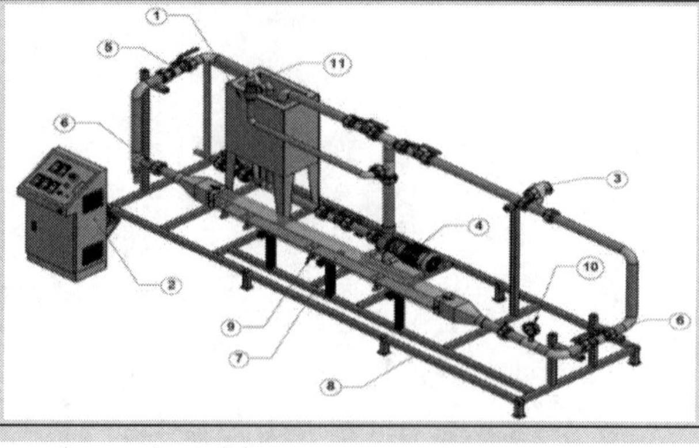

| | |
|---|---|
| 1 | Tank |
| 2 | Control Panel |
| 3 | Flow Meter |
| 4 | Pump |
| 5 | 2" Ball Valve |
| 6 | 2" Union |
| 7 | Flow Duct |
| 8 | Platform |
| 9 | Pressure Transmitter |
| 10 | Pressure Gauge |
| 11 | 1" PRV |

**Figure 1.** Shows the experimental rig.

 DOI: 10.1051/matecconf/20167501007

## 2 Experimental set-up and measuring techniques

### 2.1 Experimental Apparatus and Riblets Fabrication

The main purpose of the present work is to determine the effect of longitudinal triangular riblets on the development of the pressure characteristics over the selected models. A closed loop system was used in the present work to test the drag reduction efficiency of the chosen riblet dimensions as shown in Figure 1. The system consists mainly of one tank rectangular in shape and had a length of 500 mm width of 300 mm depth of 520 mm and capacity of 70 litters; pump with capability to drive the primary fluid at a rate of 36〗 m 〖^3/hr , PVC pipes with 2 inches in size, flow meter, pressure transducers and square channel. Tap water was used and repeatedly changed after each series of tests. Water is then pumped from the centrifugal pump, from the supply tank, into the PVC pipes, then through the transparent channel and into the supply tank again.

The square channel of the experimental rig has been designed and fabricated in such a manner that there are four transparent polycarbonate plates with dimensions (2×0.12×0.01)m, which have been used to fabricate the channel. Acrylic welding was used in fabrication of all the sections of the channel. Four pressure tabs were placed at the bottom surface of the channel that was used to measure the pressure drop. These tabs comprised the test sections, where 0.16 m was placed after two sections of 0.55 m to ensure a full turbulence flow. The system also featured six ball valves that were strategically located at the points of entry and exit of the water and were primarily present to regulate the flow of water. The pressure transmitters were connected to the first test section with a length of 0.16 m, and the pressure drop was taken for the section over the smooth plate and rib surfaces. The range of Reynolds numbers was from 13000 to 53000 based on channel hydraulic diameter. Fluctuating pressure measurements in the channel wall were taken with great care to eliminate all vibration in the experimental equipment.

Aluminum sheets with dimensions $(9.53 \times 90 \times 3000)mm$ have been used for manufacturing the riblet surface, the plate first have been cut in to the size of the test section $(8 \times 80 \times 160\,)\,mm$ with CNC machine. After prepared the plates, WEDM machine was used to fabricate the triangular groove. A Brass Wire with a diameter of 150 µm was used as an electrode to erode a wokpiece. The four triangular riblets were fabricated longitudinal to the direction of the streamwise with the characteristic dimensions listed Figure 2 that illustrates grooves that are defined by peak height (h) and peak-to-peak spacing (s) with selected dimensions. The groove peak-to-peak spacing was fixed for all fabricated riblets to 1000µm but with height to space ratio 0.4, 0.6, 0.8 and 1. The main purpose of the present study is to investigate the response of turbulent flow to longitudinal triangular grooves of with different rib height and fixed rib spaced and compares the effect of the turbulence structure over smoothed and grooved surfaces with pressure drop measurements.

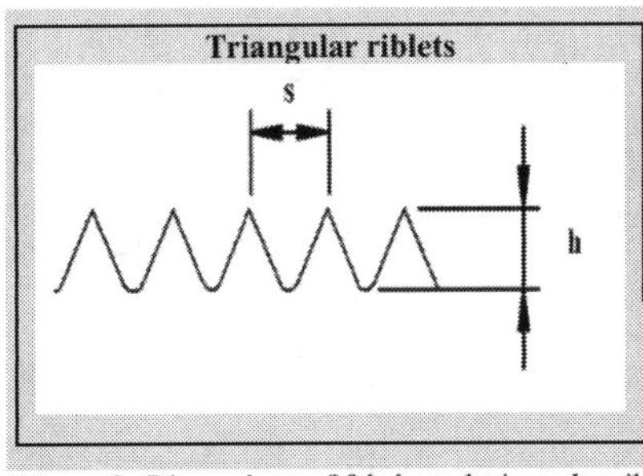


| h/s | h (µm) |
|---|---|
| 0.4 | 400 |
| 0.6 | 600 |
| 0.8 | 800 |
| 1 | 1000 |

**Figure 2.** Dimensions of fabricated triangular riblets.

### 2.2 Measurements

The The skin friction coefficients (*Cf*) and Reynolds numbers (Re) were calculated from the wall shear stress and flow rate measurements by using the following equations:

$$Cf = \frac{2\tau_w}{\rho u^2} \text{ and } Re = \frac{u\,D_h}{\upsilon} \qquad (1)$$

where $(u)$ is mean velocity $(m/\text{sec})$, $(\upsilon)$ is the kinematic viscosity of water $(m^2/\text{sec})$, $(\rho)$ is the density of water $(Kg/m^3)$, and $(\tau_w)$ is the wall shear stress in a fully developed pipe flow, as defined by Perry et al. [33]. This equation relates the wall shear stress to the pressure drop during the turbulent flow inside the pipelines.

$$\tau_w = \frac{\Delta p\ D_h}{4L} \qquad (2)$$

The height and spacing of the grooves in wall units are calculated by using Walsh [34] identification in terms of wall units as non-dimensional height ($h^+$) and spacing ($s^+$), as;

$$h^+ = \frac{h\,u}{\nu}\sqrt{\frac{C_f}{2}} \text{ and } s^+ = \frac{s\,u}{\nu}\sqrt{\frac{C_f}{2}}, \qquad (3)$$

where $(C_f)$ is local skin friction coefficient, $(s)$ is peak to peak spacing of riblets and $(h)$ is the valley to peak height of riblets.

The percentage drag reduction (DR %) is defined as:

$$DR\% = \left(\frac{\Delta P_{Smooth} - \Delta P_{riblet}}{\Delta P_{Smooth}}\right) \times 100 \qquad (4)$$

The diameter used in the calculation of the Reynolds number and friction coefficient is defined as the hydraulic diameter $D_h = 4\frac{A}{p}$, where $(A)$ is the area section of the duct and $(p)$ is the wetted perimeter of the duct.

## 3 Results and Discussion

As per the theory, the effect of longitudinal grooved surfaces on wall-pressure fluctuations can be estimated by calculating the effect first on smooth flat plate and then on ribbed plate. In order to get a well-defined result, the Reynolds number and surface roughness of both the plates should be the same.

The first thing that needs to be addressed before measurement is the time series of pressure. We can analyze the time domain by plotting a graph of the sequence of data points of signal obtained. This will give an accurate description of both the time scale, rate and complexity of the flow. The pressure signal, on the other hand, depends on the random fluctuations which are roughly near the pressure mean value.

The turbulent flow relies on fluctuations in three factors, namely, pressure, acceleration and shear stress. All three are directly associated with position and time. These fluctuations directly affect the velocity and pressure terms in the momentum and energy equations which will show varied differences. Figures 3 show the pressure drop fluctuations over the smooth and riblet plates at Reynolds number $5.3 \times 10^4$.

There is a mechanism that controls the drag reduction and behavior of the grooved surfaces which needs to be

investigated in the experiment. For that, the pressure drop readings of each type of surface and the flow rate are measured at the same time at 60s. As a result, there arise two distinct groups of fluctuations that contribute to the wall pressure as per the suggestion provided by the collective body of results.

There are two groups of fluctuations that make up the wall pressure. The first one is made up of large-scale disturbances that are of low-frequency. Such disturbances come from the surrounding portions of the boundary layer and goes up to within the unsteady potential flow. The large-scale disturbances remain consistently on course with the character of the interface of the potential flow which is outside the boundary layer.

The second group of pressure fluctuation, on the other hand, is made up of small-scale disturbances that are of high-frequency and are considered to have a relation with the burst-sweep cycle of events. Such disturbances are also noted to follow the wall pressure fluctuations of large-amplitude. The large-amplitude wall pressure peaks, according to the results of the conditional sampling techniques on signals of pressure and velocity that are simultaneously acquired, are realized to have relation to the shear layer structures found in the buffer region, and have an association with the bursting process in flat-plate turbulent boundary layers, turbulent flows of pipes as well as direct numerical simulation turbulent channel flows.

The pressure drop performed on the surface showed readings that are stable while having a turbulence frequency lower than other surfaces. It was seen that the pressure drop point is below the smooth surface pressure drop but with irregular readings. Now the structured surfaces exhibit pressure readings which are higher in amplitude and smooth surface readings which are lower in amplitude. It is seen that the structured surfaces have a lower pressure drop on average but in some points, it may be higher. This is due to the fact that the low pressure drop arises when the pressure drops stabilizes for a certain time and then returns to the smooth surface readings. The same pattern repeats several times which explains the low pressure drops on average.

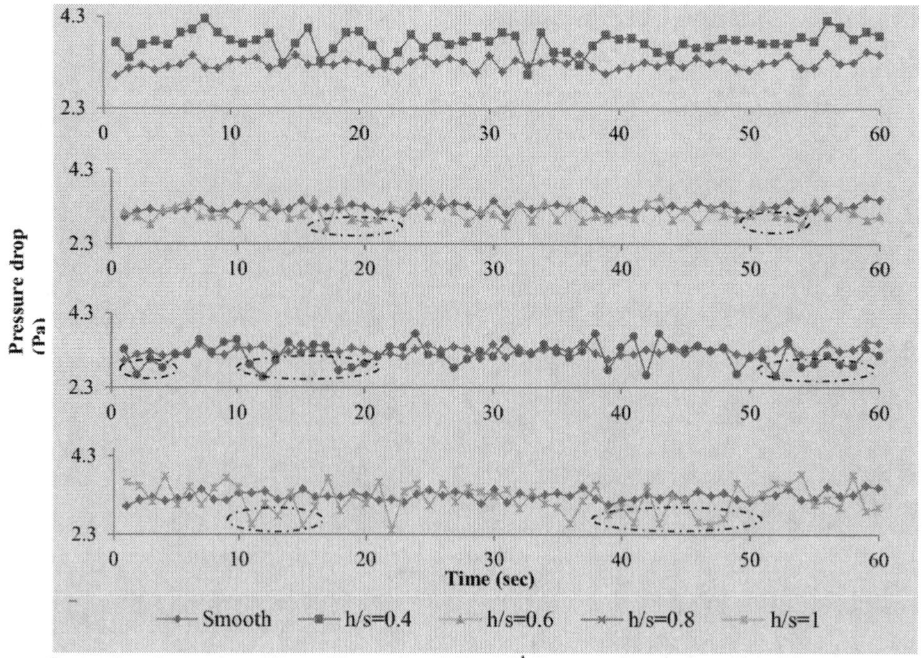


**Figure 3.** Pressure fluctuation time series at Reynolds number $5.3 \times 10^4$ for smooth and triangular riblet.

The profiles of the pressure fluctuation time series under different conditions were illustrated as in Figures 4. The turbulent characteristics in the square channel were fully embodied. Generally the pressure fluctuation profile

DOI: 10.1051/matecconf/20167501007

with different Reynolds numbers at smooth and triangular riblet, the fluctuation of pressure time series obviously increased as the Reynolds number of the inlet flow increased over riblets compared to smooth surface. But the amplitudes average values of riblet is less than amplitudes average values of smooth surface.

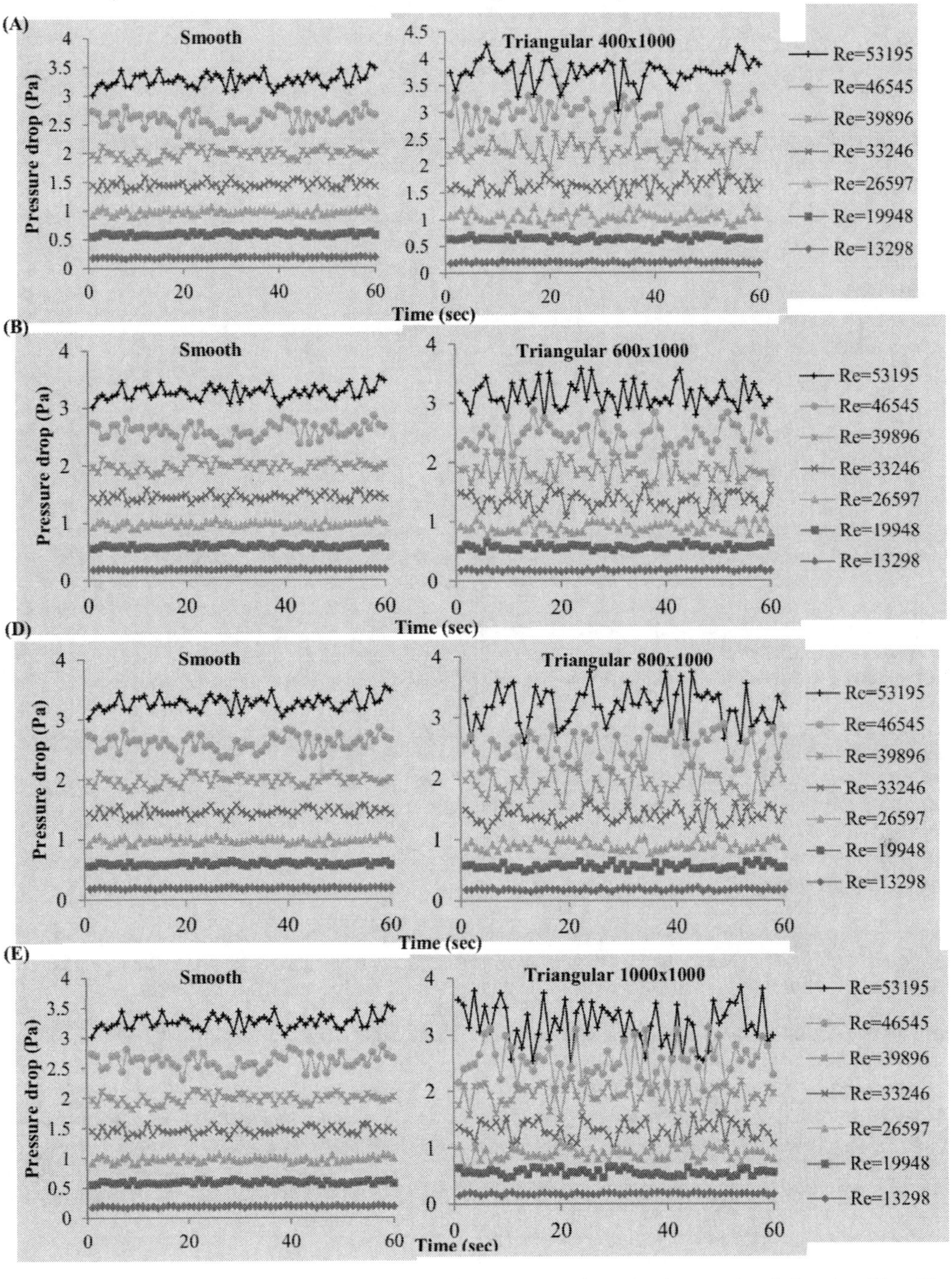


**Figure 4.** The profiles of pressure fluctuation time series under different Reynolds number for smooth and triangular riblet with selected dimensions (A) triangular riblet 400×1000μm, (B) triangular riblet 600×1000μm, (C) triangular riblet 800×1000μm, (D) triangular riblet 1000×1000μm.

These experiments deal with the Reynolds numbers between 13000 and 5000. Figure 5 discuss the pressure drop measured across the smooth plate and all triangular riblet surfaces. The initiation happens in between the region of $1.3 \times 10^4$ and $1.9 \times 10^4$. And as the Reynolds number increases, the pressure drops increase. Afterwards, in the region of $1.9 \times 10^4$ to $5.3 \times 10^4$, the zone is represented by the pressure drop which continues to rise on two characterized zones can be spotted out easily. This are the joining zone where the pressure drop of the triangular grooved surfaces and that of the smooth surface are near a Reynolds number going from $1.3 \times 10^4$ and $2.6 \times 10^4$ are conveniently used to delineate a relentless increment in the rate of pressure drop as the Reynolds number expansions.

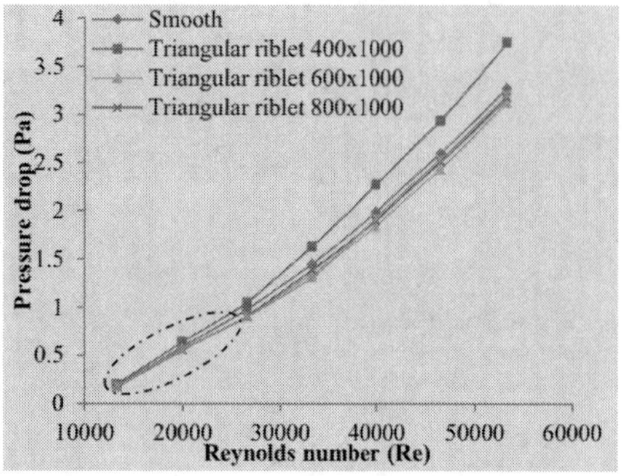


**Figure 5.** Pressure drop comparison over smooth and triangular groove.

The other defined zone is the pressure drop of the tried plates that has a method for ascending on a consistent premise and begins to separate to the Reynolds number running from $2.6 \times 10^4$ to $5.3 \times 10^4$. On the hand, it must be noted of that the trends of the three triangular riblet dimensions with height 600, 800 and 1000μm are below the pressure drop of smooth plate which show a capability to reduce the drag force but with triangular riblet 400 ×1000 show increasing in pressure drop compare to smooth surface. Figures 6 demonstrate the effect of non-dimensional space ($s^+$) and Reynolds number on the percentage of drag reduction with different triangular riblet heights. the data obtained for Reynolds number range of Reb≈13000–53000. The riblet spaces in nondimensional wall units were then estimated from equation (3) and the percentage of drag reduction was estimated from equation (4). The present measurements of riblet drag reduction shows that the percentage drag reduction increases by increasing the $s^+$ and Reynolds number for most of the tested riblets until a certain value where the percentage drag reduction start to descend. Which means increasing the turbulence spectrum and reaching a maximum point or maximum performance "which might be considered as optimum performance for this point only". After that and by a further increase in the value of the $s^+$ and Reynolds number start to decline in most cases due to the strong action of these turbulent structures over riblets, where riblets no longer could modify these structures, which is dependent on the properties of the particular boundary layer. However, the best performance for riblets was in range of $s^+$ equal to 15-20 wall unit and for Reynolds number range equal to 30000 to 40000, high drag reduction up to 10.20% was obtained based on repeatable pressure drop measurements.

## 4 Conclusions

Experimentation was carried on triangular riblets with four different sizes of grooves to determine which one shows the maximum drag reduction in a particular water square channel at different Reynolds numbers. The peak to peak spaced dimension was the same for all four tasted riblets 1000 μm with height to space ratio was chosen to be 0.4-1. After the thorough experimentation, we can conclude that the triangular riblets show an increase in pressure drop as the Reynolds number increases. The triangular riblet with 1000 x 1000 μm reach maximum drag reduction 10.2 at Reynolds number $3.3 \times 10^4$ and then start to decline at Reynolds number $4.2 \times 10^4$.

## References

1. H. A. Abdulbari, H. D. Mahammed, and Z. B. Y. Hassan, ChemBioEng Reviews 2, 185 (2015).
2. B. Dean and B. Bhushan, Philosophical Transactions of the Royal Society of London A: Mathematical, Physical and Engineering Sciences 368, 4775 (2010).
3. F. E. Fish and G. V. Lauder, Annual Review of Fluid Mechanics 38, 193 (2006).
4. M. Gad-el-Hak, Applied Mechanics Reviews 49, 365 (1996).
5. A. M. Savill, in Structure of Turbulence and Drag Reduction (Springer Berlin Heidelberg, 1990), pp. 429.
6. H. A. Abdulbari, R. M. Yunus, N. H. Abdurahman, and A. Charles, Journal of Industrial and Engineering Chemistry 19, 27 (2013).
7. M. J. Walsh, in Viscous drag reduction in boundary layers, edited by D. Bushnell, and J. Hefner (American Institute of Aeronautics and Astronautics, Washington, DC, 1990), pp. 203.
8. A. M. Savill, Flow visualization IV 1, 303 (1987).
9. S. P. Wilkinson, J. B. Anders, B. S. Lazos, and D. M. Bushnell, International Journal of Heat and Fluid Flow 9, 266 (1988).
10. P. Vukoslavcevic, J. M. Wallace, and J. L. Balint, AIAA Journal 30, 1119 (1992).
11. K.-S. Choi, in Turbulence Management and Relaminarisation, edited by H. W. Liepmann, and R. Narasimha (Springer Berlin Heidelberg, 1988), pp. 149.
12. L. W. Reidy, Flat Plate Reduction in a Water Tunnel Using Riblets, 1987.
13. L. Reidy and G. Anderson, in 26th Aerospace Sciences Meeting (American Institute of Aeronautics and Astronautics, 1988).

# Machinability and Tribological Properties of Stir Cast LM6/SiC/GR Hybrid Metal Matrix Composite

Montasser S. Tahat

*Al-Balqa' Applied University, Al-Huson University College, Mechanical Engineering Department, PO Box 50, Al-Huson 21510, Irbid, Jordan*

**Abstract.** Analysis on machining characteristics in turning of LM6/SiC/Gr hybrid metal matrix composites is made of (Al-11.8%Si)/SiC/Gr hybrid metal matrix composites. The process performances such as porosity, wear rate of the composites, tool wear, tool life, specific modulus, surface roughness and material removal rate with equal weight fraction of SiC and Gr particulates of 3%, 7%, 10% and 13% reinforcement are investigated. This experimental analysis and test results on the machinability of Al/SiCMMC will provide essential guidelines to the manufacturers. Hybird metal matrix composites reinforced with graphite particles posses better machinability and tribological properties.

## 1 Introduction

Lubricants are extensively used between the contacting surfaces to reduce friction and wear. These lubricants are usually toxic and not readily biodegradable and thus these lubricants can cause considerable damage to the environment. The use of external lubricants can be eliminated by designing self-lubricating composite materials. These composite materials have the ability to achieve low friction and wear at the contact surfaces without any external supply of lubrication during sliding. The metal matrix composites reinforced with various self-lubricating particles such as graphite are being used as self-lubricating materials for various engineering applications [1]

Nowadays manufacturing industries are constantly focused on lower cost solutions with reduced lead time and better surface quality in order to maintain their competitiveness and efficiency. The knowledge of cutting forces, tool wear and temperature developed in various machining processes under given cutting conditions is being a dominating criterion of material machinability, to both the designer manufacturer of machine tools, as well as to user.

Aluminum alloys are promising structural materials due to their high specific strength and stiffness. However, their applications are restricted because of their poor wear resistance. Aluminum alloys reinforced with SiCp particles composites are being considered for their superior mechanical and tribological properties over the conventional aluminum alloys, and therefore, these composites have gained extensive applications in automotive and aerospace industries. But Al/ SiCp are extremely difficult to machine (turning, milling, drilling, threading) due to their extreme abrasive. As the presence of hard reinforcement particles makes them extremely difficult to machine as they lead to rapid tool wear [2]. Therefore, graphite particles were added in SiC/Al composites to improve its machinability [3], [4]. Meanwhile, their damping and wear resistance properties were also improved.

Characteristics of AMCs reinforced with soft reinforcement particles of Gr, and volume fraction of reinforcements, as well as machinability and testing condition have been reported [5]-[7].

The incorporation of SiC hard reinforcing particles into the AMCs improve their mechanical and tribological behavior, but may result in deteriorated machinability together with rapid counter-face wear [2], [8], [9]

Hybrid composites containing both hard and solid lubricant materials with improved mechanical and tribological behavior have been developed to enhance the machinability of the AMCs [10]-[13].

From the published literature it is clear that Al/SiC MMCs machining is one of the major problem which counter the industry. While machining Al−SiC−Gr hybrid composites, the criteria considered for the best performance are lower surface roughness, wear rate of the composites and tool flank wear of tool, and higher tool life, material removal rate and specific modulus. The present study has been carried out to investigate the machinability of the stir cast LM6/SiCp/Gr hybird metal matrix composites at varying equal weight fraction of SiC and Gr particulates of 3%, 7%, 10% and 13% reinforcement. The influence of reinforcement volume fraction on porosity, density, surface roughness, tool life and wear, the matrix wear rate, specific modulus and material removal rate has also been investigated. This experimental analysis and test results on the machinability of Al/SiC/Gr-HMMCs will provide essential guidelines to the manufacturers.

## 2 Experimental procedure

### 2.1 Materials selected for processing composites

The main materials used in this project are LM6 aluminum alloy (Aluminum – 11.8% silicon) as a matrix material and SiC-Graphite as a particulate reinforced added in different percentages. The details of the LM6 alloy properties and composition is shown in Table 1. Silicon Carbide particles of 45 μm average size and graphite particles of 90 μm average size were used in this study.

**Table 1.** Composition of LM6 [14]

| Elements | Al | Cu | Mg | Si | Fe | Mn | Others |
|---|---|---|---|---|---|---|---|
| Weight % | 85.95 | 0.2 | 0.1 | 11.8 | 0.5 | 0.5 | 0.75 |

### 2.2 Preparation of composites

For the purpose of improving surface wettability and interface bonding between matrix alloy and graphite particles, the surface of graphite particles was coated with copper by a cementation process using aqueous and supersaturated copper sulphate solution [15]. The SiC particlues was preheated at 750-800 C for 2 hrs to make their surface oxidized to achieve better weldability and to prevent decarburization.

The stir casting technique is the simplest and the most economical process for producing particulate reinforced MMCs available for particulate reinforced metal matrix composites [16]. Two step stir casting method was used to fabricate the Al/Si-SiC-Gr hybird composites. Aluminium alloy LM6 was melted up to 750°C. Varying equal weight fraction of SiC and Gr particulates of 3%, 7%, 10% and 13% reinforcement was added with this molten LM6. It is allowed to cool to solidus temperature and stirred at 120 rpm using steel impeller. In this way 100% of the SiC particles were transferred to the metal. Simultaneously, graphite was added with Aluminum – 11.8% silicon (LM6)-SiC. After that the composite alloy was roll poured into the pre-heated (250°C) permanent mould. Composites were cast in required diameter and length. The silicon carbide particles exhibit in the form of solid crystal whereas the graphite particles appear to be flakes.

The different samples were precision weighed in an electronic balance to an accuracy of 0.1 mg. In general, the experiment data obtained from three specimens cut from the same composite were taken as the average value.

The density measurements were carried out to determine the porosity levels of the samples. The density of the samples was measured according to ISO 2738 standard. The measured density was compared to the value obtained using rule-of-mixtures so as to determine the volume fraction of porosity.

Composites wear of pin-on-disc tests were undertaken under the normal load of 20 N (resulting in a normal pressure of 1 MPa), the sliding velocity of 0.5 ms-1 and different sliding distance of 250, 500, 750, and 1000 m using a wear track diameter of 16 mm.

He machining tests were conducted using an ACE LT2 type of CNC lathe under dry turning condition. The as cast bars were machined with K10 Tungsten Carbide inserts (positive rake angle 14°, clearance angle 6°, and nose radius 0.3 mm) and MTJNL 2525M16 tool holder. The recommended levels of turning parameters to minimize surface roughness and flank wear, and to maximize material removal rate are: cutting speed (200 m/min), feed rate (0.075 mm/r) and at a constant depth of cut of 1 mm [Suresh, P, Marimuthu, K, Ranganathan, S & Rajmohan, T, 'Optimisation of machining parameters in turning of Al-SiC-Gr hybrid metal matrix composites using Grey-fuzzy algorithm', Trans. Nonferrous Met. Soc. China 24(2014) 2805−2814]. The maximum flank wear on the tool tip was measured by a SZX12-type Olympus toolmaker's microscope with measuring grids on the eye-piece. The maximum flank wear of 0.6 mm was chosen as the tool life limit.

Surface roughness values have been measured by using Mitutoyo Portable Surface Roughness Tester. The Material Removal Rate (MRR) is determined from the amount of material machined in the given period of time in minutes.

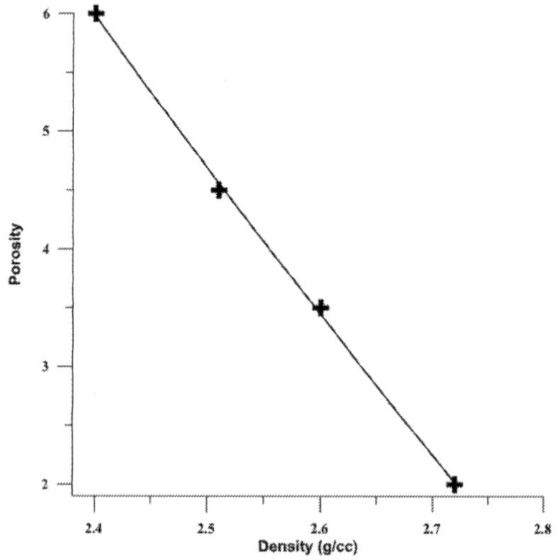


**Figure 1.** Porosity and density variations with increasing SiC/Gr contents.

## 3 Results and discussion

### 3.1. Density and porosity

Fig. 1 shows the variation of both density and porosity as a function of the SiC/Gr volume fraction. The densities of composites were found to increase linearly with the volume fraction, in accordance with the rule of mixtures. Also the percentage porosity present in the composites

DOI: 10.1051/matecconf/20167502001

reduces with the increase of the reinforcements content. Graphite is a solid lubricant material and facilitates movement and rearrangement of matrix and reinforcement particles resulting in higher densification. With relevance to characterization, the composites were found to be totally non-porous and microscopically homogeneous in the distribution of SiC particles. The volume fraction of porosity, and its size and distribution in a cast metal matrix composite play an important role in controlling the material's mechanical properties. It is necessary that porosity levels be kept to a minimum.

### 3.2 Wear of the composites

As shown in Fig. 2, with the increase of the SiC/Gr contents, the wear rate of Al/SiC/Gr HMMCs reduces. It was found that the amount of graphite released on the wear surface forms a tribofilm on the contact surfaces. This reduces the overall friction coefficient and wear rate. The presence of graphite tribolayer (or mechanically mixed layer) also increases the seizure resistance and enables to run under boundary lubrication without galling. Increasing of protrusion of SiC particles result in formation of more stable lubricating film on the tribosurface of hybrid composites.

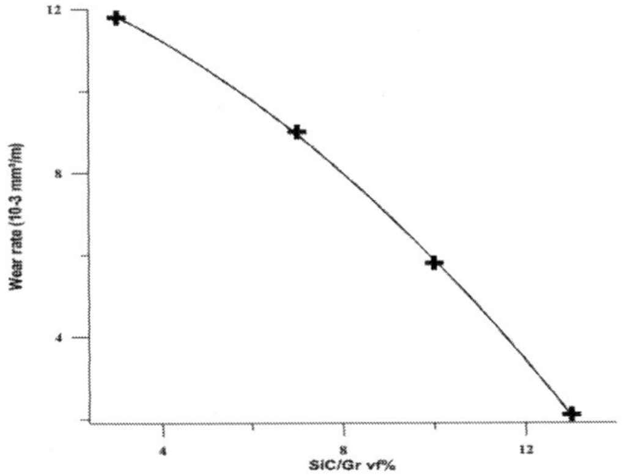


**Figure 2.** The effect of SiC/Gr addition on the matrix wear rate of the investigated composites.

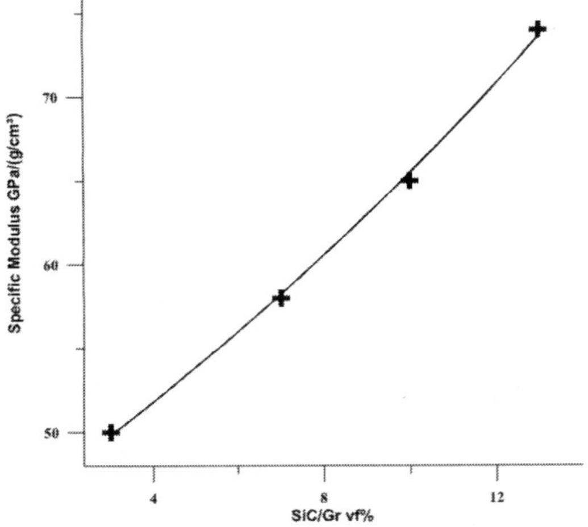


**Figure 3.** Specific moduli of the composites.

### 3.3 Specific modulus

As shown in Fig. 3, the specific moduli of 13%SiC/13%Gr/Al are approximately as twice as those of preferred inertial guidance materials. A high specific modulus would increase the dimensional stability of the materials used for the inertial instruments. Then an inertial guidance device could be made smaller and lighter. So this paper will provide a novel solution to machinable materials for precision instruments.

### 3.4 The surface roughness

The surface roughness affects the dimensional accuracy of the machined components. It is evident from the plot in Fig. 4 that the SR of Al–SiC–Gr hybrid composites decrease with increase in % reinforcement. This is due to increase in brittleness of the material and avoids build up edge formation in the tool tip.

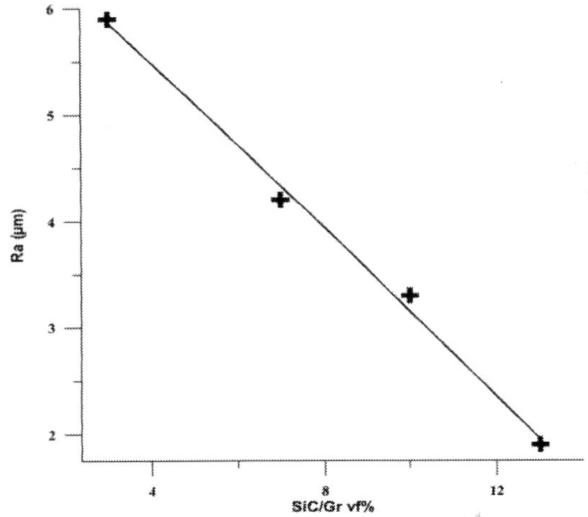


**Figure 4.** Variations in surface roughness of the composites.

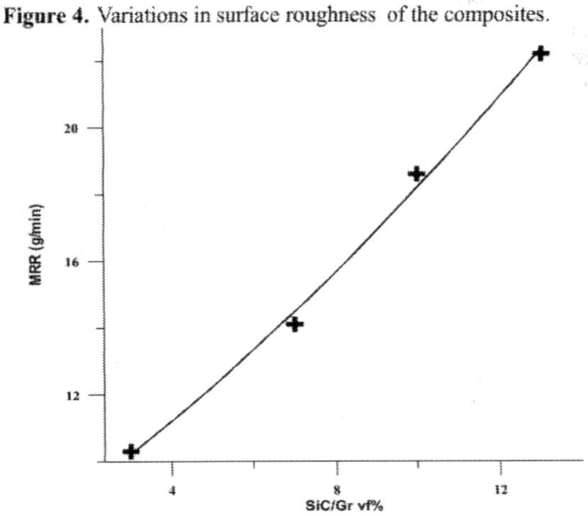


**Figure 5.** Effect of reinforcements volume fraction on material removal rate.

### 3.5 Material removal rate

MRR in is an important factor because of its vital effect on machining characteristics of the material in the industrial economy. As shown in Fig. 5, the MRR increases with increase in volume fraction of SiC/Gr. The increase in combined % reinforcement of SiC and Gr particulates results in decrease in hardness and fracture toughness of Al-SiC-Gr hybrid composites. Therefore the machining of Al-SiC-Gr hybrid composites with higher weight fraction of graphite is easy with maximum MRR and less tool wear.

### 3.6 Tool wear and tool life

From Fig. 6a, it can be observed that an increase in mass fraction of SiC–Gr decreases the flank wear of the tool. The tool life values for specimens investigated in the present work are shown in Fig. 6b. By adding 3-13% SiC/Gr particles, the tool life is prolonged by 50-200%. The tool life can reflect the machinability of the composites. The flaky graphite in SiC/Gr/Al composites promotes chip formation and the SiC particle in matrix also act as chip breaker resulting in an improved machinability. The addition of graphite particles reduces flank wear of the tool due to the formation of tribolayer. Some of the crushed or removed graphite particles trapped between flank face of the tool and machining surface, and reduced friction at the machining interface.

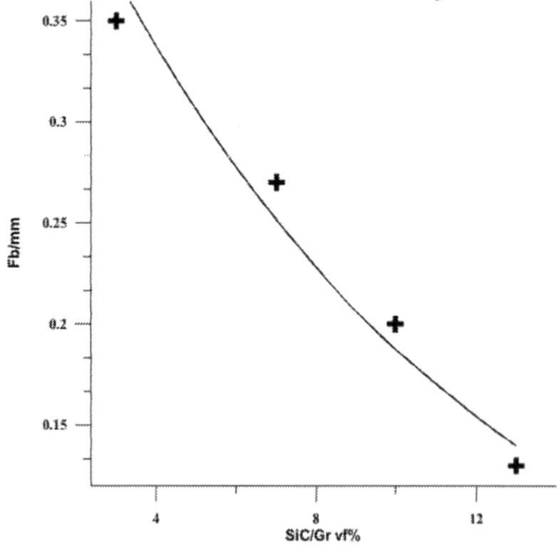


a. Effect of the SiC/Gr particles on flank wear of the tool

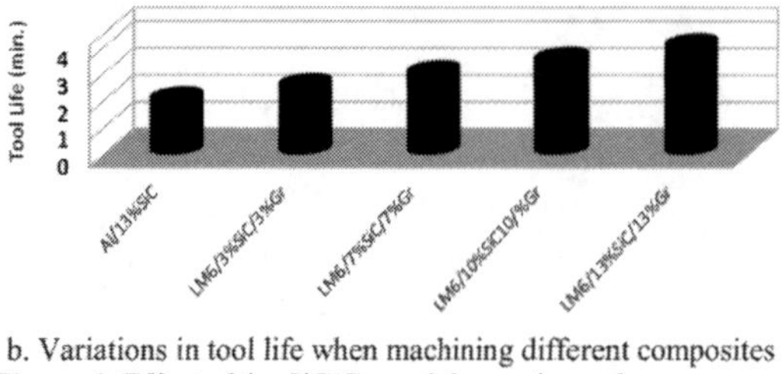


b. Variations in tool life when machining different composites
**Figure 6.** Effect of the SiC/Gr particles on the tool.

## 4 Conclusions

The following points were concluded from the conducted tests of the composites:

1. The processing employed in this paper would enable realization of electronic packages made out of Al-Si/SiCp MMCs.

2. The amount of porosity are decreased by increasing of silicon carbide and graphite contents in the composites.

3. It is obvious that the SR of Al–SiC–Gr hybrid composites decrease with increase in combined SiC-Gr % reinforcement.

4. Low wear rate was exhibited in the Al–SiC/Gr surface hybrid composite due to mechanically mixed layer generated between the composite pin and steel disk surfaces which contained fractured SiC and Gr. The presence of SiC particles serves as load bearing elements and Gr particles acted as solid lubricant.

5. Machining of Al–SiC–Gr hybrid composites with a higher mass fraction of graphite is easy with a maximum MRR and tool life and less tool wear.

6. The MRR increases with the increase in combined SiC-Gr % reinforcement. The increase in combined % reinforcement of SiC and Gr particulates results in decrease in hardness and fracture toughness of Al-SiC-Gr hybrid composites.

7. The high specific modulus of the investigated composites would increase the dimensional stability of the materials used for the inertial/ precision instruments.

## References


1. Pradeep L. Menezes, Pradeep K. Rohatgi, Michael R. Lovell, *Green Tribology*, Green Energy and Technology (2012)
2. Rabindra Behera and G. Sutradhar. Journal of Engineering and Applied Sciences **7**, (2012)
3. J Singh, D Narang, NK Batra, IInternational Journal of Engineering Science and Technology. **5**, (2013)
4. Prakash C.H. and Pruthviraj R. D, Research Journal of Chemical Sciences **6**, (2011)
5. Pardeep Sharma, Satpal Sharma, Dinesh Khanduja, Particulate Science and Technology **34**, (2016)
6. Akhlaghi, F., ZareBidaki, A, Wear **266**, (2009)
7. Zhao, H., Liu, L., Hu, W., Shen, B, Mater. Des. **28**, (2007)
8. Rao, R.N., Das, S. Mater. Des. **31**, (2010)
9. Rabindra Behera, S. Das, D. Chatterjee, G. Sutradhar, Journal of Minerals & Materials Characterization & Engineering **10**, (2011)
10. S. Mahdavi, F. Akhlaghi, Tribol Lett **44**, (2011)
11. F. Akhlaghi, S. Mahdavi, Advanced Materials Research, **264-265**, (2011)
12. Adalet Zeren , Industrial Lubrication and Tribology **67**, (2015)
13. Jinfeng Leng, Longtao Jiang, Qiang Zhang, Gaohui Wu, Dongli Sun, Qingbo Zhou, Journal of Materials Science **43**, (2008)
14. Sulaiman, S., Sayuti, M., and Samin, R. Mechanical properties of the as-cast quartz particulate reinforced LM6 alloy matrix composites. Journal of Materials Processing Technology **201(1-3)**, (2008)
15. Yang J.B., Lin C.B., Wang T.C., and Chu H.J., *Wear* **257**, (2004)
16. Umanath K, Selvamani S T, Palanikumar K, I. J. Composite Part-B Engineering **53**, (2013)

DOI: 10.1051/matecconf/20167502002

# An Experiment Study on Surface Roughness in High Speed Milling NAK80 Die Steel

Fa SU [1,2], Guofu Lian[2]

[1]Ship Engineering Department, Guangzhou Maritime Institute, Guangzhou,Guangdong, China
[2]School of Mechanical & Automative Engineering, Fujian University of Technology, Fuzhou, Fujian, China

**Abstract.** The paper introduces that the high speed milling experiments on NAK80 die steel was carried out on the DMU 60 mono BLOCK five axis linkage high speed CNC machining center tool by the TiAlN coated tools, in order to research the effect of milling parameters on surface roughness $R_a$. The results showed that the $R_a$ value increased with the decrease of milling speed $v_c$, increased with the axial depth of milling $a_p$, and feed per tooth $f_z$ and radial depth of milling $a_e$. On the basis of the single factor experiment results, the mathematics model for between surface roughness and milling parameters were established by linear regression analysis.

## 1 Introduction

NAK80 is a kind of pre-hardening plastic die steel, which is mainly made to produce some mirror polishing mold, automobile parts electrical discharge machining and so on. High Speed Machining Technology plays important part in application of mold processing, which could largely increase the producing efficiency and quality of mold.in this case, for NAK80 die steel, developing the research of high speed milling is really important. Surface roughness is one of the parameters that influence parts surface processing quality. It influences the parts' fitting property, endurance, fatigue resistance, corrosion resistance, contact stiffness and so on. Besides, it also influences parts' working life, stability and reliability.

By consulting literature, it could be deferred that scholars have researched on surface roughness of high speed milling steel: Z. Y. Hu [1] and W. W. Liu [2] have all received parameters about milling parameters' influence on surface roughness from experimental study on high speed milling GH4169 nickel-base superalloy. Meanwhile, W.W. Liu ,etc [3]created empirical formula of surface roughness based on SPSO; P. Koshy, etc [4] carried on high speed milling test on AISI D2 tool steel and researched on selection of tool wear mechanism and milling parameters, D. A. Axinte, etc [5] carried on high speed milling experimental research on AISI H13 hot die steel. From this, they received milling parameters' influence on surface roughness, micro-structure, micro-hardness and residual stress. Moreover, they established relevant experimental mathematical model; J. J. Chen, etc [6] researched on milling parameters' influence on surface roughness and established relevant surface roughness mathematical model; S. Y. Wang ,etc [7] created surface roughness predicted model based on experimental research on high speed milling 45 steel, C. F. Yao, etc [8] carried on high speed milling test on TC11 titanium alloy and analysed milling parameters' influence on surface roughness and three-dimension surface topography; F. Su ,etc [9] carried on experimental research on 718 die steel and established regressive mathematical model on surface roughness. So far, the research on surface roughness of NAK80 die steel has not been found.

The author carried on the high speed milling experiments on NAK80 die steel was carried out on the DMU 60 mono BLOCK five axis linkage high speed CNC machining center tool by the TiAlN coated tools and researched on milling parameters' influence on surface roughness and established mathematical model on relationship between surface roughness and milling parameters. Before high speed milling processing NAK 80 die steel, surface roughness could be appropriately predicted to select proper milling parameters. This importantly guide the practical production.

## 2 Experimental Tests

### 2.1 Specimens

NAK80 die steel had dimensions of 80 × 30 ×30mm$^3$, 80 × 35 ×30 mm$^3$, 80 × 40 ×30mm$^3$ and 80 × 45 ×30mm$^3$. Some main chemical component of NAK80 die steel as follows: C is 0.28~0.40%, Si is 0.20~0.80%, Mn is 0.60~1.00%, Mo is 0.30~0.55, Cr is 1.40~2.00%, Ni is 0.80~1.20%. Some main physical mechanical properties as follows: density is 7.8g/cm$^3$, elasticity modulus is 205GPa, extension strength is 1100MPa, yield strength is 980 MPa, hardness is HBS356, ductility is 0.13, Original state is anneal.

## 2.2 Milling experiments

Surface milling experiments were conducted on a high-speed machining center tool (DMU 60 mono BLOCK produced by Germany DMG company).The machine spindle is capable of running up to 18000 rpm, the maximum feed rate of (X, Y, Z) is 30m/min, the maximum power of three-dimensional Heldenhain iTNC 530 control system The power of the spindle drive motor is 35kW.

Coolant fluid was not used in the milling tests. The experimental setup is illustrated schematically in Fig.1. The cutting tools of SNNDVIK's TiAlN coating cemented carbide indexable blade were used ,whose type is 490R-140408M- PM 4220, The cutter holder type is A1B05-4022100, which is overall face milling cutter handle, its diameter is 63mm, it has five blade. The surfaces milling was carried out on the 80 × 30 mm$^2$, 80 × 35 mm$^2$, 80 × 40 mm$^2$ and 80 × 45 mm$^2$ surfaces in a up- milling mode. A series of tests were performed using different peripheral milling speeds, axial depths of milling, feeds per tooth and radial milling depths to study their influences on the milling performances. Detailed milling parameters used were listed in Table 1.

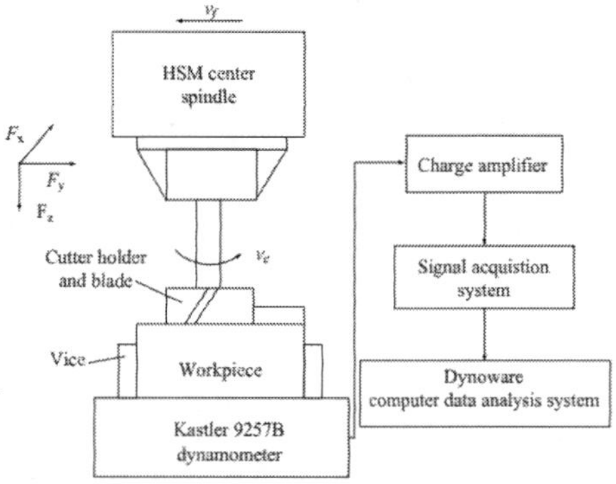


**Figure 1**. Schematic illustration of the processing.

**Table 1.** Parameters used in milling of the NAK80 die steel.

| milling parameters | Value |
|---|---|
| Peripheral milling speed $v_c$ (m/min) | 198, 296, 394, 495, 594 |
| axial depth of milling $a_p$ (mm) | 0.1, 0.2, 0.3, 0.4, 0.5 |
| feed per tooth $f_z$ (mm/z) | 0.01, 0.02, 0.03, 0.04, 0.05 |
| axial depth of milling $a_e$(mm) | 30, 35, 40, 45 |

## 2.3 Force measurement and ground sample evaluation

Milling forces were measured using a piezoelectric dynamometer (Kistler 9257B), the radial cutting depth direction $F_x$, feed direction $F_y$, axial direction $F_z$. were measured. The force signals were fed to a 5067A1 type charge-amplifier and 5697A1 type data collection system. Under each set of parameters, the grinding process was repeated at least three times to obtain the average values of grinding forces.

A surface profilometer (Mahr Products, XR20) was used to measure surface roughness of milling ground along the directions parallel and perpendicular to the milling direction, three measurements were taken at different locations, the mean values was calculated. $R_a$ (//) is used to show surface roughness that is parallel to milling direction, $R_a$ (⊥) is used to show surface roughness that is perpendicular to milling direction. The selected measurement parameters were sampling length is 0.8mm, evaluated length is 4.0mm, measured length is 5.6mm.

## 2.4 Experiment method

The single factor experiment method [10] applied to carry on high speed milling experiment. Through the analysis of experiment results, the author received the influence of milling parameters on surface roughness. Meanwhile, the surface roughness mathematical model was established. When milling speed is larger than 300m/min, it could be named high speed milling [1]. The surface roughness values of measuring were listed in Tab. 2.

**Table 2.** The measuring surface roughness values.

| Milling parameters | | | | $R_a$(//)(μm) | $R_a$(⊥)(μm) |
|---|---|---|---|---|---|
| $v_c$ | $a_p$ | $f_z$ | $a_e$ | | |
| 198 | 0.2 | 0.02 | 30 | 0.683 | 0.651 |
| 296 | 0.2 | 0.02 | 30 | 0.628 | 0.603 |
| 394 | 0.2 | 0.02 | 30 | 0.581 | 0.551 |
| 495 | 0.2 | 0.02 | 30 | 0.549 | 0.455 |
| 594 | 0.2 | 0.02 | 30 | 0.456 | 0.417 |
| 495 | 0.2 | 0.01 | 30 | 0.252 | 0.229 |
| 495 | 0.2 | 0.03 | 30 | 0.361 | 0.371 |
| 495 | 0.2 | 0.04 | 30 | 0.375 | 0.437 |
| 495 | 0.2 | 0.05 | 30 | 0.429 | 0.458 |
| 495 | 0.1 | 0.02 | 30 | 0.441 | 0.388 |
| 495 | 0.3 | 0.02 | 30 | 0.560 | 0.517 |
| 495 | 0.4 | 0.02 | 30 | 0.681 | 0.601 |
| 495 | 0.5 | 0.02 | 30 | 0.754 | 0.651 |
| 495 | 0.2 | 0.02 | 35 | 0.453 | 0.505 |
| 495 | 0.2 | 0.02 | 40 | 0.581 | 0.626 |
| 495 | 0.2 | 0.02 | 45 | 0.642 | 0.735 |

DOI: 10.1051/matecconf/20167502002

## 3 Result and discussion

### 3.1 The influence of milling parameters on surface roughness $R_a$

The influence rule curve of the milling speeds $v_c$ on measured surface roughness results of the $R_a$ (//) and $R_a$ ($\perp$) are presented in Fig.2. It could be seen that the $R_a$ values decrease with increases, It's mainly because as increases, milling forces $F$ decreases[11]. This could make frequency produced by tool system of processing far away from inherent frequency of tool system. This could decrease of eliminate vibration, thus the $R_a$ values were decreased, and the processing quality of milling surfaces could be increased. Fig 3 is the influence rule curve of the axial milling depth $a_p$ on measured surface roughness results of the $R_a$ (//) and $R_a$ ($\perp$) . It could be seen that as the $R_a$ values increase with the $a_p$ increases, the $R_a$ values are approximate linear improving. The reason is that the milling forces $F$ increase with the $a_p$ increases [11]. The friction and extrusion between flank surface of cutting tools and machined surface were intensified. Therefore, the values increase, but the milling surface quality decreases. Fig 4 is the influence rule curve of the feed per tooth $f_z$ on measured surface roughness results of the $R_a$ (//) and $R_a$ ($\perp$) . It could be seen that as the $R_a$ values increase with the $f_z$ increases,. The reason is that as increases, the milling layer' s thickness increases with $f_z$ increasing, the metal plastic deformation's volume of the finished surface increases, the milling forces $F$ increases[11].so, the surface roughness values increase with the $f_z$ increase. The influence rule curve of the axial depth $a_e$ on measured surface roughness results of the $R_a$ (//) and $R_a$ ($\perp$) is drew in Fig 5. It could be seen that the $R_a$ values increase with the $a_e$ increases .The reason is that the milling area and the milling forces increase[11], thus, the surface roughness values increase and the milling surface quality decreases ,and the $a_e$ is the most influence on the surface roughness values among milling parameters.

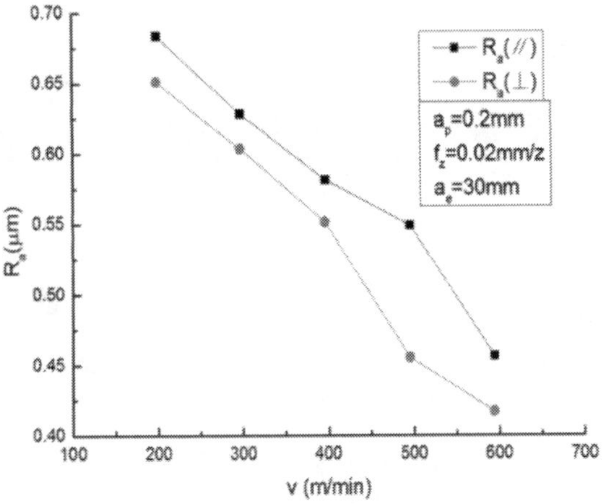

**Figure 2.** Impact of milling speed on surface roughness.

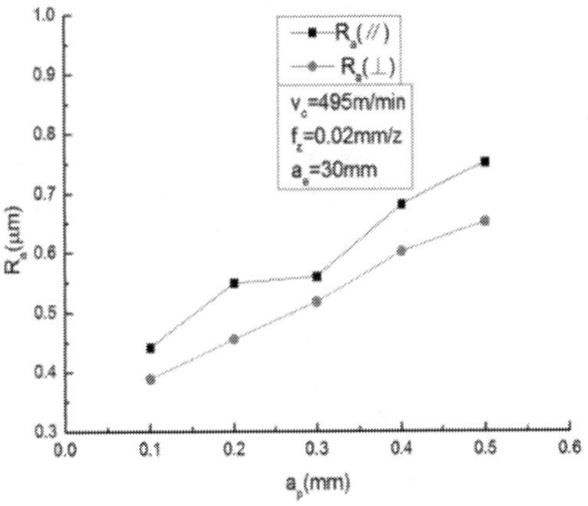

**Figure 3.** Impact of axial milling depth on surface roughness.

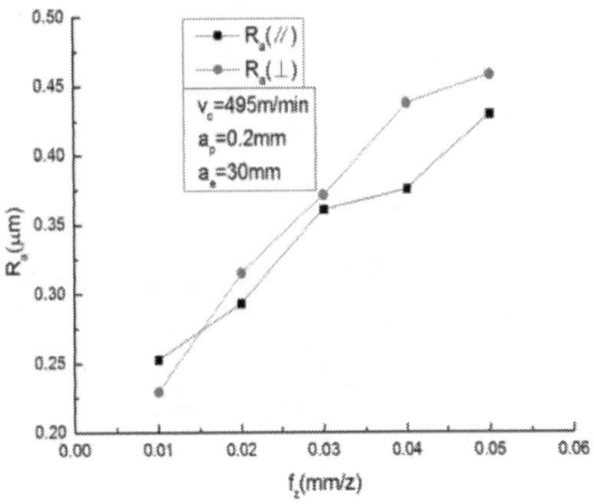

**Figure 4.** Impact of feed per tooth on surface roughness.

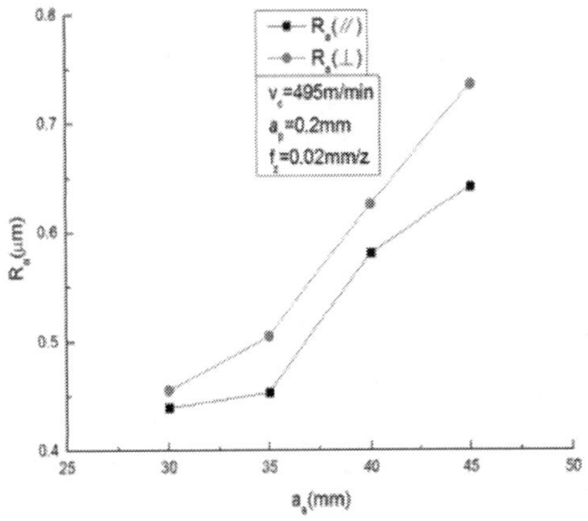

**Figure 5.** Impact of radial milling depth on surface roughness.

### 3.2 Establishment and verification of the prediction mathematical model of surface roughness $R_a$

Lots of research results show that surface roughness and milling parameters have complex index relationship, for high speed milling, mathematical expression of high speed milling could be expressed as

$$R_a = C_a a_p^{b_1} v_c^{b_2} f_z^{b_3} a_e^{b_4} \qquad (1)$$

where $a_p$ is axial depth of milling ,mm. $v_c$ is peripheral milling speed ,m/min. $f_z$ is feed per tooth, mm/z. $a_e$ is radial milling depth, mm. $C_a$ is comprehensive coefficient of milling condition. $b_1$, $b_2$, $b_3$ and $b_4$ are the influence parameters of each milling parameters， which are decided by the factors of work-piece material and cutting condition.

As for single factor experiment method, it needs to transform Eq.[1] to followed 4 mathematical expression in order to obtain Eq.[1]:

$$R_a = C_a^{a_p} a_p^{b_1} \qquad (2)$$

$$R_a = C_a^{v_c} v_c^{b_2} \qquad (3)$$

$$R_a = C_a^{f_z} f_z^{b_3} \qquad (4)$$

$$R_a = C_a^{a_e} a_e^{b_4} \qquad (5)$$

It is very difficult to get mathematical solution of exponent equation of Eq.[2], Eq.[3], Eq.[4] and Eq.[5]. The common method is to take the both sides logarithm of Eq.[2], Eq.[3], Eq.[4] and Eq.[5]. So, exponent equation could be transformed to linear equation and solve the equation. For example, for Eq.[2], the process is as follows:

$$\log R_a = \log C_a^{a_p} + b_1 \log a_p \qquad (6)$$

make $\log R_a = y$， $\log(C_a^{a_p}) = b_0$， $\log a_p = x_1$ . therefore, surface roughness exponent equation could be transformed into linear equation:

$$y = b_0 + b_1 x_1 \qquad (7)$$

The unary linear regression mathematical model was established as follows:

$$\begin{cases} y_1 = \beta_0 + \beta_1 x_{11} + \varepsilon_1 \\ y_2 = \beta_0 + \beta_1 x_{21} + \varepsilon_2 \\ \cdots\cdots \\ y_n = \beta_0 + \beta_1 x_{n1} + \varepsilon_n \end{cases} \qquad (8)$$

where $\varepsilon_i$ is random variable error, for Eq.(8), the form of matric could be expressed as:

$$Y = X\beta + \varepsilon \qquad (9)$$

where, $Y = \begin{bmatrix} y_1 \\ y_2 \\ \cdots \\ y_n \end{bmatrix}$ 、 $X = \begin{bmatrix} 1 & x_{11} \\ 1 & x_{21} \\ \cdots & \cdots \\ 1 & x_{n1} \end{bmatrix}$ 、 $\beta = \begin{bmatrix} \beta_0 \\ \beta_1 \\ \beta_2 \\ \beta_3 \\ \beta_4 \end{bmatrix}$ 、 $\varepsilon = \begin{bmatrix} \varepsilon_0 \\ \varepsilon_1 \\ \varepsilon_2 \\ \varepsilon_3 \\ \varepsilon_4 \end{bmatrix}$. The parameter $\beta$ is estimated by the least square method, hypothesis $b_0$ and $b_1$ are respectively the least square estimation of parameter$\beta_0$, $\beta_1$, thus, regression mathematics model is:

$$\hat{y} = b_0 + b_1 x_1 \qquad (10)$$

where, $\hat{y}$ is statistical variable. $b_0$ and $b_1$ are regression coefficient, and could be calculated by $b = (X'X)^{-1} X'Y$ .

According to measured surface roughness results of the $R_a(//)$ and $R_a(\perp)$ by single factor experiment, each regression coefficient could be solved by using origin software , then the each coefficient in Eq.[2], Eq.[3], Eq.[4] and Eq.[5] could be solved in return.

After solution, The mathematical models of the surface roughness $R_a$ ( // ) , $R_a$ ( $\perp$ ) between single milling parameters are as follow: $R_a$( // ) $=0.710a_p^{0.39794}$ ,$R_a$( // ) $=2.782v_c^{-0.33208}$ ,$R_a$( // ) $=0.753f_z^{0.32674}$ , $R_a$( // ) $=0.057a_e^{0.54742}$ ,$R_a$( $\perp$ ) $=0.485a_p^{0.32301}$ ,$R_a$( $\perp$ ) $=2.169v_c^{-0.31714}$ , $R_a$( $\perp$ ) $=1.072f_z^{0.44095}$ , $R_a(\perp)=0.003a_e^{1.29914}$.

The coefficient $C_a$ value in Eq.[1]could be solved according to the $C_a^{a_p}$ , $C_a^{f_z}$ , $C_a^{v}$ and $C_a^{a_e}$ value , For example, The steps of solving $C_a$ were as follow:

$$R_a = C_a^{a_p} a_p^{b_1} = C_a^1 a_p^{b_1} v_c^{b_2} f_z^{b_3} a_e^{b_4}$$

Thus $$C_a^1 = \frac{C_a^{a_p}}{v_c^{b_2} f_z^{b_3} a_e^{b_4}} \qquad (11)$$

where, $v_c$, $f_z$, $a_e$ are fixed values when $a_p$ changes.
In addition, the solve is done as follows:

$$R_a = C_a^{v_c} v_c^{b_2} = C_a^2 a_p^{b_1} v_c^{b_2} f_z^{b_3} a_e^{b_4}$$

Thus $$C_a^2 = \frac{C_a^{v}}{a_p^{b_1} f_z^{b_3} a_e^{b_4}} \qquad (12)$$

where, $a_p$, $f_z$, $a_e$ are fixed values as $v_c$ changes.

$$R_a = C_a^{f_z} f_z^{b_3} = C_a^3 a_p^{b_1} v_c^{b_2} f_z^{b_3} a_e^{b_4}$$

Thus $$C_a^3 = \frac{C_a^{f_z}}{a_p^{b_1} v_c^{b_2} a_e^{b_4}} \qquad (13)$$

where, $a_p$, $v_c$ , $a_e$ are fixed values when $f_z$ changes.

$$R_a = C_a^{a_e} a_p^{b_4} = C_a^4 a_p^{b_1} v_c^{b_2} f_z^{b_3} a_e^{b_4}$$

Thus $$C_a^4 = \frac{C_a^{a_e}}{a_p^{b_1} v_c^{b_2} f_z^{b_3}} \qquad (14)$$

where, $a_p$, $v_c$ , $f_z$ are fixed values when $a_e$ changes.

In addition, The coefficient $C_a$ value could be solved according to $C_a = \frac{C_a^1 + C_a^2 + C_a^3 + C_a^4}{4}$ , thus $R_a = C_a a_p^{b_1} v_c^{b_2} f_z^{b_3} a_e^{b_4}$ could be received. After solution, the mathematical models of surface roughness $R_a(//)$ and $R_a(\perp)$ are as follow:

$$R_a(//) = 2.711 a_p^{0.39794} v_c^{-0.33208} f_z^{0.32674} a_e^{0.54742} \qquad (15)$$

$$R_a(\perp) = 0.210 a_p^{0.39301} v_c^{-0.31714} f_z^{0.44095} a_e^{1.29914} \qquad (16)$$

The measured values and prediction values calculated by model was compared so as to verify prediction accuracy of mathematical model Eq.[15], Eq.[16]. By calculation, prediction error is 3.20%~24.60%, it shows that the mathematical models have higher accuracy and could be used to predict surface roughness of high speed milling NAK80 die steel.

## 4 Conclusion

The conclusions could be deferred from high speed milling experimental research on NAK80 die steel:

Through the analysis for test results, the influence rules of milling parameters on surface roughness could be known: the surface roughness $R_a$ increases with peripheral milling speed $v_c$ decreases, and radial milling depth $a_e$, axial depth of milling $a_p$ and feed per tooth $f_z$ increases. However, influence degree is different, and the is the most impact.

As for obtaining better finished surface quality, smaller radial milling depth $a_e$ and feed per tooth $f_z$ could be selected. Besides, higher milling speed $v_c$ and smaller axial depth of milling $a_p$ could be selected.

Based on unary linear regression analysis method, the author carried on numerical fitting analysis on experimental data and established mathematical model for relationship between surface roughness and milling parameters in case of high speed NAK80 die steel. Moreover, the author verified the model and proved the established model is available.

## Acknowledgments

This research was supported by grant no. 2015J01629 from the Natural Science Foundation of Fujian Province/China, by grant no. JK2012034 from the Fujian provincial/China special plan of university's scientific research projects.

## References

1. Z. Y. Hu, G. Y. Meng, H. T. Xia, Research on Surface Roughness of High Speed Milling GH4169 Based on the Orthogonal Experiment Method, Manufacturing Technology and Machine Tools, 1 ( 2011 )
2. W. W. Liu, F. Li, CH. F. Yao, etc, Effect of Milling Parameters on Surface Roughness in High Speed Milling GH4169, Aeronautical Manufacturing Technology, 12 (2012)
3. W. W. Liu, F. Li, X. J. Ren, etc, Research on Surface Roughness Based on SPSO in High Speed Milling of GH4169, China Mechanical Engineering, **22**, 22(2011)
4. P. Koshy, R. C. Dewes, D. K. Aspinwall, High speed end milling of hardened AISI D2 tool steel (~58 HRC), Journal of Materials Processing Technology, **127**(2002)
5. D. A. Axinte, R. C. Dewes, Surface integrity of hot work tool steel after high speed milling-experimental data and empirical models, Journal of Materials Processing Technology, 127(2002)
6. J. J. Chen, C. Long, C. Wang, Prediction Model of Surface Roughness on P20 Mold Steel by High Speed Milling, Modular Machine Tool & Automatic Manufacturing Technique, 12(2012)
7. S. Y. Wang, X. Ai, J. ZHao, etc, A Surface Roughness Prediction Model for High Speed Milling, Manufacturing Technology and Machine Tools, 8(2006)
8. CH. F. Yao, D. Q, CH. Jin, etc, Exploring Surface Roughness and Surface Morphology of High-speed Milling TC11 Titanium Alloy, Mechanical Science and Technology for Aerospace Engineering, **30**, 9(2011)
9. F. Su, Y. J. ZHan, H. Hong, The Construction of an Mathematics Model Study on Surface Roughness in High Speed Milling 718 Die Steel, Journal of Xiamen university of technology, **13**, 1(2015)
10. W. ZH. Wang, *The Design of the Test and Analysis* (Beijing: Higher Education Press, 2004)
11. F. Su, Y. J. ZHan, H. Hong, An experiment study on milling forces in high speed milling NAK80 die steel, Journal of Zhong kai University of Agriculture and Technology, **27**, 4(2014)

# Stiffness of serial and quasi-serial manipulators: comparison analysis

Alexandr Klimchik[1], Evgeni Magid[1], Stephane Caro[2,3], Kriangkrai Waiyakan[4] and Anatol Pashkevich[2,5]

[1]*Innopolis University, Universitetskaya 1, 420500 Innopolis, The Republic of Tatarstan, Russia*
[2]*Institut de Recherches en Communications et en Cybernétique de Nantes, 1 rue de la Noe, 44321 Nantes, France*
[3]*Centre National de la Recherche Scientifique (CNRS), France*
[4]*Prince of Songkla University, 15 Karnjanavanich Rd., Hat Yai, Songkhla 90110, Thailand*
[5]*Ecole des Mines de Nantes, 4 rue Alfred-Kastler, Nantes 44307, France*

**Abstract.** The paper deals with comparison analysis of serial and quasi-serial manipulators. It shows a difference between stiffness behaviours of corresponding industrial robots under external loading, which is caused by machining process. The analysis is based on the estimation of compliance errors induced by cutting forces that are applied to the manipulator end-effector. We demonstrate that the quasi-seral manipulators are preferable for large-dimensional tasks while the quasi-serial ones better suit small size tasks.

## 1 Introduction

Enhancement of industrial robot performances extend their applications from traditional pic-and-place to operations where manipulator end-effector is subjected to essential external loadings [1]. They progressively take their niche in the drilling, milling, friction stir welding and other operations [2], [3], replacing more expensive CNC machines. Nevertheless, robot positioning accuracy under external loading is still an open high-demand research issue.

To improve robot accuracy under loading, there were developed different online and off-line error compensation methods [4]-[6], which allow reducing the impact of manipulator elasto-static deformations on the machining quality. The most efficient technique is based on the online tracking of robot position and compensation of related deflections. Another group of on-line methods is based on information from internal robot sensors; these methods could be implemented relatively easily and usually do not impose any restriction on the robot workspace. However, these methods require rather accurate geometric and stiffness models of the manipulator, which should be obtained case-by-case from the dedicated experimental study [7], [8]. The most essential limitation of the on-line approach is related to necessity of the manipulator model modification in robot controller software, which is usually not completely open for end-users. In contrast, the off-line error compensation technique does not require any intervention in the controller software, it is based on the modification of a target trajectory that is obtained using either a complete or a reduced manipulator model [9].

Another approach to improve manipulator accuracy under loading is based on mechanical methods, where the manipulator stiffness enhancement is achieved by means of closed loops, i.e. transforming conventional serial robots into quasi-serial ones. The most common way here is using gravity compensators, which in fact do not affect essentially manipulator stiffness. An alternative way is to use kinematic parallelograms [10], which potentially improve robot stiffness but require additional detailed investigations.

In classical robotics, robots are usually compared from their kinematic properties point of view. However, these performance measures do not suit well a manipulator under external loading since they are not able to take into account robot elasticity and influence of the external force which is applied to the manipulator end-effector. To overcome this difficulty, in robotics other performance measures were developed that are based on the norms of the stiffness matrix [11] or manipulator deflections at a specific "test pose" [12]-[14]. Nevertheless, they cannot be applied directly to the robot architecture comparison for machining application since they ignore some important technological issues. To overcome this difficulty, this paper proposes a new approach to comparison analysis of industrial robots, which is based on the compliance errors estimation for typical machining operations.

## 2 Problem of robot comparison

Typical *serial manipulators* contain robot base, robot arm and robot wrist (Fig. 1a). The robot base defines the arm orientation with respect to the robot world frame and usually contains a single actuated joint providing rotation around z-axis. As shown in Fig. 1a, link #1 provides shifting of axis #2 with respect to axis #1. The robot arm is responsible for major movements of the robot end-effector. For 6-dof non-redundant robots, the translational movements are usually implemented using two actuated joints (joints #2 and #3). In anthropomorphic manipulators, these joints are also referred to as a robot shoulder and an elbow respectively. Corresponding links #2 and #3 define the manipulator workspace. The end effector orientation is achieved by the robot wrist. In the majority of manipulators, a robot wrist contains three revolute actuated joints (joints #4, #5 and #6) where joint axes intersect in a same point (a so-called *spherical wrist*). In a relevant geometric model, the wrist linear parameters may be omitted and included in the parameters which describe link #3 and the tool transformation. However, in practice, the robot can be equipped with a non-spherical wrist (e.g., an off-set wrist, a hollow wrist, etc.) whose geometrical model includes more parameters.

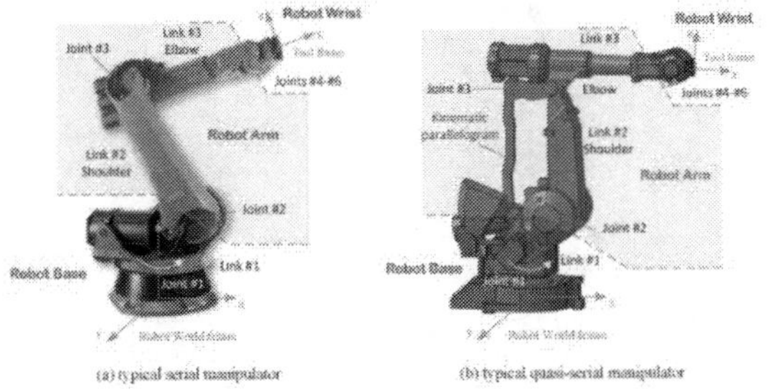

**Figure 1.** Architecture of typical industrial robots

*Quasi-serial robots* have roughly similar architecture (see Fig. 1b) but in contrast to their serial counterparts, a robot arm of a quasi-serial manipulator contains a kinematic parallelogram, which can be treated as an internal closed-loop. For this reason, such robots are often referred as quasi-serial ones. In practice, the kinematic parallelogram allows robot designers to increase robot dynamic properties by means of moving masses reduction (locating heavy actuating motor #3 on the robot base instead of a robot elbow). Usually, the parallelogram does not essentially affect manipulator control and does not change manipulator direct/inverse kinematic equations. On the other hand, the quasi-serial manipulator stiffness model essentially differs from its serial counterpart since re-location of the manipulator compliant element (actuator #3 transmission) essentially influences the stiffness behaviour [15]. For this reason, previous results, which are obtained for strictly serial manipulators cannot be directly applied in such case.

A stiffness model of a robotic manipulator (both serial and quasi-serial) describes the manipulator behavior under loading [16], [17]. In addition to the conventional robot parameters (i.e., geometric ones), it includes a number of elastic parameters which describe flexibility of the manipulator links and joints. In a number of industrial applications, the manipulator elasticity cannot be ignored since the high loading is applied to a robot, while the required positioning accuracy is rather high. For example, from our experience it is known that the end-effector deflection of heavy industrial robots under loading of 1kN may vary from 1 to 10 mm within the robot workspace, while demanded accuracy for the machining process is typically about 0.1 mm. These compliance errors can be reduced down to admissible level using both on-line and off-line error compensation techniques, which are based on the appropriate stiffness model [18], either "complete" or "reduced". The complete stiffness model of an industrial robot is complicated and takes into account all manipulator links and actuators compliances [19] (Fig. 2a). In practice, a number of manipulator components may be treated as rigid ones (e.g., some links), while the main compliance is concentrated in the actuator transmissions. This allows us to apply so-called reduced models that take into account the joint elasticities only [20]. Such models are quite common for stiffness modeling of heavy industrial robots where the links are massive and their deflections under the force of 1kN are much lower than 0.1 mm. The reduced stiffness model is also quite useful at the design stage since it gives valuable approximation of the manipulator stiffness behavior, which is required for optimization. Moreover, in most cases these models could be used to compare stiffness properties of the manipulators of different architectures. For this reason, the presented in this paper comparison study is based on the reduced stiffness model.

To make the comparison comprehensible, let us perform further simplification of the considered models. At first, in the frame of the comparison study, it is possible to exclude the robot base from the stiffness analysis. It is obvious that this component influences the stiffness properties of both serial and quasi-serial manipulators in the same way. Further, since the wrist dimensions are essentially smaller than the lengths of the links #2 and #3, we simplify the model by excluding the robot wrist components.

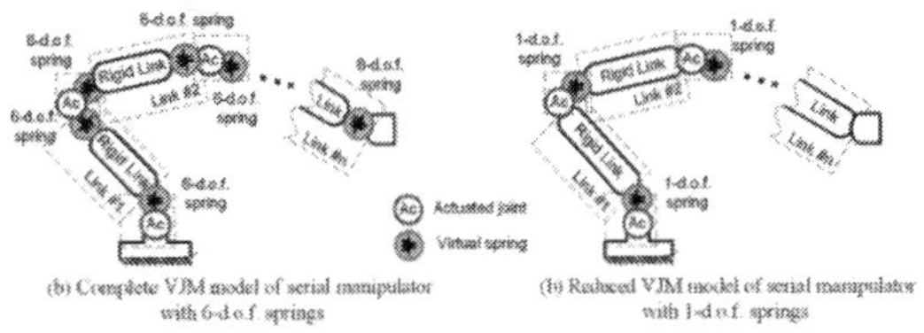

**Figure 2.** Complete and reduced stiffness models of an industrial robotic manipulator

## 3 Stiffness model of a serial manipulator

Let us obtain first the reduced stiffness model of a typical serial manipulator. The desired model takes into account the actuator compliances only, which usually are the most important elastic components of the entire manipulator. In the case of a serial manipulator, the robot arm contains two actuated revolute joints $q_2 . q_3$ and two links of lengths $l_2, l_3$ (Fig. 3). It is worth mentioning that in some robots link #3 is described with two geometrical parameters $l_{3x}, l_{3y}$ (see Fig. 3). Nevertheless, these two parameters can be easily presented as a single link length $l_3 = \sqrt{l_{3x}^{\;2} + l_{3y}^{\;2}}$ and the joint offset $\Delta q_3 = atan2(l_{3y}, l_{3x})$.

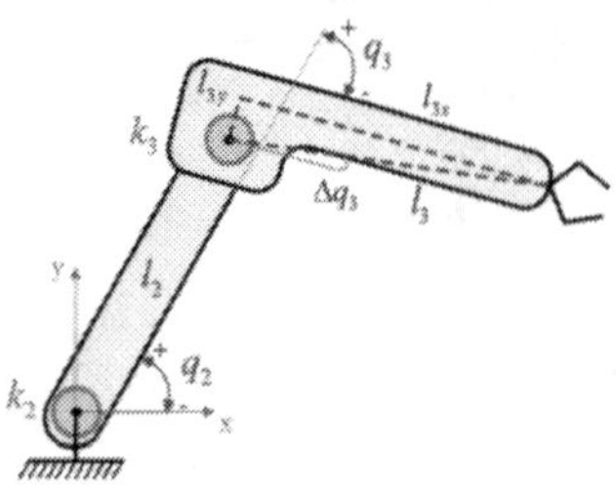

**Figure 3.** Kinematic structure and basic parameters of a serial robot arm

For the considered manipulator, the stiffness model contains two elastic parameters $k_2$ and $k_3$ describing compliances of corresponding joints. For this geometry, the Cartesian stiffness matrix $\mathbf{K}_C$ describes manipulator resistance to the external force $\mathbf{F}$, which is applied to the end-effector, and could be expressed as

$$\mathbf{K}_C = \left(\mathbf{J}_S\,\mathbf{k}_\theta\,\mathbf{J}_S^T\right)^{-1} \tag{1}$$

where the diagonal matrix $\mathbf{k}_S = diag(k_2, k_3)$ collects the joint compliances, and $\mathbf{J}_S$ is kinematic Jacobian. For the serial arm, the Jacobian matrix is presented as

$$\mathbf{J}_S = \begin{bmatrix} -l_2\cos q_2 - l_3\cos(q_2+q_3) & -l_3\cos(q_2+q_3) \\ l_2\sin q_2 + l_3\sin(q_2+q_3) & l_3\sin(q_2+q_3) \end{bmatrix} \tag{2}$$

where $l_2, l_3$ are the link lengths, $q_2, q_3$ are the actuated joint coordinates. After relevant substitution and transformations, we obtain the following expression for the serial manipulator compliance $\mathbf{k}_C = \mathbf{K}_C^{-1}$

$$\mathbf{k}_{CS} = \begin{bmatrix} k_{CS}^{(1,1)} & k_{CS}^{(1,2)} \\ k_{CS}^{(1,2)} & k_{CS}^{(2,2)} \end{bmatrix}$$

$$\begin{aligned} k_{CS}^{(1,1)} &= k_2\left(l_2\,c_2 + l_3\,c_{23}\right)^2 + k_3 l_3^{\,2}\,c_{23}^2 \\ k_{CS}^{(1,2)} &= -k_2(l_2^{\,2}\,c_2\,s_2 + l_2 l_3\,c_3 + l_3^{\,2}\,c_{23}\,s_{23}) - k_3 l_3^{\,2}\,c_{23}\,s_{23} \\ k_{CS}^{(2,2)} &= k_2\left(l_2\,s_2 + l_3\,s_{23}\right)^2 + k_3 l_3^{\,2}\,s_{23}^2 \end{aligned} \tag{3}$$

where $c_2 = \cos q_2$, $s_2 = \sin q_2$, $c_3 = \cos q_3$, $s_3 = \sin q_3$, $c_{23} = \cos(q_2+q_3)$, $s_{23} = \sin(q_2+q_3)$. As follows from our experience, the joint compliances $k_2, k_3$ are quite close to each other and may be assumed to be equal, i.e. $k_2 = k_3 = k$. Besides, it is reasonable to introduce a ratio coefficient $\mu = l_3 / l_2$ which defines the ratio between the link lengths. This allows re-writing the compliance matrix in a more compact form

$$\mathbf{k}_{CS} = k l_2^{\,2}\begin{bmatrix} (c_2 + \mu c_{23})^2 + \mu^2 c_{23}^2 & -c_2 s_2 - \mu c_3 - 2\mu^2 c_{23} s_{23} \\ -c_2 s_2 - \mu c_3 - 2\mu^2 c_{23} s_{23} & (s_2 + \mu s_{23})^2 + \mu^2 s_{23}^2 \end{bmatrix} \tag{4}$$

which is convenient for the further comparison analysis.

## 4 Stiffness model of a quasi-serial manipulator

Applying a similar technique, let us obtain the reduced stiffness model of a quasi-serial manipulator. In contrast to its serial counterpart, the robot arm of the quasi-serial manipulator contains a kinematic parallelogram whose adjacent sides could be treated as the links #2 and #3 (Fig. 4). Besides, here the actuators are usually located at the same axis (corresponding to the joint #2 of the serial manipulator). Such kinematics is quite common in heavy industrial robots since it allows reducing the robot arm inertia and improving its dynamic properties. For comparison purposes, the geometric and elastic parameters of the quasi-serial arm are assumed to be the same as for the corresponding serial case.

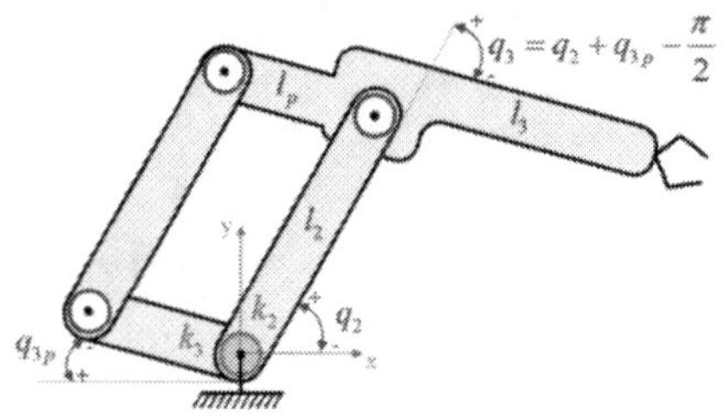

**Figure 4.** Kinematic structure and basic parameters of a quasi-serial robot arm

For the quasi-serial geometry, the desired stiffness matrix can be computed using expression (1), where the Jacobian should be replaced by

$$\mathbf{J}_P = \begin{bmatrix} -l_2\cos q_2 & -l_3\cos q_{3p} \\ l_2\sin q_2 & l_3\sin q_{3p} \end{bmatrix} \tag{5}$$

Here $q_{3p}$ is the coordinate of the actuated joint #3, which differs from $q_3$ in the serial manipulator. Using this notations, the compliance matrix of the quasi-serial robot can be expressed as

$$\mathbf{k}_{CP} = \begin{bmatrix} k_2 l_2^{\,2}\,c_2^2 + k_3 l_3^{\,2}\,c_{3p}^2 & -k_2 l_2^{\,2}\,c_2\,s_2 - k_3 l_3^{\,2}\,c_{3p}\,s_{3p} \\ -k_2 l_2^{\,2}\,c_2\,s_2 - k_3 l_3^{\,2}\,c_{3p}\,s_{3p} & k_2 l_2^{\,2} s_2^2 + k_3 l_3^{\,2} s_{3p}^2 \end{bmatrix} \tag{6}$$

where $c_{3p} = \cos q_{3p}$, $s_{3p} = \sin q_{3p}$. Following the assumptions on the stiffness coefficients (i.e., $k_2 = k_3 = k$) and introducing the link lengths ratio $\mu = l_3 / l_2$, the above matrix can be presented in the compact form as

$$\mathbf{k}_{CP} = k l_2^{\,2}\begin{bmatrix} c_2^2 + \mu^2 c_3^2 & -c_2 s_2 - \mu^2 c_3 s_3 \\ -c_2 s_2 - \mu^2 c_3 s_3 & s_2^2 + \mu^2 s_3^2 \end{bmatrix} \tag{7}$$

Using the compliance matrices (4) and (7), it is possible to estimate the manipulator end-effector deflections under the given external loading (for a given robot configuration) and also to compare the manipulator stiffness behaviour and influence the link-length ratio on the force-deflection relation.

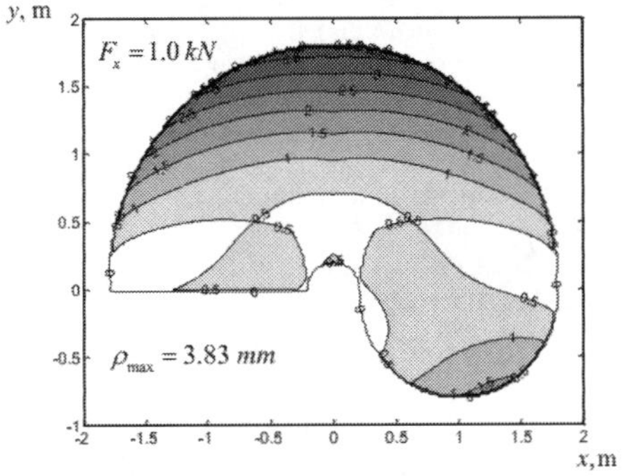


(a) Compliance map for *serial manipulator* under the external loading in x-direction

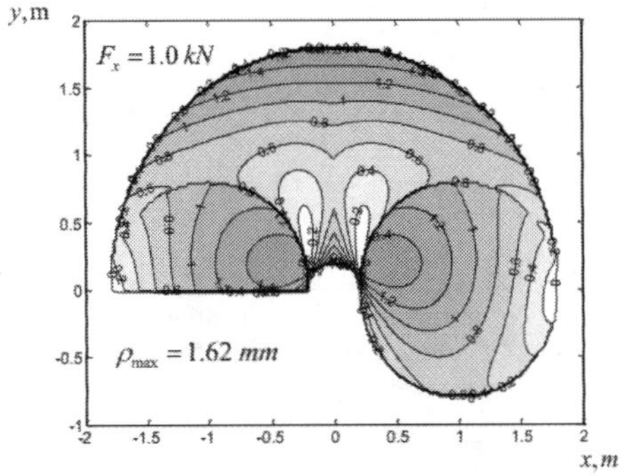


(b) Compliance map for *quasi-serial manipulator* under the external loading in x-direction

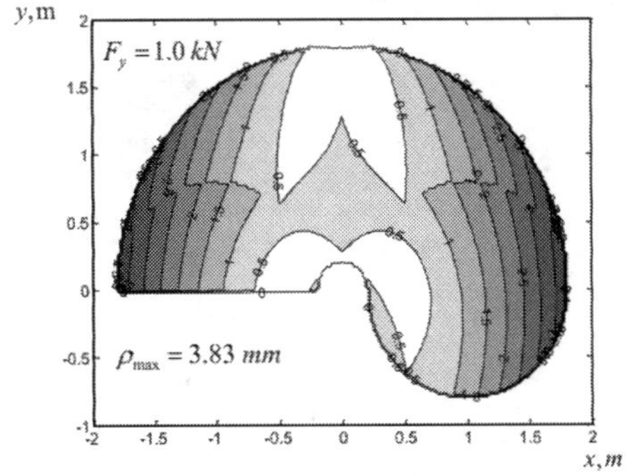


(c) Compliance map for *serial manipulator* under the external loading in y-direction

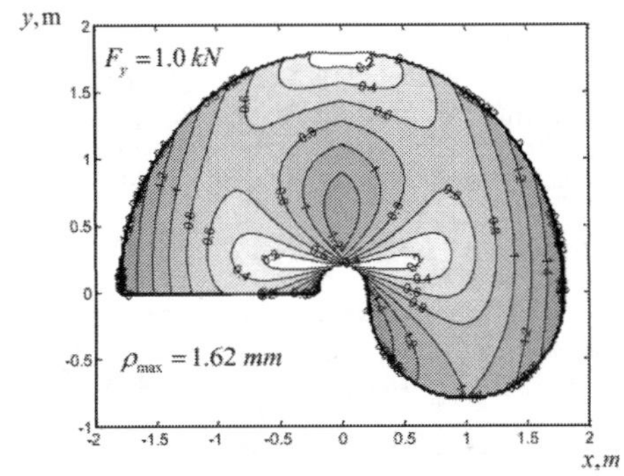


(d) Compliance map for *quasi-serial manipulator* under the external loading in y-direction

**Figure 5.** Compliance error maps for serial and quasi-serial manipulators under different external loadings ($F_x$ and $F_y$)

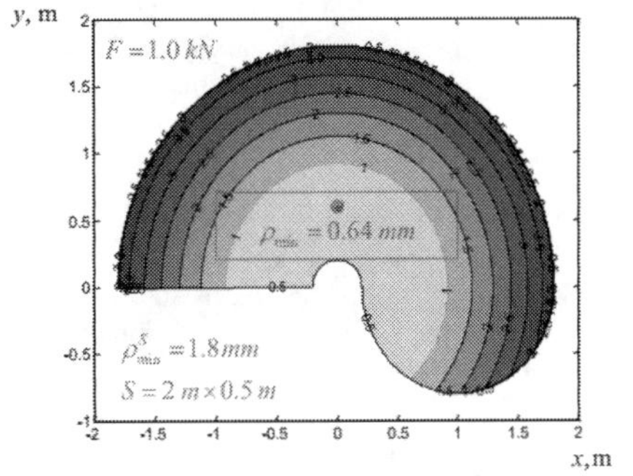


(a) Maximum compliance errors for arbitrary force directions for *serial manipulator*

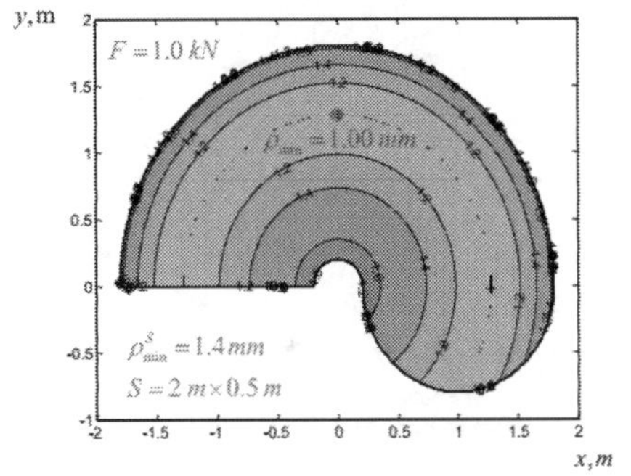


(b) Maximum compliance errors for arbitrary force directions for *quasi-serial manipulator*

**Figure 6.** Maximum compliance errors for serial and quasi-serial manipulators: case of arbitrary force direction

# 5 Comparison of compliance errors for serial and quasi-serial manipulators

To demonstrate advantages/disadvantages of both serial and quasi-serial architectures, let us present an example showing that the selection of a proper manipulator type essentially depends on the technological task dimension and external force orientation. This example deals with two manipulators (serial and quasi-serial ones) with the same geometric parameters $l_2 = 1m$ , $l_3 = 0.8m$ and identical joint compliances $k = 10^{-6}\,N m/rad$ . These values are typical for industrial robots that are used in machining applications. To compare their stiffness behaviour, let us compute the compliance errors caused by an external force 1.0 *kN* which is applied to the end-effector. Relevant results have been obtained for two different external loadings (applied in x- and y-directions). They are presented in Fig. 5, which shows the compliance error distribution within the manipulator workspace for both serial and quasi-serial robots.

As follows from the result presented in Fig. 5, the elastostatic deflections for the serial manipulator vary through the workspace and may reach up to 3.83 mm. For the quasi serial manipulator, the upper value is 1.62 mm. These results show that for the considered case study the compliance errors range does not depend on the force direction, while the compliance error maps of the two manipulators are not identical. It should be also mentioned that for both manipulators the best accuracy is achieved at the workspace boundary (i.e., in a singular configuration), which is not acceptable in the most of industrial applications. Moreover, at the corresponding workspace points, the compliance error achieves its maximum value for the orthogonal force direction.

To compare the considered architectures, let us first find the workspace point, which provides the smallest compliance errors for all possible force directions. It is clear that in the neighbourhood of this point it is reasonable to locate a technological task of relatively small dimensions. The relevant computations demonstrated that the minimum compliance errors are 0.64 and 1.00 mm for the serial and quasi-serial robots respectively (Fig. 6). Hence, the serial architecture is preferable for small-dimensional tasks.

Further, let us locate in the manipulator workspace a technological task of size $0.5\times2$ m, which can be treated as a large-dimensional one (compared to the manipulator workspace). Corresponding computational results show that the minimum compliance errors that can be achieved for all possible force directions and for all task points are 1.8 mm and 1.4 mm for the serial and quasi-serial manipulators correspondingly. Hence, the quasi-serial architecture is preferable for the large-dimensional tasks.

## 6 Conclusions

The paper deals with comparison analysis of serial and quasi-serial manipulators. The obtained results show that the quasi-serial manipulators are preferable for large-dimensional tasks, while the serial ones better suit small and medium size tasks, assuming that the machining task is optimally located within the robot workspace. When the task location is predefined and cannot be optimized, the quasi-serial manipulators should be used since they provide more homogenous compliance error distribution within the entire robot workspace. Another advantage of the quasi-serial manipulators is related to the fact that the best accuracy is ensured in the mid-workspace, while the compliance errors minimum for the serial manipulators is achieved on the workspace boundary.

## Acknowledgments

The work presented in this paper was partially funded by the project Partenariat Hubert Curien SIAM 2016 France-Thailand.

## References

1. Chérif, M., et al., Generic modelling of milling forces for CAD/CAM applications. International Journal of Machine Tools and Manufacture, 2004. 44(1): p. 29-37.
2. Vosniakos, G.-C. and E. Matsas, Improving feasibility of robotic milling through robot placement optimisation. Robotics and Computer-Integrated Manufacturing, 2010. 26(5): p. 517-525.
3. Guo, Y., et al., Vibration analysis and suppression in robotic boring process. International Journal of Machine Tools and Manufacture, 2016. 101: p. 102-110.
4. Klimchik, A., et al., Compliance error compensation technique for parallel robots composed of non-perfect serial chains. Robotics and Computer-Integrated Manufacturing, 2013. 29(2): p. 385-393.
5. Chen, Y., et al., Spatial statistical analysis and compensation of machining errors for complex surfaces. Precision Engineering, 2013. 37(1): p. 203-212.
6. Drouet, P., et al., Compensation of geometric and elastic errors in large manipulators with an application to a high accuracy medical system. Robotica, 2002. 20(03): p. 341-352.
7. Klimchik, A., et al. Identification of geometrical and elastostatic parameters of heavy industrial robots. in IEEE International Conference on Robotics and Automation (ICRA) 2013.
8. Klimchik, A., et al., Optimal Selection of Measurement Configurations for Stiffness Model Calibration of Anthropomorphic Manipulators. Applied Mechanics and Materials, 2012. 162: p. 161-170.
9. Belchior, J., et al., Off-line compensation of the tool path deviations on robotic machining: Application to incremental sheet forming. Robotics and Computer-Integrated Manufacturing, 2013. 29(4): p. 58-69.
10. Guo, Y., et al., A multilevel calibration technique for an industrial robot with parallelogram mechanism. Precision Engineering, 2015. 40: p. 261-272.
11. Sun, Y. and J.M. Hollerbach. Observability index selection for robot calibration. in Robotics and Automation, 2008. ICRA 2008. IEEE International Conference on. 2008.
12. Klimchik, A., et al., Geometric and elastostatic calibration of robotic manipulator using partial pose measurements. Advanced Robotics, 2014. 28(21): p. 1419-1429.
13. Imoto, J., et al., Optimal kinematic calibration of robots based on maximum positioning-error estimation (Theory and application to a parallel-mechanism pipe bender), in Computational Kinematics. 2009, Springer. p. 133-140.
14. Wu, Y., et al., Geometric calibration of industrial robots using enhanced partial pose measurements and design of experiments. Robotics and Computer-Integrated Manufacturing, 2015. 35: p. 151-168.
15. Yan, S.J., S.K. Ong, and A.Y.C. Nee, Stiffness analysis of parallelogram-type parallel manipulators using a strain energy method. Robotics and Computer-Integrated Manufacturing, 2016. 37: p. 13-22.
16. Klimchik, A., D. Chablat, and A. Pashkevich, Stiffness modeling for perfect and non-perfect parallel manipulators under internal and external loadings. Mechanism and Machine Theory, 2014. 79: p. 1-28.
17. Quennouelle, C. and C.m. Gosselin. Instantaneous kinemato-static model of planar compliant parallel mechanisms. in ASME 2008 International Design Engineering Technical Conferences and Computers and Information in Engineering Conference. 2008. American Society of Mechanical Engineers.
18. Pashkevich, A., A. Klimchik, and D. Chablat, Enhanced stiffness modeling of manipulators with passive joints. Mechanism and machine theory, 2011. 46(5): p. 662-679.
19. Klimchik, A., et al., Identification of the manipulator stiffness model parameters in industrial environment. Mechanism and Machine Theory, 2015. 90: p. 1-22.
20. Nubiola, A. and I.A. Bonev, Absolute calibration of an ABB IRB 1600 robot using a laser tracker. Robotics and Computer-Integrated Manufacturing, 2013. 29(1): p. 236-245.

 DOI: 10.1051/matecconf/20167502004

# Humic Acid Degradation by ZnO Photocatalyst

Putri A. Sekartaji and Sandhya Babel

*School of Biochemical Engineering and Technology, Sirindhorn International Institute of Technology (SIIT), Thammasat University, Thailand*

**Abstract.** Humic acid (HA) is universally present in soils and natural water resources in a yellow-brown form. HA can react with chlorine during drinking water treatment and produce disinfection byproducts (DBPs), such as trihalomethanes (THMs) and haloacetic acids (HAAs), which are harmful for health. Therefore, HA has to be eliminated from water environment. The photocatalysis is an effective alternative solution for the degradation of HA in a water environment. This research aims to degrade HA from water environment. The rapid degradation of HA, using zinc oxide nanoparticles, irradiated by ultraviolet light (ZnO/UV), is investigated. The optimum conditions of pertinent factors, which include the light wavelength (UV-A and UV-C), and light intensity, HA concentration, ZnO dose, and contact time are investigated at neutral pH conditions, considered for drinking water treatment. HA degradation efficiency reached more than 80% after 60 min for both types of irradiation in optimum conditions of 0.3 g/L ZnO dose in 180 min of contact time. Comparisons for degradation efficiency under UV-A and UV-C irradiation indicate that UV-C has higher efficiency, up to 150 min of contact time. The reusability of catalyst is performed for three reuses and still revealed effective for beneficial commercial applications.

## 1 Introduction

Humic acid (HA) commonly exist in surface and ground water. Chemical structure is complex and contributes to natural organic matter (NOM). HA sources are from animal and plant materials break down. This can affect drinking water quality. HA can react with chlorine during drinking water treatment and produce disinfection byproducts (DBPs), such as trihalomethanes (THMs) and haloacetic acids (HAAs), which can cause potential health impacts [1], [2]. Hepplewhite et al reported that HA is present in drinking water at concentration levels of between 2 and 15 mg/l [3].

Recently, ZnO is used as an effective photocatalyst for organic pollutants degradation, due to its high photochemical reactivity, low cost and nontoxic [4]. ZnO showed a good performance in promoting generation of reactive oxygen species (ROS) under irradiation with energy [5]. Oskoei et al. studied HA degradation at lower initial concentrations under UV mercury lamp up to 60 minutes of contact timeusing ZnO [6]. Kim et al. analysed for higher initial concentrations of HA, using UV-C/NZLC [7]. However, behavior of HA degradation under higher range of contact time and under different intensities of UV-A and UV-C lamps are not comparatively investigated.

Thus, in this study, the potential of applying ZnO as catalyst in photocatalytic process to degrade HA from water environment under UV-A and UV-C is evaluated. Effect of pertinent factors affecting degradation are also evaluated.

## 2 Materials and methods

### 2.1 Preparation of synthetic HA solution

HA was obtained from Sigma-Aldrich. A stock solution of HA was prepared by using laboratory-grade humic acid-sodium salt solution. HA concentration of 5, 8, 11 and 14 mg/L were prepared from stock solution using serial dilution with deionized water. A calibration curve was plotted using standardized HA solutions.

### 2.2 ZnO as catalyst

Zinc Oxide nanoparticle with a size of 6-12 nm, hexagonal crystal structure, 99.8% purity, was used. Since ZnO nanoparticles are more difficult to disperse in organic solvent, ultrasonication for 5-10 min is required for dispersion in solution.

### 2.3 Experimental set up

Experimental setup is shown in Fig. 1. Photocatalytic degradations were performed in sealed conditions. The whole system was covered by aluminium foil. One lamp was installed at the top of the reactor and two lamps on each side of the container. Two reactors each for UV-A (315-400 nm) and UV-C (100-280 nm) lamps, were used. During the experiment, the HA solution was mixed by using a magnetic stirrer.

### 2.4 Methodology

Photocatalytic experiment was done in a Pyrex beaker placed on the top of magnetic stirrer containing 100 ml of

HA solution at the concentration of 5, 8, 11 and 14 mg/L. The ZnO doses were added at range 0.1 – 1 g/L. The suspensions were irradiated for various contact time ranging from 30 – 180 min in interval of 30 min. After irradiation, the suspensions were filtered using 0.45 μm syringe filters. HA concentrations was analyzed by using UV-visible spectrophotometer Thermo Scientific, Genesys at 276 nm wavelength. Effect of light intensity was investigated by varying the number of 8W UV-A lamps i.e. 0.131 mw/cm$^2$, 0.167 mw/cm$^2$ and 0.236 mw/cm$^2$, respectively, and 8W UV-C lamps 0.196 mw/cm$^2$, 0.233 mw/cm$^2$ and 0.284 mw/cm$^2$, respectively. Light intensity was measured at liquid surface. All experiments were performed at room temperature (25 ± 3°C) and at neutral pH conditions. The reusability experiment of ZnO as a catalyst was done by centrifuging the sample for 20 min at 5000 rpm. It is followed by drying the catalyst at 60°C for at least 12 h.

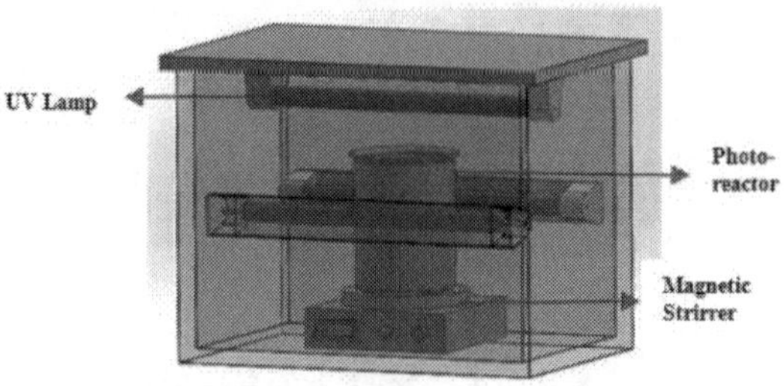


**Figure 1.** Schematic of the photocatalytic reactor

This experiment was done for UV-A and UV-C at light intensity 0.236 mw/cm$^2$ and 0.284 mw/cm$^2$, respectively, at optimum condition (14 mg/L of HA, ZnO dose 0.3 g/L, contact time 180 min).

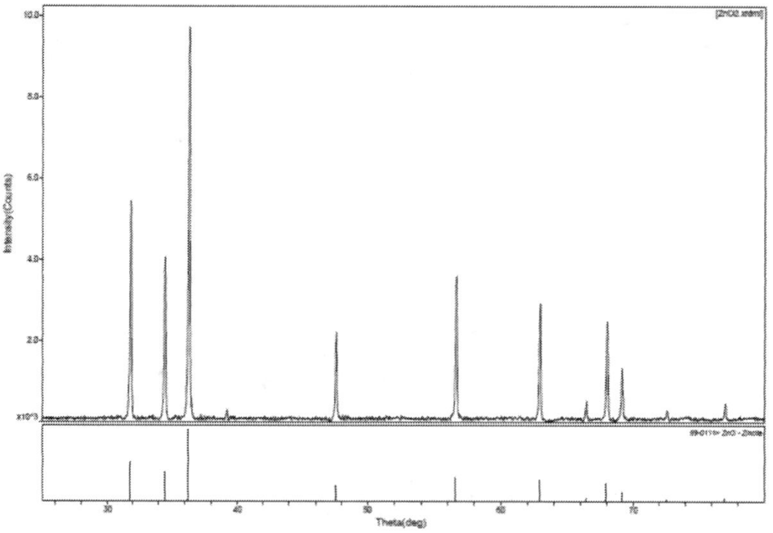


**Figure 2.** X-ray diffraction pattern of ZnO

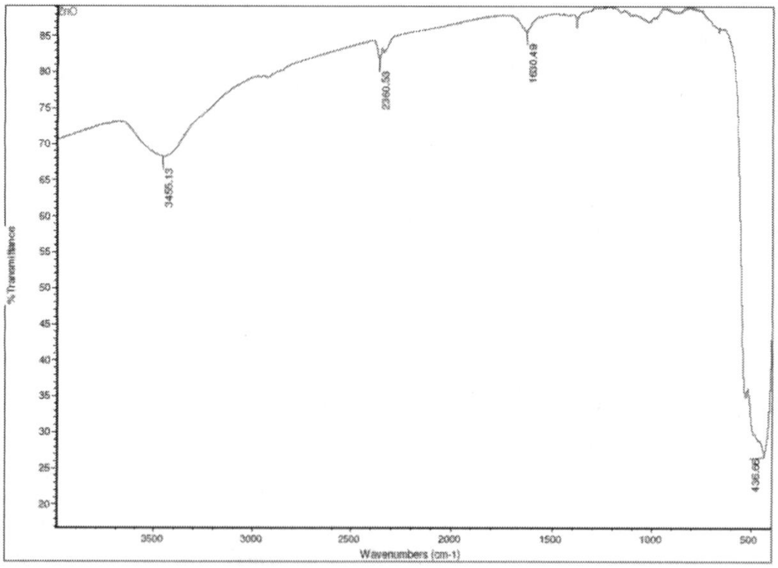


**Figure 3.** FT-IR spectra of ZnO

DOI: 10.1051/matecconf/20167502004

# 3 Results and discussions

## 3.1 ZnO characterization

From XRD pattern of ZnO in Fig. 2, the facets (2h value) at 31.7, 34.3, 36.3, 47.6 and 56.7 were assigned to (100), (002), (101), (102) and (110) reflection lines of hexagonal ZnO, respectively. It indicates that the characteristic peaks represent the ZnO with hexagonal phase [8].

FT-IR spectra of the ZnO nanoparticles in Fig. 3 showed the fundamental mode of vibration at 3000-3600 $m^{-1}$ which corresponds to the O-H stretching vibration, 2360.53 $m^{-1}$ which corresponds to C-H stretching vibration and 1365 $m^{-1}$ corresponds to C=O asymmetric C=O stretching vibration, 1630.49 $m^{-1}$ corresponds to C=O symmetric stretching vibration. O-H bending of the hydroxyl group at 436.66 $m^{-1}$ is observed [8].

## 3.2 Effect of HA concentration

The effect of HA concentration of was carried out at 0.1 g/L ZnO dose, for 60 min. HA concentrations was varied from 5 to 14 mg/L. From Fig. 4, it can be seen that increasing the initial concentration of HA decreases the degradation efficiency. The efficiency of HA degradation varied from 95% to 51% for UV-A and 97% to 47% for UV-C, respectively.

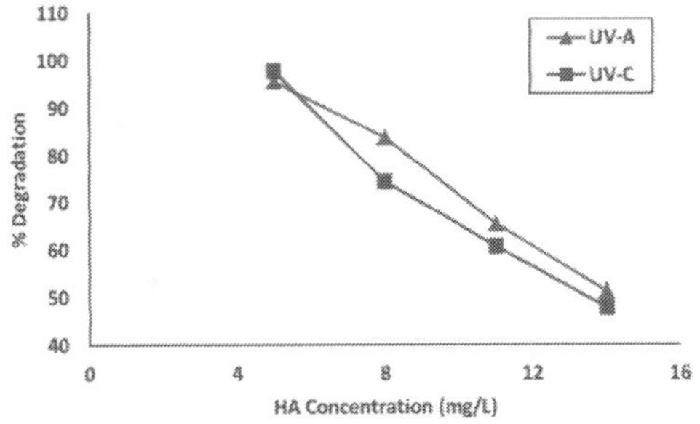


**Figure 4.** Effect of HA concentration on degradation by ZnO photocatalyst (ZnO dose 0.1 g/L, contact time 60min)

HA degradation decreased as the initial concentration increases and repulsive forces between molecules occurred, which effected in reduction in the number of hydroxyl radicals generated on the surface of catalyst by UV irradiations. Futhermore, a solution with a low concentration of HA with the same amount of hydroxyl radicals will have a higher degradation rate than a solution with a high concentration [9-11]. Previous studies on trichloroethylene (TCE) reported that the reaction by-products generated during the photocatalytic reaction can also decrease the degradation rate [11].

## 3.3 Effect of ZnO dose

In order to avoid the use of excess catalyst, it is necessary to find out the optimum loading for efficient removal of pollutant. Fig. 5 shows that increasing the ZnO dose from 0.1 to 1 g/L increased degradation of HA from 51% to 94% for UV-A and 48% to 94% for UV-C, respectively, with a contact time of 60 min. This is due to an increase in the active sites available for absorption of UV irradiations and adsorption of pollutant molecules which effects in producing large amounts of hydroxyl radicals. The result showed that rapid increase of HA degradation up to 0.5 g/L ZnO dose. The increase subsequent to 0.5 g/L seemed gradual.

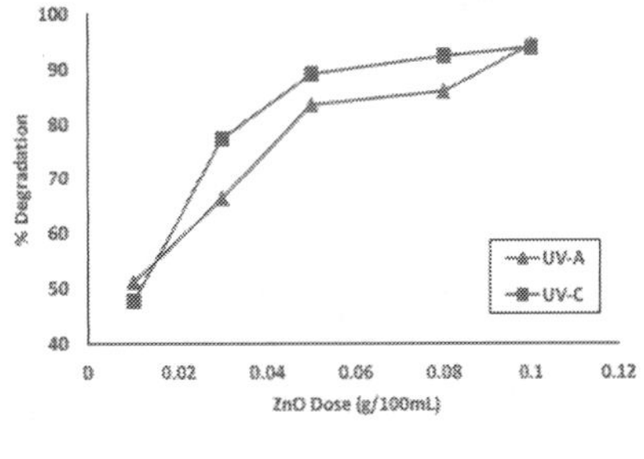


**Figure 5.** Effect of ZnO dose on degradation of HA (HA conc. 5 mg/L, contact time 60 min)

## 3.4 Effect of contact time

From the ZnO dose results, 0.5 g/L seemed to be the optimum dose. However, experiments were done for both 0.3 g/L and 0.5 g/L. It was observed that significant results for 60 min of contact time, could be obtained from 0.3 g/L. Therefore, 0.3 g/L was selected as the optimum of ZnO dose. For experiments conducted here onwards.

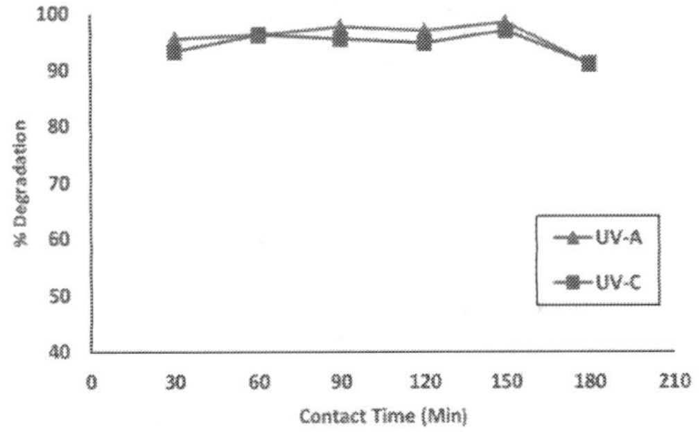


**Figure 6(a).** Effect of contact time on degradation of HA by ZnO photocatalyst (HA conc. 5 mg/L, ZnO dose 0.3 g/L)

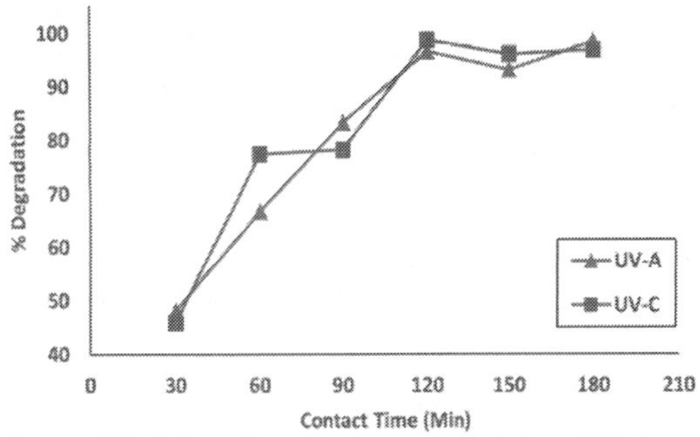


**Figure 6(b).** Effect of contact time on degradation of HA by ZnO photocatalyst (HA conc. 14 mg/L, ZnO dose 0.3 g/L)

Fig. 6(a) and (b) show the effect of contact time for HA concentration of 5 mg/L and 14 mg/L, respectively . It shows that increasing the contact time from 30 min to 180 min increased degradation efficiency. The result has shown insignificant variation of degradation efficiency for lower concentration at 5 mg/L, with time. The ZnO dissolution might have occurred due to this condition. The results show rapid increase of degradation efficiency for higher concentration at 14 mg/L, especially at low contact time. After 120 min of contact time, degradation efficiency remained nearly constant in this condition. The degradation efficiency was rapid at the beginning because there were a large number of active sites available on the surface of the adsorbent. The number of sites decreased over time with the gradual occupation of sites by HA. This results in increasing repulsive forces between the adsorbed molecules leading to decreased degradation [6].

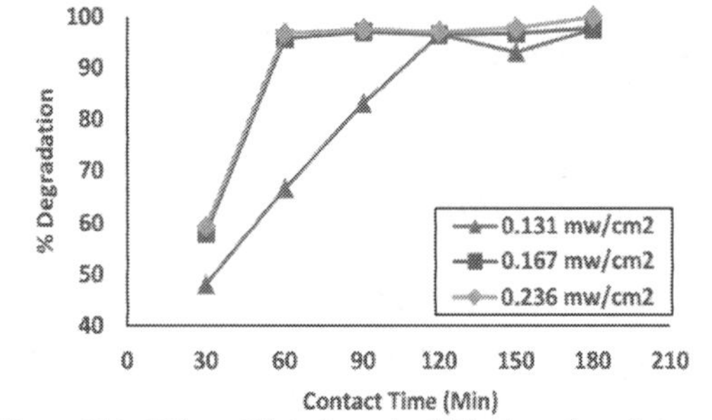


**Figure 7(a).** Effect of light intensity (UV-A) on degradation of HA by ZnO photocatalyst (HA conc. 14 mg/L, ZnO dose 0.3 g/L, contact time from 30 – 180 min)

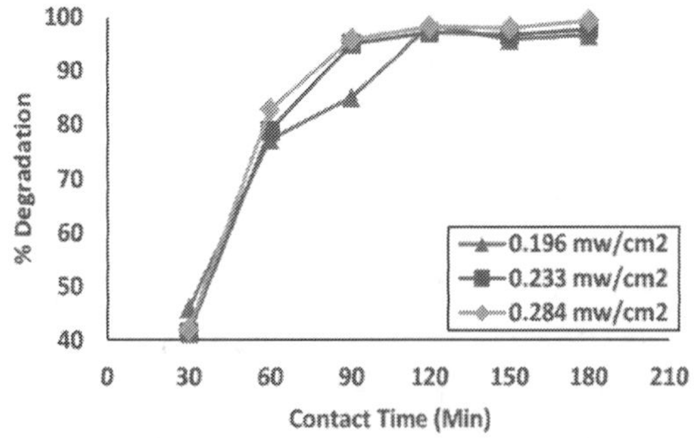


**Figure 7(b).** Effect of light intensity (UV-C) on degradation of HA by ZnO photocatalyst (HA conc. 14 mg/L, ZnO dose 0.3 g/L, contact time from 30 – 180 min)

### 3.5 Effect of light intensity

Fig. 7(a) and Fig. 7(b) show the effects of light intensity of 14 mg/L HA concentration for UV-A and UV-C, respectively, with time. In this study, contact time is varied from 30-180 min when finding the effect of light intensity. Fig. 7(a) shows that under UV-A irradiation, different behaviours are observed in different light intensities. The lower intensity required 120 min of contact time to reach more than 90% of HA degradation. On the other hand, higher intensities required only 60 min of contact time. Fig 7(b) shows lower intensity of UV-C irradiation required 120 min to reach more than 90% of HA degradation. For higher intensities required 90 min of contact time. These graphs indicate that in terms of contact time, UV-A required lower contact time than UV-C to reach 90% of HA degradation. However, UV-C seemed to have higher degradation efficiency up to 150 min. Similar observation was obtained in a previous study by Bak et al, in an experiment done for trichloroethylene [11]. Higher intensity conditions will lead to higher electron-hole pair generation rate. Therefore, higher degradation efficiency is obtained, due to higher photocatalytic activity [12].

Fig. 8 indicates that more than 50% of HA degradation is dominated by photodegradation. The activity of ZnO showed better degradation of HA under irradiation condition. Due to that result, the photodegradation using ZnO provides higher concentration of hydroxyl ions to react with holes to form hydroxyl radicals. Hydroxyl radical is considered to be responsible for oxidation decomposition of organic compound. However, agglomeration and accumulation of ZnO might be occurred in the absence of irradiation.

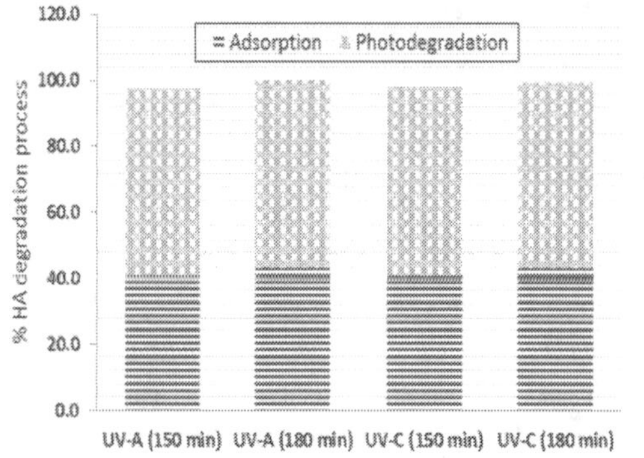


**Figure 8.** The ratio of adsorption and photodegradation to total of HA degradation percentage.

### 3.6 Reusability of Catalyst

In order to study the reusability of the catalyst, centrifugation separation method was used. Photocatalytic experiments were performed at 14 mg/L HA and 0.3 g/L ZnO, with 180 min of contact time. The ZnO was reused up to three cycles with consistent activity. Similar studies were done on $TiO_2$ for five cycles [13]. After three cycles of reusability, the percentages of degradation obtained were 88% for UV-A and 76% for UV-C, respectively. With higher number of cycles, the percentage of catalyst recovery decreased. The decrease of HA degradation might occur from the photocorrosion effect, which affected catalyst to a large extent. L.Huang et al, showed similar result on $Cu_2O$ where the photocorrosion rate was rapid in catalyst as against their micron sized counterparts [14]. Therefore, three cycles of reusability is suggested to be acceptable from the experiments of this study.

## 4 Conclusions

In this study, HA degradation by ZnO photocatalyst was carried out. The photocatalytic degradation of HA was studied by varying parameters that can be used to degrade HA in drinking water. Increasing the initial concentration of HA decreases the degradation efficiency due to the reduction of generated hydroxyl radicals. As the ZnO dose increases, degradation efficiency increases. HA degradation under high intensity UV-A irradiation required lower contact time, to reach 90% degradation. HA degradation efficiency under UV-C has higher efficiency, up to 150 min of contact time. The reusability of ZnO was demonstrated with consistent activity, and is still effective at three cycles. The use of catalyst creates a technical problem in retrieving the catalyst from the treated water. The authors suggest an immobilized catalyst for commercial applications for future works. Moreover, using the results of this study, efficiencies of different photocatalysts are to be comparatively investigated to remove HA from drinking water.

## Acknowledgment

This research was supported by Sirindhorn International Institute of Technology, Thammasat University and The Asian Development Bank.

## References

1. De Wuilloud J.C, Sadi B.B. and Caruso J.A., "Trace humic and fulvic acid determination in natural water by cloud point extraction/pre-concentration using non-ionic and cationic surfactants with FI-UV detection". *The Analyst* **128**, 453-458, (2003).
2. Liu S, Lim M, Fabris R, Chow C, Drikas M, and Amal R., "$TiO_2$ Photocatalysis of Natural Organic Matter in Surface Water: Impact on Trihalomethane and Haloacetic Acid Formation Potential" in *Environ. Sci. Technol.* **42**: 6, pp. 218-223, (2008).
3. Hepplewhite C, Newcombe G and Knappe D, "NOM and MIB, who wins in the competition for activated carbon adsorption sites?" in *Water Sci. Technol.* **49**, pp. 257-267, (2004).
4. Hongbo Ma, Phillip L. Williams, Stephen A. Diamond, Ecotoxicity of manufactured ZnO nanoparticles – a review, *Environ. Pollut.* **172**: 76–85, (2013)
5. Ravichandrika, K., Kiranmayi, P. & Ravikumar, "Synthesis, characterization and antibacterial activity of ZnO nanoparticles", *International Journal of pharmacy and pharmaceutical Science*, **4**, ISSN-0975-1491, p. 336-338, (2012).
6. V. Oskoei, M.H. Dehghani, S. Nazmara, B. Heibati, M. Asif, I. Tyagi, Shilpi Agarwal and Vinod Kumar Gupta, "Removal of humic acid from aqueous solution using UV/ZnO nano-photocatalysis and adsorption", *Journal of Molecular Liquids* **213**: 374-380, (2016).
7. Kim, J.K,. Alajmy, J., Borges, A.C., Joo, J.C., Anh, H., and Campos, L.C, "Degradation of Humic Acid by Photocatalytic Reaction Using Nano-sized ZnO/Laponite Composite", *Water Air Soil Pollut.* **224**:1749, (2013).
8. G.E. Fryxell, and G. Cao, "Environmental Applications of Nanomaterials: Synthesis, Sorbents and Sensors", *World Scientific*, (2007).
9. M. A. Behnajady, N. Modirshahla, and R. Hamzavi, "Kinetic study on photocatalytic degradation of C.I. acid yellow 23 by ZnO photocatalyst," *Journal of Hazardous Materials*, **133**, no.1–3, pp. 226–232, (2006).
10. H. Yang, G. Li, T. An, Y. Gao, and J.Fu, "Photocatalytic degradation kinetics and mechanism of environmental pharmaceuticals in aqueous suspension of $TiO_2$: a case of sulfa drugs," *Catalysis Today*, **153**, no. 3-4, pp. 200–207, (2010).
11. Bak S, Song M, Nam I, Lee WG, "Photocatalytic Oxidation of Trichloroethylene in Water Using a Porous Ball of Nano-ZnO and Nanoclay Composite", *Journal of Nanomaterials*, Article ID 160212, 8 pages, (2015).
12. G.E. Fryxell, G. Cao, Environmental Applications of Nanomaterials: Synthesis, Sorbents and Sensors, *World Scientific*, (2007).
13. Sekartaji, Putri Ardyarini and Sandhya Babel. Degradation of humic acid by using $TiO_2$ photocatalyst. In *Proceedings of the 10th GMSARN International Conference 2015 on Smart Energy, Environment and Community Development in GMS*, (2015).
14. L.Huang, F.Peng, H.Yu and H.Wang, *Solid State Sci.* **11**: 129–138, (2009).

# Classification of Textures Using Filter Based Local Feature Extraction

Veysel Gokhan Bocekci [1] and Kazim Yildiz [2]

[1]*Marmara University, Technology Faculty, Electric and Electronic Engineering, 34722 Istanbul, Turkey*
[2]*Marmara University, Technology Faculty, Computer Engineering, 34722, Istanbul, Turkey*

**Abstract.** In this work local features are used in feature extraction process in image processing for textures. The local binary pattern feature extraction method from textures are introduced. Filtering is also used during the feature extraction process for getting discriminative features. To show the effectiveness of the algorithm before the extraction process, three different noise are added to both train and test images. Wiener filter and median filter are used to remove the noise from images. We evaluate the performance of the method with Naïve Bayesian classifier. We conduct the comparative analysis on benchmark dataset with different filtering and size. Our experiments demonstrate that feature extraction process combine with filtering give promising results on noisy images.

## 1 Introduction

The feature extraction and identification of the texture is a challenging duty in computer vision, especially number of features. It is known texture analysis is a kind of problem in literature. In feature extraction process, so many methods were designed like filter banks [1], co-occurrence statistics [2] and hidden Markov models [3]. These all application are based on two dimensional surfaces. To overcome of the problem like rotation, scale and illumination, the texton based approaches were designed [4-6]. Images are served as histograms of the discrete vocabulary of features.

Local binary pattern (LBP) is most well-known and popular descriptor because of its computational simplicity, robustness of the gray scale changes and promising performance [7]. There are also some limitations to using it, for instance the descriptor is not invariant to rotation, the size of the features are growing exponentially and increase the time complexity of the algorithm. The other reason is structural information which is taken from texture is limited [8]. For improving of the algorithm performance and overcome of the limitations, there are some modifications on it. LBP descriptors are designed and the local orientation is completely removed [7]. LBP-Histogram Fourier based algorithm is designed by Zhao and friends [9]. The mentioned algorithm interested in magnitude spectrum of the transform. Multiresolution information is taken from images with the help of LBP, various granularities can be captured [10]. Liao and friends are designed dominant LBP to choose the outlier features from images [11]. Guo proposed a method to overcome of the feature increasing problem which is fisher criterion based algorithm [12]. We designed filter based LBP to extract features from images. The Brodatz dataset is used for evaluate the performance of the method. It is one of the first and common datasets which have been used in texture analysis. Naïve Bayesian algorithm is used in classification process. The classification algorithm is effective on complex problems when comparing the neural network classifier that is the reason why we choose it as a classifier. The paper is organized as follows: In Section 2, the MATERIALS and method are given. Results of the experiments and finally conclusion part is given. Finally conclusion is given.

## 2 Materials and method

In this part we give information about the details of proposed approach and using algorithms. First LBP is described, then median filter and classification process are introduced.

### 2.1 Feature extraction with LBP

Central pixel and its neighbors difference is taken by LBP which operates in a local circular [13]. It is referred in 1;

$$LBP_{R,P} = \sum_{p=0}^{p-1} s(g_p - g_c)2^p, s(g_p - g_c) = \begin{cases} 1\ g_p \geq \\ 0\ g_p < \end{cases} \quad (1)$$

Central pixel and its neighnbors are represented with $g_c$ and $g_p$ which are the gray value. Neighbor index is determined by p and radius of the circular neighborhood is R. Number of the neighbors are identified by P. The coordinates of uniformly spaced circular neighbourhood are determined as (x+Rcos(2πp/P), y−Rsin(2πp/P))(x+Rcos(2πp/P), y−Rsin(2πp/P)) for p=0,1,2…P−1p=0,1,2…P−1. [8].

The Figure 1 shows how the LBP works. Label's partitions is computed after the thresholding process which can be used as a texture descriptor. Local primitives are encoded via each histogram partition.

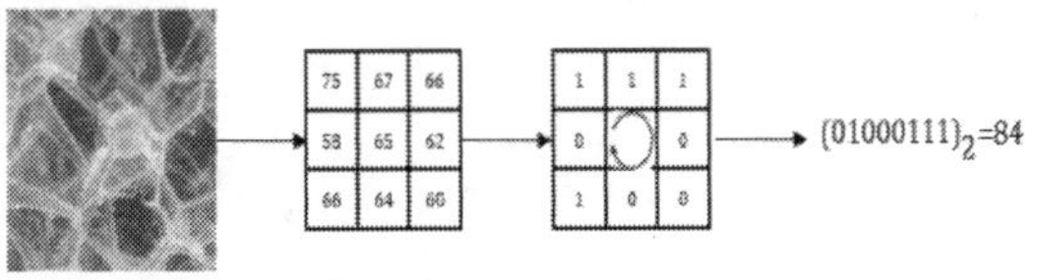


**Figure 1.** Scheme of local binary pattern.

### 2.2 Median filter

Median filter is used as a non-linear filter. Generally it is used for removing noise [14]. In this filter, the median value is used for the value of corrupted pixel in noisy image of the corresponding window. This reference pixel describes the value which is in the middle position of any sorted matrix [15].

For example the gray levels of any pixel value, in any window ($w_x$) of size nxn are represented by x1, x2, x3,...xn. After sorting it should be ascending or descendingxi1 ⩾ xi2 ⩾ xi3⋯ ⩾xin. Equation 2 value of the median;

$$M_x = Median(w_x) = \begin{cases} X_{\frac{i(n+1)}{2}}, & n \ is \ odd \\ \frac{1}{2}\left[ X_{i(\frac{n}{2})} + X_{i(\frac{n}{2})+1} \right] & n \ is \ even \end{cases} \quad (2)$$

There is so many different median filter based image denoising algorithm in literature. In this paper we used MATLAB median filter to remove the noise and outlier points from image.

### 2.3 Wiener filter

Wiener filter builds an optimal estimate of original image. The difference between original image and estimate is aimed to enforce a smallest mean square error [16]. It is used as an optimum filter which minimize mean square error. It has capability of handling both degradation function and also removing the noise [17]. In general the best way to reconstruction of a noisy signal is using wiener filter because of its optimal and adaptive ability. The aim of the filter is minimize the least square error is referred in (3) [18];

$$\min E\left\{ \left[ \hat{f} - f \right]^2 \right\} \quad (3)$$

where $f$ represents the original image and $\hat{f}$ is the restored image.

### 2.4 Naïve Bayes classifier

In pattern recognition, the classification process is an important issue. There are so many different algorithms such as nearest neighbor, decision tree and neural network. Naïve Bayes (NB) is based on Bayesian theorem which has probabilistic model [19]. NB takes into account each features to subscribe independently to probability. For parameter estimation, NB uses maximum likelihood method [20]. For a classifier, a conditional model is referred in (4);

$$p(CIX_1, .....X_n) \quad (4)$$

where C is a class and X is a feature variable. Feature can take large number values, when n is grow then basic model should be infeasible. It is referred in (5);

$$p(CIX_1, .....X_n) = \frac{p(c)p(X_1....X_n IC)}{p(X_1,....X_n)} \quad (5)$$

## 3 Results and discussion

All the codes and experiments were developed using Matlab2015a. In this part, information of the dataset and obtained results are given.

### 3.1. Data source

There are so many number of texture databases on the net for comparison the algorithms performance. In this paper we used Brodatz texture album. It has been used much and one of the first dataset which includes textures of natural scenes and materials such as bark, sand and grass. The original images of the dataset are not digital format [21]. The images in this paper should be find in [22]. We used thirteen classes from texture album. Figure 2 shows the dataset which includes classes

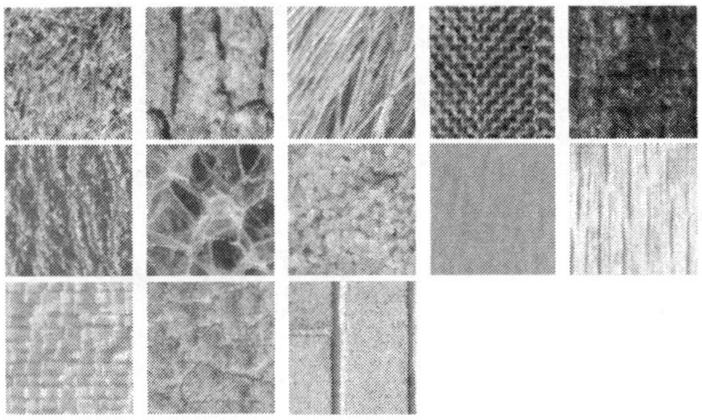

**Figure 2.** Brodatz texture album which has 13 classes.

### 3.2 Comparison of the performance

We proposed two different approaches. The dataset for the experiment is divided two part. First we used half of the dataset as train and the rest of it for test. For second approach to show the robustness of the proposed approach we decreased the number of train images. 52 images are used for train 156 images are used for test. Also we test the algorithm performance to add three different noises to both train and test images before filtering part which are Gaussian, speckle and Poisson. After adding noise, filtering part is applied to the images. The feature extraction part come after this. Obtained feature vector is used for the classification process. Figure 3 shows the example of adding noises to image.

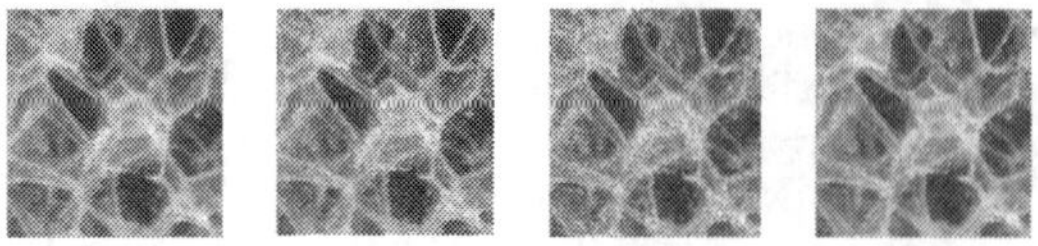


**Figure 3.** Original image (a) Gaussian noise added (b) speckle noise added (c) Poisson noise added (d)

After noise adding, for removing the noise we apply the median and wiener filter to the images respectively. The example of the application is shown in Figure 4. From images it can be seen the effective of the noises. Median and wiener filters improvement stage also is shown. Because of the its adaptive structure wiener filter does more soft effects to remove noises from images.

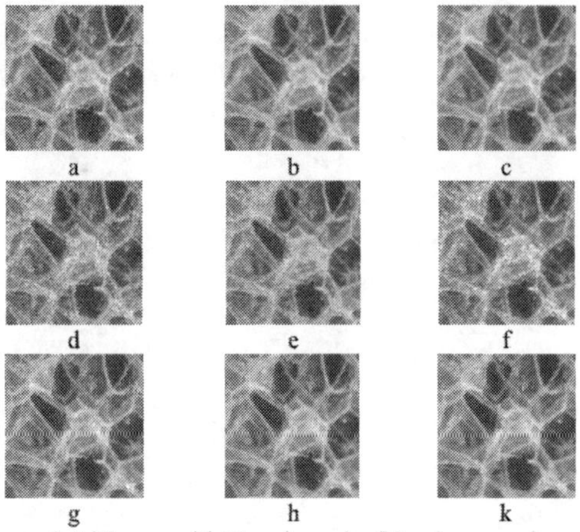


**Figure 4.** (a) Image with Gaussian noise (b) noise removing with wiener filter (c) noise removing with median filter (d) image with speckle noise (e) noise removing with wiener filter (f) noise removing with median filter (g) image with Poisson noise (h) noise removing with wiener filter (k) noise removing with median filter

We try our proposed work on different conditions. Five different train and test images are selected for experimental process. The classification accuracy on image dataset with Gaussian noise (GN) is shown in Table 1. For comparing the algorithm performance five different train and test image subset is selected. Also different size of the filter is applied to the images before classification to compare the effect on it. Averagely wiener filter with the size 5x5 is the most powerful filter to remove the noise before the classification when comparing the median filter (MF). The average performance of the classification is 99.232 with wiener filter (WF).

**Table 1.** The classification accuracy with Gaussian noise and filter.

| | GN | MF 3x3 | MF 5x5 | MF 10x10 | WF 3x3 | WF 5x5 |
|---|---|---|---|---|---|---|
| Case 1 | 92.31 | 94.23 | 96.15 | 97.12 | 91.35 | 98.08 |
| Case 2 | 83.65 | 93.27 | 96.15 | 98.08 | 96.15 | 100 |
| Case 3 | 88.46 | 92.31 | 94.23 | 92.31 | 97.12 | 99.04 |
| Case 4 | 87.5 | 93.27 | 97.12 | 94.25 | 99.04 | 99.04 |
| Case 5 | 87.5 | 93.27 | 95.19 | 95.19 | 96.15 | 100 |
| Average | 87.88 | 93.27 | 95.76 | 95.39 | 95.96 | 99.23 |

Table 2 shows the classification accuracy on image dataset with speckle noise (SN). The original classification with noise images is averagely 88%. Wiener filtering help the process with ten point increase. It is really so effective of the removing noises from images because of the adaptive structure. The obtained results with median filter is just about giving four point improvement. The average performance of the classification is 98.078 with wiener filter.

**Table 2.** The classification accuracy with speckle noise and filter.

| | SN | MF 3x3 | MF 5x5 | MF 10x10 | WF 3x3 | WF 5x5 |
|---|---|---|---|---|---|---|
| Case 1 | 86.54 | 90.38 | 92.31 | 91.35 | 93.27 | 98.08 |
| Case 2 | 88.46 | 86.54 | 91.35 | 93.27 | 94.23 | 97.12 |
| Case 3 | 91.35 | 93.27 | 93.27 | 91.35 | 99.04 | 100 |
| Case 4 | 88.46 | 91.35 | 95.19 | 93.27 | 92.31 | 99.04 |
| Case 5 | 84.62 | 88.46 | 91.35 | 89.42 | 94.23 | 96.15 |
| Average | 87.88 | 90 | 92.69 | 91.73 | 94.61 | 98.07 |

Table 3 shows the classification accuracy on image dataset with Poisson noise (PN). Actually with Poisson noise the classification is about the 94% averagely. It is not so effective to disturb the structure of the image. However filtering process should avoid these situation. And the classification is goes up to 99%. The average performance of the classification is 98.078 with wiener filter.

**Table 3.** The classification accuracy with Poisson noise and filter.

| | PN | MF 3x3 | MF 5x5 | MF 10x10 | WF 3x3 | WF 5x5 |
|---|---|---|---|---|---|---|
| Case 1 | 96.15 | 98.08 | 99.04 | 99.04 | 100 | 99.04 |
| Case 2 | 93.27 | 99.04 | 98.08 | 99.04 | 99.04 | 99.04 |
| Case 3 | 94.23 | 96.15 | 97.12 | 98.08 | 99.04 | 100 |
| Case 4 | 93.27 | 99.04 | 98.08 | 98.08 | 99.04 | 99.04 |
| Case 5 | 96.15 | 99.04 | 99.04 | 98.08 | 100 | 98.08 |
| Average | 94.61 | 98.27 | 98.27 | 98.46 | 99.42 | 99.04 |

To compare the effectiveness of the algorithm on the process, we reduce the number of training images. 52 images are selected for training stage randomly and the rest of them are used for testing. For all process the wiener filter so effective for removing noise. The averagely obtained results are about 88% accuracy for classification which can be seen in Table 4. Our proposed approach makes improvement when comparing the non-filtered results.

**Table 4.** The classification accuracy with small training set usage.

| | No filter | MF 3x3 | MF 5x5 | MF 10x10 | WF 3x3 | WF 5x5 |
|---|---|---|---|---|---|---|
| Gaussian noise | 63.64 | 78.21 | 83.22 | 76.92 | 79.71 | 90.97 |
| Speckle noise | 86.3 | 88.31 | 88.44 | 86.54 | 91.39 | 85.92 |
| Poisson noise | 73.68 | 64.9 | 80 | 85.26 | 72.19 | 86.45 |

## 4 Conclusion

In this paper we described a framework which is a filtering based feature extraction process. The main idea of the paper to show the robustness of the feature extraction and classification process with additive noise. We choose different types of noise to compare the algorithm performance. For feature extraction process the common method which is LBP used. LBP is local texture feature extraction method and it's commonly used in texture feature extraction process. Brodatz texture images are used to test proposed model efficiency. The dataset is divided into two parts.

For removing noise, the wiener and medina filter is used respectively to compare the efficiency of them. Both of them known as effective on noise and outlier pixels on images. In this work wiener filter is so effective to remove the noise form images because of its adaptive structure. Naïve Bayes algorithm is used in classification process. Averagely 99% classification result is taken with wiener filter usage during removing noise process. Non filtering classification accuracy is decreased when using training set is reduced. Averagely filtering classification accuracy is about 88%. Our proposed approach shows promising results for both situation. For future work we should try our method on different types of datasets. To show the filtering part efficiency it will design new types of filters and it is going to try under different noise conditions.

## References

1. Randen, T. and Husoy, J.H., *Filtering for texture classification: A comparative study.* Pattern Analysis and Machine Intelligence, IEEE Transactions on, (1999). **21**(4): pp. 291-310
2. Haralick, R.M., *Statistical and structural* approaches *to texture.* Proceedings of the IEEE, (1979). **67**(5): pp. 786-804
3. Chen, J.-L. and Kundu, A., *Rotation and gray scale transform invariant texture identification using wavelet decomposition and hidden Markov model.* Pattern Analysis and Machine Intelligence, IEEE Transactions on, (1994). **16**(2): pp. 208-214
4. Lazebnik, S., Schmid, C., and Ponce, J., *A sparse texture representation using local affine regions.* Pattern Analysis and Machine Intelligence, IEEE Transactions on, (2005). **27**(8): pp. 1265-1278
5. Mehta, R. and Egiazarian, K., *Texture Classification Using Dense Micro-block Difference (DMD)*, in *Computer Vision--ACCV 2014.* (2014), Springer. pp. 643-658.
6. Varma, M. and Zisserman, A., *A statistical approach to texture classification from single images.* International Journal of Computer Vision, (2005). **62**(1-2): pp. 61-81
7. Ojala, T., Pietikäinen, M., and Mäenpää, T., *Multiresolution gray-scale and rotation invariant texture classification with local binary patterns.* Pattern Analysis and Machine Intelligence, IEEE Transactions on, (2002). **24**(7): pp. 971-987
8. Mehta, R. and Egiazarian, K., *Dominant Rotated Local Binary Patterns (DRLBP) for texture classification.* Pattern Recognition Letters, (2016). **71**: pp. 16-22
9. Zhao, G., et al., *Rotation-invariant image and video description with local binary pattern features.* Image Processing, IEEE Transactions on, (2012). **21**(4): pp. 1465-1477
10. Qian, X., et al., *PLBP: An effective local binary patterns texture descriptor with pyramid representation.* Pattern Recognition, (2011). **44**(10): pp. 2502-2515
11. Liao, S., Law, M.W., and Chung, A., *Dominant local binary patterns for texture classification.* Image Processing, IEEE Transactions on, (2009). **18**(5): p. 1107-1118
12. Guo, Y., et al., *Descriptor learning based on fisher separation criterion for texture classification*, in *Computer Vision–ACCV 2010.* (2010), Springer. pp. 185-198
13. Tai, S.-C., Chen, Z.-S., and Tsai, W.-T., *An automatic mass detection system in mammograms based on complex texture features.* Biomedical and Health Informatics, IEEE Journal of, (2014). **18**(2): pp. 618-627
14. Gupta, V., Chaurasia, V., and Shandilya, M., *Random-valued impulse noise removal using adaptive dual threshold median filter.* Journal of Visual Communication and Image Representation, (2015). **26**: pp. 296-304
15. Gonzalez, R.C. and Woods, R.E., *Digital image processing.* (2002): Prentice hall Upper Saddle River
16. Jog, A. and Halbe, S., *Multiple Objects Tracking Using CAMShift Algorithm and Implementation of Trip Wire.* International Journal of Image, Graphics and Signal Processing, (2013). **5**(6): p. 43
17. Mane, S.A. and Kulhalli, K., *Mammogram Image Features Extraction And Classification For Breast Cancer Detection.* (2015)
18. Jin, F., et al., *Image Restoration Based on Wiener Filtering* Journal of communication university of china (Science and Technology), (2011). **4**: p. 005
19. Zhou, X., et al., *Detection of pathological brain in MRI scanning based on wavelet-entropy and naive bayes classifier*, in *Bioinformatics and biomedical engineering.* (2015), Springer. pp. 201-209
20. Platt, J.C., Cristianini, N., and Shawe-Taylor, J. *Large Margin DAGs for Multiclass Classification.* in *nips.* (1999)
21. Tuceryan, M. and Jain, A., *Textures: a photographic album for artists and designers.* (1966), Dover Publications
22. The USC-SIPI, I.d. *http://sipi.usc.edu/database/.* (1977) [cited 2016 November]

DOI: 10.1051/matecconf/20167503002

# Multi-Language Sentiment Analysis for Hotel Reviews

Maleerat Sodanil[a]

*Faculty of Information Technology, King Mongkut's University of Technology North Bangkok, Thailand*

**Abstract.** Touristes and traveler use avariety of information sources (e.g. travelportals, blogs, or social networkingsites like twitter) to help themdecide for a hotel room. Thesesources allcontainhighly subjective textthat expresses theopinions of many. Wetook a preliminaryview on user generatedhotelreviewsfromtwotravelportals in English andThai. Wedeveloped a taxonomy of features and specificallyinvestigated how accuratelytheycanbepredictedwiththreeclassification methods. The resultsindicatethat support vector machines perform best for thisspecificdomain.

## 1 Introduction

People clearly enjoy traveling. Travel and tourism is continuously on the rise over the past six decades and has become one of biggest and fastest-growing economic sectors on the planet, regardless any economic crisis. According to the World Tourism Organization UNWTO, international tourist arrivals have increased from 25m globally in1950, to 278m in 1980, 527m in 1995, and have now reached an impressive 1133m in 2014. Likewise, tourist spending has risen from US$ 2b in 1950 to US$ 1245b in 2014, based on the 2015 edition of the World Tourism Highlights report [1]. The amount of information about hotels has increased with it and there was never a time where travellers had better and easier access to hotel reviews than today. This however also means that travellers have to base their decisions on a lot more detailed information about hotels and an even larger amount of generally very subjective information from previous guests about their quality. Decision making has therefore become more challenging and information systems use sentiment analysis to assist in the analysis of human opinion to help users make a sensible choice. In [4], for example, supervised learning with two types of information (frequency and TF-IDF) are used to realize polarity classification of reviews.

In this paper we took a preliminary view on user generated hotel reviews from two travel portals. We developed a taxonomy of features for these reviews and specifically investigated how accurately they can be predicted with three different classification methods.

We selected reviews for English, as the major spoken language in the travel domain, and Thai as a local comparison. After reviewing related work in the next section, we highlight our evaluation methodology and describe our data set in Section 3. Results are presented and discussed in Section 4 before reaching our conclusions.

## 2 Related work

This section describes an overview of sentiment analysis related disciplines such as Text mining, Feature Extraction/Identification, Sentiment classification techniques and also performance evaluation Text mining is conventional data mining done using text features. Text features are usually keywords which extracted based-on the language model. In this paper, the keyword means the word that can be used to classify reviews in terms of polarity. The accuracy of classification depends on the text features which extracted from the reviews [2]. The goal of sentiment analysis is to detect opinion, identify the sentiments people express by classifying the contextual polarity of a given text. The polarity can be positive, negative or neutral. In general, there are three different levels in sentiment analysis: document, sentence, and feature/aspect level respectively. Document level in sentiment analysis is applied to detect the polarity of the whole product or service reviews [3], [8]. Sentence level is applied to define a given text usually a sentence or phrase while feature/aspect level determine the opinion expressed on different feature or aspects (attribute or component) of entities [3]. The results of sentence level sentiment analysis are dependent on the definition of subjectively used when annotating texts [5]. The advantage of feature-based sentiment analysis is the possibility to extract about object of interest and determine an opinion expressed on each feature/aspect is positive, negative or neural [4].

In this paper, we applied three classification approaches, Naive Bayes (NB) [9], Decision Trees (DT) [10] and Support Vector Machine (SVM) [11], for the experimental comparison for Thai and English. These

[a] Corresponding author: msn@kmutnb.ac.th.

DOI: 10.1051/matecconf/20167503002

methods are selected as a common technique for this paper as they represent a very common and universally applicable approaches in an attempt to get initial results on our data.

## 3 Evaluation methodology

This section explained the main module that used to evaluate a reviews: One is reviews weight analysis, Second is classifier module which compared from three type of classification algorithms such as Naïve Bayes, Decision tree and Support Vector Machine (SVM).Third is testing module which the best classifier is applied.

### 3.1 Proposed framework

There are three main parts of sentiment analysis of hotel reviews framework. Part 1 is a reviews analysis which an input is both Thai and English reviews and the output is a reviews separated into positive and negative. Weight calculation is applied in this part in order to assign weight for all properties (Room, Location, Service, Price and Facilities) Part 2 is a training module which reviews separated used as input and output is the best model of classification process. Part 3 is testing module which model is applied to classify a hotel review in to positive and negative as shown in Fig. 1.

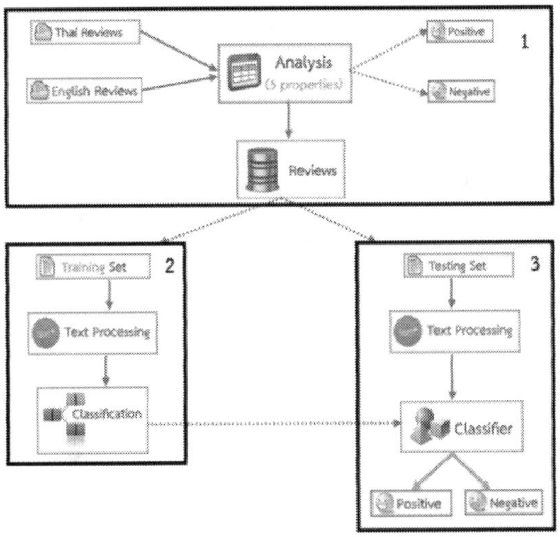


**Figure 1.** Proposed framework of sentiment analysis.

**Table 1.** Taxonomy of categories and sub-categories.

| Property | Sub-property | Property | Sub-property |
|---|---|---|---|
| **Room** | Clean<br>Room size<br>Room location<br>Other satisfaction | **Location** | Location<br>Public services<br>Environment |
| **Service** | Receptionist<br>Check-in, check-out | **Price** | Room price<br>Accessories price |
| **Facilities** | Breakfast<br>Other facilities<br>Internet access | | |

### 3.2 Data collection and process

We collected a total of 3,000 hotel reviews for English and Thai from the two well-known websites TripAdvisor[a] and Agoda[b] between January and March 2015. We had an equally sized set for each languages (1500 reviews each). The data was collected with the tool import.io[c], and free web site scraping tool, to transform the review pages into text. A domain expert then assigned the text to the five categories Room, Location, Service, Price, and Facilities where each was labelled as a class of polarity (Positive and Negative). Then we manual separate and labelled polarity from each reviews into five categories: Room, Location, Service, Facilities, and Price. All reviews are labelled as a class of polarity (Positive and Negative) by experts. Data is spitted into three sets of training, validating and testing with 60%, 20% and 20% respectively. Then three classifier were tested based-on the same data in order to compare the performance in terms of accuracy. Fig. 2 and Fig. 3 show an example of Thai reviews and English reviews respectively.

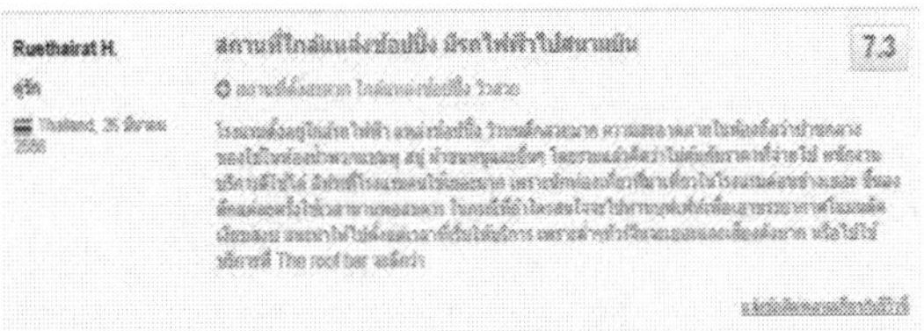


**Figure 2.** Thai reviews from agoda.com/th-th.

**Figure 3.** English reviews from Tripadvisor.

### 3.3 Weight calculation

In order to calculate weight property, five properties (Room, Location, Service, Price and Facilities)and fourteen sub-property need to be extracted from the reviews. The weight for all sub properties under the core property are sum into 1. The weight ratio of sub-property can be calculated as follows:

$$\%\ \text{mentioned properties} = \frac{\text{No. of mentioned properties} * 100}{\text{Total of reviews}} \quad (1)$$

Then summarize the total weight for each properties as:

$$W_{total} = W_1 + W_2 + \ldots + W_n \quad (2)$$

[a]www.tripadvisor.com
[b]www.agoda.com/th-th
[c]https://import.io

where $W_{total}$ is a main property weight, $W_i$ is sub-property weight where i= 1,2,..n, and n is a number of sub-property, $W_i$ equal $(+)W_i$ if the review is positive and $W_i$ equal $(-)W_i$ if the review is Negative. In ourexperiments, weight which calculated by equation 1 and 2 are illustrâtes in Table 2 and Table 3.

**Table 2.** Property weight for Thai reviews.

| Propert y | Sub-property | %Reviews | Weight (+/-) |
|---|---|---|---|
| Room | Clean | 38.33 | 0.3/ -0.3 |
| | Room size | 19.73 | 0.1/ -0.1 |
| | Room location | 12.13 | 0.1/ -0.1 |
| | Other Satisfaction | 50.47 | 0.5/ -0.5 |
| Location | Location | 18.60 | 0.1/ -0.1 |
| | Public service | 40.13 | 0.4 / -0.4 |
| | Environment | 22.73 | 0.2/ -0.2 |
| Service | Receptionist | 38.93 | 0.3/ -0.3 |
| | Check-in, check-out | 47.00 | 0.4 / -0.4 |
| Facilities | Breakfast | 22.8 | 0.2/ -0.2 |
| | Otherfacilities | 7.27 | 0.07/ -0.07 |
| | Internet access | 3.93 | 0.03/ -0.03 |
| Price | Room price | 16.13 | 0.1/ -0.1 |
| | Accessoriesprice | 6.47 | 0.06/ -0.06 |

**Table 3.** Property weight for English reviews.

| Property | Sub-property | %Reviews | Weight (+/-) |
|---|---|---|---|
| Room | Clean | 30.13 | 0.3/ -0.3 |
| | Room size | 27.93 | 0.2/ -0.2 |
| | Room location | 18.33 | 0.1/ -0.1 |
| | Other Satisfaction | 49.93 | 0.4/ -0.4 |
| Location | Location | 48.73 | 0.4/ -0.4 |
| | Public service | 46.20 | 0.4/ -0.4 |
| | Environment | 20.27 | 0.2/ -0.2 |
| Service | Receptionist | 67.13 | 0.6/ -0.6 |
| | Check-in, check-out | 43.00 | 0.4/ -0.4 |
| Facilities | Breakfast | 53.60 | 0.5/ -0.5 |
| | Otherfacilities | 41.07 | 0.4/ -0.4 |
| | Internet access | 11.60 | 0.1/ -0.1 |
| Price | Room price | 70.13 | 0.7/ -0.7 |
| | Accessoriesprice | 31.00 | 0.3/ -0.3 |

### 3.4 Evaluation measure

In order to evaluate classifier model, accuracy is used to measure how accurate of class that each classifier method performed. This paper compare the performance of the three classification models: Naïve Bayes, Decision Tree and Support Vector Machine. Accuracy is defined as:

$$\text{Accuracy} = (TP + TN)/(TP + FP + FN + TN) \quad (3)$$

where
TP is number of correct items for positive polarity,
TN the number of correct items for negative polarity,
FP the number of incorrect items for positive polarity and
FN the number of incorrect for negative polarity.

## 4 Results and discussion

The classification results are summarized in Table 4 and Fig 4. The result show the accuracies of the correct classification as described in Table 1. We have separated them by properties and the base classifier and highlighted the language for each group in the last column. The highest accuracy for each partition and language is highlighted in bold. The performance in terms of accuracy depends on the text features used in part of text processing. The accuracy in Thai language is less than English in all properties because in Thai reviews, there is a little comment and not all sub-property are mentioned while in English language reviews, there are almost cover all sub-property which we used to calculate weight statistical.

**Table 4.** Performance comparison in terms of accuracy.

| Properties | Naive Bayes | Decision Tree | SVM |
|---|---|---|---|
| Room | **90.12** | **91.40** | **92.61** |
| Location | 88.52 | 86.56 | 89.91 |
| Service | 81.70 | 87.70 | 90.39 |
| Facilities | 80.37 | 85.47 | 88.75 |
| Price | 82.02 | 80.91 | 85.50 |
| Room | 80.36 | **84.72** | 88.34 |
| Location | 83.20 | 80.93 | **86.77** |
| Service | **85.48** | 81.72 | 88.19 |
| Facilities | 81.54 | 84.40 | 85.92 |
| Price | 86.37 | 85.63 | 88.25 |

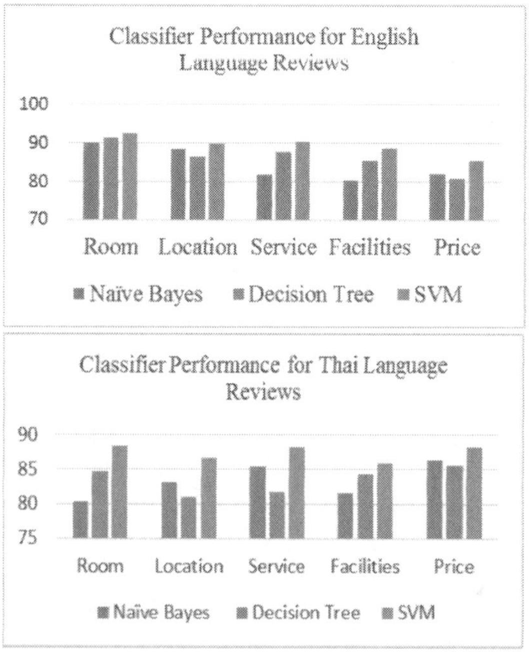


**Figure 4.** Performance comparison from three classifier of Thai and English language.

## 5 Conclusion

This paper proposed sentiment analysis based on statistic weight that is calculated from hotel reviews in both Thai and English language. We applied three classification methods — naïve Bayes, decision tree, and support vector machine have been used to classify the hotel reviews into positive and negative polarities. The experiments were conducted using hotel reviews obtained from the TripAdvisor and the Agoda portals. The results revealed that support vector machines given the best accuracy, outperforming naïve Bayes and decision tree classification. However, properties weight which calculated using training data effected the performance of the system in terms of accuracy.

## References

1. WTO. UNWTO Tourism Highlights, 2015 Edition. WTO, (2015)
2. J. Yao, H. Wang, and P. Yin. Sentiment feature identification from Chinese online reviews. In H. Tan and M. Zhou, editors, Advances in Information Technology and Education, volume 201 of Communications in Computer and Information Science, pages 315–322. Springer Berlin Heidelberg, (2011)
3. M. Hu and B. Liu. Mining and summarizing customer reviews. In Proceedings of the Tenth ACM SIGKDD International Conference on Knowledge Discovery and Data Mining, KDD '04, pages 168–177, New York, NY, USA, (2004). ACM
4. B. Liu, M. Hu, and J. Cheng. Opinion observer: Analyzing and comparing opinions on the web. In Proceedings of the 14th International Conference on World Wide Web, WWW '05, pages 342–351, New York, NY, USA, (2005). ACM
5. B. Pang, L. Lee, and S. Vaithyanathan. Thumbs up?: Sentiment classification using machine learning techniques. In Proceedings of theACL-02 Conference on Empirical Methods in Natural Language Processing - Volume **10**, EMNLP '02, pages 79–86, Stroudsburg, PA, USA, (2002). Association for Computational Linguistics
6. H.-X. Shi and X.-J. Li. A sentiment analysis model for hotel reviews based on supervised learning. In Machine Learning and Cybernetics (ICMLC), 2011 International Conference on, volume **3**, pages 950–954, (July 2011)
7. F. Su and K. Markert. From words to senses: A case study of subjectivity recognition. In Proceedings of the 22Nd International Conference on Computational Linguistics - Volume **1**, COLING '08, pages 825–832, Stroudsburg, PA, USA, (2008). Association for Computational Linguistics
8. P. D. Turney. Thumbs up or thumbs down?: Semantic orientation applied to unsupervised classification of reviews. In Proceedings of the40th Annual Meeting on Association for Computational Linguistics, ACL '02, pages 417–424, Stroudsburg, PA, USA, (2002). Association for Computational Linguistics
9. Y. Huang and L. Li, "Naive Bayes classification algorithmbased on smallsample set," *2011 IEEE International Conference on Cloud Computing and Intelligence Systems*, Beijing, (2011), pp. 34-39
10. M. T. Cazzolato and M. X. Ribeiro, "A statisticaldecisiontreealgorithm for medical data streammining," *Proceedings of the 26th IEEE International Symposium on Computer-BasedMedicalSystems*, Porto, (2013), pp. 389-392
11. R. Varghese and M. Jayasree, "Aspect based Sentiment Analysisusing support vector machine classifier," *Advances in Computing, Communications and Informatics (ICACCI), 2013 International Conference on*, Mysore, (2013), pp. 1581-1586

 DOI: 10.1051/matecconf/20167503003

# Implementation of Edge Detection Digital Image Algorithm on a FPGA

Issam Bouganssa, Mohamed Sbihi and Mounia Zaim

*Laboratory of System Analysis, Information Processing and Integrated Management, High School of Technology SALE, Mohammed V University in Rabat, Morocco*

**Abstract.** This paper presents the implementation of an adaptive contour detection filter on field programmable gate array (FPGA) using a combination of hardware and software components. The proposed system locates contours from a calculation of Gradient in preferred directions while quantifying the importance of the outline with a wise thresholding. The dedicated algorithm is implemented in an FPGA Xilinx Spartan 6, and results are displayed in a VGA monitor. The FPGA offers the necessary performance for image processing and video in real time, while maintaining the flexibility of the system to support an adaptive algorithm. Simulation results and synthesis proposed edge detection processor on FPGA chip demonstrate the efficiency of proposed architecture.

## 1 Introduction

The goal of edge detection is to produce something like a line drawing of an image. In practice we will look for places in the image where the intensity changes quickly. Observe that, in general, the boundaries of objects tend to produce sudden changes in the image intensity.

For example, different objects are usually different colors or hues and this causes the image intensity to change as we move from one object to another. In addition, different surfaces of an object receive different amounts of light, which again produces intensity changes. Thus, much of the geometric information that would be conveyed in a line drawing is captured by the intensity changes in an image. Unfortunately, there are also a large number of intensity changes that are not due to geometry, such as surface markings, texture, and specular reflections. Moreover there are sometimes surface boundaries that do not produce very strong intensity changes. Therefore the intensity boundary information that we extract from an image will tend to indicate object boundaries, but not always.

The detected edges are materialized by the intensity breakdown in the image in a given direction. Several methods exist to detect the break. Some of them are more or less complex. Others are more or less intensive computing.

In this work, we use the gradient measurement method for images followed by a judicious thresholding which allow to isolate the edge of the rest of the image. The algorithm is implemented in a FPGA device which is providing good performance of integrated circuit platform for research and development. Indeed, the FPGA technology has become an alternative for the implementation of software algorithms.

The results of edge detection algorithm implemented in FPGA are displayed on a VGA screen continuously. The image will be stored in a block memory in FPGA. Then, the algorithm reads the image of memory and stores the processed image back into memory. VGA controller designed reads the processed image from memory and displaying it.

We choose in our work the Xilinx Spartan6 circuit which offers the performance required in the real-time processing of image sequences, while retaining the flexibility of the system to support an adaptive algorithm.

## 2 Contour definition

In the image, a contour can be considered in different ways. Here we describe three main ways to consider a contour:

First, a contour can be seen as an abrupt change in the intensity of the image particularly for images in gray level [1], there are several types of variations (Figure 1).

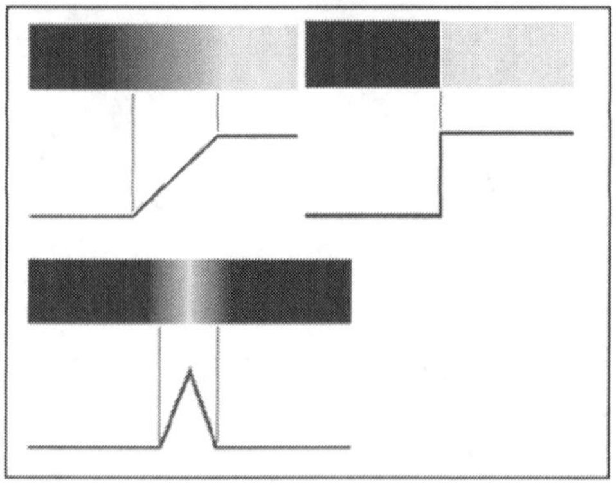

**Figure 1.** Contour profiles, ramp, doorstep, roof.

Secondly, a way very similar to that mentioned above, to consider the contour as a difference on color.

Thirdly, if we consider the image as a 2D signal, we can go in the frequency domain (Fourier transform or wavelet for example) [2]. In this case, a contour can be represented as the high frequency signal.

## 3 Edge detection principle with gradient

A local variation of intensity is a primary source of information in image processing. It's measured by ***the gradient*** [3,4] vector function of the pixel [i, j].

### 3.1 The gradient in a discrete image

In an orthogonal coordinate system (Oxy) where (Ox) is the horizontal axis and (Oy) the vertical axis, the image ***gradient*** (or rather the ***luminance f***) at any point or pixel coordinates (x, y) is denoted by:

$$Grad\, f = \nabla f = \begin{pmatrix} \frac{\partial f}{\partial x} \\ \frac{\partial f}{\partial y} \end{pmatrix} \quad (1)$$

The ***module of the gradient*** quantifies the importance of the contour highlighted, that is to say the magnitude of the jump intensity observed in the image:

$$\|\nabla f\| = \sqrt{(\frac{\partial f}{\partial x}^2) + (\frac{\partial f}{\partial y}^2)} \quad (2)$$

The ***direction $\alpha_o$ of the gradient*** determines the present edge in the image. Indeed, the gradient direction is orthogonal to that of the outline:

$$\alpha_o = \arctan(\frac{\partial f/\partial y}{\partial f/\partial x}) \quad (3)$$

The principle of ***edge detection by the use of the gradient*** is to calculate,in the first time, the gradient of the image in two orthogonal directions, then the gradient module. The next step is to make a selection of the most marked contours, that is to say the points of stronger contrast with adequate thresholding (see below).

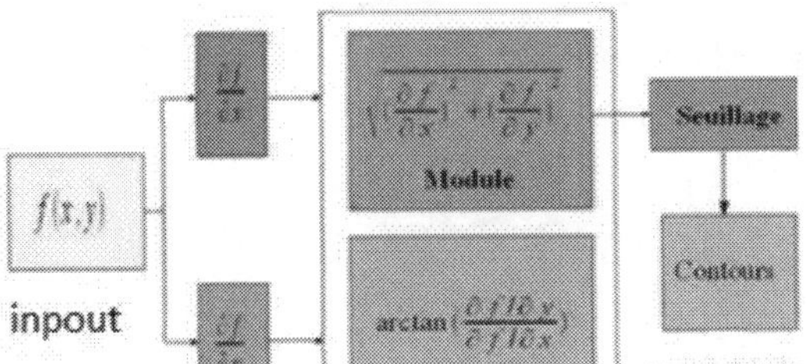


**Figure 2.** principle of edge detection.

### 3.2 Mask with two elements

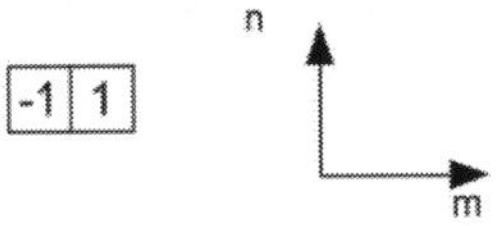


The correlation of the mask with a luminance image [5], f (i, j) can be written:

$$\sum_{m=0}^{1}\sum_{n=0} h(m,n)\, f(m+i, n+j)$$
$$= h(0,0\, f(i,j) + h(1,0)\, f(i+1,j)$$
$$= f(i+1,j) - f(i,j) \quad (4)$$

### 3.3 Mask with three elements

The gradient calculation is achieved by means of two masks, the first performing a horizontal gradient, the second vertical gradient. The second mask is derived from the first by a rotation of $\frac{\pi}{2}$.

$$[\text{-}1\ 0\ 1]\ \text{and}[-1\ 0\ 1]^T$$

Compared to previous, [6], [7] this mask has the advantage of producing two effects. To calculate the gradient in one direction, and performs a smoothing in the orthogonal direction. This smoothing makes it a little less sensitive to noise than the previous mask. The origin of this mask is always the center pixel. The value of the constant C can take 1 or 2 to increase the effect of smoothing.

The output pixel obtained after filtering is:

$$\sum_{m=-1}^{1}\sum_{n=-1} h(m,n)\, f(m+i, n+j)$$
$$= \frac{1}{c+2}\{h(-1,1)\, f(i-1,j+1) + h(0,1)\, f(i,j+1)\}$$
$$+ \frac{1}{c+2}\{h(1,1)\, f(i+1,j+1) + h(-1,-1)\, f(i-1,j-1)\}$$
$$+ \frac{1}{c+2}\{h(-1,0)\, f(i-1,j) + h(-1,1)\, f(i-1,j+1)\}$$

$$\sum_{m=-1}^{1}\sum_{n=-1} h(m,n)\, f(m+i, n+j)$$
$$= \frac{1}{c+2}\{\, f(i-1,j+1) + c\, f(i,j+1)\, f(i+1,j+1)\}$$
$$- \frac{1}{c+2}\{\, f(i-1,j-1) + c\, f(i-1,j)\, f(i-1,j+1)\} \quad (5)$$

The equation reveals the double action with a horizontal middle ages of three pixels on the lines above and below the central pixel and calculate the vertical gradient between the two lines.

For the detection of vertical edges, other mask is used. The output pixel obtained after filtering can be put in the form:

$$\sum_{m=-1}^{1}\sum_{n=-1} h(m,n)\, f(m+i, n+j) = \frac{1}{c+2}\{f(i+1,j+1) - f(i-1,j+1)\} + \frac{1}{c+2}\{c\, f(i+1,j) - c\, f(i-1,j)\} + \frac{1}{c+2}\{f(i+1,j-1) - f(i-1,j-1)\} \quad (6)$$

DOI: 10.1051/matecconf/20167503003

This equation shows the dual action: calculate on three lines the horizontal gradient then applied a vertical smoothing.

Note:

The main interest of these masks is the ease of implementation and the speed of processing. Their disadvantage is their sensitivity to noise. However the detected edges are often quite large. According to the filters of three elements are the most used in industrial applications requiring real-time constraints.

## 4 Requirements image processing material

Treatment of live image requires high computing power. For example, the standard image.jpg with a dimension of 640*480 approximately 0.25 mega pixels per image, with a calculated power of 100 frames per second for a 25Mhz processor. The size of the image can be greater [8,9], the amount of processing required per pixel depends on the image processing algorithm.

The edge detection algorithm by the concentration gradient is a combination of four steps, the horizontal gradient, the vertical gradient, their modules and finally a suitable thresholding.

Depending on the size of the masks, the processing requirements for the first two convolution steps may to be change. Using the assumption of a mask size of 2*1 or 3*1 for the partial derivative, both steps require more operations per pixel. This implementation requires less complexity than most typical 2-D convolution process because of the symmetry characteristics and separable Gaussian masks. The steps of the calculation module and thresholding requires fewer operations per pixel.

Such as high resolution images become more frequent, the processing requirements will increase. The high resolution images of the standards generally have ten times more pixels per image. The computational load is approximately ten times higher. They require more DSP or a single, very expensive high-end DSP. In this scenario, FPGAs offer a real-time alternative platform image processing. FPGA effectively supports high levels of parallel processing data flow structures (Figure 4), which are important for the effective implementation of image processing algorithms.

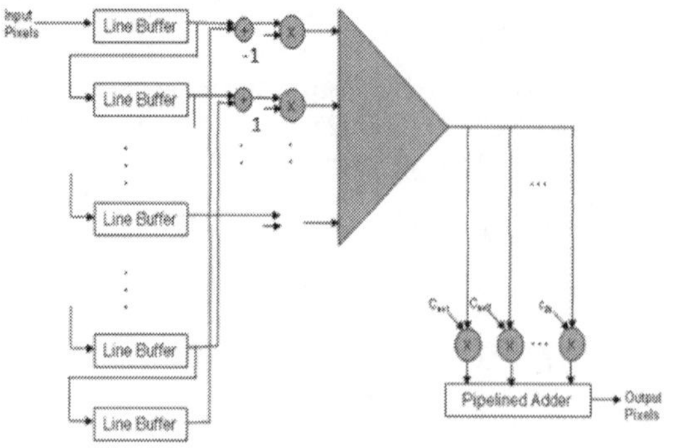


**Figure 3.** processing of data streams in parallel.

## 5 Hardware implementation on FPGA of the Xilinx family

For the implementation of our algorithm, we used the platform of Xilinx nexys3 Spartan-6 FPGA and VGA monitor to display results.

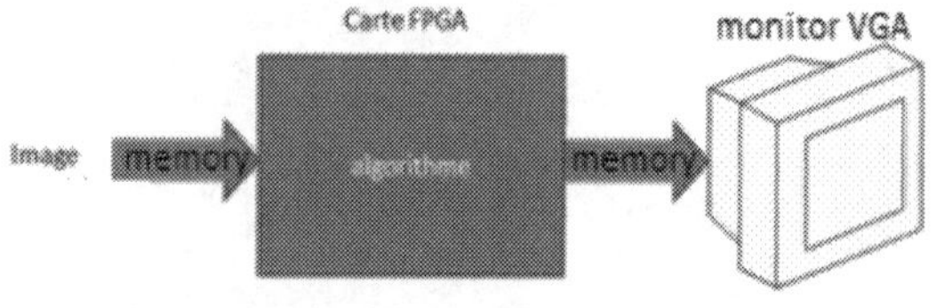


The algorithm was developed on Xilinx ISE interface and all the blocks are programmed in VHDL with the following block diagram:

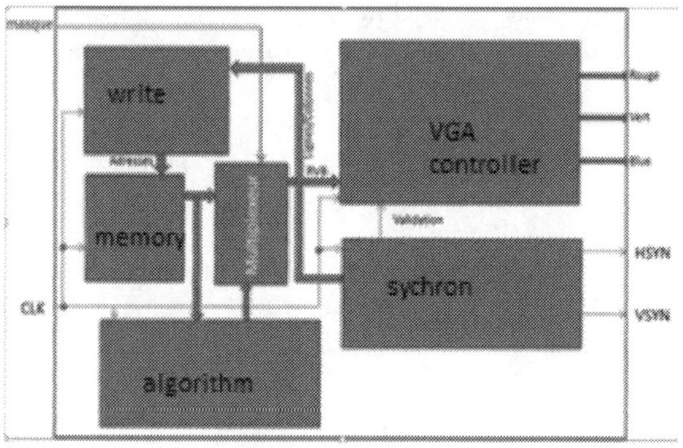


**Figure 4.** block diagram of programs.

Our project consists of several blocks, which are other than VHDL programs:

### 5.1 Block program "Memory"

The ***"Memory"*** block that exists in the party IP (intellectual property) to the Xilinix software is configured according to the size of the image that will store, each memory cell contains 8 bits (8 bits is the value which is coded each RGB pixel).

### 5.2 Block program "Memory Playback"

Since the image is not necessarily the same size as the screen, the program is necessary for the correct positioning of the image on the screen.

### 5.3 Block program "VGA Screen"

The program's role to define. The VGA monitor is controlled by five 10-bit coded signals: red (3 bits), green (3 bits), blue (2 bits), horizontal sync and vertical sync 1 bit 1 bit. The three color signals, collectively known as the RGB signal are used to control the color of a pixel at a location on the screen. In order to produce other colors, each color analog signal is to be supplied with a voltage between 0.7 and 1.0 volts for varying the intensities of the colors.

### 5.4 Block program "Synchronization"

The program lets you synchronize the scanning of pixels on the screen of a horizontally and vertically. Signals are used to control the synchronization of the scanning speed. The horizontal synchronizing signal determines the time to scan one line, while the vertical synchronizing signal determines the time to scan the entire screen. By manipulating these signals, the images are formed on the monitor screen.

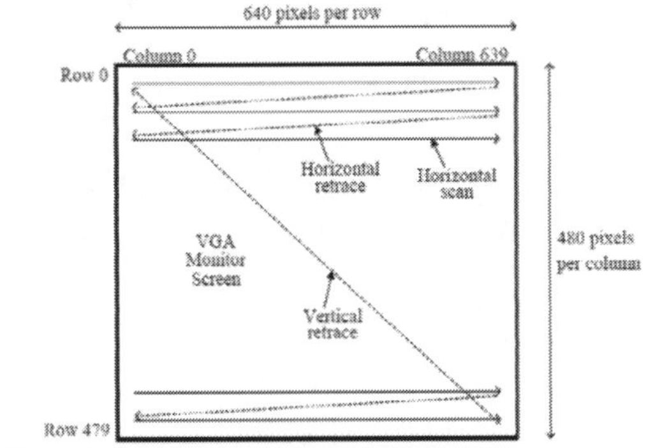


**Figure 5.** Operating principle of VGA controller.

### 5.5 Block program "algorithm"

It's the same principle previously defined in Part 3 for edge detection. The pixels after the conversion are always less than the total number of pixels. The upper limit guarantees that the operations depending on the FIFO list can be completed during the real-time processing period of a frame. FIFOs are implemented using multiple integrated M4K memory blocks, which are a size of 4 kbps each. The image of the edge detection result is stored in the blocks of RAM built in the Spartan-6 device. These large blocks of integrated RAM can be used to store temporarily the edge image when 2 bits are allocated to each pixel location.

Following diagram shows the different blocks implemented on the Xilinx ISE software:

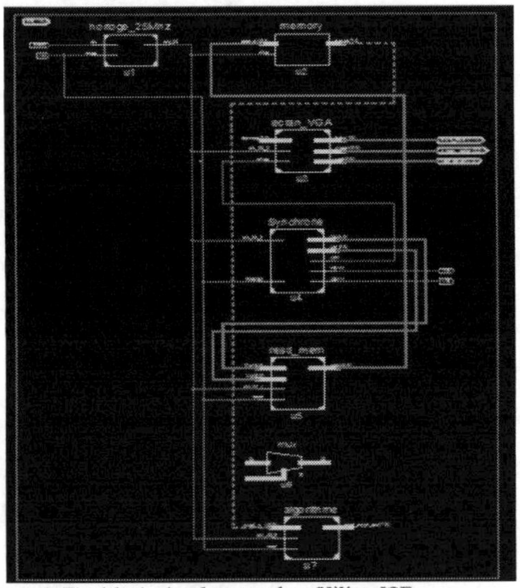

**Figure 6.** The blocks implemented on Xilinx ISE.

Preliminary results show the FPGA design clocked at 100MHz Spartan-6 targeting (Figure 7).

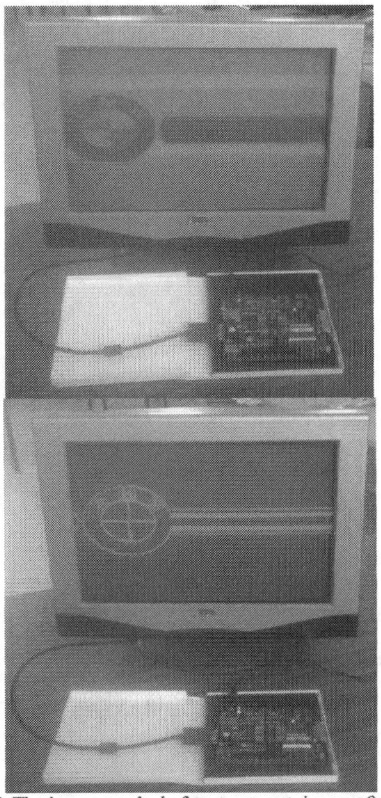

**Figure 7.** The image results before treatment, image after treatment.

We used a test image with a resolution of 200x200. The module is fully pipelined where a resulting pixel is calculated at each clock cycle [10]. With this rate and the clock frequency, it is possible to treat more than 400 image frames at 200x200 resolutions.

The design is scalable to handle high-resolution images while maintaining the clock frequency used to treat standard resolution video at 30 frames per second, as well as computer vision applications at high speeds, which require more 80 fps.

It is possible to increase the frequency above 100 MHz by introducing more pipeline stages at the expense of increased resource utilization [11]. But the uses of high resolution images require a more powerful processor that exceeds the range Spartan, in talking about the Virtex range, high performance at frequencies exceeding 500MHz.

## 6 Conclusion

In this paper, we propose implementation of edge detection algorithms on a FPGA. The presented method is based on detection of the maximum gradient of the horizontal and vertical image, on the module of them and on the thresholding to select the strongest contours.

The results obtained show is the clear presence of all the contours that exist in the images (horizontal and vertical contours, circular forms and writings), the use of an FPGA-circuit spartan6 show its ability to manage images high resolution VGA output, while maintaining the clock frequency used (100Mhz) for processing image sequences with high resolution at 30 frames per second, as well as computer vision applications at high speeds, which require more than 80 fps.

## References

1. Marimont, D., Rubner Y., "Un probabiliste Frameword pour Edge Détection et Échelle de sélection", Actes de la Conférence internationale de l'IEEE sur la vision par ordinateur, 207-214, Janvier (1998)
2. T. Lindeberg, "EdgeDetection et détection Ridge avec la sélection Echelle automatique", International Journal of Computer Vision, vol. **30**, numéro 2, 117-154, (1998)
3. Maar, D., E. Hildreth, "Théorie de la détection de bord", Actes royale Soc. Londres, vol. **207**, 187-217, (1980)
4. Technique Advanced Edge Detection: (Techniques de Vision informatique)
5. J. Malik, S. Belongie, T. Leung and J. Shi, "Contour and texture analysis for image segmentation", IJCV, vol. **43**, no. 1, (2001)
6. Duncan D-Y Po et Minh N. Do. - "Directionnel multiscale modeling of images using the contourlet" - Coordinated Science Lab and Beckman Institute, University of Illinois at Urbana-Champaign - Urbana IL 61801, (2003)
7. M. N. Do et M. Vetterli, "The contourlet transform: an efficient directional multiresolution image representation". IEEE Transactions Image Processing, (2003)
8. http://www.xilinx.com/products/design-tools/ise-design-suite.html
9. http://store.digilentinc.com/fpga-programmable-log
10. Wang, R., Cohen, G., Stiefel, K. M., Hamilton, T. J., Tapson, J., and van Schaik, A. An FPGA Implementation of a Polychronous Spiking Neural Network with Delay Adaptation. frontiers in Neuroscience, (2013)
11. Nazari, S., Amiri, M., Faez, K., and Amiri, M. Multiplier-less digital implementation of neuron–astrocyte signalling on FPGA. Neurocomputing, (2015)

MATEC Web of Conferences **75**, 03004 (2016) DOI: 10.1051/matecconf/20167503004
*ICMIE 2016*

# Image Processing Methods Usable for Object Detection on the Chessboard

Ladislav Beran[a], Pavel Chmelar and Lubos Rejfek

*Department of Electrical Engineering, Faculty of Electrical Engineering and Informatics, University of Pardubice,Pardubice, Czech Republic*

**Abstract.** Image segmentation and object detection is challenging problem in many research. Although many algorithms for image segmentation have been invented, there is no simple algorithm for image segmentation and object detection. Our research is based on combination of several methods for object detection. The first method suitable for image segmentation and object detection is colour detection. This method is very simply, but there is problem with different colours. For this method it is necessary to have precisely determined colour of segmented object before all calculations. In many cases it is necessary to determine this colour manually. Alternative simply method is method based on background removal. This method is based on difference between reference image and detected image. In this paper several methods suitable for object detection are described. Thisresearch is focused on coloured object detection on chessboard. The results from this research with fusion of neural networks for user-computer game checkers will be applied.

## 1 Introduction

Image segmentation is the process of dividing pixels in an image into categories based on shared visual characteristics like a shared edge, colour, texture etc. The image segmentation is an important first step in object detection. Although many algorithms for image segmentation have been invented, there is no simple algorithm usable for all cases of image segmentation and object detection. For this detection, it is necessary to use combination of several methods for example colour segmentation and background removal. In case of colour segmentation, it is necessary to have precisely determined colour of searched object. In many cases, it is necessary to determine colour manually. The next problem in image segmentation and object detection is an added noise to signal by sensors. This noise, in case of camera withpoor lighting of the scene, can be added. In this paper several method suitable for object detection on chess board are described.

## 2 Usable methods for detection

### 2.1 Background removal

The background removal method is very simply method for object detection. This method is based on deduction of frame without any object from frame with object. The result is areas with segmented object. For using of this method, it is necessary to use an object with different color than background. This method is mathematically very simply and algorithm is described using equation (1), (2) and (3). The equation (1) describes a case when object is fitted on black squares (brighter object than background) and equation (2) describes case, when object is fitted on white squares (darker object than background).

$$I[F(n)] = I[I(n)] - I[BG] \quad (1)$$

$$I[F(n)] = I[BG] - I[I(n)] \quad (2)$$

where: *I[F(n)]* is an image without background, *I[I(n)]* is an captured image by camera, *I[BG]*is first captured image – background.

An alternative of this method is possible to use an image difference. For this case alternative equation is used (3)

$$|I[F(n)] - I[F(n-1)]| > Tr \quad (3)$$

where: *I[F(n-1)]* is a previous captured image, *Tr* is user selected threshold.

The main problem in case of background removal applied to chessboard is necessary to have pair of images. The first image is for case, when the object is fitted on black squares, the second for case, when object is fitted on white squares. It is not possible use universal algorithm, because when object is fitted on white squares, the maximal value will be less than 255 in case of 8 bit colour channels. If first algorithm is used, the object will be lost in saturation. For this case, it is necessary to use a subtraction captured image from template. The

[a] Corresponding author: ladislav.beran@student.upce.cz

background removal method and its implementation are described in detail in 1.

### 2.2 Colour segmentation

The colour segmentation is based on extraction similar shared visual characteristic – colour. The several types of colour spaces exist in the world. The main colour space is called RGB space. This space is based on red, green and blue component. Every from these component partially carry an information about brightness. Also, RGB is suitable for colour LCD display, but very bad for colour segmentation and analysis because of the high correlation among the R, G, and B components. From this case, it is necessary to use different colour space. In our case, we are using a HSV colour space. The conversion from RGB to HSV model is provided using equations (4) to (10).

$$R_n = \frac{R}{2^b}; G_n = \frac{G}{2^b}; B_n = \frac{B}{2^b} \tag{4}$$

where $R_n$, $G_n$ and $B_n$ are normalized RGB components (Result is in range from 0 to 1), the b is a bit length of each colour component.

$$C_{max} \quad = \max(R_n, G_n, B_n) \tag{5}$$

$$C_{min} \quad = \min(R_n, G_n, B_n) \tag{6}$$

$$\Delta = C_{max} - C_{min} \tag{7}$$

where $C_{max}$ is maximal value from normalised RGB components, $C_{min}$ is minimal value from RGB normalises components

$$\text{Hue} = \begin{Bmatrix} 0^\circ & \text{if}\,\Delta = 0 \\ 60^\circ \cdot \left(\frac{G_n - B_n}{\Delta} \bmod 6\right) & \text{if} C_{max} = R_n \\ 60^\circ \cdot \left(\frac{B_n - R_n}{\Delta} + 2\right) & \text{if} C_{max} = G_n \\ 60^\circ \cdot \left(\frac{R_n - G_n}{\Delta} + 4\right) & \text{if} C_{max} = B_n \end{Bmatrix} \tag{8}$$

$$\text{Saturation} = \begin{Bmatrix} 0 & \text{if} C_{max} = 0 \\ \frac{\Delta}{C_{max}} & \text{if} C_{max} \neq 0 \end{Bmatrix} \tag{9}$$

$$\text{Value} = C_{max} \tag{10}$$

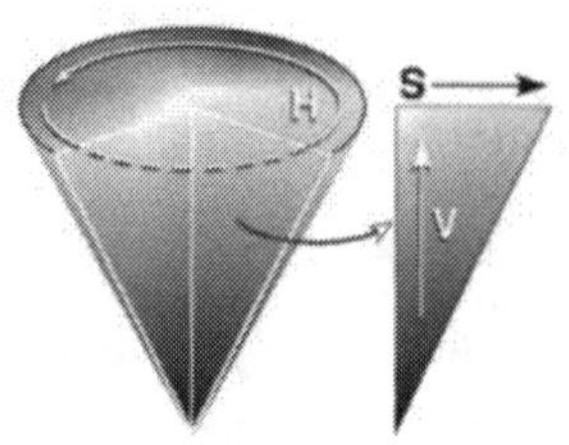


**Figure 1.** HSV color space.

The main advantage of HSV over the RGB is separated colours information. This information is called Hue, Saturation and Value 2, 3.

## 3 Proposed approach

### 3.1 Background removal

In our research, background removal algorithm was tested first. The background removal flowchart is shown at Figure 2.

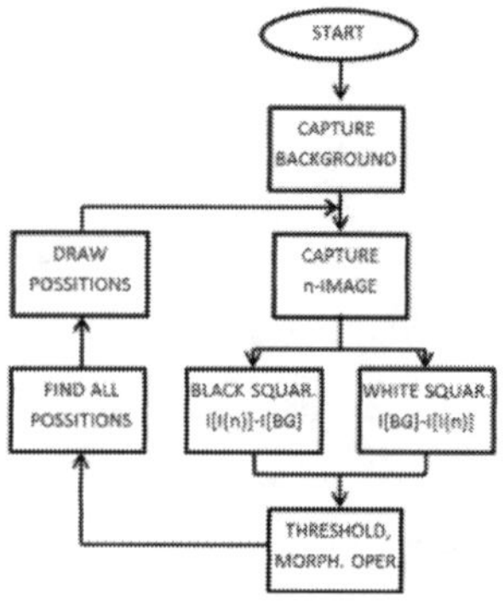


**Figure 2.** The background removal flowchart.

At first, algorithm capture template which is used as reference scene. After this process, algorithm starts periodically capturing the scene. The segmented object using the equations (1) and (2) is calculated and results are shown at Figure 3. At Figure 3 left side, segmented black squares is shown, at right side, segmented white squares is shown.

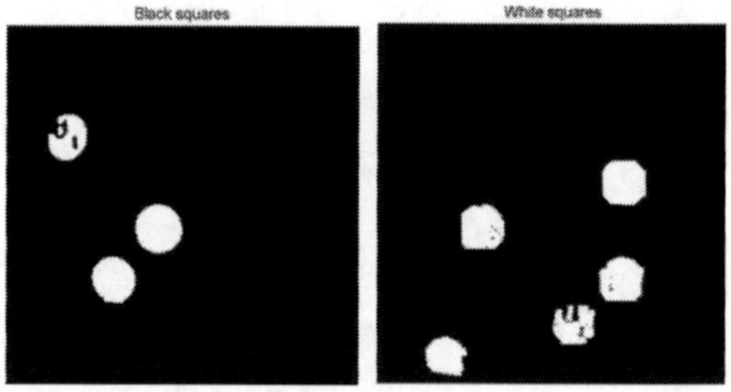


**Figure 3.** Segmented objects: black squares (left), white squares (right).

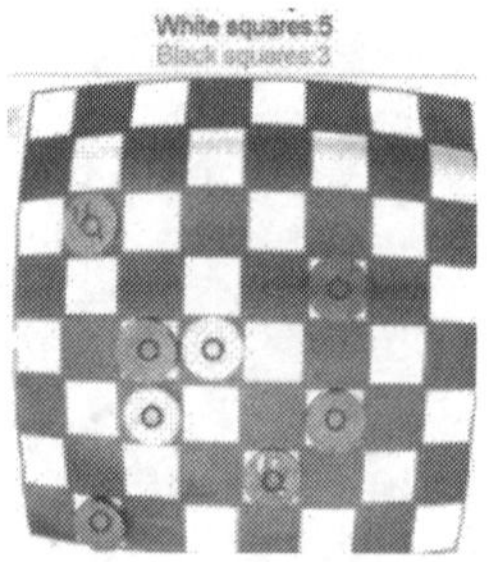


**Figure 4.** Final view of segmented and detected objects.

In this method, it is the main problem in case of wrongly selected threshold. Also, threshold can be selected correctly, but if lighting scenes can have small changes, which can cause totally wrong results. The next problem in background removal method is problem with similar colors – object and square. In this case, the object

DOI: 10.1051/matecconf/20167503004

is lost in the background and detection is not possible. The example of this problem is shown at Figure 5 and Figure 6.

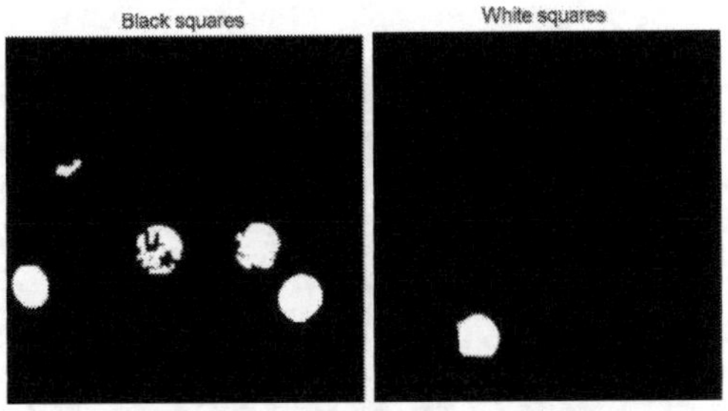


**Figure 5.** Wrongly segmented objects.

**Figure 6.** Wrongly segmented objects.

At Figure 6 the wrongly segmented object is shown. In case of grey object (undetected object in the lower part of image), the colour is mostly identic and algorithms are not able to detect this object. In case of blue object (undetected object on the top of image), is also problem with colour and hue but for this case, it is necessary to use extended algorithm for object detection like as circular object detection. Possible algorithms are described in the next chapters and in papers 4, 5, 6.

### 3.2 Colour segmentation and object detection

The colour detection is based on detection an object with same/similar colour in the image. The models used during all word are RGB and HSV colour model. The RGB model is mainly used for displays. This colour model is based on Red, Green and Blue component. Every from component partially carry an information about brightness. Also, RGB is suitable for colour display, but very bad for colour scene segmentation and analysis because of the high correlation among the R, G, and B components. The result of colour segmentation is shown at Figure 7.

Using the results shown at Figure 7, the disadvantage of RGB colour model is possible to read. For case of red colour is possible to found maximal values (on red channel) which is shown in this cannel as a white colour, but additional information about colour is also written in another component. This colour model can be used only for strictly red, green and blue colours. For this case it is necessary to use different colour model called HSV – Hue, Saturation and Value. This colour model has very big advantage comparison with RGB model – the colour component is not correlated. The RGB colour model for object detection with fusion of depth map in 7 was used.

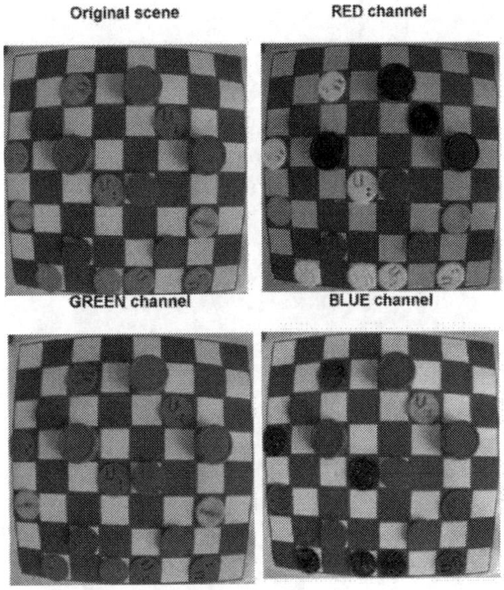


**Figure 7.** Example of RGB color channels.

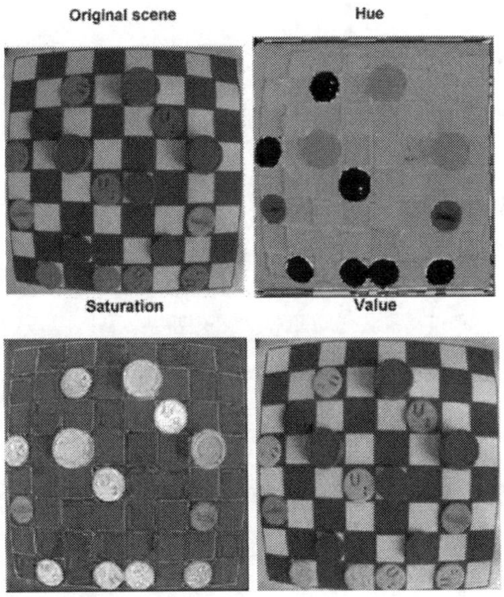


**Figure 8.** HSV color model.

At Figure 8 main difference with comparison RGB color model shown at Figure 7 is possible to shown. In case of HSV, the all information about colour in hue and saturation channel is written. The information from these channels are shown at Figure 9. This colour model is suitable for colour detection of objects. Also, from Figure 7 and Figure 8 it is possible to recognize problem with grey objects on a chessboard. These object is not possible to found using every known image processing methods, for case, when these object are fitted on the black squares. For our purpose it is neccary to use colours diiferent from black or white colours like a yellow, red, green, etc.

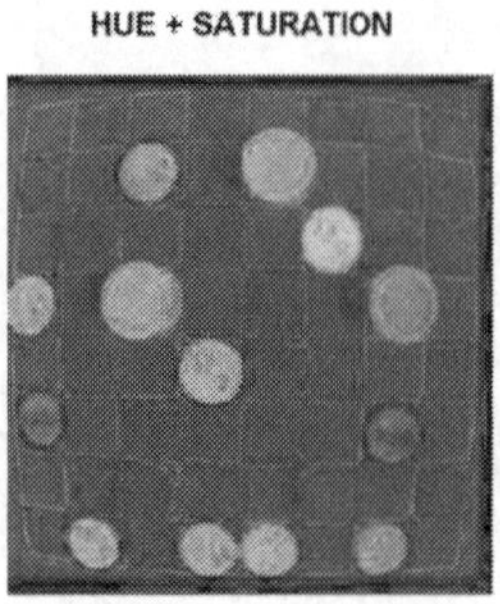


**Figure 9.** Hue and Saturation channel.

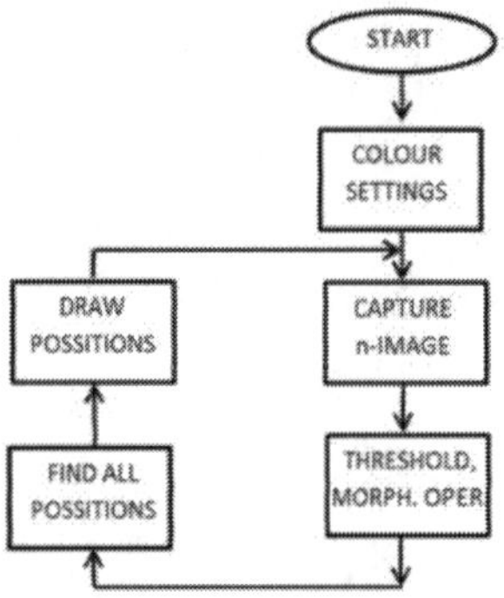


**Figure 10.** The color detection flowchart.

At Figure 10, the colour detection flowchart is shown. The algorithm after start capture scene which contain colour of the object which will be searched using algorithm. This sample is located in defined position. From this defined areas are calculated statistical values like average and standard deviation for all colour channels. These statistical values are calculated using equations (11) and (12). After this process, algorithm periodically captures a scene, calculate a threshold using statistical values, find all position of object and draw these positions to the original image. The colour segmentation algorithm is described using equation (13).

$$\overline{x} = \frac{1}{N}\sum_{i=1}^{N} x_i \quad (11)$$

$$\sigma = \sqrt{\frac{1}{N}\sum_{i=1}^{N}(x_i - \overline{x})^2} \quad (12)$$

$$\begin{Bmatrix} x = 1 \text{ when:} & x_{HUE} \in \overline{x}_{HUE} \pm \sigma_{HUE} \\ \cap & x_{SAT} \in \overline{x}_{SAT} \pm \sigma_{SAT} \\ x = 0 & \text{Otherwise} \end{Bmatrix} \quad (13)$$

where $x$ is a pixel value in scene on defined position, $x_{HUE}$ and $x_{SAT}$ are a Hue/Saturation values of a pixel x, $\bar{x}_{HUE}$ and $\bar{x}_{SAT}$ are an average Hue/Saturation values calculated at start of the algorithm from the template, $\sigma_{HUE}$ and $\sigma_{SAT}$ are a standard deviations calculated from the template.

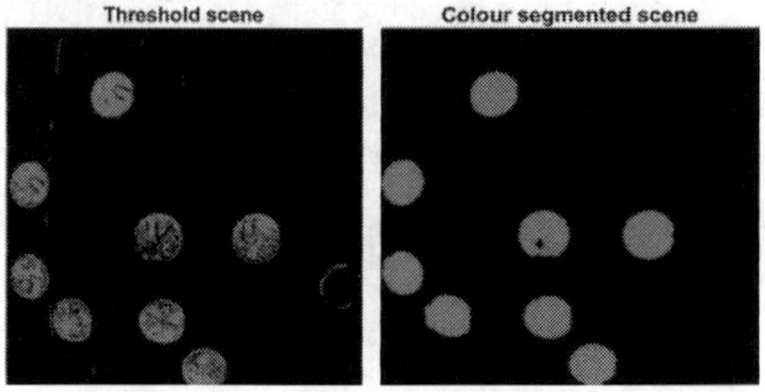


**Figure 11.** Threshold (left) and segmented (right) scene - Red objects.

At the Figure 11, threshold and segmented scene is shown.

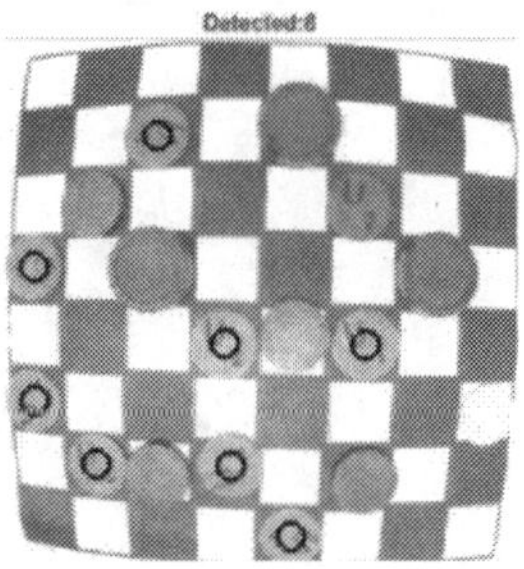


**Figure 12.** Detected red objects in the scene.

## 4 Results and discussion

At the Figure 13 our testing platform is shown. This platform is based on Basler acA-2500-14uc USB3 camera.

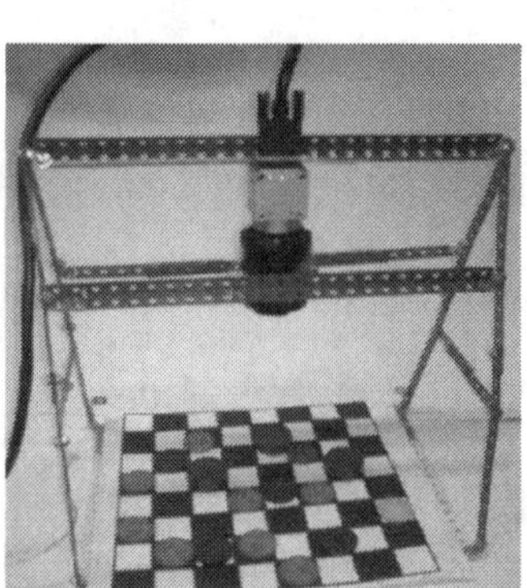

**Figure 13.** Testing platform.

In our research a several settings was tested for each method. During our testing, several advantages and disadvantages of suitable methods were found. These parameters of suitable methods are written in the Table 1 and Table 2.

**Table 1.** Background removal parameters.

| **Background removal** | |
|---|---|
| **Advantages** | Easy implementation |
| | Fast calculation |
| | Colour independence |
| **Disadvantages** | Need template before calculations |
| | Totally out when template is moved |

**Table 2.** Color segmentation parameters.

| **Colour segmentation** | |
|---|---|
| **Advantages** | Easy implementation |
| | Colour selection |
| **Disadvantages** | Problems with lighting of scene |
| | Need colour template before calculations |
| | Convert RGB -> HSV colour model |
| | Higher time consumption |

**Table 3.** Time consumption of both methods.

| **Method** | **Average Time consumption (s)** |
|---|---|
| **Background removal** | 0.14 |
| **Colour segmentation** | 0.45 |

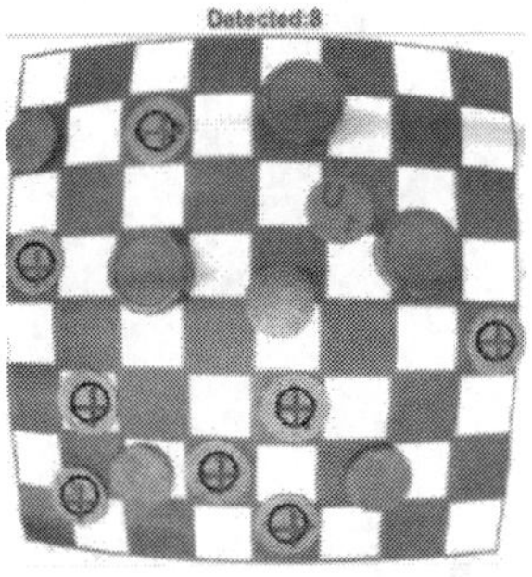


**Figure 14.** Detected red objects using background removal and color detection.

At Figure 14 the detected red objects using colour detection and background removal are shown. The detected centres using background removal using a red cross and black circles marks centres detected using colour segmentation are shown. The results are synthesis of detected centre using both methods. When calculated centres are located in defined interval between each detected centre, this point is shown in the result scene.

## 5 Conclusion

In this paper, methods for detection object placed on the chessboard are described. In this paper method of background removal was described. This method has satisfactory results and this method will be applied to the final system. The method of background removal is very sensitive for correctly selected threshold. The alone method detects all types of objection shapes and find a centroid of these object. For our purpose it is necessary to implement method of detection circular objects. Also, in this paper method based on color segmentation was described. In this method is necessary to know all parameters of colour – hue, value and saturation. These parameters from template are calculated. This method is based on detection specific colour and without change this method is not possible to detect another colour using one setting. At present, the colour of the object is selected manually. In this project will be defined two colours for players –for example green and red. In future, algorithm after start will be calibrated using special template with fitted game coins. In the future, algorithm from defined places records colour of game coins and if the colours will be different, game starts. For trouble-free game playing, it will be necessary to implement fusion of described methods – background removal and colour segmentation. The final research will be constructed as fusion of background removal, colour segmentation and circuit detection. These methods are implemented yet, but it is necessary to implement these methods with circular detection. This method is based on circular Hough transformation. At present, the both methods are implemented as complementary system, when if both detected centres are similar (in defined range), the object is marked as detected. In the future this rule will be modified. The object will be marked in case, if minimal two from three methods correctly detect centre of object. The results from this research will be applied to the user-computer game called checkers.

## Acknowledgment

The research was supported by the Internal Grant Agency of University of Pardubice, the project No. SGSFEI_2016_022.

## References

1. P. Chmelar; A. Benkrid, "Efficiency of HSV over RGB Gaussian Mixture Model for fire detection," in Radioelektronika 2014 24th International Conference, **24**, 1-4 (2014)
2. S. Sural; Qian Gang; S. Pramanik., "Segmentation and histogram generation using the HSV color space for image retrieval," in Image Processing-. Proceedings. 2002 International Conference on, **2**, II-589-II-592 (2002)
3. P. Chmelar, L. Beran, N. Kudriavtseva , The Laser Color Detection for 3D Range Scanning Using Gaussian Mixture Model, 25th International Conference Radioelektronika 2015, **25**, 248-253 (2015)
4. W.E.L Grimson and D.P. Huttenlocher, On the sensitivity of the Hough transform for object recognition. IEEE Transactions on Pattern Analysis and Machine Intelligence, **12**, 3. 255-274 (1990)
5. S. Umer; B.C. Dhara; B. Chanda, "A Fast and Robust Method for Iris Localization," in Emerging Applications of Information Technology (EAIT),

DOI: 10.1051/matecconf/20167503004

2014 Fourth International Conference of, **4**, 262-267 (2014)
6. M. Smereka; I. Duleba, Circular object detection using a modified hough transform, Int. J. Appl. Math. Comput. Sci. **18**, 1 85–91 (2008)
7. P.Jetensky, S. Karamazov - Accessible Chessboard For Blind That Is Mouse And Keyboard Free. In International Conference on Applied Electrical Engineering and Informatics (AEI 2012), 72-76 (2012)

# The Depth Map Construction from a 3D Point Cloud

Pavel Chmelar[a], Ladislav Beran and Lubos Rejfek

*Department of Electrical Engineering, Faculty of Electrical Engineering and Informatics, University of Pardubice, Czech Republic*

**Abstract.** A depth map transforms 3D points into a 2D image and gives a different view of an observed scene. This paper deals with a depth map construction. It describes the whole process, how to transform any 3D point cloud into a 2D depth map. The described method uses 3D rotation matrixes and the line equation. This process allows to create the desired view from arbitrary point and rotation in an exploration space. Using of a depth map allows to apply image processing methods on depth data to get additional information about an ambient space.

## 1 Introduction

Depth maps (RGB-D) in mapping systems and robot navigation [1]-[4] are nowadays often used, but also in medical area for the respiratory motion detection [5], and in other areas for the motion and object detection.

For the 3D range scanning we are using a mobile platform with a laser and a camera, described in [6]. A laser spot is dispersed to a vertical line which allows measurement in the height range. In our previous work we introduced the new modelling of a colored laser line by using the laser Gaussian Mixture Model (GMM) [7]. The last paper describes an elimination of the barrel distortion correction [8]. These works contribute to the increase of the measurement points amount and measurement precision. The described 3D range scanning system is a part of the bigger project called ARES (Autonomous Research Exploration System) [9-11]. Our system is able to create a 3D point cloud, but up to now was not possible to create the depth map, which is useful for mapping systems mentioned above.

This paper describes the new algorithm for the depth map construction form a point cloud. This will give more information about depth data and it will allow to apply image processing methods 2D depth data.

## 2 Optical range finder

Our 3D range scanning system for the autonomous mapping, see Figure 1, is based on the triangulation method. The principle of the distance determination and the 3D point cloud constructions is introduced in [6].

The whole measuring device is pivotally carried by tripod. For the image capturing a high-quality Basler color camera with a resolution of 2590x1942 pixels is used. The used laser diode has an output power 200 mW, especially for good recognition after vertical swapping. The laser spot is dispersed by an optical filter with an angle of 90 degrees. The 360 degrees rotation of the measuring head provides a powerful and fast stepping motor. The maximal measured distance is 15 m.

Figure 1. 3D range scanning system.

### 2.1 3D point cloud

The Figure 2 gives measurement results of one frame, which includes 80 measurement points.

**Figure 2.** One measurement frame example.

The green laser line is segmented by the GMM, described in our previous work [7].

After one 360 degrees scan a 3D point cloud is created. An example of the 3D point cloud is in Figure 3.

[a] Corresponding author: pavel.chmelar@student.upce.cz

DOI: 10.1051/matecconf/20167503005

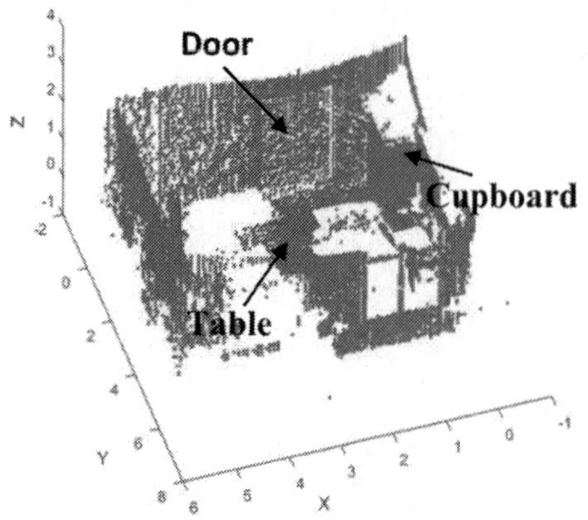


**Figure 3.** 3D point cloud.

This measured point cloud is used for the description of the depth map construction process in the following Section 3.

## 3 Depth map construction

For a depth map construction there are several initial parameters. The first is the camera position

$$P_C = (P_X, P_Y, P_Z)^T \quad (1)$$

The vector $\mathbf{P_C}$ (1) is a position from the depth map will be created. The second parameter is the camera field of view: horizontal $H_{FOV}$; vertical $V_{FOV}$,

$$C_{FOV} = \left( \frac{H_{FOV}}{2}, \frac{H_{FOV}}{2}, \frac{V_{FOV}}{2}, \frac{V_{FOV}}{2} \right)^T \quad (2)$$

The expression (2) is the field of view in the top, bottom, left and right area. It gives the used camera view range, which is able to cover. The last important parameter is the camera orientation defined by the following vector

$$C_O = (C_\alpha, C_\beta, C_\gamma)[°]^T \quad (3)$$

This allows "to look" in the space in any desired orientation. The main advantage of the described algorithm is the possibility to set these parameters according our requirements, to create the desired resultant depth map.

For next described steps the two mathematical tolls are needed. For points rotations and transformations, the rotation matrixes in the 3D space (4), (5) and (6) are needed, described in paper [9]. To express the view range, the line equation (7) is used.

$$R_x(\alpha) = \begin{bmatrix} 1 & 0 & 0 \\ 0 & \cos(\alpha) & \sin(\alpha) \\ 0 & -\sin(\alpha) & \cos(\alpha) \end{bmatrix} \quad (4)$$

$$R_y(\beta) = \begin{bmatrix} \cos(\beta) & 0 & -\sin(\beta) \\ 0 & 1 & 0 \\ \sin(\beta) & 0 & \cos(\beta) \end{bmatrix} \quad (5)$$

$$R_z(\gamma) = \begin{bmatrix} \cos(\gamma) & \sin(\gamma) & 0 \\ -\sin(\gamma) & \cos(\gamma) & 0 \\ 0 & 0 & 1 \end{bmatrix} \quad (6)$$

Symbols $\alpha$, $\beta$ and $\gamma$ are rotation angles in corresponding axes *X*, *Y* and *Z*.

$$y = \tan(\alpha) \cdot x + b \quad (7)$$

The line parameter offset *b* is zero, *tan(α)* is the line angle and *x* express the point where *y* is calculated.

### 3.1 3D point cloud initial preparations

In the initial stage the 3D point cloud is rotated according the camera orientation $\mathbf{C_O}$ (3). Although the input 3D point cloud is rotated, this rotation simplifies data selection process which describes the next Section 3.3. For the algorithm illustration $\mathbf{C_O}$ = (0, 0, 90) is used. In Figure 4. are points rotated in $\mathbf{C_O}$ angles.

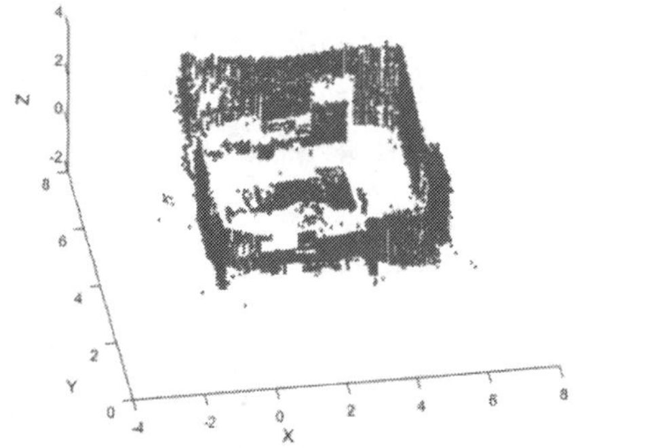


**Figure 4.** Rotated point cloud in angle of 90 degrees (Z axis)

### 3.2 Range view construction

The first algorithm step is the view range construction by using the line equation (7) and rotation matrixes in Y (5) and Z (6) axis. The line point step $q_L$ is 1 cm. Firstly, the "V" (consist from two lines) with the $V_{FOV}$ angle is created at the level Z = 0, subsequently the "V" is rotated in the angle of 90 degrees around Y axis (5). This rotated "V" is further rotated in angle of $H_{FOV}/2$ around the Z axis (6) in both sides (left and right). The whole range view is shifted to the camera position $\mathbf{P_C}$ = (1.79, 3.3, 1.2). The result is shown in Figure 5. There are also orientation angles $\mathbf{C_O}$ (3) marked.

By using of the constructed view range, see Figure 6, the points for the depth map creation can be selected.

### 3.3 Range view data segmentation

To get the point of data selection describes Figure 7. There are marked four grey rectangles, which shows stages of the segmentation process.

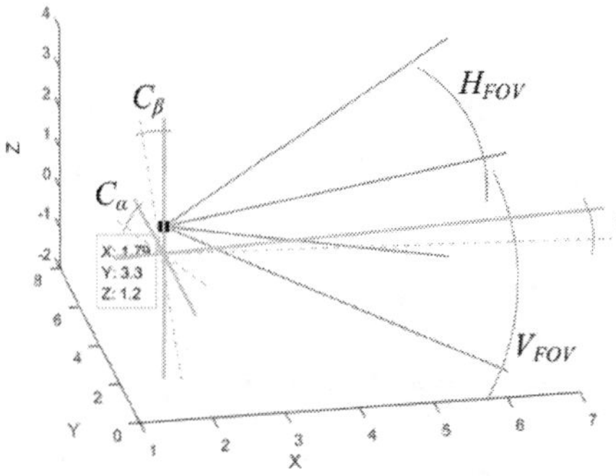


**Figure 5.** Constructed range view with rotation angles.

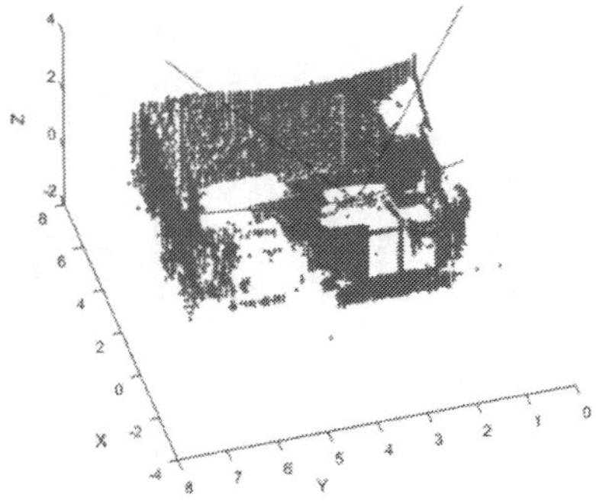

**Figure 6.** Constructed range view.

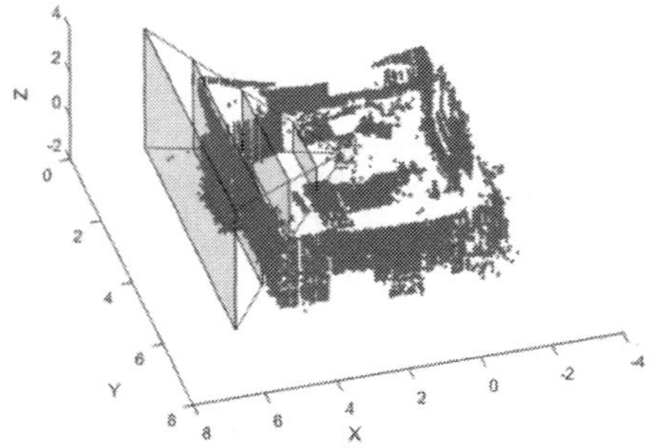

**Figure 7.** Points segmentation illustration.

The equation (8) describes the points which lies in grey rectangle $R_{SYZ}$ in Y and Z axis

$$R_{SYZ}(n) = (PC(y) > R_S(n, y_{\min}))\&(PC(y) < R_S(n, y_{\max}))\& \\ \&(PC(z) > R_S(n, z_{\min}))\&(PC(z) < R_S(n, z_{\max})) \quad (8)$$

where $n$ is the view range line point from total $N_L$ points (grey rectangle), $PC$ is the point cloud. Values $y$ and $z$ means points in the appropriate axis Y and Z. Parameter $R_S$ is segmentation rectangle and indexes *min* and *max* means the *min* position value in the appropriate axis and vice versa. The final segmentation equation (9) by using equation (8) is the following

$$\bigcup_{n=1}^{N_L} PC(x) > R_S(n, x)\, \&\, R_{SYZ}(n). \quad (9)$$

The last unknown value is $x$ which are points in the axis X. The whole segmentation process is logical sum of all points which comply the condition (9) (all points inside grey rectangles). The next Figure 8. shows the segmentation result with the marked view range.

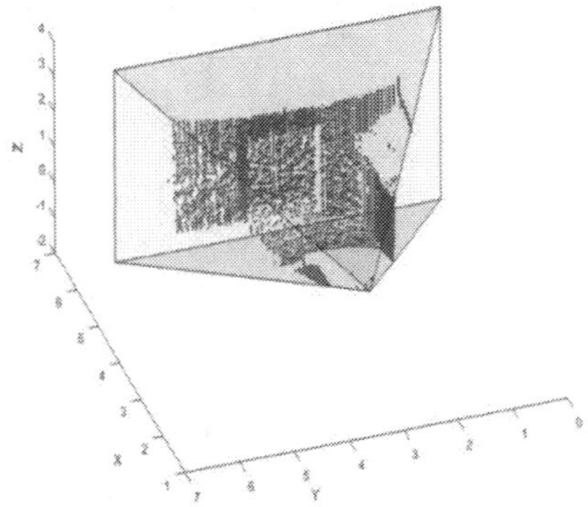

**Figure 8.** Segmented points from input point cloud.

### 3.4 Points depth image transformation

The final step is to transform points into the depth plane. The following Figure 9. illustrates points transformation. There are three red transformation lines for the demonstration.

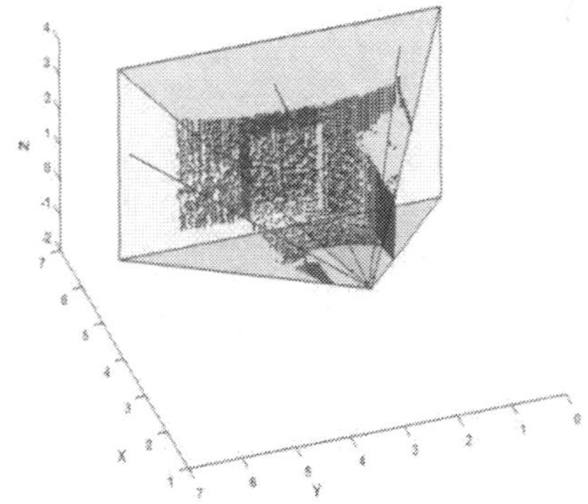

**Figure 9.** Illustration of point transformation.

Each point form the segmented point cloud has to be transformed into the final depth plane. The detailed transformation line description is in Figure 10, where the important transformation parameters are marked.

The distance parameters in Figure 10 marked by the uppercase letter means the direct distance to the depth plane and the parameters with the lower case letter are differences between two points. The parameters are the following: $D(x, y)$ is the transformation point $\mathbf{P}_t$ position in the depth plane; $D_{PD}$ is the direct distance to depth plane from the camera position $\mathbf{P_C}$; $d_y$ and $d_z$ is the distance offset from the depth plane centre; angles $\varphi_y$ and $\varphi_z$ expresses the transformation line orientation; $D_P$, $d_{Pz}$, $d_{PH}$ are parameters of the point triangle; $D_{PD}$, $d_z$, $D_{PDH}$ constitute the second triangle (to the depth plane) and $W$ and $H$ is the depth map resolution.

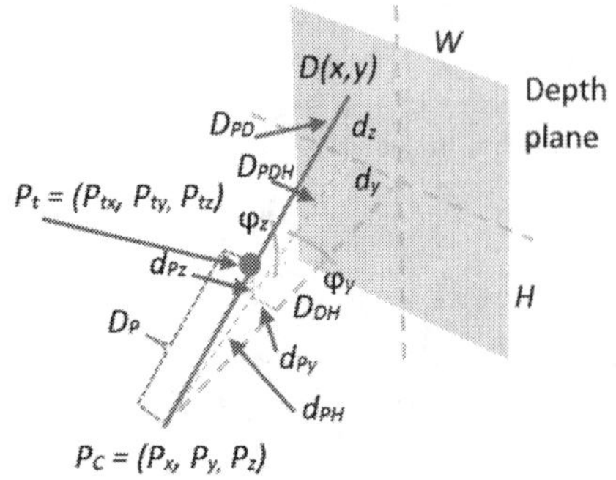


**Figure 10.** Point transformation.

The depth plane size is the last grey rectangle in Figure 7. The depth map resolution is possible to select by parameter $q_{DM}$, which express the depth map quantization. For the testing example parameter $q_{DM}$ is equal to 1 cm, and then the final depth map horizontal resolution $W$ is 724 px and the vertical resolution $H$ is 662 px for the used point cloud.

The direct distance to the depth plane $D_{DH}$ is known from the data range in X axis. Parameters $d_{Pz}$ and $d_{Pz}$ are the subtracted vectors $\mathbf{P_t}$ and $\mathbf{P_C}$. The distance $D_P$ to the measured point $\mathbf{P_t}$ is the point norm of vector in $\mathbf{R}^3$

$$D_P = \sqrt{\sum \left| P_t - P_c \right|} \tag{10}$$

Angles $\varphi_y$ and $\varphi_z$ are calculated by using of goniometric functions. From known parameters is estimated the red line angle and calculated the distance $D_{PDH}$. This allows to express the red line point $D(x,y)$ and get exact position $x$ and $y$ in the depth plane. This process is done for all points in point cloud. If two points are projected on the same position the closer point is used.

The depth map pixel intensity is equal to point distance $D_P$ quantized by 16 bit. The maximal intensity $2^{16}$ is equal to the maximal rangefinder range 15 m.

It is possible to assign all points directly in the depth plane from their position in space, but the main goal is to create the depth map from the camera view point $\mathbf{P_C}$. For this purpose, it is necessary to use the line transformation, showed in Figure 10. The point cloud typically contains points which can be covered by other objects like tables and etc. and these points is necessary to suppress in the resulting depth map. In this depth map construction process the Z-buffer method it is used [12]. The Z-buffer is known from computer graphics. When an object is rendered the Z-buffer ensure that the hidden points are not shown in the output frame. Practically the Z-buffer is the depth map. The point cloud does not give information about surfaces or points connection. The resulting depth map is therefore checked for the farther points in all points neighbourhood. If in the neighbourhood area of 10 px ($q_{DM}$ = 1 ~ 10 cm) lies a pixel with the distance farther than 0.5 m, these pixels are erased. These are estimated like covered by the closer pixel.

### 3.5 Resulting depth map

If the parameter $q_{DM}$ is small like in this example 1 cm, it is possible to use the morphological operation "dilate" to increase the depth map point size. In the resulting depth map Figure 11. the dilatation by the element size 10 is used.

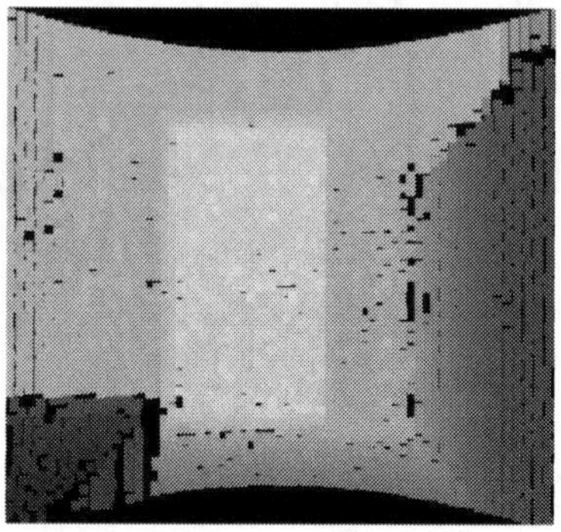

**Figure 11.** Resulting depth map.

The black pixels in top and bottom part of the resulting depth map are caused by the used input point cloud. There are missing points to display them in the created depth map. The darker pixels are closer to the camera position PC. The fineness and compact image is in farther distances where is the higher density of measurement points.

## 4 Results evaluation

The resulting depth map it is possible display also in surf graph, see Figure 12, which allows to "look" on the depth map in different view. The table, door and cupboard is obvious in this representation.

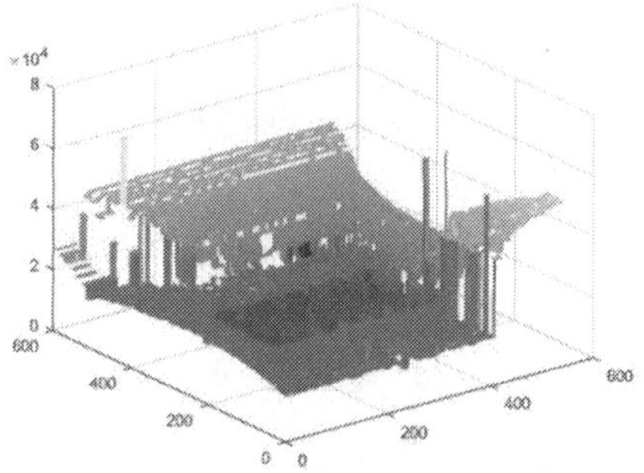


**Figure 12.** Surf depth map interpretation.

The next figures (Figure 13. 14. and 15.) show the depth map with the different camera position and

DOI: 10.1051/matecconf/20167503005

orientation angles $\mathbf{C_O}$ = (0,20,50). These figures shows that it is possible to use the different rotation and the position. The result quality is similar like the introduced example.

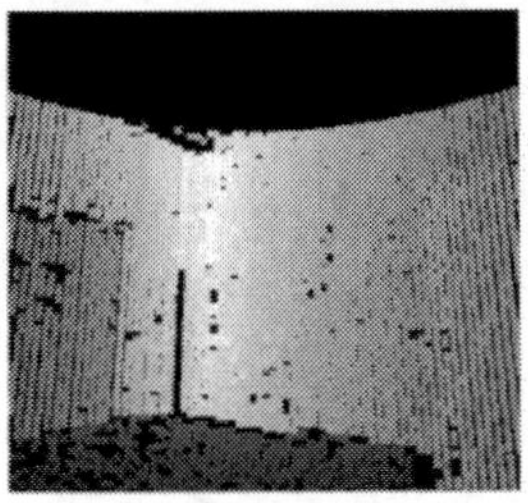

**Figure 13.** Depth Map in different orientation.

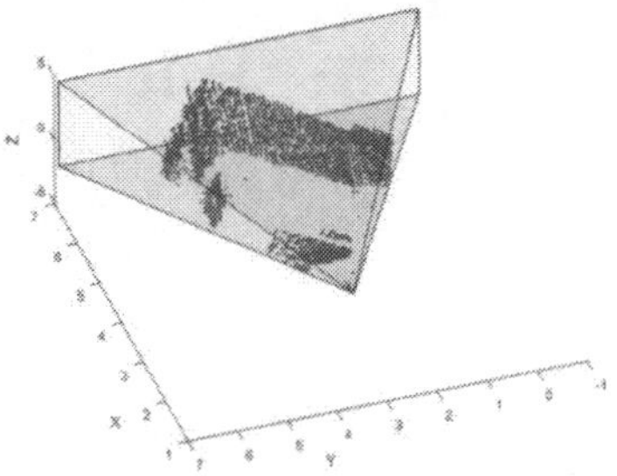

**Figure 14.** Segmented points for second depth map.

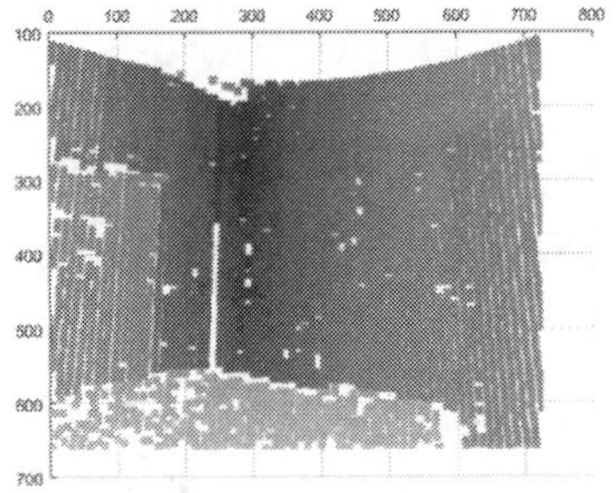


**Figure 15.** Surf interpretation for second depth map.

The Figure 15. shows the increasing distance to the edge of the room. The fineness effect is more obvious here. Points discontinuities in the closer distance are caused by the lower density of input point cloud. This is possible to solve by increasing the parameter $q_{DM}$, which has influence on depth map details.

## 5 Conclusion

In this paper we presented the algorithm for the depth map construction. We described the whole process with our optical rangefinder parameters. The main advantage of this algorithm lies in possibility to create a depth map from any point cloud with any camera position and any camera orientation.

There can be some black parts and holes in the resulting depth map. This disadvantage mainly depends on the used point cloud and the selected detail parameter $q_{DM}$, which is necessary to choose with the deliberation.

## Acknowledgement

The research was supported by the Internal Grant Agency of University of Pardubice, the project No. SGS_2016_022.

## References

1. B.M.F. Silva, L.M.G. Gonçalves, A Fast Vis. Odom. and Mapp. Sys. for RGB-D Cam., *SBR LARS Robocontrol*, 55-60 (2014)
2. A. Aldoma, Z. Marton, F. Tombari, W. Wohlkinger, C. Potthast, B. Zeisl, R. Rusu, S. Gedikli, M. Vincze, Tutor.: Point cl. lib.: Three-dim. obj. recog. and 6 DOF pose est., *Robotics Automation Magazine IEEE*, **19**, 80-91 (2012)
3. H. Du, P. Henry, X. Ren, M. Cheng, D. Goldman, S. Seitz, D. Fox, Interact. 3D model. of ind. Enviro. with a consum. depth cam., *Proc. of the Int. Conf. on Ubiquitous Computing*, (2011)
4. K. Yousif, A. Bab-Hadiashar, R. Hoseinnezhad, 3D Reg. in Dark Environ. Using RGB-D Cam., *Digital Image Computing: Techniques and Applications (DICTA)*, 1-8 (2013)
5. C. Plagemann, V. Ganapathi, D. Koller, Real-tim. Identif. and loc. of bod. parts from depth imag., *IEEE International Conference on Robotics and Automation (ICRA)*, (2010)
6. P. Chmelar, M. Dobrovolny, The fus. of ultras. and optic. measurem. dev. for autonom. mapp., *Radioelektronika 23rd Int. Conf.*, pp.292-296, 2013
7. P. Chmelar, L. Beran, N. Kudriavtseva, The Las. Col. Det. for 3D Ran. Scan. Us. GMM, *Radioelektronika 25th Int. Conf.*, 248-253 (2015)
8. Chmelar P., Beran L., Rejfek L. N. Kudriavtseva, An Eff. Lens Distort. Correct. for 3D Range Scann. Syst., *In 57th Int. Symp. ELMAR*, 37-40 (2015)
9. L. Beran, P. Chmelar, M. Dobrovolny, Navig. of robot. platf. with us. inert. measure. unit and Dir. Cos. M, *ELMAR 56th Int. Symp.* 87-90 (2014)
10. L. Beran, P. Chmelar, and L. Rejfek, "Navigation of Robotics Platform using Monocular Visual Odometry," Radioelektronika, 2015 25th International Conference
11. L. Beran, P. Chmelar, L. Rejfek, Navig. of Robot. Platf. Us. Adv. Img. Proc. Navig. Meth., *VIPIMAGE*, (2015)
12. T. Theoharis, G. Papaioannou, E.-A. Karabassi, The Mag. of the Z-Buf.: A Surv., *Proc. of 9th International Conference on Computer Graphics, Visualization and Computer Vision WSCG*, (2001)

# Theoretical aspects of aerospace image processing in quasi two-dimensional spectral space

B.V. Kostrov[1], D.V. Grigorenko[1], V.N. Ruchkin[2] and V.A. Fulin [2]

[1]*Ryazan State Radio Engineering University, Ryazan, Russian Federation*
[2]*Ryazan State University, Ryazan, Russian Federation*

**Abstract.** The article regards digital images as mathematical objects and uses a co-variation matrix of element brightness values to analyze an autocorrelation matrix of images – a dispersion matrix, to define all significant properties of images with the help of mathematical statistics and matrix computing. We suggest theoretic and methodological generalizations in the field of aerospace photo processing, basing on choice of the most advantageous variant of constructing basic functions among various Vilenkin-Chrestenson functions (VCF) in a non-trigonometrical minimally available construction form as Walsh functions with an original 'carcass' of Rademacher functions. The methodology of this research is based on a set of theorems that are different from corresponding theorems of classical spectral analysis [5]. The conclusions of the theorems allow constructing effective algorithms of filtration and correlation analysis of aerospace images when noises and distortions appear.

## 1 Introduction

An image can be regarded as a discrete signal, defined at a finite interval of its coordinates in a plane. The best results in aerospace image processing are now obtained via algorithms in spatial coordinates, where space and frequency properties of images are ignored. Information about using classical harmonic analysis in image processing is available in [1]. However, the complexity of implementation in Fournier transforms prevents these methods from becoming popular tools in aerospace image processing. For solving tasks of spectral analysis in general cases we can use any systems that contain a necessary number of orthogonal functions. Choosing the system functions will be based on convenient computing and ultimately optimize labour input necessary for implementation of a required transform. The application of alternative basic function systems requires a methodological analysis of their use for solving such tasks.

The paper generalizes the theory and methodology of previous research by both authors, which allows expanding the methodology of using orthogonal transformations in the field of aerospace image processing. In this way, a theoretical and methodological basis is made for using systems of Vilenkin-Chrestenson functions (VCF) in a non-trigonometrical minimally available construction form as Walsh functions, with an original 'carcass' of Rademacher functions. The resulting theoretical and methodological statements are necessary for development of algorithms of filtration and correlational analysis of aerospace images when noises and distortions appear.

The paper is organized as follows. Section 2 treats an aerospace image as a mathematical object, a co-variation matrix of brightness values, and autocorrelation matrix of images – dispersion matrix – all significant properties of images are described with the help of mathematical statistics and matrix computing. Section 3 defines the main idea and theoretical pre-requisites for solving the tasks with the help of Vilenkin-Chrestenson functions (VCFs) in a non-trigonometrical minimally available construction form as Walsh functions, with an original 'carcass' of Rademacher functions. Section 4 presents a methodology of using such transforms, it is based on a set of theorems that are different from corresponding theorems of classical spectral analysis [5]. In Section 5, the authors describe some experimental results of using theorems 1 and 5 for filtration of aerospace images with synchronous multiplicative noises and provide recommendations for software and hardware implementation of spectral representation of aerospace images in programmable logic integral circuits (PLICs).

## 2 Principal results

As far as the principles and tools of aerospace image photography (ASI) [2] are concerned, all video sensors in radio systems can be divided into 5 groups:

Group 1 includes sensors that are constructed on the basis of charge coupled devices (CCDs);

Group 2 includes single-beam sensors with conical or flat-bed scanning;

Group 3 includes scanning sensors;

Group 4 consists of radio-locator stations with various bases and operational principles;

Group 5 can include thermal imagery devices and televisions, as well as systems that are typically mounted on flying apparatuses and on on-ground stations.

All the listed methods and systems create panchromatic video data that can be fixed or saved in a number of ways. They all have a certain geometrical shape, and their content is represented as densities of blacker areas in video; these data can be transferred via various systems, and they all share properties that allow referring to them as aerospace images (ASIs).

To solve tasks in analysis and processing of ASIs, it is necessary to define 'image' as a mathematical object possessing certain analytical properties. Currently, performing mathematical operations on an image presupposes that it exists in digital form, as a *digital image* [2], represented as a discrete array of numbers, forming a matrix of elements, $b_{ij}$ with parameters $(i,j)$, where $1 \leq i \leq n$ and $1 \leq j \leq n$. Elements $b_{ij}$ are squares (apertures, pixels), defined by inequalities $i\text{-}1 \leq x < i$ and $j\text{-}1 \leq y < i$ ($x$ and $y$ are spatial co-ordinates of a continuous image). Then any $[\ b_{ij}\ ]$matrix of a $n \times n$ can be assigned an image, whose brightness value squared $[\ i\text{-}1 \leq x < i\ ] \times [\ j\text{-}1 \leq y < i\ ]$ equals a constant value $\bar{b}_{ij}$ for each pair $(i,j)$. A digital image element can be represented as a random value $\tilde{b}_{ij}$, and the $[\ \tilde{b}_{ij}\ ]$ matrix of random values. As the $b_{ij}$ values can be taken as full representatives of random values $\tilde{b}_{ij}$, we transpose their value onto the area of the whole element.

The sum of random values of the image element brightnesses leads to a value of $\Sigma_B = M\{(B - \bar{B})(B - \bar{B})'\}$, where $M$ is an operator of mathematical expectation, $\bar{B}$ corresponds to the vector average, $(B - \bar{B})$ represents the column matrix, and $(B - \bar{B})'$ stands for a row matrix.

The obtained matrix contains $n^2 \times n^2$ elements, the diagonal elements are dispersions of specific random values, and all the other elements correspond to co-variations of two random values $b_{ij}$, measured at different values of $i$ and $j$. A co-variation matrix is completely symmetrical.

The utility of an image can be defined via analysis of a self-correlation image matrix, also called a dispersion matrix, as is characterizes the image element dispersion.

$$S_B = \begin{bmatrix} M\{b_{11}^2\} & \dots & M\{b_{11}b_{nn}\} \\ \dots & \dots & \dots \\ M\{b_{nn}b_{11}\} & \dots & M\{b_{nn}^2\} \end{bmatrix} . \quad (1)$$

In some cases, the $\Sigma_B$ matrix is more conveniently represented via correlation coefficients, defined as $r_{ij} = \sigma_{ij}^2 / \sigma_{ii} \sigma_{jj}$. Then $\Sigma_B$ will equal $\Sigma_B = \Gamma_\eta R_K \Gamma_\eta$, where $\Gamma_\eta$ is a diagonal matrix of mean square deviations $\sigma_{ij}$, and $R_K$ is a matrix of correlation coefficients $r_{ij}$.

$$\Sigma_B = \sigma^2 R_K', \quad R_K' = \begin{bmatrix} 1 & r_1 & \dots & r_{n-1} \\ r_1 & \dots & \dots & \dots \\ \dots & \dots & \dots & \dots \\ r_{n-1} & \dots & \dots & \dots \end{bmatrix} \quad (2),$$

where is a correlation matrix for rows and columns, and $r_i$ are correlation coefficients for rows and columns.

Thus, using tools and methods of mathematical statistics and matrix computing, we can define all significant properties of images.

ASIs develop specific distortions due to their formation principles and conditions of transfer, and these distortions can be classified:

– synchronous distortions due to changing transfer properties of the pixel brightness formation tract: these are rigidly connected to the law of scanning ASIs (specific 'synchronous stripes' along the rows or columns);

– distortions as group noises that can behave absolutely destructively (shown as missing pixels along ASI rows);

– non-synchronous distortions not related to the pixel brightness formation tract of the scanning law (shown as specific periodic 'non-synchronous stripes' located at a random angle to the ASI columns).

Removal of described distortions is the primary task of preliminary processing of compression algorithm picture compression, to form prerequisites for using the images).

## 3 Main idea and theoretical pre-requisites for solving the tasks

The existence of an image digital form defines mathematical tools available for execution of processing operations. These are mostly discrete transformations defined for finite intervals. Introducing the notion of finiteness, we can avoid the contradictions that arise from the Fourier transformation, and conduct spatial-spectral analysis of images that are in a general way described by a non-stationary arbitrary process. In this respect, using the principal statements and conclusions of the theory of discrete signal at finite intervals is quite justified. This theory is based on choosing the best acceptable option of constructing a basic function system from the whole range of VCFs, for which signal shift is defined as bitwise addition of numbers in a certain module. The notion of VCF includes Walsh function systems (based on binary arithmetic) as an individual case.

The methodology of applying this system at an interval is based on the following properties:

1. All functions of the system are valid for definition interval $N = 2^n$.

2. As the system function values are limited to +1 and –1, the basic operations for expansion in a Walsh system are addition and subtraction.

3. The Walsh function remains orthogonal at the definition interval of $N$, and the Hadamard matrix built according to Walsh functions is symmetrical.

4. As the Hadamard matrix is sized $N \times N$, it includes $N$ orthogonal functions and therefore no new orthogonal functions can be added. This means that such a function system is complete and can be used for unitary transformations of non-harmonic spectral analysis. The computational complexity of such transformations will be minimal, as all the operations will be replaced with addition of real numbers, and not of discrete exponential functions, where all numbers are complex [3].

Alternatively, VCFs can also be represented by Rademacher functions (complex) for the same interval $N = m^n$:

$$R_i(x) = e^{j(2\pi/m)x_i} \quad (3)$$

where the $i$ index stands for the order of a Rademacher function.

Then, in accordance with (1), VCFs can be written as

$$F(p,x) = \prod_{i=0}^{n-1} [R_i(x)]^{<P_i>} \quad (4)$$

For the case when $m = 2$, the Rademacher function can be set like this:

$$r_i(x) = (-1)^{<x_i>} \quad (5)$$

where $< x_i >$ is the $i$ order of a binary representation of the $x$ variable.

In such a task, all Rademacher functions are valid and uneven functions for the $N$ interval. The system that they constitute in not complete. Its completion results in a Walsh function system:

$$wal(w,x) = \prod_{i=1}^{n} [r_i(x)^{<w_i>}] \quad (6),$$

where $< w_i >$ is the value of the $i$ order of a Rademacher function number, represented in a Gray code: $i = 1,2,3,..., n$.

If the number of the Rademacher function contains only one unit, the Walsh function will coincide with the Rademacher function having the same number.

Thus, for definition interval $N = 2^n$ a Walsh function system can be divided into $n$ groups (here, the zero order function is not considered). If these groups are numbered $k = 1,2,3,..., n$, each group begins with Rademacher function $r_{n+1-k}$, and each includes $2^{n-k}$ functions (including the Rademacher function proper). Thus, a Rademacher function system forms a 'carcass' on which the Walsh system is built; and this circumstance has to be considered for spectral analysis of signals.

The methodology of this research is based on a set of theorems that are different from corresponding theorems of classical spectral analysis [5]. The conclusions of the theorems allow constructing effective algorithms of filtration and correlation analysis of aerospace images when noises and distortions appear.

# 4 Methodology of aerospace image processing

**2.** Theorem on dyad convolution

If $\{X_n\}$ and $\{Y_n\}$ are digital sequences set for an $N$ *interval*, the sequence of convolution (correlation) $\{Z_S\}$:

$$Z_S = \frac{1}{N} \sum_{n=0}^{N-1} X_n Y_{n \oplus s} \quad (7)$$

will be as follows:

$$Z_s = \sum_{u=0}^{N-1} N(C_u^X \times C_u^Y), \quad (8)$$

where $C_u^X$ are spectral coefficients of the $\{X_n\}$ sequence, and $C_u^Y$ are spectral coefficients of the $\{Y_n\}$ sequence.

The theorem gives an insight into the mechanism of formation of signal elements following their filtration in a spectral space.

**2.** Theorem on realnumber-dyad convolution (RDC)

If $\{X_n\}$ and $\{Y_n\}$ are digital sequences set for an N interval, the sequence of convolution (correlation) $\{Z_s\}$

$$Z_s = \frac{1}{N} \sum_{n=0}^{N-1} X_n Y_{n-s} \quad (9)$$

can be computed as follows:

$$Z_S = \sum_{u=0}^{N-1} C_u^X (C_u^Y)_s \quad (10)$$

where $C_u^X$ are spectral coefficients of the $\{X_n\}$ sequence, and $(C_u^Y)_s$ are spectral coefficients computed after a sequence shift $\{Y_s\}$.

This theorem allows constructing complex spectral filters and effective algorithms for correlation analysis. The proof of the theorem and methods of construction of a correlation function for two images can be found in [4]. Physical foundations of RDF functioning are given in [5, 6].

**3.** Theorem on invariance of thinned basis

If $\{V_n\}$ is a digital sequence obtained from sequence $\{X_n\}$ through a dyad shift by value of $l$, i.e. $V_n = X_{n \oplus l}$,

and $C^V = \frac{1}{N} \mathbf{H_w V}$, and $C^x = \frac{1}{N} \mathbf{H_w^0 X}$, (11)

where $C^V$ are coefficients of the Walsh sequence spectrum, $\{V_n\}$ $C^x$ are coefficients of the sequent spectrum of the $\{X_n\}$ sequence, $H^0_w$ is an arbitrarily thinned Hadamard matrix, and $\mathbf{H_w}$ is the initial matrix, i.e. $(C^v)^2 = (C^x)^2$.

This theorem allows certain filtration types while obtaining spectral presentations, which is not typical of other basic function systems. Proof of this theorem is given in [7].

**4.** Theorem on restriction of a non-trigonometrical spectrum

If $n$ is the order of a Rademacher function system, and $k$ is the number of a function groupы in a Walsh system, restriction of the spectrum is applied at the level Rademacher functions, number of $n+1-k$, then in the newly formed image the element brightness $b_{ij}^{[N_s]}$ will be:

$$b_{ij}^{[N_s]} = \frac{1}{S^2}\sum_{g=1}^{s}\sum_{p=1}^{s} b_{gp}, \quad (12)$$

where $N_s = \frac{N}{S}$, $s = 2^{k-1}$, and $i, j = \overline{1, N_S}$. Here, the $i(s); j(s); g(i); p(j)$ dependencies are ignored for simplification of the formula.

Intuitive use of this property allows constructing a two-level algorithm of image alignment, requiring computational costs that are tenfold lower as compared to algorithms described in [5]. An implementation of this approach is discussed in [8, 15].

**5.** Theorem on energetic completeness of a quasi-two-dimensional spectral presentation

If $b$ is the matrix of a digital image sized $M \times N$, and $C$ are coefficients of its quasi-two-dimensional spectral representation, this equality is true:

$$\frac{1}{MN}\sum_{i=0}^{M-1}\sum_{j=0}^{N-1} b_{ij}^2 = \frac{1}{M}\sum_{i=0}^{M-1}\sum_{v=0}^{N-1} C_{iv}^2 \quad (13).$$

Expression (2) is a Parseval equation for a quasi-two-dimensional presentation of two-dimensional signals. Its proof does not require any commentary, as it results from certain properties of spectral analysis. Using this theorem allows twofold reduction of computing effort and several other benefits, through removal of certain distortion types [9].

**6.** Theorem on a constant component of spectrum

If we assume that $b(x,0) = 0$ is a concrete value of an image element in every row, it is not necessary to transfer the spectral constant component via a communications channel, because it can always be recovered with this formula:

$$B_0 = -\sum_{v=1}^{N-1} B(x, v-1)(-1)^{\sum_{i=1}^{n}\langle (v-1)_i x_i \rangle} \quad (14)$$

In this case, the transferred spectrum does not have a constant component.

## 5 Experimental results of image filtration

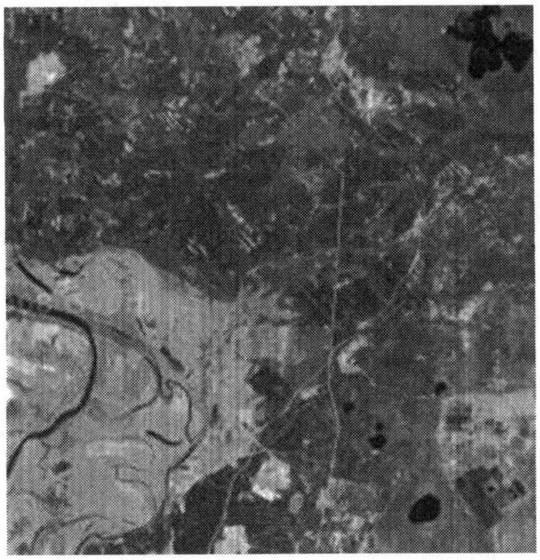

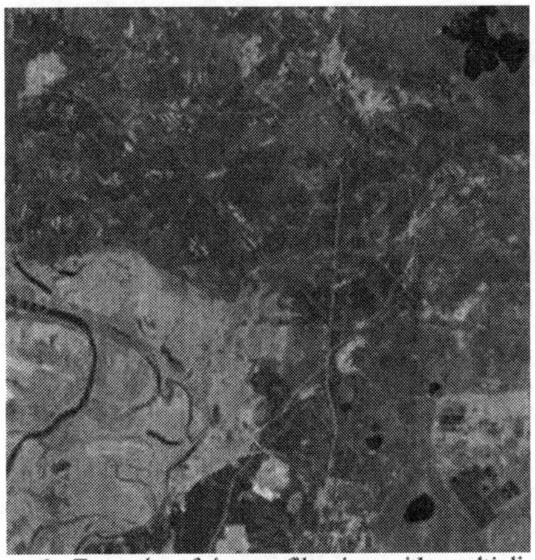

**Figure 1.** Example of image filtration with multiplicative synchronous noise: (a) is the initial image MSD=7.4184; (b) is the image transferred MSD=2.77

Fig. 1 shows the result of using Theorems 1 and 5 for filtration of images with synchronous multiplicative noises [10]. Filtration is performed through diagonal filtration in a quasi-two-dimensional spectral Walsh space. Image recovery follows this formula:

$$\mathbf{B} = \mathbf{H}_w \mathbf{H}_{uu} \mathbf{C} \quad (15)$$

where **B** is the matrix of the recovered image; **C** is the matrix of spectral coefficients; $\mathbf{H}_w$ is a Walsh-Hadamard matrix; $\mathbf{H}_{uu}$ is a matrix whose rows are made up of coefficients of row $h_{uu}$ diagonal filters:

$$h_{uu} = \frac{(\sum_G)_{uu} - (\sum_\eta)_{uu}}{(\sum_G)_{uu}} \quad (16)$$

where $(\sum_G)_{uu}$ and $(\sum_\eta)_{uu}$ are co-variation matrices of image rows with noises and of the noise itself, correspondingly, computed in a spectral space.

## 6 Conclusion

Removal of distortions that emerge during formation and transfer of aerospace images is the principal task of preliminary processing of compression algorithm image compression, to form prerequisites for using the images.

Large volumes of data to process make us constantly look for methods and tools of acceleration, and improve the mathematical toolset for processing them as discrete transformations at their finite intervals. Introducing the notion of finiteness, we can avoid the contradictions that arise from the Fourier transformation, and conduct spatial-spectral analysis of images that are in a general way described by a non-stationary arbitrary process. In this respect, using the principal statements and conclusions of the theory of discrete signal at finite intervals is quite justified. This theory is based on choosing the most acceptable option of constructing a basic function system from the whole range of VCFs, for which signal shift is defined as bitwise addition of numbers in a certain module. Using this method considerably reduces computing expenditure (tenfold or more), while the quality of preliminary processing of aerospace images considerably increases. Examples of using the described approaches in development of systems that are to implement practical tasks are given in [15].

## References

1. Gonzalez R.C., Woods R.E. Digital Image Processing: Prentice Hall. New Jersey. 2002. P. 1072.
2. Kostrov B.V. The Features of Aerospace Image Forming by Radar Facilities // Proektirovaniye i tekhnologiya elektronnykh sredstv. 2011. Vol. 1. Pp. 41-43.
3. Kostrov B.V. Correlation-extremum method of digital signal detection // Tsifrovaya obrabotka signalov. 2011. № 2. Pp. 46-50.
4. Zlobin V.K., Kolesenkov A.N., Kostrov B.V., Correlation-extremum methods of aerospace image alignment // Vestnik Ryazanskogo gosudarstvennogo radiotekhnicheskogo universiteta. 2011. № 37. Pp. 12-17.
5. Zlobin V.K., Kostrov B.V., Sablina V.A. Place and role of sequential analysis methods in processing of aerospace images // Radiotekhnika. 2012. № 3. Pp. 64-72.
6. Kolesenkov A.N., Kostrov B.V. Method of thinning of basic functions in correlation-extremum algorithms // Voprosy radioelektroniki. – 2010. – Issue 1. – Series OT. Pp. 176 – 184.
7. Zlobin V.K., Kostrov B.V., Sablina V.A. Algorithm of sequential filtration of group noises in images // Vestnik Ryazanskogo gosudarstvennogo radiotekhnicheskogo universiteta. 2009. № 30. Pp. 3-7.
8. Kostrov B.V., Sablina V.A. Adaptive filtration of images with structural distortions // Tsifrovaya obrabotka signalov. 2008. № 4. Pp. 49-53.
9. Kostrov B.V., Upakova A.G. Quasi-two-dimensional filtration of synchronous noises in images // Proektirovaniye i tekhnologiya elektronnykh sredstv. 2012. № 1. Pp. 32-35.
10. Kostrov B.V., Grinchenko N.N., Stepanov D.S., Upakova A.G. Algorithm of image transfer with recovery of the constant component // Izvestya Tulskogo gosudarstvennogo universiteta. Tekhnicheskiye nauki. 2013. № 9-1. Pp. 244-249.
11. Ahmed N., Rao K. R. Orthogonal Transforms for Digital Signal Processing.: Springer-Verlag Berlin, Heidelberg New York. 1975. P. 250.
12. Kostrov B.V., Kostrov V.V. Method of fold computation in a non-trigonometrical finite basis // Radiotekhnicheskiye i telekommunikatsionnye systemy. 2015. № 2 (18). Pp. 66-73.
13. Kolesenkov A., Kostrov B., Ruchkina E., Ruchkin V. Anthropogenic situation express monitoring on the base of the fuzzy neural networks // Proceedings – 2014 3rd Mediterranean Conference on Embedded Computing, MECO 2014 – Including ECyPS 2014 3. 2014. Pp. 166-168.
14. Ruchkin V.N., Kostrov B.V., Kolesenkov A.N., Ruchkina E.V. Algorithms of Fire seat Detection, Modeling Their Dynamics and Observation of Forest Fires via Communication Technologies // Proceedings – 2015 4th Mediterranean Conference on Embedded Computing. BUDVA. 2015. Pp. 254 - 257.
15. Kolesenkov A.N., Kostrov B.V., Ruchkin V.N. Emergencies monitoring and preventing // Proceedings – 2013 2nd Mediterranean Conference on Embedded Computing, MECO – 2013. 2013. Pp. 263-265.

MATEC Web of Conferences **75**, 03007 (2016) DOI: 10.1051/matecconf/20167503007
*ICMIE 2016*

# Application and Optimization of Kalman Filter for Baseband Signal Processing of GPS Receivers

Yanpin He, Weimin Li [a] and Bo Bi

*Beijing Microelectronics Technology Institute, Beijing, 100076, China*

**Abstract.**High sensitivity tracking in GPS receiver is required in many weak signal circumstances. The key of improving sensitivity is the optimization of the loop filter in tracking. As Kalman filter is the most optimized linear filter, it is used in many engineering fields. This article introduced the application of Kalman filter as the loop filter of the carrier tracking loop in GPS receiver, to improve tracking sensitivity. The traditional loop filter is replaced. Simulation results show that the new structure improves the tracking sensitivity by 6dB and can make the tracking loop more robust when the navigation signal is languishing. The optimization of theKalman filter is also analysed, which further improves the sensitivity by 4dB.

## 1 Introduction

The navigation signal is very weak when it arrives at earth surface from the satellites. The nominal power of the signal is -130dBm [1]. But in practical circumstances, it is further blocked by buildings, bridges, forests, etc. The signal power can be under -140dBm, or even lower. How to make the receiver module keep working in low carrier-to-noise ratio (CNR) is an important issue for the navigation receiver design.

After the radio frequency module converts the radio-frequency signal to intermediate-frequency signal, the acquisition module starts working to get the rough estimate of carrier frequency and code phase, which is searching along frequency dimension and pseudo random noise (PRN) code phase dimension. In tracking step, the acquisition result is used to generate local signal. The received signal is multiplied by the local signal to strip off the carrier and the PRN code. Because there is error in frequency and code phase from the acquisition result, the multiplied result has residual frequency and code phase. Since code loop is more robust and simple than carrier loop, so code loop won't be discussed here. A phase detector and a frequency detector are used to get phase error and frequency error, respectively. The errors are filtered by loop filter to remove noise. Then the filtered result is used to adjust local numerically controlled oscillator (NCO) to generate the signal of the next moment with the updated frequency. From the analysis above, appropriate loop filter method is important to the noise performance of a loop.

Kalman filter is originally used in positioning algorithm to improve position precision, which is in post navigation part. But just taking into account its properties that it updates its estimates considering the measurement noise and the uncertainty in its current estimates, it also performs well in previous signal processing of navigation [2]. Difference is that the total physical model should be rebuilt according to the signal properties.

For higher Signal to Noise Ratio (SNR) during processing, coherent integration is adopted before phase detection and frequency detection. At the beginning of the tracking step, 1ms coherent integration is adopted because the bit synchronization has not realised. Longer coherent integration will increase the possibility of the integration crossing bit edges, which impairs the integration result if there is a bit transition. Short coherent integration means low integration gain. So if the CNR is low, noise may be too large to keep tracking. In this article, the application circumstance is that the tracking has just started following acquisition, so 1ms coherent integration is employed. The rest of the article will compare traditional loop filter and Kalman filter in noise performance. The final section will analyse the optimization of the Kalman filter.

## 2 The traditional tracking structure

### 2.1. Structure overview

Fig. 1 describes the traditional tracking loop structure with a third-order phase-locked loop assisted by a second-order frequency-locked loop. The transfer functions of the second-order frequency-locked loop filter and the third-order phase-locked loop filter are expressed by (1), (2) respectively, which can be described in Fig. 2 after bilinear transformation.

$$F_f(s) = \frac{1}{K_f}\left(a_2\omega_n + \frac{\omega_n^2}{s}\right) \quad (1)$$

[a] Corresponding author: liwm@mxtronics.com

$$F_p(s) = \frac{1}{K_p}\left(b_3\omega_n + \frac{a_3\omega_n^2}{s} + \frac{\omega_n^3}{s^2}\right) \tag{2}$$

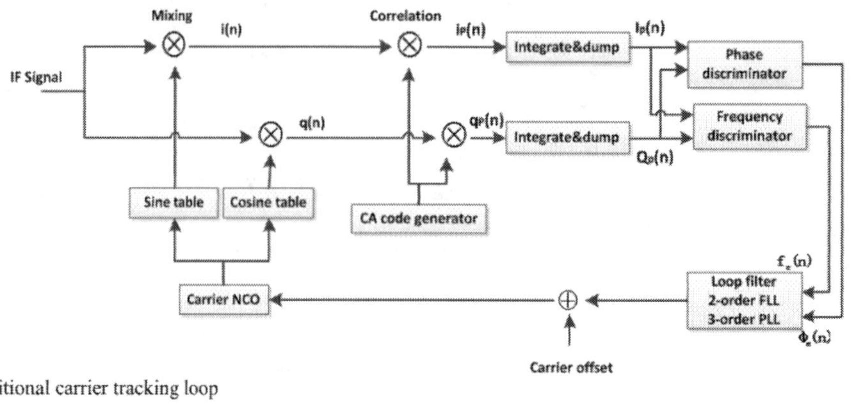


**Figure 1.** Traditional carrier tracking loop

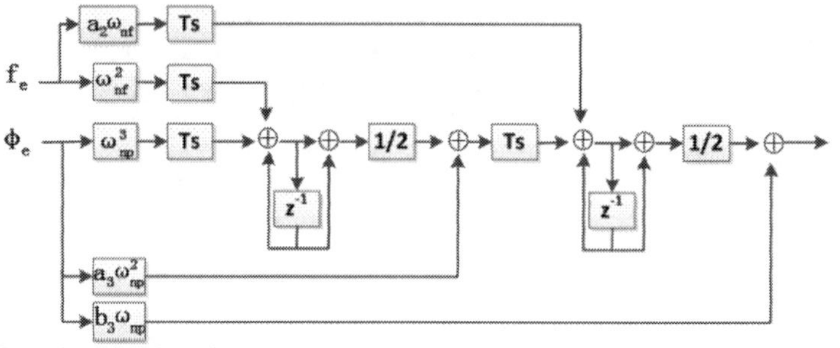


**Figure 2.** Bilinear transformation of the loop filter.

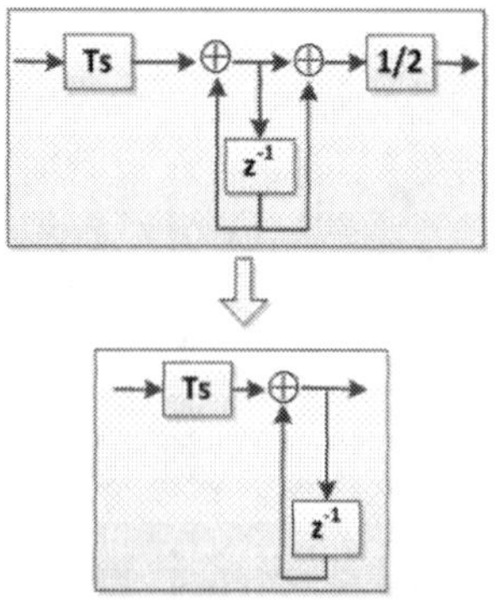


**Figure 3.** The integrator's transform

PLL can track carrier phase closely, but its dynamic tolerance is poor. It will lose lock when noise power is strong or the loop bandwidth is wide. FLL tracks signal frequency, the integral of carrier phase. So it is more robust with good dynamic performance and it can keep locked in lower SNR circumstances. But its tracking is not quite close, inducing large tracking error. Thus, combining PLL and FLL together is a reasonable solution for both high sensitivity and high dynamics. Frequency error should be integrated to become phase error, so FLL order is one order less than that of PLL. The bilinear transformation of the integrator can be equivilant to a rectangle integrator as in Fig. 3 [3]. The filter paramaters are set with empirical values as in Table 1 [1].

**Table 1.** Empirical values of filter parameters.

| **Loop order** | **parameters** |
|---|---|
| Second-order | $a_2 = 1.414$<br>$B_L = 0.53\omega_n$ |
| Third-order | $a_3 = 1.1$<br>$b_3 = 2.4$<br>$B_L = 0.7845\omega_n$ |

A critical step in filter design is the selection of the loop bandwidth. Large bandwidth has fine dynamic performance while small bandwidth has good anti-noise performance. The different needs in actual situations should be taken into consideration. For application on

 DOI: 10.1051/matecconf/20167503007

vehicles, bandwidth can be set to 20Hz and this is also the simulation condition in this article.

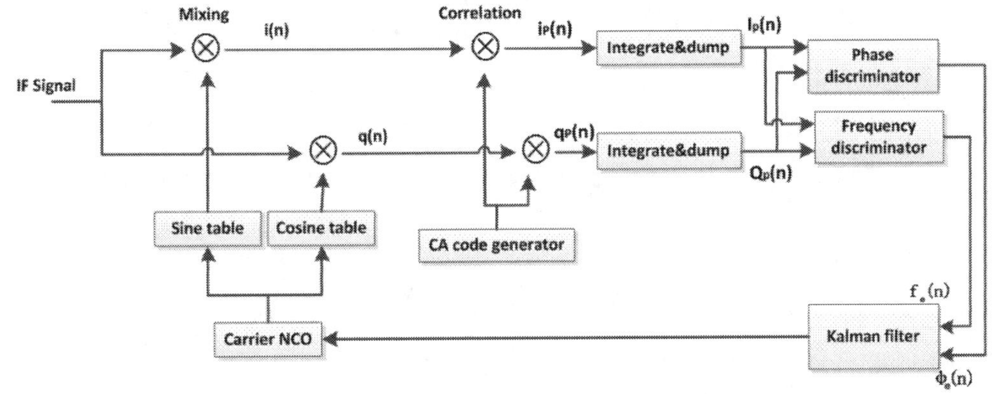


**Figure 4.** Kalman carrier tracking loop.

### 2.2 NCO tuning

The phase of the NCO output cannot jump. Sohow to transfer the desired phase adjustment amount to frequency amount? Simple in concept, the integration of delta frequency equals to the adjustment amount of the initial phase [1]. An example can be employed to explain this conclusion. Set initial phase $\theta_0 = 0$. A sine signal first runs for 1ms with frequency $f_0 = 2Hz$, then runs for 1ms with frequency $f_1 = 2.001Hz$.The phase after 2ms is $\theta_2 = \theta_0 + f_0 * 360^{\circ} * 0.001s + f_1 * 360^{\circ} * 0.001s = 1.44036^{\circ}$ . The integration of delta frequencyis $(f_1 - f_0) * 360^{\circ} * 0.001s = 0.00036^{\circ} = delta_{phase}$ . Thus $\theta_2$ can also be calculated by $\theta_2 = delta_{phase} + f_0 * 360^{\circ} * 0.001s * 2$.This means the phase after 2ms equals to the phase that the sine signal runs for 2ms with frequency $f_0 = 2Hz$, and initial phase $\theta_0{}' = 0.00036^{\circ}$.

Asthe filter output is $\frac{delta\_phase}{T_c}$, so

$$local\ freq_{next} = f_0 + filter\ output \quad (3)$$

where $f_0$ is the acquisition result frequency.

## 3 Tracking structure using Kalman filter

### 3.1 Structure overview

All modules remain the same except that the filter is replaced by Kalman filter, as in Fig. 4. The inputs of the Kalman filter are from phase discriminator andfrequency discriminator. The filter output is the updated state vector$X$. There is no need to additionally add a carrier offset. The filter output is directly used by NCO.

### 3.2 Kalman filter modelling

Use absolute phase and frequency as state vector $X$.

$$X = \binom{\theta}{\omega} \quad (4)$$

*3.2.1 Forward-prediction*

$$\hat{X}_k^- = \Phi\hat{X}_{k-1} + W, \Phi = \begin{pmatrix} 1 & T_c \\ 0 & 1 \end{pmatrix} \quad (5)$$

where $W$ is the process noise; $T_c$ is the coherent integration time. $\hat{X}_k^-$ is used to generate the local carrier at next moment.

*3.2.2 Mean square error (MSE) of forward-prediction*

$$\hat{P}_k^- = \Phi\hat{P}_{k-1}\Phi^T + Q \quad (6)$$

where $Q$ is the covariance matrix of the process noise $W$ and it can be simplified to 0 in the simulation where there is no crystal jitter. $\hat{P}_{k-1}$is the MSE matrix of the posteriori estimates of the last moment. $\hat{P}_k^-$ is the MSE matrix of the priori estimates of the present moment.

*3.2.3 Kalman gain*

$$K = \hat{P}_k^- H^T\left(H\hat{P}_k^- H^T + R\right)^{-1},\ H = \begin{pmatrix} 1 & 0 \\ 0 & 1 \end{pmatrix} \quad (7)$$

where $H$ is the relationship matrix between the state vector and the observation vector. $R$is the covariance matrix of the observation noise. In this scenario, $R$is the phase resolution and the frequency resolution from acquisition.

*3.2.4 Update*

$$\hat{X}_k = \hat{X}_k^- + K(Z_k - H\hat{X}_k^-) \quad (8)$$

$$\hat{P}_k = (I - KH)\hat{P}_k^- \quad (9)$$

where $Z_k$ is the observation vector at present moment.So the observation equation is

$$Z_k = \binom{\theta_k}{\omega_k} = H\hat{X}_k^- + V \quad (10)$$

where $V$ is the observation noise.Define residual error as $Res$:

$$Res = Z_k - H\hat{X}_k^-$$

$$= \begin{pmatrix} output\ of\ phase\ discriminator \\ output\ of\ frequency\ discriminator \end{pmatrix} \quad (11)$$

$$\hat{\boldsymbol{X}}_k = \hat{\boldsymbol{X}}_k^- + \boldsymbol{K} \cdot \boldsymbol{Res} \quad (12)$$

So Kalman filter comprehensively considers the prediction value and the observation value, adding them together with reasonable weight to get the most reliable estimation.

### 3.3 NCO tuning

The updated phase $\hat{\boldsymbol{X}}_k(1)$ and the updated radian frequency $\hat{\boldsymbol{X}}_k(2)$ are the unbiased estimations of the signal carrier phase and the signal carrier frequency of the *kth* moment. Compared to the *kth* local carrier, which is generated based on the *kth*priori estimate$\hat{\boldsymbol{X}}_k^-$, the phase adjustment amount is [4]

$$delta_{phase} = \hat{\boldsymbol{X}}_k(1) - \hat{\boldsymbol{X}}_k^-(1) \quad (13)$$

The forward-prediction based on $\hat{\boldsymbol{X}}_k$ is used to generate the *k+1th* local carrier, using both the phase adjustment amount and the frequency adjustment amount. According to the analysis in section 2, the integration of delta frequency equals to the adjustment amount of the initial phase, i.e.

$$delta_{phase} = T_c * delta_{freq} \quad (14)$$

$$local\ freq_{k+1} = \hat{\boldsymbol{X}}_{k+1}^-(2) + delta_{phase}/T_c \quad (15)$$

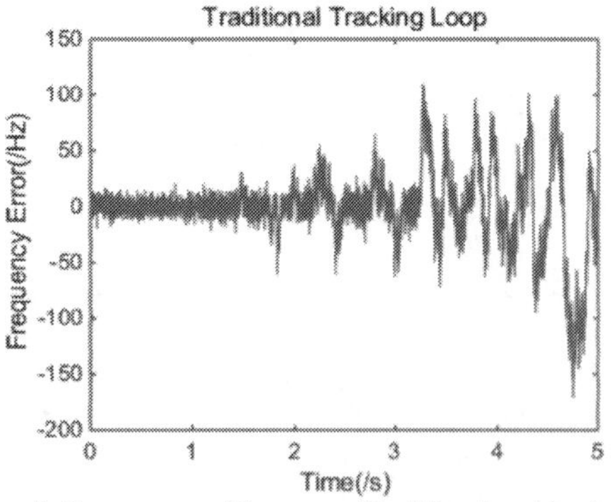


**Figure 5.** Frequency tracking error of traditional tracking loop running for 5s with CNR decreasing 2dB every 1s from 35dBHz.

## 4 Simulations

### 4.1 Traditional tracking loop

Assume the frequency error from acquisition is -12Hz and the phase error is 25°. The PLL bandwidth and the FLL bandwidth are both 20Hz. The signal CNR start with 35dBHz and is decreased by 2dB every 1s for 5s. So the CNR is 35dBHz, 33dBHz, 31dBHz, 29dBHz, 27dBHz. From Fig. 5, it is seem that at 35dBHz and 33dBHz, the loop can converge and the frequency error is hopping between +10Hz and -10Hz. When CNR decreases further, the tracking error is gradually increasing, making the tracking invalid. Considering the thermal noise MSE equations of PLL and FLL, this phenomenon can be explained [5], [6]:

$$\sigma_{tPLL} = \frac{180^\circ}{\pi}\sqrt{\frac{B_L}{C/N_0}\left(1 + \frac{1}{2T_c \cdot C/N_0}\right)} \quad (16)$$

$$\sigma_{tFLL} = \frac{1}{2\pi T_c}\sqrt{\frac{4FB_L}{C/N_0}\left(1 + \frac{1}{T_c \cdot C/N_0}\right)} \quad (17)$$

The general value of F is 1when CNR is high. It is 2 when CNR is low and the tracking is approaching the threshold.

When CNR=29dBHz, F=2, $\sigma_{tFLL}$ =107Hz. When CNR=27dBHz, F=2, $\sigma_{tFLL}$=155Hz. It can be seen that the simulation result corresponds to the theoretical calculation results.

### 4.2 Kalman tracking loop

For Kalman filter, importance should be attached to the setting of the measurement noise $\boldsymbol{R}$ and the initial values of the state vector$\hat{\boldsymbol{X}}_0$ and the MSE matrix$\hat{\boldsymbol{P}}_0$. The covariance matrix of the observation noise$\boldsymbol{R}$ is a 2*2 diagonal matrix related to the error variances of the phase discriminator and the frequency discriminator. Remember that equation (16) and (17) are measurement variances of the whole loop. According to [5], the independent variances errors of the phase discriminator and the frequency discriminator are:

$$\sigma_{PD}^2 = \frac{1}{2C/N_0 T_c}\sigma_{FD}^2 = \frac{1}{C/N_0 T_c^3} \quad (18)$$

$$R(1,1) = \sigma_{PD}^2 R(2,2) = \sigma_{FD}^2$$

$\boldsymbol{R}$is constant as long as CNR doesn't change. The measured phase and radian frequency at t=0 is used as $\hat{\boldsymbol{X}}_0$[7]. In the application of this article, the initial phase and radian frequency are the results from acquisition. The MSE matrix $\hat{\boldsymbol{P}}_0$ is a 2*2 diagonal matrix reflecting the MSE of $\hat{\boldsymbol{X}}_0$, which corresponds to the search resolution in acquisition. Set:

$$\hat{P}_0(1,1) = \left(\frac{\pi}{180^\circ} \cdot 25^\circ\right)^2 \hat{P}_0(2,2) = (2\pi \cdot 12Hz)^2$$

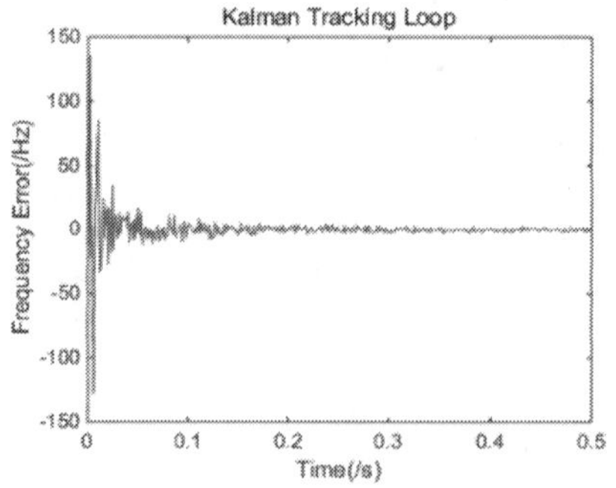


**Figure 6.** Frequency tracking error of Kalman tracking loop running for 500ms with CNR= 27dBHz.

$\boldsymbol{P}$is changing during iterations, decreasing gradually when converging [7].The initial frequency error and

DOI: 10.1051/matecconf/20167503007

phase error settings are the same as that of traditional tracking loop. Fig. 6 shows the loop tracking error converging to 0Hz within 500ms at CNR=27dBHz.

Another simulation scenario is that the signal CNR start with 27dBHz and is decreased by 2dB every 1s for 5s, with $\boldsymbol{R}$ changing accordingly. So the CNR is 27dBHz, 25dBHz, 23dBHz, 21dBHz, 19dBHz. It can be seen in Fig. 7 that as long as the loop converges at 27dBHz, when CNR decreases further, loop can still keep locked.

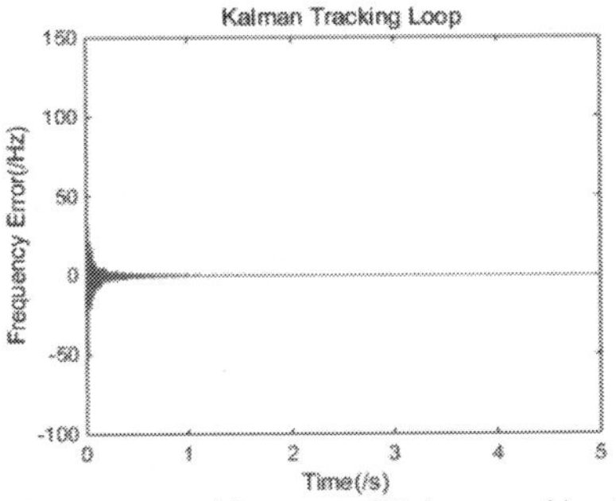


**Figure 7.** Frequency tracking error of Kalman tracking loop running for 5s with CNRdecreasing 2dB every 1s from 27dBHz.

## 5 Optimization of Kalman filter

When CNR is under 27dBHz, the loop is likely to be locked at a steady error not equals **0**. The reason is that theMSE matrix $\boldsymbol{P}$ over converges, making the values in the second column of the Kalman gain matrix $\boldsymbol{K}$ too small [8]. So the frequency error can't be reflected in the updated state vector. This phenomenon is reflected in Fig. 8 with CNR=23dBHz.In fact, according to other engineering experiences, Kalman filter can decrease the error by 70%~80%. Athreshold can be set for $\boldsymbol{P}$ to keep its value large enough so that the values in the second column of the Kalman gain matrix $\boldsymbol{K}$ are large enough. This actually sacrifices tracking precision for the correct lock. The converging speed will also slow down.If the threshold is set too large, the tracking precision will be poor and thefiltering effect is lost. If the threshold is set too small, false lock will still appear. Considering the engineering experiences, set:

$$P_{2,2\,Threshold} = \frac{P_0(2,2)}{5^2}$$

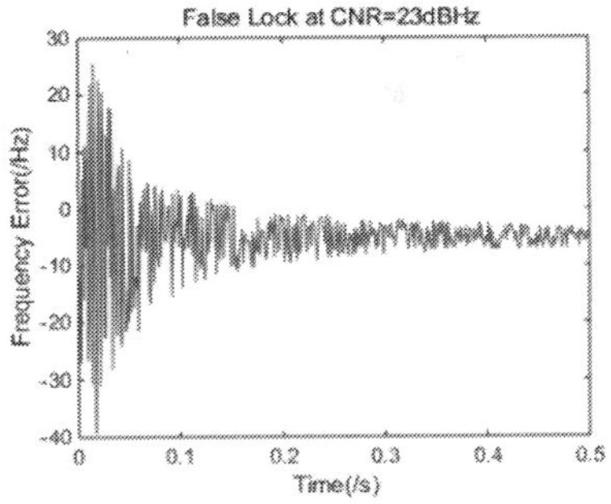


**Figure 8.** Frequency tracking error of Kalman tracking loop running for 500ms with CNR=23dBHz.

That means the maximum convergence is that the frequency MSE$\boldsymbol{P}(2,2)$ become 1/25 of its initial value. Fig. 9 shows the effect of this optimization with CNR=23dBHz and time=5s.It can be seemed more obviously with the frequency error of the updated Kalman state vector at 23dBHz in Fig. 10.

## 6 Conclusions

The articledescribed in details about the application of Kalman filter in tracking navigation signals. Simulation results have shown that the tracking loop using Kalman filter exceeds traditional tracking loop by 6dB in sensitivity. For ultra-weak signal tracking, which is under 27dBHz, an optimization method is introduced to sacrifice tracking precision for correct lock. Future research should pay attention to make Kalman tracking loop be suitable for more complex circumstances.

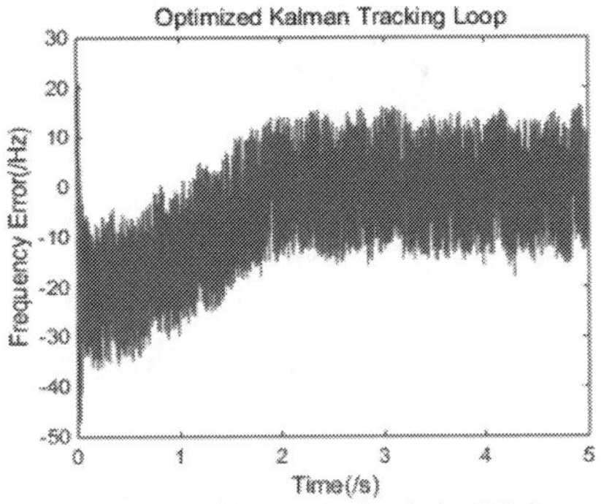


**Figure 9.** Frequency tracking error of optimized Kalman tracking loop running for 5s with CNR=23dBHz.

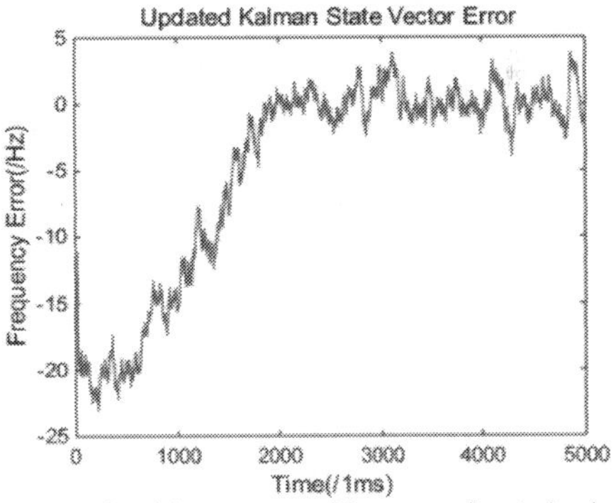


**Figure 10.** Updated frequency tracking error of optimized Kalman tracking loop running for 5s with CNR=23dBHz.

## References

1. G. Xie, *Principles of GPS and Receiver Design*, pp. 380-381 (2011)
2. N. Ziedan, *GNSS Receivers for Weak Signals*, pp. 129-136 (2006)
3. A.D. Men, F. Su, L. Wang, H.Y. Wang, J.J. Li, *Digital Signal Processing, 2nd ed.*, 2009, pp. 124-127.
4. T.T. Ren, *A Standalone Approach for High-Sensitivity GNSS Receivers*, A thesis for the

PhD, Dept. GeomaticsEngineering, Calgary,Canada, 2014.
5. E. Kaplan, *Understanding GPS: Principles and Applications, 2nd ed.*, pp. 135-141 (2006)
6. B. Parkinson, J. Spilker, P. Axelrad, P. Enge, *Global Positioning System: Theory and Applications*, American Institute of Aeronautics and Astronautics, 1996.
7. R. Kalman, *Transactions of the ASME - Journal of Basic Engineering*, **82**, pp. 35-45 (1960)
8. R. J. Fitzgerald, *IEEE Transactions on Automatic Control*, **AC-16**, pp. 736-743 (1971)

# Image Haze Removal Using Dark Channel Prior and Inverse Image

Lei Shi [1], Xiao Cui [1], Li Yang[1], Zhigang Gai [1], Shibo Chu[1] and Jing Shi [2]

[1]*Institute of Oceanographic Instrumentation, Shandong Academy of Sciences, Qingdao, China*
[2]*Luoyang Institute of electro-optical devices, China Aviation Industry Corporation, Luoyang, China*

**Abstract.** Image haze removal using dark channel prior is prone to encountering color distortion in sky and brightness region. To solve the problem, we proposed an improved method based on inverse image and dark channel prior. Firstly, we applied inverse image to estimating a new transmission map. The new transmission map can be used to modify original transmission map in order to avoid color distortion. Next, fast guider filter was applied to refining transmission map. Using the transmission map, we can recover a high quality haze-free image. Experimental results showed that the proposed method is feasible, and visibility can be enhanced.

## 1 Introduction

The contrast and visibility of the haze images are all decreased, which directly affects the air, water and road transport. Thus, simple and effective image haze removal algorithm is of great value to improve the robustness of the visual system.

In order to remove haze effect, many methods required additional information about the scene, such as multiple images captured in different weather conditions [1] or polarization angles [2]. Also, Kopf et al. [3] employed the depth knowledge of known 3D models for haze removal.

Recently, single image haze removal has made great progress. Many single image haze removal algorithms have been developed. Tan et al. [4] maximized the contrast of a hazy image, assuming that the haze-free image has a higher contrast than the hazy image. The results are visually compelling but may not be physically valid. Fattal et al. [5] decomposed the scene radiance of an image into the albedo and the shading, and then estimated the scene radiance, assuming that the shading and the object depth are locally uncorrelated. It can remove haze locally but cannot restore densely hazy images. Kim et al. [6] proposed a cost function produced by the contrast cost and the information loss cost. By optimizing the cost function, the proposed algorithm can enhance the contrast and preserve the information optimally.

He et al. [7] proposed a single image haze removal based on Dark Channel Prior (DCP). He randomly selected about 5000 images for DCP. The algorithm produced impressive result; however, the method may suffer from a limitation. It is that color distortion will appear in bright region (such as sky region). In this paper, we proposed an improved method based on [7] to deal with this case.

## 2 Single image haze removal using dark channel prior

The observed brightness of a hazy image can be modeled based on the atmospheric optics [7] via

$$I(x)=J(x)t(x)+A(1-t(x)) \qquad (1)$$

where, $I(x)$ is the observed haze image, $J(x)$ is the scene radiance(Haze-free Image), $A$ is the global atmospheric light, and $t(x) \in [0,1]$ is the medium transmission describing the portion of the light that is not scattered and reaches the camera.

He et al. [7] found that the minimum intensity in the non-sky patches on haze-free outdoor images should have a very low value, which is called dark channel prior(DCP). His observation told us that except for the sky region, the intensity of dark channel image $J_{dark}$ is low and tends to be zero. According to the DCP, the transmission can be estimated simply by:

$$t_f(x)=1-\min_c(\min_{y\in\Omega(x)}(\frac{I_c(y)}{A^c})) \qquad (2)$$

A constant parameter $\omega(0<\omega<1)$ was introduced into Equation (2) to keep a small amount of haze for the distant objects:

$$t_f(x)=1-\omega\min_c(\min_{y\in\Omega(x)}(\frac{I_c(y)}{A^c})) \qquad (3)$$

In the next step, He et al. [7] refined the transmission map using soft matting method [8]. For reducing the computational complexity, soon he applied guided image filter [9] to refine the transmission map. After the transmission map is estimated, the scene radiance $J(x)$ can be recovered according to Equation (1). The transmission $t_f(x)$ is restricted by a low bound to, which means that a small amount of haze is preserved in very dense haze regions. *to* is set as 0.1. The final scene radiance $J(x)$ is recovered by

$$J(x)=(I(x)-A)/t(x)+A, \quad t_f(x)=max(t_f(x),to) \tag{4}$$

In [7], He analyzed about 5000 out-door images. Then, He told us that about 75% of the pixels in the dark channels have zero values, and the intensities of 90% of the pixels are below 25. This statistic gave very strong support to DCP. The algorithm can perform very well in most cases. Most pixels satisfied DCP. But, for bright pixels such as pixels in sky region, the method can not perform very well because these pixels cannot satisfy DCP. We presented an improved method to solve the problem.

## 3 Image haze removal based on dark channel prior and inverse Image

If there are large-area sky region and brightness object in out-door image, color distortion may appear in these regions when He's method [7] is applied, as described in Figure 2 under section 4. According to our observations, these regions will be brighter than others in most cases, and these differences between RGB channels values are very small. In general, the pixel values of RGB channels may be close to atmospheric light value $A$. DCP is not established in these regions. According to Equation (1), we obtained:

$$t_{actual}(x)=\frac{1-\min\limits_{c}(\min\limits_{y\in\Omega(x)}(I_c(y)/A))}{1-\min\limits_{c}(\min\limits_{y\in\Omega(x)}(J^c(y)/A))} \tag{5}$$

where, $t_{actual}(x)$ donates actual transmission value.

According to Equation (5), if $Ic(y)$ value is close to atmospheric light value $A$, the term $\min\limits_{c}(\min\limits_{y\in\Omega(x)}(J^c(y)/A))$ is not equal to 0.0 but close to 1.0, and $1-\min\limits_{c}(\min\limits_{y\in\Omega(x)}(J^c(y)/A))$ is close to 0.0. Therefore, $t_{actual}(x)$ is far above $t_f(x)$ based on DCP. The calculated transmission $t_f(x)$ value by Equation (3) is very small, and $J(x)$ value by Equation (4) is not accurate.

If $to$=0.1 is chosen, max($t_f(x)$, $to$) may be also very small. If $I_c(x)$ value is small to a certain extent, the absolute value of $(I(x)-A)$/max($tf(x)$, $to$) will probably be amplified to a considerable value. At this time, $J(x)$ value will be very small in this channel. If other two RGB channel values are bigger than $A$, color distortion will appear.

Jiang et al. [11] added a controllable parameter $K$ in in order to adjust the transmission value of each pixel .When $|I(x)-A|$ is smaller than $K$, the pixel may belong to brightness region(such as sky). At the time, He will multiply the transmission value value $tf(x)$ by max($K/|I(x)-A|$, 1). The value of $K$ is suggested to be 50 by Jiang. When $|I(x)-A|$ is bigger than $K$, the transmission value is max($t_f(x)$, $to$). Sun et al. [12] adjusted the transmission values based on $|I(x)-A|$. Wang et al. [13] applied sky segmentation based on gradient threshold, then set the transmission values in sky region directly into a fixed value. Deng et al. [14] separated firstly sky region from background image, then applied histogram equalization. Kim et al. [6] enhanced visibility and contrast by minimizing energy function. *to* was equal to 0.3. All of the above algorithms can eliminate color distortion in some cases. These algorithms both depend on the experience value. But, there are also some situation is that they can't deal with it. Besides that, if we have to separating sky region from background image, this will result in burdensome calculation.

In [15], Ancuti et al. applied semi-inverse image to image haze removal. The semi-inverse image can be obtained by replacing the RGB values of each pixel on a per channel basis by the maximum of the initial channel value and its inverse. In many cases, the methods can do it well. In [6], Kim built cost function by two constraints, as described in Equation (6) and Equation (7):

$$t_f(x)\geq\min_{c}(\min_{y\in I(x)}(1-\frac{I_c(y)}{A})) \tag{6}$$

$$t_f(x)\geq\max_{c}(\max_{y\in I(x)}(1-\frac{255-I_c(y)}{255-A})) \tag{7}$$

Equation (7) denoted inverse image can be used to haze removal. Kim's cost function included two portions. Besides Equation (6) and Equation (7), he also added the information loss cost. In [16], Tang et al. proposed similar constraints to [15].

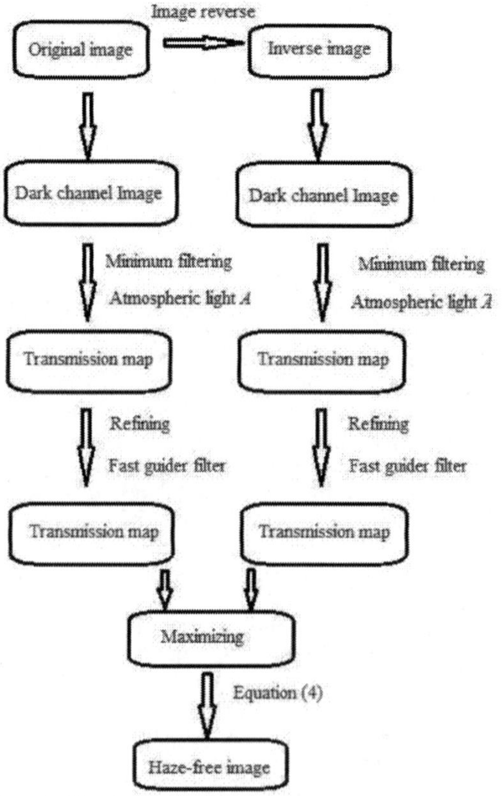


**Figure 1.** The flow chart.

The three works [6], [15]-[16] told us that inverse image is very useful for haze removal. But, only it can not deal with all cases. We presented an improved method to overcoming color distortion using inverse image. The algorithm can be described as follows:

1) Calculated inverse image of original image, then calculated their dark channel images, and applied minimum filtering.

2) Calculated global atmospheric light value respectively.

3) Calculated initial transmission map respectively.

4) In order to reduce the complexity of calculation, we refined the two transmission map by fast guide filter [10] respectively.

5) Determined the maximum between the two refined transmission maps.

6) Recovered free-haze image by Equation (4).

The flow chart is described as in Figure 1.

## 4 Experiment and analysis

All of these pictures in this paper are from the internet. In experiments, we applied minimum filtering, and the patch size was set to 9×9. In Equation (3), $\omega$ was equal to 0.95. The parameters were $r = 40$, $\varepsilon = 0.001$ for the fast guided filter [10]. In order to compare these algorithms, we didn't enhance these results after processing. The program was written in C++ using Microsoft Visual 2008. We installed OpenCV 2.4.1 Library for programming. Our CPU was Intel(R) Core(TM) i5-3450 CPU @3.10GHz. Computer memory was 8G. Operating system was Windows 7.

**Figure 2.** Comparison with He's work [7]. From left to right: Input image, He's results, our results

We selected the four pictures. All pictures both included large-area sky region. In experiments, we compared our method with He's method [7], [10]. These results can be shown in Figure 2 and Table 1. According to Figure 2, we observed that these samples encountered color distortion when there is large-area sky region and utilizing He's method [7], [10]. Only our proposed method can deal with color distortion in all cases.

**Table 1.** Consuming time.

| Sample Number | He's method by Guider filter(ms) | Our method by Guider filter(ms) | Our method by Fast Guider filter(ms) | Picture size (pixels) |
|---|---|---|---|---|
| 1 | 733 | 1451 | 406 | 690×517 |
| 2 | 172 | 328 | 93 | 320×240 |
| 3 | 359 | 687 | 187 | 500×333 |
| 4 | 515 | 967 | 266 | 585×413 |

In Table 1, we gave the time consumption comparison with He et al. [7], [10]. As can be seen, our approach will spend more time if the transmission map was refined by guider filtering [9]; but, aided by fast guider filtering [10], our approach is much faster than others and may achieve the real-time requirement.

## 5 Conclusion

In the paper, we presented an improved method based on He's work [7], [10] to deal with color distortion. The method firstly determined the inverse image and corresponding dark channel images, and then applied minimum filtering. Next, we calculated the transmission maps and refined it by the fast guider filter [10]. Then, we modified original transmission map based on it. The final transmission map can be used to obtain haze-free image. In final, we can deal with color distortion and enhance the visibility. In future work, we will extend our work to real-time video dehazing.

## Acknowledgment

We thank He Kaiming for generating the results of his method [7], [10]. The authors were supported by the Shinan District Science and technology development fund of Qingdao City under Grant No. 2014-14-013-ZH and Qingdao pioneering and innovative leading talent project under Grant No. 13-CX-23.

## References

1. S. G. Narasimhan, S. K. Nayar, Contrast restoration of weather degraded images. IEEE Transactions on pattern Analysis and Machine Intelligence (PAMI). **25**:713–724 (2003)
2. Y. Y. Schechner, S. G. Narasimhan, and S. K. Nayar, Instant dehazing of images using polarization. Processing of 2006 IEEE Conference on Computer Vision and Pattern Recognition CVPR. 1:325 (2006)
3. Kopf, J., Neubert, B., Chen, B., Cohen, M., Cohen-Or, D. et al., Deep photo: Model-based photograph enhancement and viewing. Processing of the 1st ACM SIGGRAPH Conference and Exhibition on Computer Graphics and Interactive Techniques in Asia. **27**(5):110.1-110.6 (2008)

4. Tan R T., Visibility in bad weather from a single image. Processing of 2008 IEEE Conference on Computer Vision and Pattern Recognition (CVPR). **1**:1-8 (2008)
5. Fattal R, Single image dehazing. Processing of the 1st ACM SIGGRAPH Conference and Exhibition on Computer Graphics and Interactive Techniques in Asia. **27**(3):1-9 (2008)
6. Jin-Hwan Kim, Won-Dong Jang, Jae-Young Sim., O Chang-Su Kim: Optimized contrast enhancement for real-time image and video dehazing. Journal of Visual Communications and Image Representation. **24**(3): 410-426 (2013)
7. He Kaiming, Sun Jian, Tang Xiaoou, Single image haze removal using dark channel prior. Processing of 2009 IEEE Conference on Computer Vision and Pattern Recognition (CVPR). 1956-1963 (2009)
8. Levin A, Lischinski D, Weiss Y, A closed form solution to natural image matting. Processing of 2006 IEEE Conference on Computer Vision and Pattern Recognition (CVPR). **1**: 61-68 (2006)
9. He Kaiming, Sun Jian, Tang Xiaoou, Guided Image Filtering. IEEE Transactions on Pattern Analysis and Machine Intelligence. **35**(6):1397-1409 (2013)
10. He Kaiming, Sun Jian, Fast Guided Filter. http://arxiv.org/abs/1505.00996. (2015)
11. Jiang Jianguo, Hou Tianfeng, Qi Meibin, Improved algorithm on image haze removal using dark channel prior. Journal of Circuits and Systems. **16**(2):7-12 (2011)
12. Sun Xiaoming, Sun Junxi, Zhao Lirong, Cao Yonggang, Improved algorithm for single image haze removing using dark channel prior. Journal of Image and Graphics, **19**(3):381-385 (2014)
13. Wang Gangyi, Ren Guanghui, Jiang Lihui, Quan Taifan, Single image dehazing Algorithms based on sky region segmentation, Information Technology Journal. **12**( 6):1168-1175 (2013)
14. Deng Yue, Wang Yanjie, Li Jing-yu, Chen Xiaoxi, Zhang Ye, Improvement of enhancement algorithm for aerial image. Laser & Infrared. **42**(9):1080-1085 (2012)
15. Ancuti C O, Ancuti C, Hermans C, et al, A Fast Semi inverse Approach to Detect and Remove the Haze from a Single Image. 10th Asian Conference on Computer Vision (ACCV 2010). Part II. LNCS, **6493**, pp. 501–514 (2014)
16. Tang Ketan, Yang Jianchao, Wang Jue, Investigating Haze-relevant Features in A Learning Framework for Image Dehazing. IEEE Conference on Computer Vision and Pattern Recognition (CVPR). 2995–3002 (2014)

# Estimation for torques applied to the master side in a construction robot teleoperation system

Lingtao Huang , Yanan Li, Tao Ni and Haidong Huang

*Jilin University, School of Mechanical and Engineering, 130022 ChangChun, Jilin, China*

**Abstract.** The purpose of this research is to develop a method that measures a torque which an operator applies to the joystick without using a torque/force sensor on a joystick. We dealt with a construction robot teleoperation system which is comprised of two joysticks as the master, and an excavator with four degrees of freedom consisting of fork glove, swing, boom, and arm as the slave. It is necessary to know the torque (force) exerted on the master side in force control. Because the joystick does not have a force/torque sensor in our lab, it is not possible to directly obtain the torque. To give prominence to the simple and practical construction robot teleoperation system in our lab, we proposed a method that measures the torque exerted on the joystick to existing equipment. Its effectiveness was verified by a reaction torque experiment.

## 1 Introduction

Teleoperation systems are employed to sense and operate objects at a distance. They are composed of a slave system and a master system. The slave side is composed of sensors to perceive remote environment, actuators to interact with it, and network channels to communicate with the master side [1]. In the master side, an operator commands the motion of the remote slave robot via a local master joystick [2]. The teleoperation system where the master side operator directly controls the remote robot according to information reflected from the remote environment is more effective and reliable [3].

We can have different classifications to joysticks which are applied to teleoperation systems according to different standards. The most common joysticks are movement-sensing joysticks because they respond to the tilt applied to the joystick's post [4]. On the other hand, isometric joysticks, whose post are rigid and do not deflect, respond to the forces applied to their transducers and theoretically require less strength and dexterity for transduction [5]. Since the mechanical abrasion is inevitable, the noncontact electronic joystick based on magnetic technology is much more precise and reliable than the mechanical joystick [6-8]. A lot of joysticks with 2-DOF [9] or 3-DOF [10] are implied in industrial applications. In teleoperation systems, there are a variety of feedback signals that flow from the remote environment to the master side [1], such as visual feedback [11] and force feedback [3].

In order to provide force feedback for the operator, an AC servo motor is used. And in order to measure the force which the operator applies to the joystick, a force sensor is attached to the rotational axis of each AC servo motor [12]. Recently, joysticks are commonly employed in many construction and industrial applications, such as cranes [13], excavators [14], and power wheelchairs [5]. It is necessary to know a torque (force) which an operator applies to a joystick in force control method. A common method is to install a force/torque sensor on the joystick.

In our lab, two joysticks, which are Microsoft Co., Ltd. manufactured and named as Side Winder Feedback 2, are introduced in our master-slave system. Because the joystick does not have a force/torque sensor, it is not possible to directly get the information of the torque applied to the joystick. To give prominence to the simple and practical construction robot teleoperation system in our lab, a method is proposed to measure the torque which an operator applies to the joystick to existing equipment in this research.

## 2 Master side - joystick

**Figure 1.** Joystick in a construction robot teleoperation system.

A typical teleoperation system consists of a master and a slave side. In our research, the master–slave system is composed of two joysticks, which act as the master, and a construction robot (a commercially available excavator modified based on Hitachi LandyKID-EX5) is the slave which has four links: fork glove, arm, boom and swing.

The joysticks enable movements forward/reverse and right/left, as in Figure 1. Its displacements are detected by position sensors. Each joystick is composed of two direct current motors that enable the operator to feel the sense of grasping an object by the fork glove and the work reaction torque (force sense) generated by the construction robot.

## 3 Torque analysis and calculation

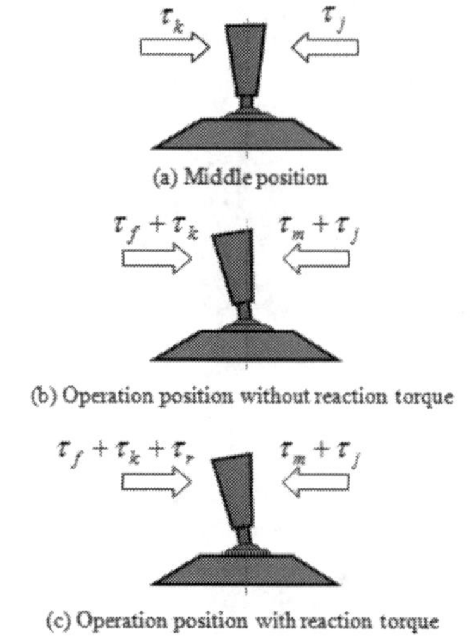


**Figure 2.** Joystick position and torques in horizontal placement.

Figure 2 shows the postures of the joystick and torques condition when the operator controls it in different operation position. In Figure 2, (a) shows the case that the joystick automatically places in middle position (zero position) when an operator does not manipulate it. In this case, a torque caused by handle's spring and damp effect $\tau_k$ overcomes a torque caused by handle's gravity $\tau_j$. (b) expresses the case that the joystick places in operation position without reaction torque generated by the construction robot $\tau_r$. The operator applies an operation torque $\tau_m$ to the joystick, and $\tau_f$ is a torque caused by friction. (c) means that the joystick places in operation position with $\tau_r$, and there is a equilibrium equation for the torques $\tau_f + \tau_k + +\tau_r = \tau_m + \tau_j$. In this research, a non-dimensional quantity of torque 1 means 0.57 Nm.

To calculate the torque that the operator applied to the joystick, three problems should be examined and measured. First, the relationship between output torques measured by a torque meter and command torques to the joystick is examined. Second, the influence of handle's gravity is investigated and its mass is estimated. Third, $\tau_f$ is measured when the joystick is controlled and moved by the operator.

### 3.1. Command torques and output torques

To avoid the influence of handle's gravity, the joystick is placed vertically. Command torques is set to increase 5% per 10 seconds. Output torques is measured by a torque meter. Torque meter is exerted on the center place of the joystick where the operator manipulates the handle. Experiment data is shown as Figure 3. The horizontal axis is the command torque, and vertical axis is the output torque. Due to the effect of friction, the output signal is zero though the command torques act on the joystick in range of $\pm 0.1$. We can see that the relationship between the command signals and output signals is described as linear except ain the initial stage of input signals (near 0.1), and the fitting line is expressed by

$$y = 0.9882x - 0.0076 \tag{1}$$

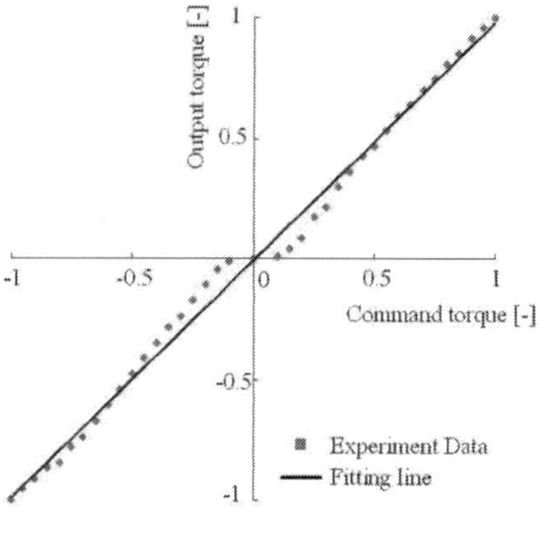


**Figure 3.** Command torques and output torques.

### 3.2. Gravity Compensation

As shown in Figure 2(b), there is the influence of handle's gravity when the joystick does not place in middle position. To calculate and measure the gravity of handle, experiment which measures the relationship between reaction torque to joystick and joystick displacement, is conducted when the joystick is placed horizontally and vertically, respectively.

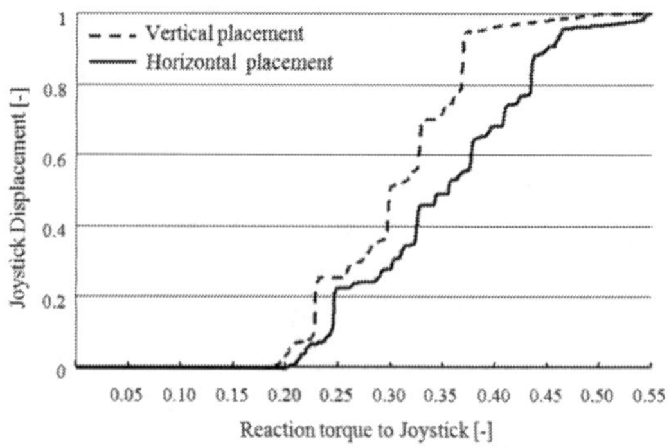


**Figure 4.** Reaction torque to joystick and joystick displacement.

When a control computer connected with the joystick send a command signal (reaction torque signal) to the joystick, the torque signal slowly and continuously increases 5% per 10 s. The relationship between torque to the joystick and the joystick displacement is shown as Figure 4. The horizontal axis expresses the torque to joystick, and the vertical axis shows the joystick displacement, which is non-dimensional quantity and 1 denotes 0.06 m. The dash line and the black line express the experiment data when the joystick was placed in vertical and horizontal placement, respectively. Because of the influence of the handle's gravity, a slightly significant discrepancy is observed between the red line and black line.

From Figure 2(b), it is known that the torque caused by handle's gravity can be expressed as follows:

$$\tau_j = mg \times \sin(\theta) \times L \tag{2}$$

Where $m$ is the mass of handle, and $L$ is the length between the center of rotation and a point that the operator applies a force to the handle. $\theta$ means the angle between the handle and vertical direction.

$\tau_j$ can be calculated in term of the data in Figure 4. It is known that there are two torques while the joystick places in a displacement, and the difference between the torques is considered as the action of $\tau_j$. So if angle $\theta$ is known, the handle's gravity can be calculated by fitting. Before calculating the gravity, it is necessary to find out the relationship between the joystick displacement and its angle.

### 3.3. Relationship between joystick displacement and angle

Experiment was conducted with the condition that the joystick was placed in vertical, and an angle senor was attached on the handle. Command torques are sent to the joystick and the joystick is moved slowly in experiment. The joystick displacement and angle are shown as Figure 5. The horizontal axis expresses the joystick displacement, and the vertical axis shows the angle of the handle, whose maximal absolute value is 16 degree. The red dots mean experiment data. In term of the experiment data, a fitting curve consists of three lines, which is shown as black lines and the mathematic expression of the curve is shown as follows:

$$y = \begin{cases} 19.7484x + 2.71992 & x \leq -0.34 \\ 10.241x - 0.12867 & |x| < 0.34 \\ 19.2728x - 3.09612 & x \geq 0.34 \end{cases} \tag{3}$$

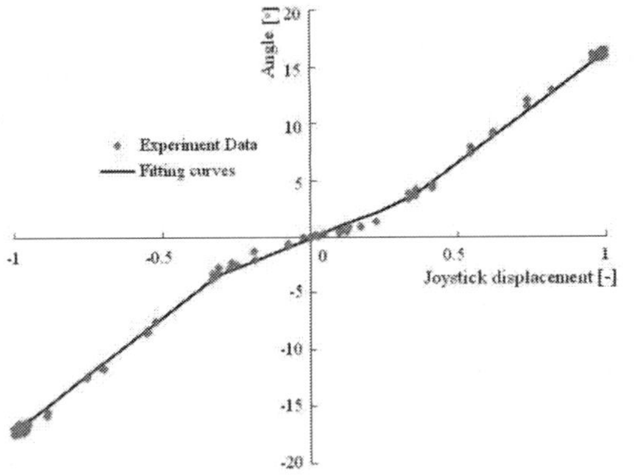


**Figure 5.** Joystick displacement and angles of the handle.

### 3.4. Torque Caused by the gravity of handle

Figure 6 shows the experiment data (black dots) and a fitting line, which are expressed as equation (4). The horizontal axis is the sin($\theta$ ), and the vertical axis shows the torques. In term of the equation (1), since the parameters ($mgL$) are constant values, there should be a proportional relationship between sin($\theta$ ) and torque. Since the influence of friction and measurement error, it is considered that 0.0071 Nm is a part of $\tau_f$, and then the value of $mgL$ is 0.1565 Nm.

$$y = 0.1565x + 0.0071 \tag{4}$$

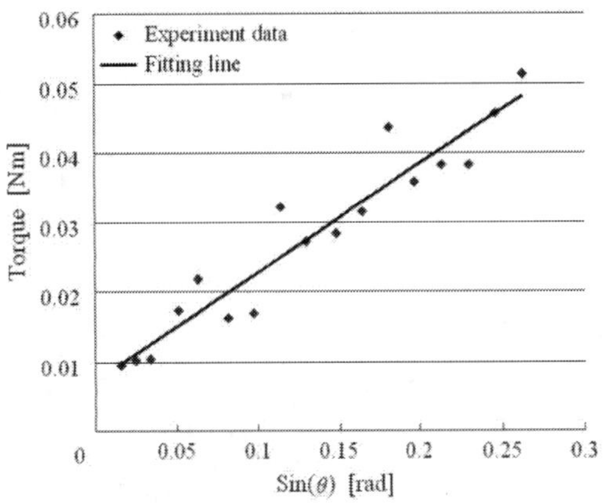


**Figure 6.** Estimation for $mgL$.

### 3.5. Relationship between joystick displacement and torque to joystick

A torque meter is used to slowly push the handle of the joystick to simulate the case that an operator grasps the joystick and moves it. Experiment is conducted under the joystick in horizontal placement on a table. Experiment data are collected when the joystick is moved from zero position to maximal position by the torque meter.

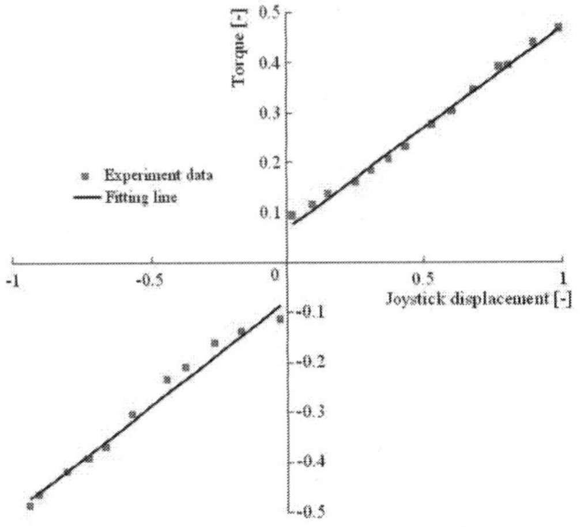


**Figure 7.** Relationship between torque to joystick and joystick displacement.

Red dots in Figure 7 show the data recorded by the torque meter, which are just like the torques applied to the joystick by an operator. $\tau_j$ can be calculated in term of equation (2), then the value $\tau_f + \tau_k$ equals to $\tau_m + \tau_j$ in term of Figure 2(b), which is shown as red dots in Figure 7. By fitting, equation (5) can be obtained and shown as follows:

$$y = \begin{cases} 0.4010x + 0.0681 & x > 0 \\ 0.4202x - 0.0757 & x < 0 \end{cases}. \quad (5)$$

In general, $\tau_r$ which is the reaction torque generated by the construction robot, is changed to an actual torque by equation (1). The torque $\tau_j$ caused by handle's gravity, is calculated by equation (2) and (3) in term of joystick displacement. The sum of $\tau_f$ and $\tau_k$ can be obtained by equation (5). Then torques exerted on the joystick by an operator $\tau_m$ is calculated by the equation $\tau_f + \tau_k + +\tau_r = \tau_m + \tau_j$.

## 5 Experiment

Reaction torque experiment was conducted as follows: firstly, $\tau_r$ was set to a constant reaction torque – 0.5, the torque meter pushed the joystick from zero position to maximal position and collected experiment data, which are shown as "Fixed torque" in Figure 8. The red line is the result in term of equations (1), (2) and (5), blue dots are experiment data collected by torque meter. Secondly, $\tau_r$ was set to be increased from 0 to 1 and changed 5% per 10 s, and the joystick was pushed to 0.528391 position by the torque meter. Experiment data are "Fixed position" in Figure 8, the black line is the fitting result according to equations (1), (2) and (5), pink dots are the data collected by the torque meter. From two ground data, we can see that the fitting result almost equals to the measured result. It proofs that it is effective to calculate torques applied to the joystick by our proposed method.

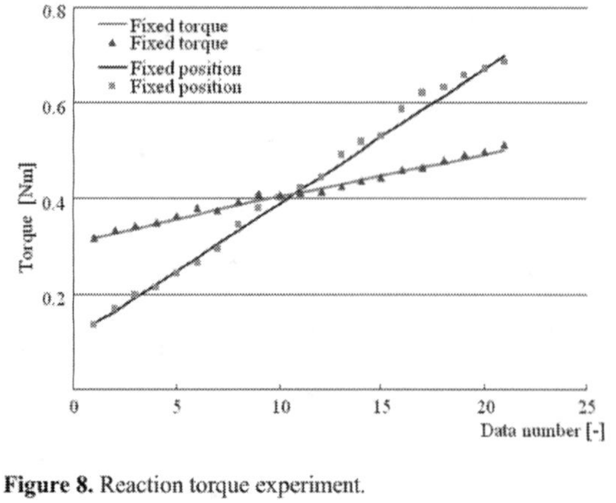


**Figure 8.** Reaction torque experiment.

## 6 Conclusion

The torque applied to a joystick by an operator must be known, when force control is introduced in a construction robot teleoperation system. In this research, a method was proposed by the authors, which measures a torque which an operator applies to the joystick without using a torque/force sensor on the joystick. In this research, we firstly examined the relationship between command torques controlled by a computer and output torques measured by a torque meter, then estimated the gravity of the handle by fitting, next, investigated and calculated relationship between $\tau_k$ and joystick displacement. Finally, reaction torque experiment proved the feasibility and validity of using the proposed control method in the construction robot teleoperation system.

## Acknowledgment

The authors would like to thank the financial support from the National Natural Science Foundation of China (nos. 51575219 and nos. 51305153), Key Scientific Research Projects of Jilin Province (nos. 20130206034GX).

## References

1. C. Pacchierotti, L. Meli, F. Chinello, M. Malvezzi, D. Prattichizzo. Int J Robot Res. 34, 1773-1787 (2015)
2. G. Niemeyer, J. J. E. Slotine. Int J Robot Res. 23, 873-890 (2004)
3. S. K. Cho, H. Z. Jin, J. M. Lee, B. Yao. Ieee-Asme T Mech. 15, 17-26 (2010)
4. B. E. Dicianno, H. Mahajan, R. A. Cooper. Pm&R. 7, 555-561 (2015)
5. H. Mahajan, D. M. Spaeth, B. E. Dicianno, D. M. Collins, M. L. Boninger, R. A. Cooper. Arch Phys Med Rehab. 92, 1298-1304 (2011)
6. H. Jin, H. Lu, S. Cho, J. Lee. IEEE-ASME T Mech. 12, 549-556 (2007)
7. M. Kayal, F. Burger, R. S. Popovic. IEEE Sens J. 4, 759-763 (2004)
8. R. A. Cooper, D. K. Jones, S. Fitzgerald, M. L. Boninger, S. J. Albright. IEEE T Bio-Med Eng. 47, 902-910 (2000)
9. W. H. Li, B. Liu, P. B. Kosasih, X. Z. Zhang. Sensor Actuat a-Phys. 137, 308-320 (2007)
10. H. Zangl, T. Bretterklieber, G. Steiner, K. Riedmuller. *2007 IEEE Sensor*. 1069-1072 ( Atlanta, GA, 2007)
11. L. Zhang, Q. Huang, Y. P. Lu, T. Xiao,J. P. Yang, M. Keerio. *2006 IEEE/RSJ International Conference on Intelligent Robots and Systems*. 1110-1115 (Beijing, CHINA, 2006)
12. A. Takemoto, K. Yano, T. Miyoshi, K. Terashima. *13th IEEE International Workshop on Robot and Human Interactive Communication*. 533-538 (Kurashiki, JAPAN, 2004)
13. M. A. Huysmans, M. P. De Looze, M. J. M. Hoozemans, A. J. Van der Beek, J. H. Van Dieen. Ergonomics. 49, 1021-1035 (2006)
14. Y. J. Nam, M. K. Park. J Mech Sci Technol. 29, 397-407 (2015)

# A Quarter Active Suspension System Based Ground-Hook Controller

Arjon Turnip

*Technical Implementation Unit for Instrumentation Development, Indonesian Institute of Sciences, Bandung, Indonesia*

**Abstract.** An alternative design technique for active suspension system of vehicle using a developved ground-hook damping system as a reference is proposed. The controller parameters are determined using Lyapunov method and can be tuned to precisely achieve the type of desired response which given by reference model. The simulation result show that the designed active suspension system based ground-hook reference model is able to significantly improve the ride comfort and the road holding compared with semi-active suspension.

## 1 Introduction

In recent years there has been a significant growth in the design of suspension systems by which should reduce vibration due to the external disturbances to the maximum possible degree [1-16]. Two criteria of good vehicle suspension performance are typically their ability to provide good road handling and increased passenger comfort. It is difficult to meet both requirements, since this involves opposing tendencies in the selection of the main parameters of suspension, i.e. stiffness and damping. Suspension should be rigid enough to effectively carry the static load, and soft enough to ensure the good isolation of vibrations. These two factors are related to the body vertical acceleration and the suspension deflection. If the suspension system is unable to absorb the vibration caused by the unevenness of the road surface, a significant deflection of the suspension will occur resulting unsafe condition of the vehicle. Moreover, the sprung mass of the vehicle will also exhibit a large vertical acceleration which can be the reason of a ride discomfort. In order to achieve a good ride quality, the performance of the suspension system plays an important role.

Since most of the vehicle suspensions do not operate with a single well defined roadway input, real suspensions are rarely optimal for any particular road and speed. On smooth roads the suspension will be stiffer than necessary and on rough roads softer that desirable. This leads to the concept of an adaptable suspension which can change its characteristics bases on measured performance. By using an adaptation algorithm, it is expected that the optimal performance can be maintained over the wide range of input conditions typically encountered by a vehicle. In this paper, an alternative design technique for active suspension system of vehicle using adaptive controller based skyhook damper model is proposed. The reference model and the controller systems are designed and determined using skyhook model and Lyapunov method, respectively. The control design was applied to two different suspension models, namely the body vertical acceleration and the suspension deflection models, each representing comfort and safety factors.

## 2 Dynamic of suspension system

### 2.1 A Quarter model of suspension system

In most cases, dynamic systems are stimulated by the movement of anchorages. The suspension system of car is stimulated as a result of moving on the ruggedness of the road. A car is moving on a bumpy road in which some forces are exerted on the passengers as the result of the change of the length of springs and dampers. The schematic diagram of a quarter-vehicle suspension system under investigation can be presented as in Fig. 1, where ms and mu each represents the sprung mass and unsprung mass respectively, $c_a$ represents absorber coefficient, $k_a$ represents spring constants, $k_t$ represents the tire equivalent constants, $z_r$ represents surface roughness of road, $z_s$ represents vehicle body position and $z_u$ represent the tire vehicle position. The variable Q represents an applied actuator force for the active suspension. The values of the parameters in this model, which are taken from a typical passenger car, are collected in Table 1. However, the value of $k_a$ is unknown therefore it can be assumed that there is no limitation in generating the actuator force $Q$. Using Newton's second law, the equations of motion for the model in Fig. 1 are given by

$$m_s\ddot{z}_s = -k_a(z_s) - z_u - Q \qquad (1)$$

$$m_u\ddot{z}_u = k_a(z_s - z_u) - k_t(z_u - z_r) + Q\,. \qquad (2)$$

By defining $x_1 = z_s$, $x_2 = \dot{z}_s$, $x_3 = z_s - z_u$, and $x_4 = \dot{z}_s - \dot{z}_u$, then the equations (1) and (2) can be expressed in the state space form as follows:

$$\dot{X} = AX + BQ + Cz_r \tag{3}$$

where, the matrices A, B, and C are given by.

$$A = \begin{bmatrix} 0 & 1 & 0 & 0 \\ 0 & 0 & \frac{k_a}{m_s} & 0 \\ 0 & 0 & 0 & 1 \\ \frac{k_t}{m_u} & 0 & -\left(\frac{k_a}{m_s} + \frac{k_a + k_t}{m_u}\right) & 0 \end{bmatrix},$$

$$B = [0 \quad -1/m_s \quad 0 \quad -(1/m_s + 1/m_u)]^T,$$

$$C = [0 \quad 0 \quad 0 \quad -k_t/m_u]^T.$$

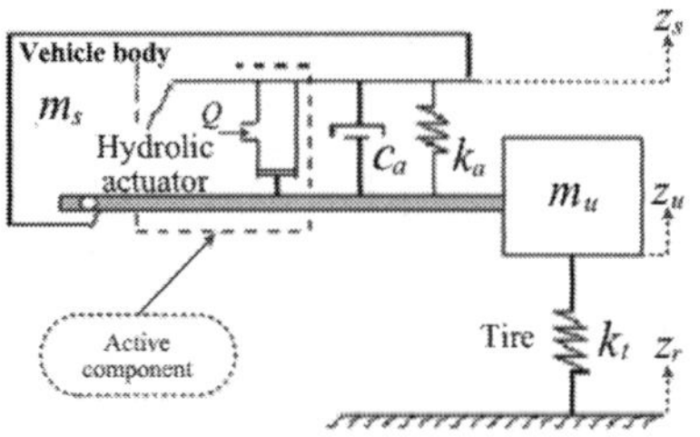


**Figure 1.** Schematic diagram of a quarter-vehicle model.

**Table 1.** System parameters.

| Parameter | Value |
|---|---|
| Sprung mass ( $m_s$ ) | 412.5 kg |
| Unsprung mass ( $m_u$ ) | 51 kg |
| Tire stiffness ( $k_t$ ) | 210 N/m |
| Rattle space ( $z_s - z_u$ )max | 91 mm |

### 2.2 A ground-hook damper as reference model

In order to specify the desired performance of the suspension system, the reference model is derived from quarter-vihicle ground-hook model (see Fig. 2). Ground-hook proposed the control method Ground-hook that in which an assumed amortize has been incorporated between the sprung mass and the ground in order to reduce dynamic force of tire and to improve the conductibility of the car. In Fig. 2, The linear equation of the state for reference model Ground-hook from the order of four has been presented as follows:

$$m_s \ddot{z}_s = -k_b(z_s - z_u) - c_b(\dot{z}_s - \dot{z}_u) \tag{4}$$

$$m_u \ddot{z}_u = k_b(z_s - z_u) - c_g \dot{z}_u + c_b(\dot{z}_s - \dot{z}_u) - k_{t1}(z_u - z_r) \tag{5}$$

where, $c_b$ and $c_g$ are indicate a variable damper and a coefficient that connected between the unsprung mass and the fixed fictitious frame on the ground, respectively. $\dot{z}_s$, $\dot{z}_u$, $\dot{z}_s - \dot{z}_u$ are the velocities of sprung mass, unsprung mass, and relative velocity between sprung mass and unsprung mass. Theoretically, ground-hook damper model will improve the responses of the unsprung system. The ideal ground-hook damper model is given by: if $-\dot{z}_u(\dot{z}_s - \dot{z}_u) \geq 0$ then $f = c_g \dot{z}_u$, if $-\dot{z}_u(\dot{z}_s - \dot{z}_u) < 0$ then $f = 0$, where $f$ is ground-hook damper model force. The generated forces by the ground-hook model are added to the passive vehicle model which used as a reference on active suspension controller design.

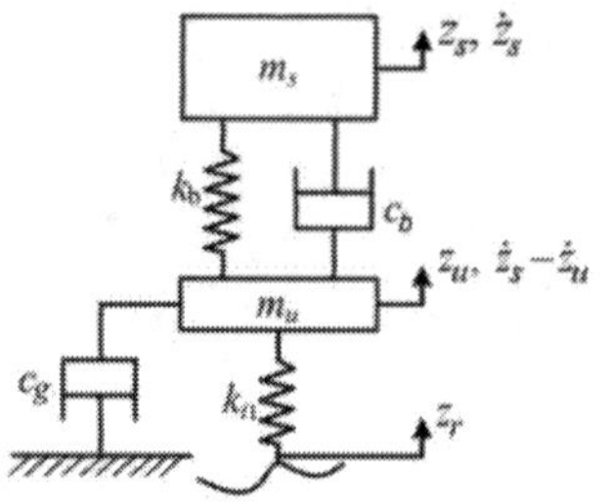


**Figure 2.** Ground-hook reference model of a quarter-vehicle suspension system.

In the state space form, equations (4) and (5) is written as

$$\dot{X}_d = A_d X_d + C_d z_r \tag{6}$$

where the matrices $A_d$ and $C_d$ are given by

$$A_d = \begin{bmatrix} 0 & 1 & 0 & 0 \\ 0 & -c_g/m_s & -k_b/m_s & -c_b/m_s \\ 0 & 0 & 0 & 1 \\ \frac{k_{t1}}{m_u} & -\left(\frac{c_g}{m_s} + \frac{c_g}{m_u}\right) & -\left(\frac{k_b}{m_s} + \frac{k_b + k_{t1}}{m_u}\right) & -\left(\frac{c_b}{m_s} + \frac{c_b}{m_u}\right) \end{bmatrix}$$

$$C_d = [0 \quad 0 \quad 0 \quad -k_{t1}/m_u]^T.$$

## 3 Controller design

One of the main approach often used in adaptive control design is the model-reference adaptive system. Here, the desired performance is expressed in terms of a reference model, which gives the desired response to a command signal. The desired output of the model is compared to the output of plant, where its error is used to generate an actuating signal. One of most critical point in the design is how to make the error as small as possible. This will depend on the model and the command signal. If it would be possible to make the error equal to zero for all command signals, then perfect model-following would be achieved. Choosing the control law as follows:

$$Q = \theta X, \tag{7}$$

where $\theta = [\theta_1 \quad \theta_2 \quad \theta_3 \quad \theta_4]$ is an adjustable controller gain matrix, then equation (3) becomes

$$\dot{X} = (A + B\theta)X + Cz_r. \tag{8}$$

Assumed that $C = C_d$, then the error and its dynamic between the states can written as

$$e = X - X_d. \tag{9}$$

$$\dot{e} = \dot{X} - \dot{X}_d = A_d e + (A - A_d + B\theta)X \tag{10}$$

In order to make $e$ goes to zero, then there must be exist a set of unknown constant gain $\theta^*$ such that

$$A + B\theta^* = A_d. \tag{11}$$

By substituting (11) into (10), we have

$$\dot{e} = A_m e + B\Phi X \tag{12}$$

where $\Phi = \theta - \theta^*$. In order to drive the adjustment law of the parameter, the Lyapunov function is introduced as

$$V = \frac{1}{2}\left(\eta e^T P e + \Phi^T \Phi\right) \tag{13}$$

$$\dot{V} = \frac{1}{2}\left(\eta \dot{e}^T P e + \eta e^T P \dot{e} + 2\Phi^T \dot{\Phi}\right) \tag{14}$$

or

$$\dot{V} = \frac{1}{2}\eta e^T\left(A_d^T P + P A_d\right) + \eta \Phi^T B^T P e X^T + \Phi^T \dot{\Phi}. \tag{15}$$

By chosing $\dot{\Phi} = -\eta B^T P e X^T$, then equation (15) can be simplify as

$$\dot{V} = -\frac{1}{2} e^T \Psi e. \tag{16}$$

Then the parameter adjustment law is obtained as

$$\dot{\theta} = -\eta B_p^T B^T P e X^T. \tag{17}$$

## 4 Discussions

To investigate and observe the performance of the designed adaptive controller to changing dynamics of the vehicle suspension systems, two different types of input signals (i.e., sinusoid and random signals) representing different types of road to stimulate the dynamics of suspension of the vehicle have been identified. It is desired that when the vehicle runs on the three different road types, the controller will choose the coefficients of controller θ such that the closed-loop output response of each model will be close to the model-reference response for any changes of the dynamic suspension system.

Revisiting the ground-hook scheme in Fig. 2, the variables $c_b$ and $c_g$ define the control gains that are used in computing $u$ in proportion to $\dot{z}_s - \dot{z}_u$ and $\dot{z}_u$. Fig. 3 shows that, as the damping coefficients $c_1$ and $c_2$ (to indicate $c_b$ and $c_g$, respectively) increases, the peaks of the accelerations of the sprung and the unsprung masses at their natural frequencies decrease, while the acceleration levels in the frequency range 4-8 Hz and in the frequency range 12 Hz and higher increase. Decreasing $c_b$ while increasing $c_g$ results in an increase of the amplitude of the unsprung-mass natural frequency, with the gain of the reduction of the amplitude of other frequency ranges. Another notable observation is that the transmissibility to the sprung-mass acceleration with constant $c_b$ in the higher frequency range of over 12 Hz is constant for any value of $c_g$. The dynamic response of the system for different values of the damping coefficients is clearly a weighted combination of the damping coefficients of the skyhook control. The controlled skyhook then is used as reference model in the adaptive controller design.

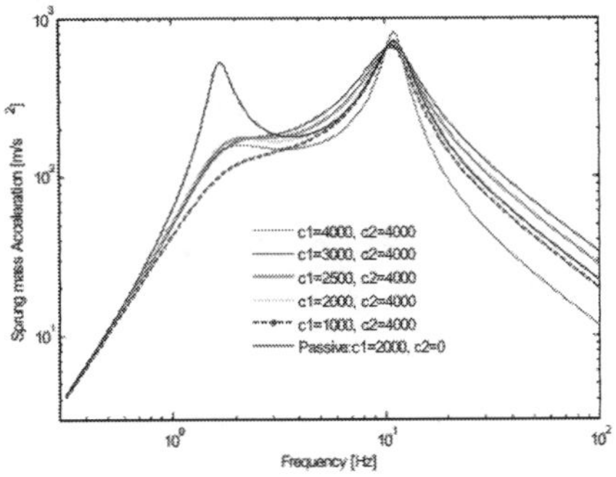


**Figure 3.** Comparison of the magnitude plots of the frequency responses of $\ddot{z}_s(s)/z_r(s)$ for various damping coefficients.

The magnitudes of the frequency responses of the sprung-mass acceleration with respect to the road disturbance are shown in Fig. 4. The solid line and the dotted line represent the magnitude plots of the frequency responses of the semi-active (with ground-hook controller) and the active. In Fig. 4, the dashed line depicts a large improvement according to the other values. However, the dashed line shows a deterioration of ride comfort in the frequency range 4-8 Hz and the deterioration of noise & harshness in the frequency range of 12 Hz and higher, when compared with the passive suspension.

If the results between semi-active and active suspension systems are compared in Figs. 5-9. It can be seen that, for the same input road disturbance in the form of random, the output response of semi-active suspension will follow the disturbance, whereas for the active suspension using the designed model reference adaptive technique then the output response only slightly react accordingly to the disturbance. The velocities and the accelerations of the sprung and unsprung masses (Figs. 6 – 9), respectively, of the active suspension were highly reduce compate to the only semi-active suspension control design. The performance of the sprung mass displacement and sprung mass veocity were improved about 20% and 35%, respectively. The response of the sprung mass acceleration with the active one was poor at a certain point. This condition suspected as a accumulation of the sprung and unsprung masses amplitude. However, another performance improvement are achieved in the unspung mass responses (velocity and acceleration). At the same condition with the sprung mass response, the velocitiy and the acceleration with active one were highly reduces which indicate that the road holding performance is improved. This means that the designed active suspension performs better than the semi-active one when the vehicle is subject to different road disturbance.

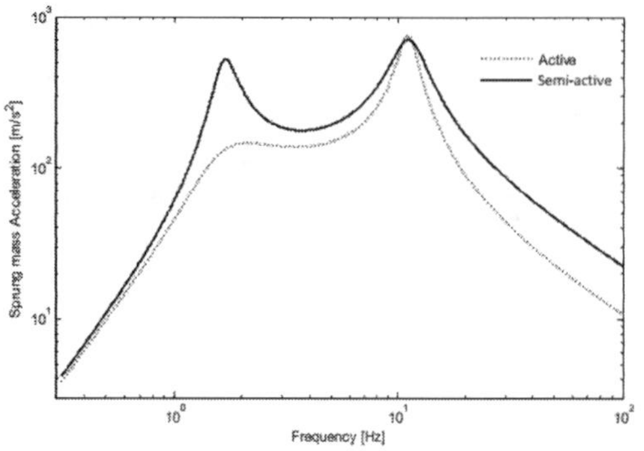


**Figure 4.** Comparison of the frequency responses of $\ddot{z}_s(s)/z_r(s)$.

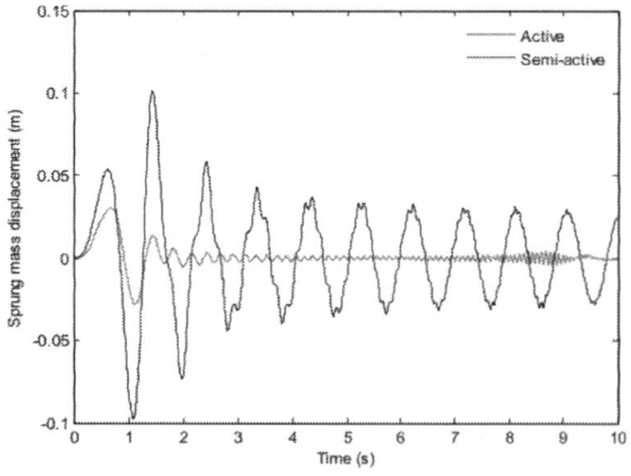


**Figure 5.** Sprung mass displacement response.

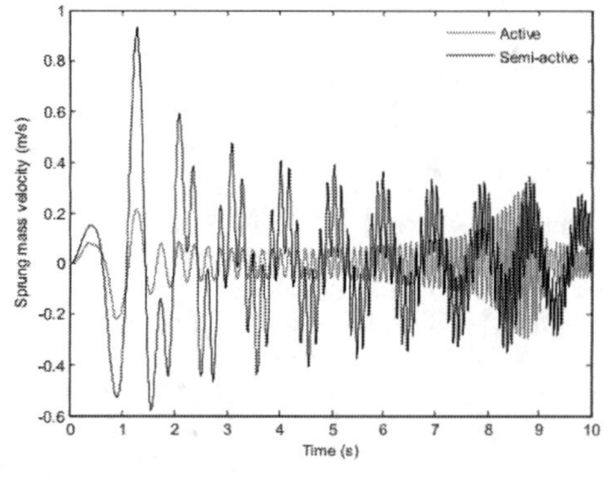


**Figure 6.** Sprung mass velocity response.

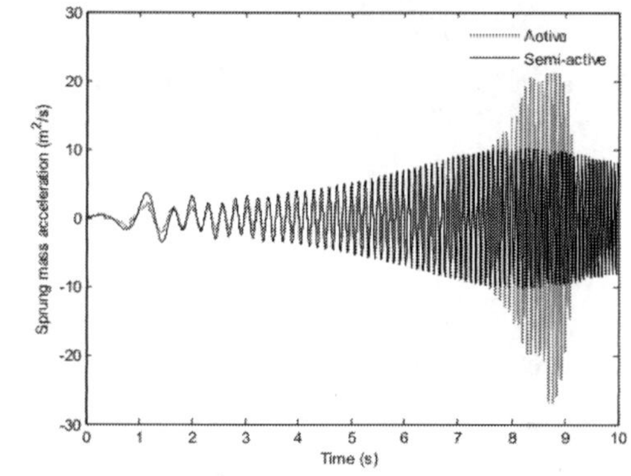


**Figure 7.** Sprung mass acceleration response.

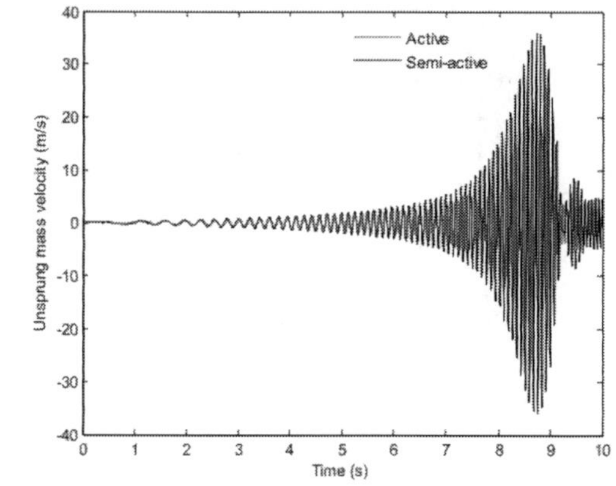


**Figure 8.** Unsprung mass velocity response.

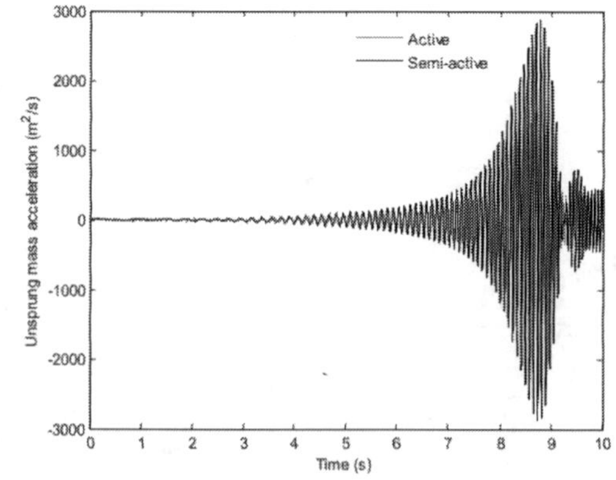


**Figure 9.** Unsprung mass acceleration response.

## 5 Conclusions

The controller based Lyapunov method and ground-hook reference model for suspension system is designed. The effects of the road excitation frequency and road roughness in association with the ride comfort and the road holding of the vehicle were studied. The simulation result show that an active suspension system with the proposed control strategy is able to improve the ride comfort and the road holding, significantly compared with the conventional passive suspension systems as well with semi-active suspension.

## Acknowledgment

This research was supported by the thematic program through the Bandung Technical Management Unit for Instrumentation Development (Deputy for Scientific Services) and the excellent program through the Research Center for Physics (Deputy for Engineering Sciences) funded by Indonesian Institute of Sciences, Indonesia.

## References

1. E. Guglielmino, T. Sireteanu, C. W. Stammers, G. Ghita and M. Giuclea: Semiactive suspension control, Springer Verlag London Limited, (2008)
2. A. Turnip, K.-S. Hong and S. H. Park: Modeling of Hydraulic Engine Mount for Active Pneumatic Engine Vibration Control Using the Extended Kalman Filter, Journal of Mechanical Science and Technology, **23**, 1 229-236 (2009)
3. D. Karnopp: Active damping in road vehicle suspensions, Vehicle System Dynamics, **12**, 291–316. (1983)
4. D. Karnopp, M. J. Crosby and R. A. Harwood: Vibration control using semi-active for generators, SME Journal of Engineering for Industry, **96**, 619–626 (1974),
5. D. Karnopp and G. Heess: Electronically controllable vehicle suspensions, Vehicle System Dynamics, **20**, 207–217 (1991)
6. A. Turnip and K.-S. Hong: Road-frequency based optimization of damping coefficients for semi-active suspension systems, International Journal of Vehicle Design, **63**, 1 84-101 (2013)
7. M. Ahmadian, X. Song and S. C. Southward: No-jerk skyhook control methods for semi-active suspensions, Journal of Vibration and Acoustics, **126**, 4 580–584 (2004)
8. A. Turnip, S. H. Park and K.-S. Hong: Sensitivity Control of a MR-Damper Semi-Active Suspension, International Journal of Precision Engineering and Manufacturing, **11**, 2 5-13, (2010)
9. M. Ahmadian and A.P. Christopher: A Quarter-Car Experimental Analysis of Alternative Semi-Active Control Methods, Journal of Intelligent Material Systems and Structures, **11**, 604-612 (2000)
10. A. S. Cherry, R. P. Jones and T. E. C. Potter: The Use of Multibody System Modeling and Multivariable System Decoupling Technique in Vehicle Ride Control," ASME Transactions, Journal of Dynamic Systems, Measurement, and Control,. **121**, 3 479-486 (1999)
11. A. Turnip, D. Soetraprawata, Hariyadi and D.E. Kusumandari: Design of an adaptive intelligent controller in a semi-active suspension systems, Teknologi Indonesia, **36**, 1 9-15 (2013)
12. S. B. Choi, S. K. Lee and Y. P. Park: A Hysteresis Model for the Field-Dependent Damping Force of a Magnetorheological Damper, Journal of Sound and Vibration, **2**, 375-383 (2011)
13. F. D. Goncalves and M. Ahmadian: A Hybrid Conrol Policy for Semi-Active Vehicle Suspensions, Shock and Vibration, **10**, 59-69 (2003)
14. A. Turnip and H. Fakhrurroja: Estimation of the wheel-ground contact tire forces using extended Kalman Filter, International Journal of Instrumentation Science, **2**, 2 34 (2013)
15. K. S. Hong, D. S. Jeon, W. S.Yoo, H. Sunwoo, S. Y. Shin, C. M. Kim and B. S. Park: A New Model and an Optimal Pole-placement Control of the Macpherson Suspension System, SAE International Congress and Exposition, Detroit, MI, SAE paper No. 1999-01-1331, 267-276 (1999)
16. Y. Liu, T. P. Waters and M. J. Brennan: A Comparison of Semi-Active Damping Control Strategies for Vibration Isolation of Harmonic Disturbances, Journal of Sound and Vibration, **280**, 21-39 (2005)

MATEC Web of Conferences **75**, 04003 (2016) DOI: 10.1051/matecconf/20167504003
*ICMIE 2016*

# Design of EEG Signal Acquisition System Using Arduino MEGA1280 and EEGAnalyzer

Debyo Saptono[1], Bambang Wahyudi[2] and Benny Irawan[3]

[1]*Electrical Engineering Dept, Gunadarma* University,Jakarta - Indonesia
[2]*Informatics Engineering Dept,* Gunadarma University,Jakarta - Indonesia
[3]*Informations System Dept,* Gunadarma University,Jakarta - Indonesia

**Abstract.** This study integrates the hardware circuit design and software development to achieve a 16 channels Electroencephalogram (EEG) system for Brain Computer Interface (BCI) applications. Signals obtained should be strong enough amplitude that is usually expressed in units of millivolts and reasonably clean of noise that appears when the data acquisition process. The process of data acquisition consists of two stages are the acquisition of the original EEG signal can be done by the active electrode with an instrumentation amplifier or a preamplifier and processing the signal to get better signals with improved signal quality by removing noise using filters with IC OPAMP. The design of a preamplifier with high common-mode rejection ratio and high signal-to-noise ratio is very important. Moreover, the friction between the electrode pads and the skin as well as the design of dual power supply. Designs used single-power AC-coupled circuit, which effectively reduces the DC bias and improves the error caused by the effects of part errors. At the same time, the digital way is applied to design the adjustable amplification and filter function, which can design for different EEG frequency bands. The next step, those EEG signals received by the microcontroller through a port Analog to Digital Converter (ADC) that integrated and converted into digital signals and stored in the RAM of microcontroller which simultaneously at 16 ports in accordance with the minimal number of points of data collection on the human scalp. Implementation results have shown the series of acquisitions to work properly so that it can be displayed EEG signals via software EEGAnalyzer.

## 1 Introduction

An electroencephalogram (EEG) evaluates electrical activity produced by the brain, which can signify or rule out certain conditions, most commonly seizure disorders. Electroen-cephalography is the neurophysiological measurement of electrical activity in the brain as recorded by electrodes placed on the scalp or, in special cases, subdurally or in the cerebral cortex. The resulting traces are known as an EEG and represent a summation of postsynaptic potentials from a large number of neurons. When the wave of ions reaches the electrodes on the scalp, they can push or pull electrons on the metal on the electrodes. Since metal conducts the push and pull of electrons easily, the difference in push or voltage between any two electrodes can be measured by a voltage. Recording these voltages over time gives us the EEG. This study mainly designs and fabricates portable electroencephalograph (EEG) machine 16 CH with single-power supply[1].

Avariety of new signal processing methods have been applied to EEG signal processing over the past fifteen years . These new methods require new tools to allow routine processing of EEG data, and also make possible the analysis of multimodal data collected using more complex experimental designs than previous analysis methods allowed [2], [3].

## 2 Research method

### 2.1 EEG signal processing

Amplitude of EEG is in the range of microvolt up to 100 μV when it is taken over the scalp and in the range of milivolt when measured directly from surface of brain (in open operation) and its frequency is approximately up to 100Hz. There are five common frequency bands in EEG signals which their characteristics are shown in Table 1.

**Table 1.** Characteristics of EEG rhythms [4].

| rhythm | Frequency tipically(Hz) | Amplitude Tipically (μV) |
|---|---|---|
| Alpha | 8-13 | 20-200 |
| Beta | 13-30 | 5-10 |
| Delta | 0.5-4 | 20-200 |
| theta | 4-8 | 10 |
| Gamma | >30 | 5-10 |

As is shown in Table 1 amplitude of any rhythm has certain range. Intensity or value of amplitude is depended to some factor such as state of patient as well as location of electrode on the scalp. Amplitude of this rhythm varies from 5µv to 200 µv[4] .

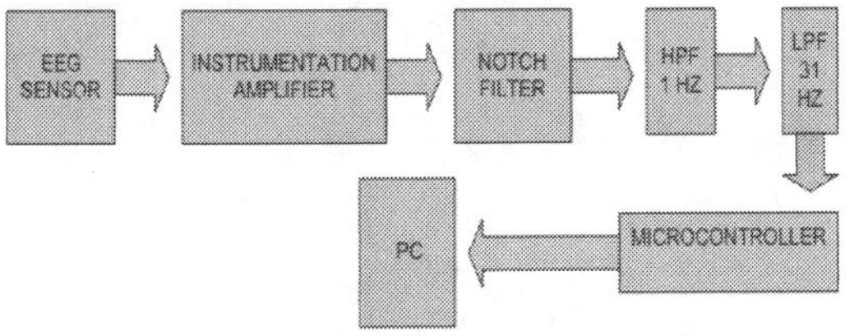


**Figure 1.** EEG acquisition model proposed

In Fig. 1, after signal acquisition phase (eeg sensor), signals are to be preprocessed. Signal preprocessing is also called as Signal Enhancement. In general, the acquired brain signals are contaminated by noise and artefacts.

After obtaining the noise-free signals from the signal enhancement phase(notch filter,HPF 1 Hz, LPF 31 Hz), the noise-free signal will be sent by serial communication into microcontroller, and next essential features from the brain signals were extracted in feature extraction in PC. For feature extraction from EEG signals use methods Fast Fourier Transformations (FFT). It extracts the signal features by transforming the signals from time domain to frequency domain. It works well for stationary signals and, linear random processes. It cannot measure both the time and frequency. With prior assumptions some of Forier Transform (FT) techniques may exhibit better performance in other cases. In this frequency analysis the signals are divided into one-second windows overlapping a half second window. This half second overlaps results in large amount of data for the training of the classifier that forms classes. This frequency based analysis is named as Discrete Fourier Transformation (DFT) and also termed as Power Spectral Density (PSD) [5].

PSD is the one of the most important tool for digital signal processing (DSP). It helps us to know how the strength of a signal is distributed in the frequency domain and its unit is energy per frequency. PSD demonstrates the strength of the variations of energy of a signal as a function of frequency. Hence, it gives us an idea about at which frequencies variations of energy are strong and at which frequencies they are weak.

### 2.2 Microcontroller arduino MEGA1280

The Arduino Mega is a microcontroller board based on ATmega1280. It has 54 digital input/output pins (of which 14 can be used as PWM outputs), 16 analog inputs, 4 UARTs (hardware serials ports), a 16 MHz crystal oscillator, a USB connection, a power, an ICSP header, and a reset button [6].

### 2.3 GUI (Graphical User Interface) EEG analyzer

To use this application, the user must first understand the design of the system being designed. For the first step, select the "new design", choose one of the existing menu on the menu bar "insert element". Having chosen one, it will appear in the form of input-output blocks display in the windows application design and settings dialog box selected menu. If the settings of the selected menu accordingly, the user can add another block output input on the design window by selecting another menu.All procedures is shown in Fig. 2.

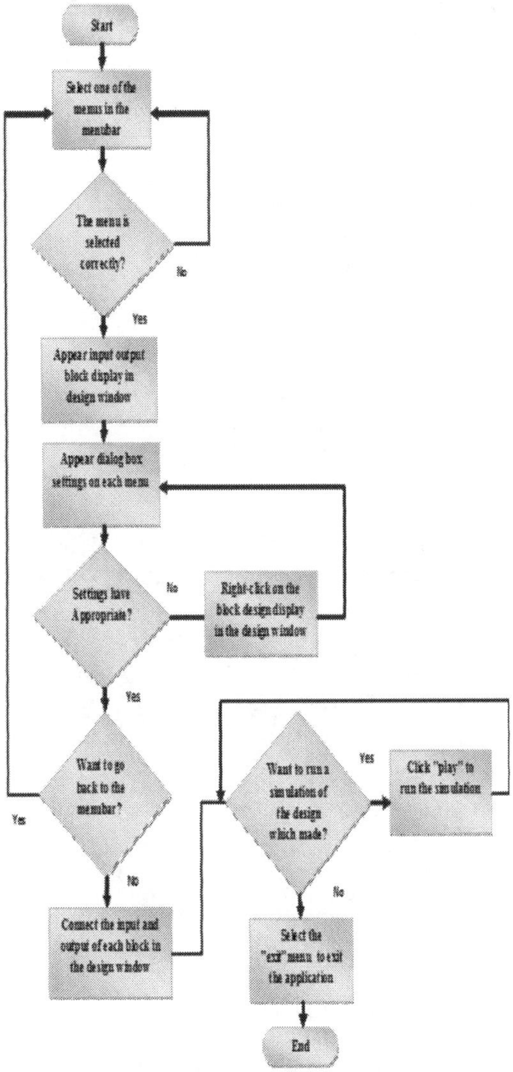


**Figure 2.** Flowchart of EEGAnalyzer

## 3 Implementation

### 3.1 Active electrode with instrumentation amplifier

The instrumentation amplifier also has some useful features like low offset voltage, high CMRR (Common mode rejection ratio), high input resistance, high gain etc. The circuit diagram of a typical instrumentation amplifier using opamp is shown in Fig. 3.

DOI: 10.1051/matecconf/20167504003

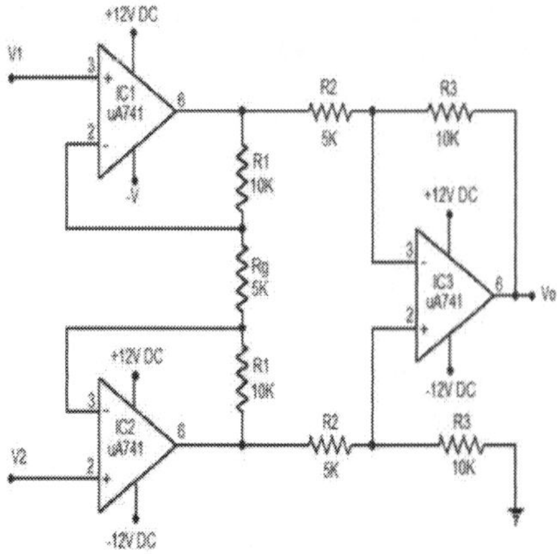


**Figure 3**. Instrumentation Amplifier for Active Elektrode

### 3.2 Notch filter of 50 Hz

Digital approaches have been widely adopted to solve the power line interference problems including proper grounding and electrical shielding in analog recordings, however, the preprocessing is necessary because power line interference which overwhelms the desired signals is actually all around. This design modifies the classical twin-T notch filter to be Q-factor tunable as shown in Fig. 4.

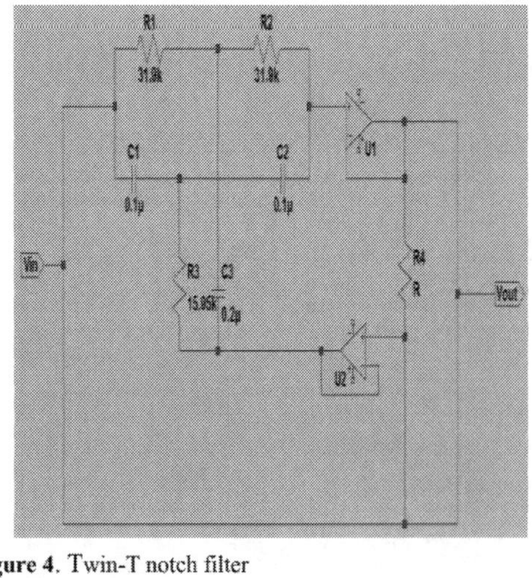


**Figure 4**. Twin-T notch filter

As differensial mode signal input only, the midpoint voltage then the output of U2 is zero, as differential-mode signal and common-mode signal input simultaneously, the output of U2 only contains the common-mode part the input signals. It means that the common-mode signal could input to the instrumentation amplifier directly instead of being translated to differential-mode disturbance for the mismatch of the components to RC coupling circuit.

### 3.3 High pass filter (HPF) 1 Hz

In the Fig. 5, the beginning of this circuit contains a quick HPF of cutoff frequency 1 Hz(Fc=1/(2*pi*R3*C2), just for some extra attenuation of unwanted noise. On the other end, the resistor and capacitor in parallel provide some extra filtering of high frequencies (Fc = 1/(2*pi*10nF*100kΩ) = 160Hz on a low-pass filter).

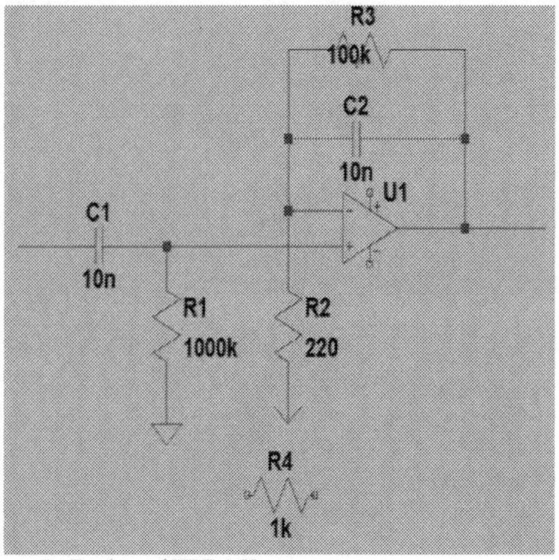


**Figure 5**. Design of HPF 1 Hz

The main purpose of this section, however, lies below this, with the 220Ω resistor and potentiometer (pot for short). This op-amp is a non-inverting amplifier, and so has a gain of G = 1 + R3/(R2+R4), (ignoring the 10nF capacitor, as it's a small value and won't contribute much to the gain). The potentiometer is a variable resistor - when the input is connected to the first pin and the output to the second, turning the wiper changes its resistance linearly between 0 and 1000 ohms. This means that when the pot is turned all the way to the left, the gain of this circuit is G = 1 + R3/(R2 + 0) = 1 + 100k/(220 + 0) = 455. When it is turned all the way to the right, the gain is G = 1 + R3/(R2 + 1000) = 1 + 100k/(220 + 1k) = 83.

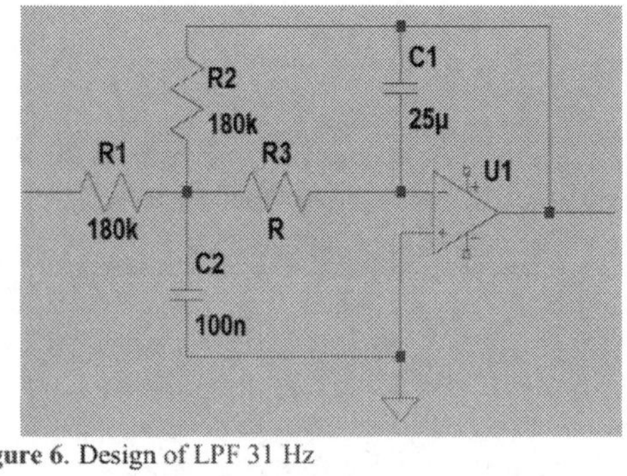


**Figure 6**. Design of LPF 31 Hz

### 3.4 Low pass filter (HPF) 31 Hz

Next up, we want to filter out data above the frequencies we are interested in. More specifically, as bheta wave information stops out at 30Hz, we want to get rid of anything above that, as combined it can contribute a good amount of noise to our data. The circuit design is very similar to the high pass filter from stage 3 - it has a gain of .71 at 31.23Hz, and decreases from there at a rate such that by 300Hz it has attenuated the data by about a factor of 100. Design of HPF-31 Hz circuit as shown in Fig. 6.

DOI: 10.1051/matecconf/20167504003

### 3.5 EEG Analyzer for EEG signal feature extraction

EEGAnalyzer has developed from BrainBay opensource application [7]. In this applications has been added DWT algorithm modul for signal processing.

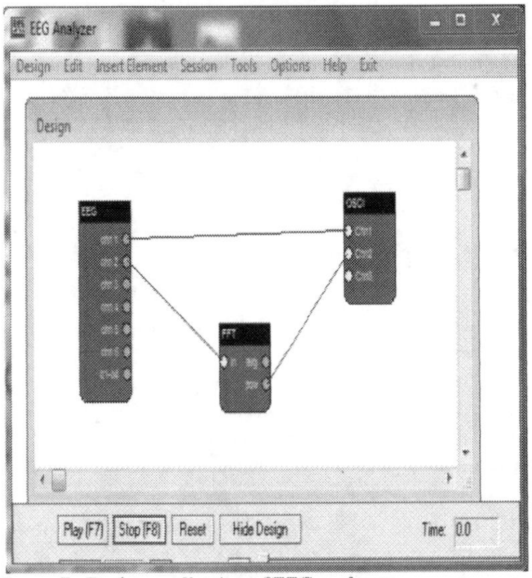


**Figure 7**. Design application of EEG analyzer

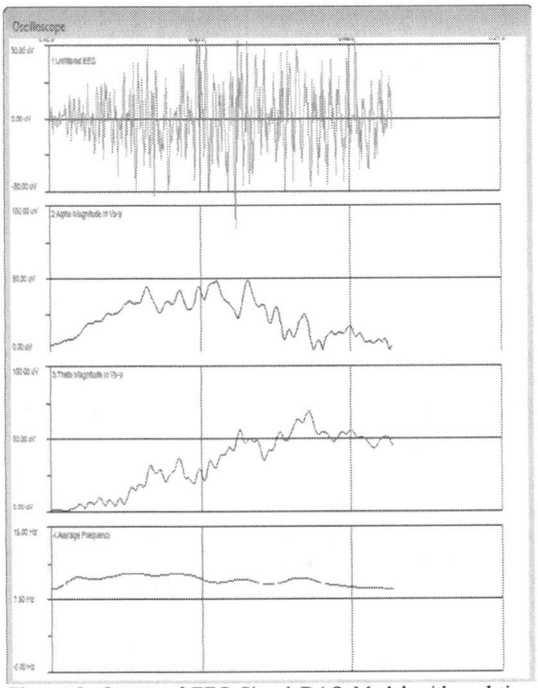


**Figure 8**. Output of EEG Signal DAQ Model with real-time application in EEG Analyzer

The purpose of this dialog box is for EEG module configuration settings that will be used in the design of the block to be created, at the first is source (Generic Bio signal-Amplifier with OpenEEG and Modular P2-Packet format with modification for ATMEGA1280), next is Processing, in this menu will display such as Discrete Wavelet Transform and Fast Fourier Transform. The last is Target, this menu will display 4 kids menu that will be used as input output block to be added to the system design including Spectrum-Analyzer and Oscilloscope as shown in Fig. 8. There are unfiltter EEG Signal, Alpha Magnitude, Theta Magnitude, and average frequency.

## 4 Conclusion

This research has produced instrumentation equipment to take the form of active electrode EEG signals are equipped with instrumentation amplifier and digital filters that work well by eliminating signal noise and missed only the expected frequency of 1 Hz - 30 Hz. The next signal is sent to the microcontroller Arduino MEGA1280 through ADC port and immediately forwarded to the serial port to the PC and then the data were analyzed by EEG Analyzer to be separated between the signals alpha, bheta, and theta. In the future, we hope to integrate the EEG acquisition system in FPGA platform.

## References

1. B. R. Myung and S. K. Yoo, Development of 16-channels Compact EEG System Using Real-time High-speed Wireless Transmission. http://www.scirp.org/journal/eng
2. M. Rajya Lakshmi, Dr. T. V. Prasad, Dr. V. Chandra Prakash, Survey on EEG Signal Processing Methods, International Journal of Advanced Research in Computer Science and Software Engineering, **4**, 1, January (2014).
3. C.-S. Wang, Design ofa 32-Channel EEG System for Brain Control Interface Applications, Journal of Biomedicine and Biotechnology, (2012).
4. A. S. Al Mejrad, Human Emotions Detection using Brain Wave Signals: A Challenging, European Journal of Scientific Research, **44**, 4, (2010).
5. M.R. Azim, M.S. Amin, M.S. Haque, M.N. Ambia, M.A. Shoeb, Extraction of Human Sleep EEG Signals using Wavelet transform and Fourier transform, International Conference 2nd Signal Processing Systems (ICSPS) 2010 , 5-7 July (2010).
6. N. Sathurappan, S.P. Potty, Measurement of environmental conditions and biomedical parameters using robotic-gripper, International Conference Advanced Computing and Communication System(ICACCS) 2013 , 19-21 Dec (2013).
7. C. Veigl, An Open-Source System for Biosignal and Camera-Mouse Applications, Young Researchers Consorsium of 10th Internasional Conference on Computers Helping People(CCHP) 2006 ,10-14 July (2006).

MATEC Web of Conferences 75, 04004 (2016) DOI: 10.1051/matecconf/20167504004
*ICMIE 2016*

# Development of logistic support for space equipment on the base of the "Sail-BMSTU" midget spacecraft

Alla Brom, Evgeniya Gorlacheva

*Bauman Moscow State Technical University, 105005, 2-ya Baumanskaya,5, Moscow, Russian Federation*

**Abstract.** The paper envisages the application of integrated logistic support conception (ILS) for space equipment on the base of the example of the student's «Sail BMSTU» midget spacecraft (MS). The peculiarities of space equipment logistic support in operation phase are considered. The special focus is done to the problem of decrease in production expenses of spacecrafts. The paper suggests that the solution of this problem has to be based on tools commonly used in engineering fields – functional analysis and FMECA. The fragment of FMECA is presented. Due to FMECA it is clear what products in spacecrafts should be calibrated in accordance with quality requirements of military class and what ones should be calibrated in accordance with quality requirements of commercial and industrial classes. Each failure mode of midget spacecraft, identified within FMECA, is studied by assessing of criticality, severity and probability of emergence. The paper describes the main procedures of integrated logistic support on the base of the student's «Sail MGTU» midget spacecraft. Recommended guidelines providing reliability of electro radio products are elaborated. The practical application of integrated logistic support in aerospace industry is reasonably presented.

## 1 Introduction

The concept of Product Lifecycle Management (PLM) covering all the processes of the life cycle (LC) technology products is actively developed in the last decade. In the field of engineering PLM tools have quite well and successfully developed and in operation of the equipment, the main approach has become an integrated logistics support (ILS) of a product. Integrated Logistics Support presents a methodology to cost optimization of products life cycle, taking into account the best operability of a technological item to exploitation support.

## 2 Main Part

The operability to exploitation support is a correspondence of product design characteristics and its information and logistic system to a demand of permanent readiness to work [1]. Therefore in ILS procedures, methods for monitoring and control of compliance with the actual performance in real operating conditions at various stages of product life cycle design requirement are very relevant. Taking into account the peculiarity of spacecraft the possibility of using ILS technologies in this sector is of undoubted concern, especially in view of the emerging world trend towards the optimization of engineering costs.

The development of midget spacecraft (MS) is a new scientific and technological direction in the satellite industry emerged in the last fifteen years. While the satellite industry has traditionally been maintained by large, high-performance satellites, the new direction has opened up new opportunities for space missions and its share is growing. Midget spacecrafts provide a drastic reduction of the cost, as the development, production and testing, and start-up: the timing of development projects are reduced from 4-5 years to 1 year, the requirements for test and production equipment are also dramatically reduced due to the reduced dimensions, in connection with this the total cost of the projects is also reduced.

All objects of space systems have common peculiarities that affect the structure of the logistic processes of production and operation:

- the complexity and high cost of production;
- an one-off production, therefore, a unique product;
- the diversity and complexity of processes that require prior experimental and research work;
- continuous improvement of the requirements for quality, reliability, product life cycle;
- the requirement to ensure maximum reliability and quality in the pre-production and production stages without further maintenance and repair work on the next stages of the life cycle;
- as a rule, long service life of equipment (the effective operation of space systems should be provided for a limited period of its useful life). For the moment it ranges from 3 to 20 years for different specimens of space systems;
- limited capacity of utilization that should be carried out by means of space systems themselves.

 DOI: 10.1051/matecconf/20167504004

Thus, the provision of maximum reliability of a device while minimizing the cost of its development is one of the most important tasks of the space industry. The implementation of integrated logistics support can yield tangible results in solving these problems.

## 3 Proposed method

At first sight complete implementation of ILS concept is impossible, as it has been mentioned the possibility of technical maintenance is strongly limited for space systems items. However, the elements of ILS concept can and should be applied to space systems, and one could refer to the Manual «International procedure specification for Logistics Support Analysis», published by the European Association of Aerospace Equipment Manufacturers (ASD) [2].

The appliance of ILS standardized methodologies – functional analysis and FMECA – to space systems items will solve the following tasks:

- Provision of "functional completeness and non-redundant set" in an item;
- Provision of maximum reliability of an item while operating;
- Provision of necessary and sufficient spectrum and amount of material and technical support;
- Provision of mission item performance;
- Provision of product life cycle cost reduction.

Unlike civilian products for space systems items the focus on the value of the product is transferred to its reliability and dependability. The primary task is to ensure trouble-free performance in the active operation of the machine process, only in this case the goal of creating a space vehicle will be made and a scientific experiment will be carried out (in the case of scientific apparatus), the cost of production will be repaid (in the case of a commercial vehicle).

The cost reduction of the product is possible due to implementation of electronic components designed for commercial and industrial use. The foreign space systems manufacturers use more often electronic components designed for commercial and industrial use.

The usage of the electronic components could reduce the production costs increasingly. The fact that up to the 2000s the space systems manufacturers use items designed only for space application. They are characterized by a high degree of reliability. Reliability characteristics of most high-end products (space applications) are different from those in the commercial is about 80-100 times. The difference in electronic components cost for space and commercial application is estimated approximately by the same figures.

The application of electronic components of industrial class instead of expensive and inaccessible foreign ones in space systems is possible due to the implementation of a vast program of rejected, super rejected and diagnostic tests on the base of non-destructive examination methods, as well as the structural control and physical analysis of representative samples from each batch. [3].

Detailed selection procedure can significantly improve the quality and reduce breakdown rates in batch in 10-30 times on the average due to exclusion of electronic components with hidden defects.

The use of functional analysis and FMECA algorithms for space systems allow to determine where it is possible to use electronic components of industrial class and where one can use more reliable components.

The application of above mentioned algorithms will be shown on the base of student's "Sail BMSTU" midget spacecraft (fig.1).

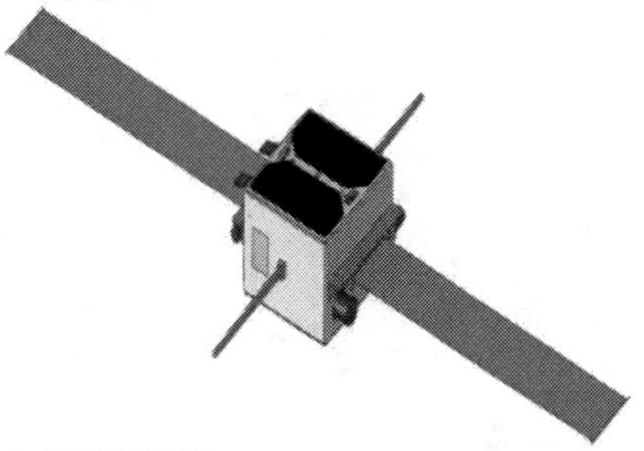

**Figure 1.** Sail BMSTU

"Sail BMSTU" was launched in September 2009 in Bauman Moscow State Technical University in academic and research youth space center. The main goal of this center is the elaboration of a midget spacecraft with a solar sail. The main idea of this project is the usage as a light-reflecting surface of the solar sail two narrow strips, which can simply be rolled up on the reel [4].

At the moment the project is adopted by the Federal Space Agency (Roscosmos) to the program of scientific experiments at the International Space Station (ISS). The project team has had to elaborate experimental design of "Sail BMSTU" midget spacecraft.

During the space experiment it is planned firstly to disclose a thin film structure in the form of two strips under the influence of centrifugal forces from the board of the midget spacecraft in the close proximity of the ISS and then test the onboard systems to flight mode for the duration of the active existence of the spacecraft.

This experiment has to achieve the following objectives:

- deploying technology demonstration of the thin film long strips by centrifugal forces;
- demonstration of the on-board equipment based on electronic components designed for non specialized purposes (microcontrollers, temperature sensors, pressure and angular velocity, radios, batteries) in space flight conditions;
- verification of the mathematical model of the thin-film structure dynamics during space flight.

"Sail BMSTU" midget spacecraft has a non-standard way to start - manual start of an astronaut during extravehicular activity on the International Space Station. Therefore, in orbital flight "Sail BMSTU" midget spacecraft is in two states: a state in the storage part of the ISS and the active work of the state, after the launch of the ISS.

## 4 Analysis

Because of the limited scope of the article it is difficult to describe a fully developed procedure for "Sail BMSTU" midget spacecraft – functional analysis and FMECA. In order to present the logic of the conducted work there are fragments in this paper as these procedures are standardized methodologies and the problem is not the algorithms essence but in their application to specific objects. The fragment of functional analysis is presented in Table 1.

**Table 1.** Functional analysis of "Sail BMSTU" midget spacecraft.

| Functions | List of subsystems |
|---|---|
| **F0: Deployment of a thin-film structure and downlink** | **"Sail BMSTU" midget spacecraft** |
| **F1: Deployment of studied construction of a solar sail** | Deploying system of a frameless thin-film construction |
| F11: strip storage | Tape reel |
| F12: deployment of a strip | Stepper engine |
| F13: midget spacecraft spin-up | Balance-wheel engine |
| F131: Electric energy conversion into rotation mechanical energy | Brushless engine |
| F132: Rotation mechanical energy storage | Balance-wheel |
| F14: generation of commands in experiment management | Micro controller |
| **F2: getting and storing data** | Onboard complex control system |
| F21: rotation sensing of midget spacecraft in three axel groups | MEMS gyro sensor |
| F22: Photo registration of deployment process | Photo camera |
| F23: information output about current time | Real-time clock |
| F24: temperature measuring of component parts of midget spacecraft | Temperature sensor |

Taking into account the results of the conducted functional analysis of MS one could have concluded that the optimal product design and the possibility of its further development. The phone is no duplication of functions in different elements, all functions are blocked structural elements. As a result of the functional analysis obtained composition features and design structure of products for FMECA procedure.

Quantitative evaluation of the reliability of the element can be carried out, using as an indicator of the probability of failure-free operation (PFFO). PFFO is calculated by the following formula:

$$P(t) = e^{-\lambda t} \quad (1)$$

where $\lambda$ – failure rate level of an element; 1/hour; $t$ – running time of an element, hour. At that the probability of failure initiation is a reciprocal magnitude of probability of non-failure operation.

In FMECA each structure element has a targeted magnitude of failure rate level. Failure rate is calculated separately for each type of element. The magnitudes of failure rate level for electronic components are given in reference literature and manufacturer's specifications, so it is reasonable to carry out FMECA up to the level of each separate electronic component. Since MS will be founded in two states, it is necessary to calculate the resulting PFFO. Taking into consideration that successive events are independent, one can use the formula:

$$P(t) = \prod_{i=1}^{n} P_i(t) \quad (2)$$

where $n$ - number of states of the system.

The algorithm of FMECA for MS is presented on fig.2.

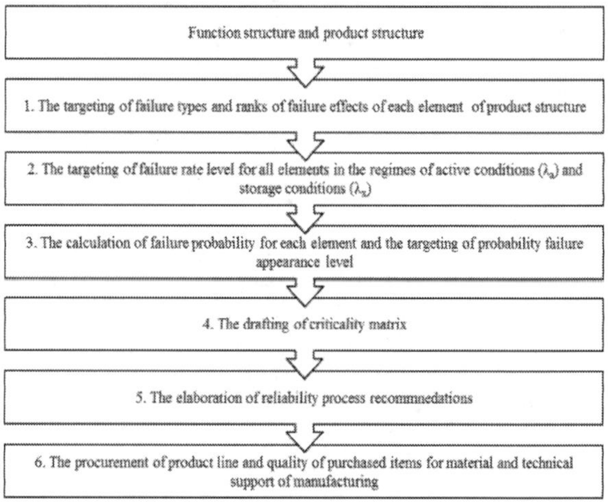


**Figure 2.** The Algorithm of FMECA

Then for MS the authors of the present article have elaborated the classification of failure critical levels – ranks of failure effects (table 2) and probability level of failure appearance (table 3).

**Table 2.** The classification of failure critical levels.

| Critical level | Rank | Failure effects |
|---|---|---|
| 1 | Catastrophic | Element failure leads to the inability of performing two tasks by a satellite |
| 2 | Critical | Element failure leads to the inability of performing one task by a satellite |
| 3 | Significant | Element failure leads to a decrease in the volume of scientific data obtained |

| | | |
|---|---|---|
| 4 | Non-significant | Failure element doesn't lead to the lack of system exploitation characteristics, but causes the decrease of a reliability level ( reserve rate) |

**Table 3.** Probability level of failure appearance.

| Probability level of failure appearance | Description |
|---|---|
| A | Frequent failure. The probability of failure appearance for a predetermined time exceeds $6*10^{-3}$ |
| B | Probable failure. The probability of failure appearance for a predetermined time ranges from $5*10^{-3}$ to $6*10^{-3}$ |
| C | Possible failure. The probability of failure appearance for a predetermined time ranges from $3*10^{-3}$ to $5*10^{-3}$ |
| D | Rare failure. The probability of failure appearance for a predetermined time ranges from $1*10^{-3}$ to $3*10^{-3}$ |
| E | Remote failure. The probability of failure appearance for a predetermined time is lower than $1*10^{-3}$ |

Then MS elements are recorded in the criticality matrix, where they are ranked according to the probability level of failure appearance and ranks of failure effects.

Fig.3 presents the criticality matrix for "Sail-BMSTU" midget spacecraft.

| | Ranks of failure effects | | | |
|---|---|---|---|---|
| Probability level of failure appearance | 4 | 3 | 2 | 1 |
| A | Converter UART-USB | Photo camera | Controller of battery charge/discharge, Micro assembly – receiver/transmitter | Microcontroller |
| B | | | | Line power stabilizer |
| C | | MEMS gyro sensor | USB-connector | |
| D | Temperature sensor | Real time clock, memory EEPROM | Tape reel, stepper engine, relay switch, body frame | Circuit plate |
| E | Receiving and technological aerial, light emitted diode | | Transmitting dipole aerial, film heater, brushless engine, balance-wheel | Charge battery, solar element, toggle switcher, micro commutator |

**Figure 3.** Criticality matrix of "Sail BMSTU" midget spacecraft

The dark grey area indicates the first priority; the light grey area is of the second priority; the white one is the third priority area.

Based on the FMECA results, the recommendations to ensure the reliability of the MS in accordance with the received priority areas have been elaborated (Fig. 4).

In case of first priority area it is necessary to conduct calibration trials to obtain reliability of elements to the military class [5]. Electronic components of industrial reliability level have the quality factor K = 10. They can be brought to the level of quality corresponding to military equipment (K = 1), using complex of additional calibration trials (Table 4).

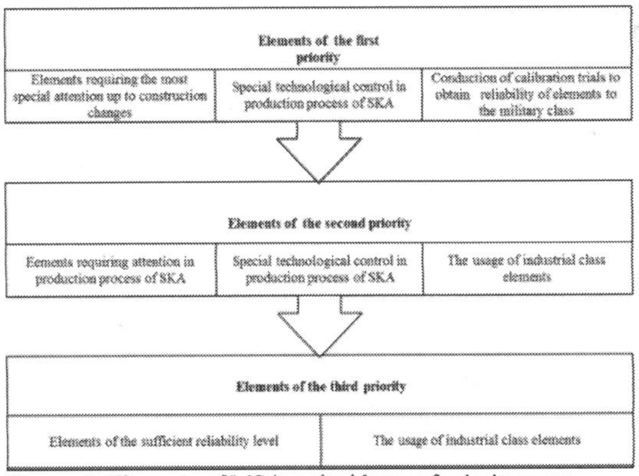

**Figure 4.** Elements of MS in coincidence of priority areas

**Table 4.** Calibration trials.

| Recommended additional calibration trials | The expected increase of quality for electronic components |
|---|---|
| Measurement of electrical parameters for increased severe standards for normal, positive and | 1,9 |

| | |
|---|---|
| negative temperature | |
| 10 thermo cycles with the measurement of electrical parameters | 1,6 |
| Heat test in period of no less than 168 hours. With maximum working temperature for a given element | 2 |
| Measurement of informative-bearing parameters after heat test | 1,5 |

The authors have calculated the quality of purchased items of first priority area. We'll use the formula of determining the sample size which would allow to evaluate the rejection proportion in the production lot (10,000 units), up to 2% at a confidence figure $P = 0,95$ That is, if the rejection proportion is the $k\%$, then with probability of $0,95$ we could consider to get rejection share $k'\%$ in the range of $k\% + 2\% < k'\% < k\% - 2\%$ .

$$n = \frac{t^2 pq}{\Delta^2} = \frac{2^2 * 0,01 * 0,09}{0,02^2} = 9 \quad (3)$$

where $n$ - elements number in the sample,

$t$ – the confidence coefficient. It is determined by the table of the Laplace function $F(t)$, in case the confidence figure is known to the researcher

$p$ - the rejection proportion in the sample.

$q$ - the proportion of qualitative items in the sample.

$\Delta$ - given accuracy.

The magnitude of rejection share has been calculated as an average arithmetic mean of rejection shares in electronic components tests for different products. These tests are fulfilled by JSC "Russian Scientific Institute "Elektronstandart", specializing calibration trials and increasing of electronic components reliability level. For the other elements of the number of items purchased is equal to one piece.

## 5 Conclusion

Thus, having fulfilled the elaborated algorithm of ILS tools application, we get a particular product line and quality of purchased items for material and technical support of manufacturing for spececraft. Thus it is possible to achieve a significant reduction in product prices due to the use of electronic components of industrial class reliability while ensuring the required level of reliability of the product.

The application of flexible technique such as FMECA has provided the separation of priority elements to military and civil classes. Thus by means of FMECA it is clear what elements should be calibrated. That is the essence of logistic support conception in aerospace industry.

## References

1. E. Sudov, A. Levin, A. Petrov, E.Chubarova. Technologies of integrated logistic support of engineering products. – Moscow, Publishing House "InformByro" Ltd., (2006).
2. International procedure specification for Logistics Support Analysis. S3000L. ASD, (2010).
3. N. Danilyn Inforation technologies and certification of element base in new Russian aerospace telecommunications. – Moscow, (2000).
4. D. Rachkin, S. Tenenbaum, A. Dmitriev, N. Nerovniy, O. Kotsur, A. Vorobyov. 2-blades deploying by centrigugal force solar sail experiment (IAC-11.E2.3.8) // Proceedings of 62nd International Astronautical Congress. - Cape Town, SA, (2011).
5. N. Evstifeeva, A. Brom, V. Mayoriva, D Rachkin. Special issues of integrated logistic support of an aerospace object on the base of on the example of midget spacecraft - «Sail BMSTU» // Humanitarian Herald of BMSTU. #10 (12). Access mode: http://hmbul.bmstu.ru/catalog/econom/log/111.html (2013)

# Multipurpose monitoring system for icebreakers: Development, implementation, and testing

Andrey Zhirnov, Maxim Anufriev, Nikolay Pozhar, Konstantin Stepanov, Anton Chernutsky, Ilya Makhrov, Evgeniy Nesterov, Dmitriy Shelestov, Kirill Koshelev, Aleksey Fedorov, Valeriy Karasik, and Alexey Pnev

*Bauman Moscow State Technical University, 105005 Moscow, Russia*

**Abstract.** We demonstrate the results of a project on implementing a complete and practical multipurpose monitoring system for icebreakers. We describe the basic principles of operating the developed monitoring system. We present a 3D finite element analysis of the ship model to optimise the sensor arrangement. We also present the benchmarks and test results of the system during an expedition on the Russian scientific research vessel *Akademik Tryoshnikov*.

## 1 Introduction

The development of solutions for the structural health monitoring of complex objects such as buildings, bridges, and ships is a challenging problem [1]. It has been the subject of active study for the last several decades [2]-[5].

In particular, monitoring systems are very important for icebreakers and seagoing ships. Because the cost of technical service of such objects is significantly high, problems related to the design of monitoring systems for accident prevention have attracted a great deal of interest. Note that obtaining accurate data regarding ship structural health is key for route planning optimisation. Currently available solutions are not universal because they are used to control a limited number of parameters only [6-12]. Light Structures AS (Norway), *SENSFIB project,* monitors the ship hull in real time. It analyses strain continuously and generates emergency messages with advices to the stuff when safe limits are exceeded. BMT Scientific Marine Services Inc. (USA), *Structural Monitoring System*, designed mainly for monitoring of offshore structures (offshore drilling platforms, including ice-resistant) and also mooring systems. Weir-Jones Ltd (Canada), *Ice Force IDEAL*, designed for monitoring of the interaction between ice and ship,measuring its parameters and for the protection of offshore platforms from the ice loads. Some gravity type ice-resistant platforms in Caspian Sea are equipped with this system. MCA Consulting (USA), *Hull Monitoring System*, designed mainly for monitoring of offshore structures. HULLMOS systems was installed on a five oil tankers of the project *Sakhalin-1*. Sea Structure Technology Co. Ltd (Korea), *Naviscan*, and Strainstall UK Ltd (UK), *StressAlert II*, designed for monitoring of the interaction between ice and ship.

We also note that there are no multipurpose systems for monitoring of the hull structure and the shafting and steering module altogether.

A number of technologies under active development, such as Supervisory Control and Data Acquisition (SCADA) [3] and wireless sensor networks [4], have attracted interest as a basis for monitoring systems. The aim of the current project was to develop a system for multipurpose monitoring of an object with the following advantages:

1) It contains all the latest achievements of the analogues.

2) Gives a hall monitoring parameter set from four hull monitoring subsystems and two shafting and steering module monitoring subsystems.

3) Uses fiber optic sensors for increasing the number of control points and decreasing the connection cable length.

4) The ability of system scaling for large and small structures.

5) For Russian companies it is more rapid and comfortable communication with the developer then in case of foreign implementer.

## 2 Multipurpose monitoring systems

The monitoring system provides complex control of the state of the object by measuring a number of parameters. The system consists of subsystems. Fig. 1 presents a complete and detailed decomposition of the monitoring system into subsystems as well as a scheme of the sensor arrangement.

We can decompose our monitoring system into six subsystems. Three subsystems control local (fibre Bragg sensors) and global (GPS/GLONASS, gyroscopes, accelerometers) deformations. The fourth subsystem using acoustic emissions to monitor cracks in the object.

DOI: 10.1051/matecconf/20167504005

The fifth subsystem controls the shafting and steering module. Finally, the last subsystem integrates with pre-installed monitoring systems in the object. This subsystem allows general information about the situation to be visualised on the operator desk during an expedition. It typically includes control of the ice around the ship, a meteorological summary, information about navigation, dynamical parameters (such as speed), fuel consumption, and timeframes for the operation of different systems on the ship.

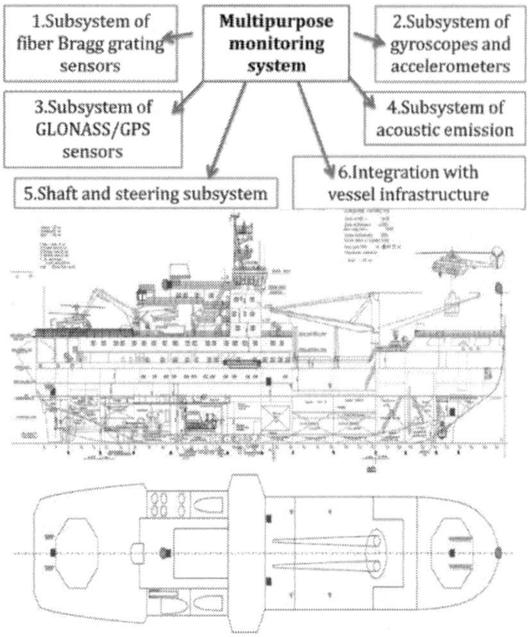


**Figure 1.** (Upper) Decomposition of the system. (Bottom) Structure and arrangement schemes of different types of sensors. Green circles: GLONASS/GPS sensors, violet squares: gyroscopes and accelerometers, yellow triangles: strain sensors.

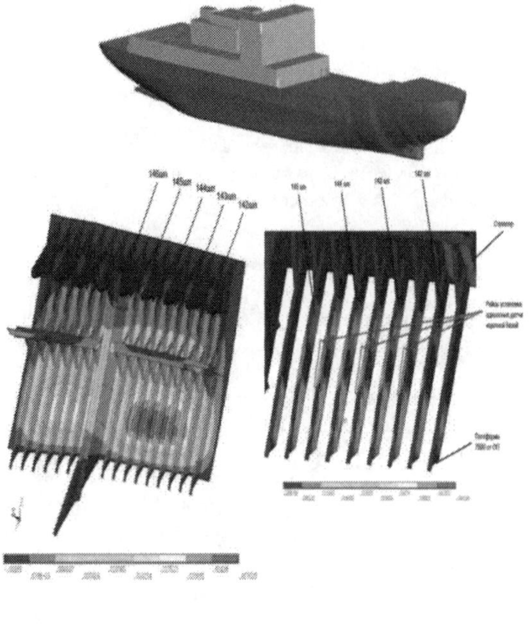

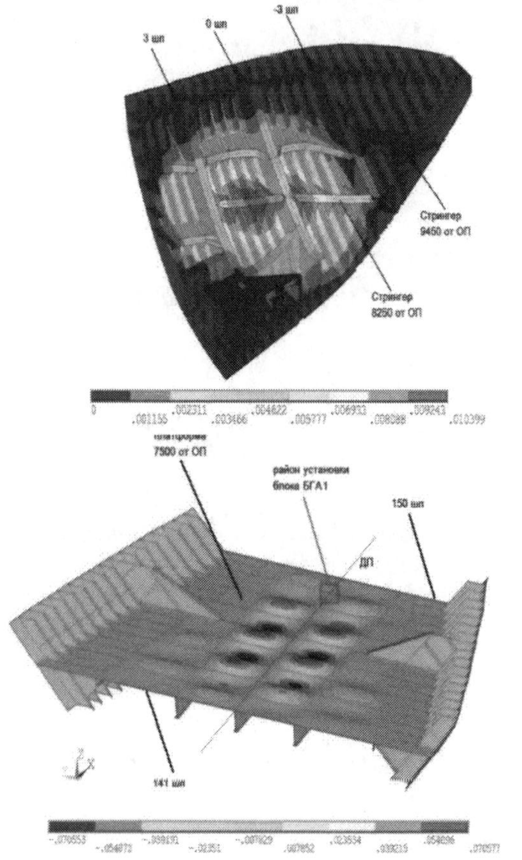

**Figure 2.** 3D numerical simulation results for different parts of the scientific research vessel *Akademik Tryoshnikov*.

Special attention should be paid to the monitoring of the hull construction. The monitoring workflow is based on the work of several subsystems. The gyroscope and accelerometer subsystem measures global deformations of the hull, such as displacements and rotations of the bow. The acoustic subsystem controls the construction integrity, i.e. it prevents crack formation by measuring acoustic pictures. The fibre-based sensor subsystem controls local deformations in different parts of the object. The GPS/GLONASS subsystem monitors longitudinal deformations.

## 3 Numerical simulations

The first steps of the project involved determining the optimal position for all types of sensors. In order to do this, we performed numerical simulations of the ship. The developed model of the object, which was chosen for an expedition in later stages of the project (see Section 5), allows all subsystems to be optimally located.

The model was designed in two basic stages. First, a geometric (CAD) three-dimensional model was created and analysed with the engineering software ANSYS. Second, a geometrical model of the object was analysed in ANSYS with the finite-element method.

DOI: 10.1051/matecconf/20167504005

**Table 1.** Model characteristics before optimisation.

| | |
|---|---|
| Number of elements | 2 645 259 |
| Number of nodes | 2 636 014 |
| Degrees of freedom | 15 816 084 |

**Table 2.** Model characteristics after optimisation.

| | |
|---|---|
| Number of elements | 797 974 |
| Number of nodes | 740 359 |
| Degrees of freedom | 4 442 154 |

Special attention was paid to the problem of model optimisation. A large number of intersection lines for internal partitions of the vessel and the outer skin resulted in a large number of surfaces and boundaries with complex shapes. This complicated the creation of a finite element mesh with acceptable quality. Note that we excluded the vessel superstructure and its internal partitions, and some stiffening ribs were reduced to a common stiffness coefficient.

These assumptions allowed us to successfully address this problem without significant computational resources. Tables 1 and 2 present the achieved results.

To construct the finite-element model, three- and four-node shell elements having six degrees of freedom at each node (three translational and three rotational) were used. Each element was about 0.2 m in size. Fig. 2 presents the results of the numerical simulation.

## 4 Implementation

Each subsystem has its own processing block, which transmits data to the central server of the monitoring system. This server queries all measurements and allows any PC or notebook connected to the system network to access them.

Because the system was installed on the vessel, we had to follow some special requirements. First, all processing blocks were designed in accordance with the IP66 Rating Relationship to Humidity Tolerance.

**Figure 3.** Part of the equipment in IP66 unit cases before being mounted on the vessel.

Note that the fibre-optic cables were halogen-free and flame-retardant, even though this is less safe for fibre splices. Finally, because an icebreaker strongly shakes during its expedition, all devices, sensors, and connection cables needed to be firmly fixed to the installation sites.

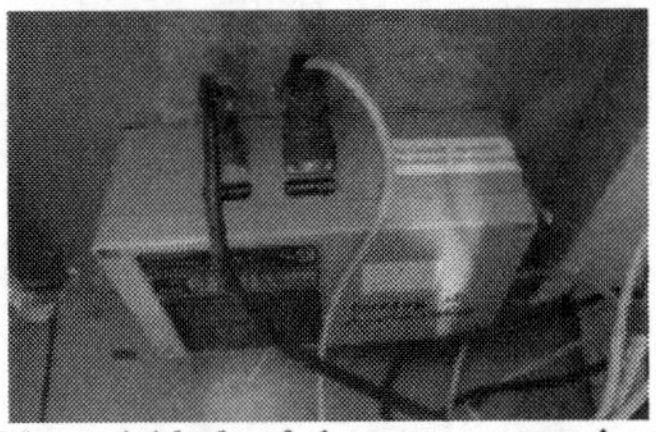

**Figure 4.** Mounted block of the gyroscope and accelerometer subsystem.

**Figure 5.** Mounted fibre Bragg grating strain sensor with the temperature compensator.

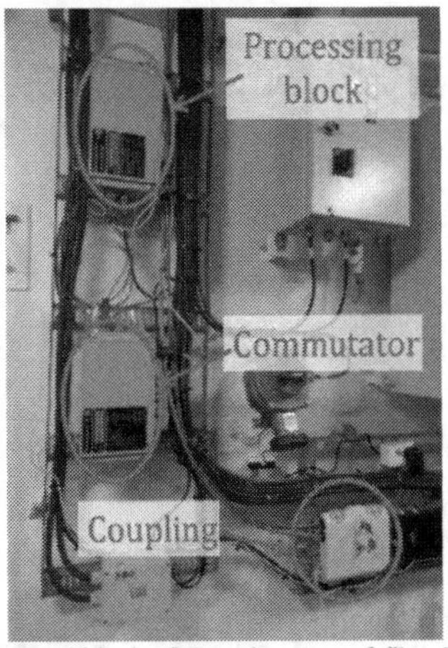


**Figure 6.** Mounted block of the subsystem of fibre Bragg grating strain sensors.

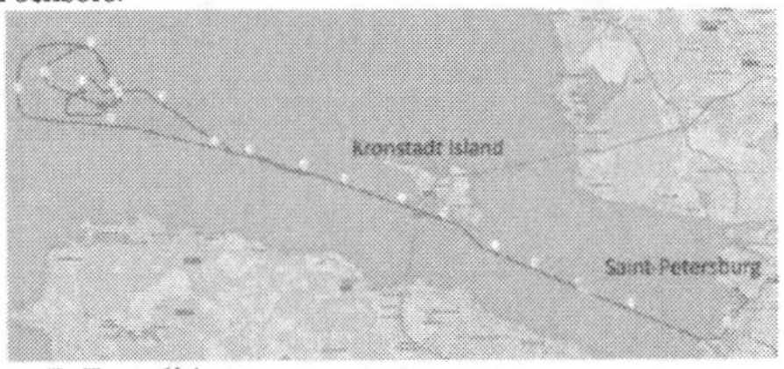


**Figure 7.** Expedition map.

## 5 Expedition: Benchmarks and tests

The above monitoring system was installed on the Russian scientific diesel-electric research vessel *Akademik Tryoshnikov*. All subparts of the monitoring system were installed (see Fig. 1) in the nose and tiller parts (strain gauges, accelerometers, gyroscopes), on the decks (antennas GLONASS/GPS), and tween-decks (strain gauges, accelerometers, gyroscopes, and sensors for acoustic-emission diagnostics).

To collect benchmarks of multipurpose monitoring systems, the systems were tested over a two-day expedition (29 February–1 March 2016) on *Akademik Tryoshnikov*. Fig. 7 presents a map of the expedition.

The vessel was tested under different driving modes: forward and reverse, direct and circular vessel movement, various ice conditions and clean water, and different speeds.

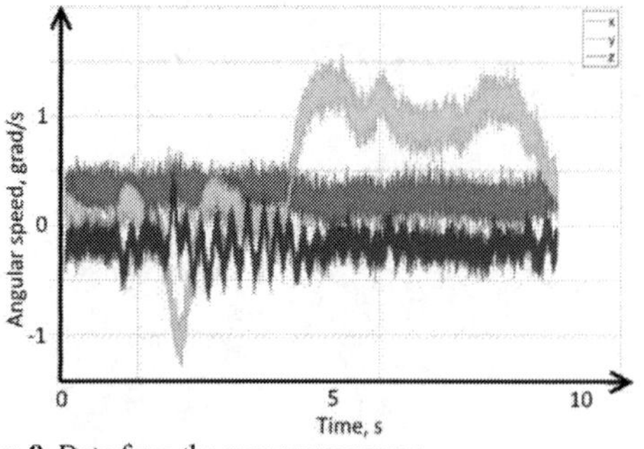


**Figure 8.** Data from the gyroscope sensor.

Fig. 8 shows an example signal from the gyroscope sensor from the tween-decks, which clearly shows the vessel incline during circulation. Fig. 9 shows the results of the GLONASS/GPS subsystem. A possible accuracy of 2 cm was confirmed. The moment of the ship impacting the pier was clearly recorded at 19:30. Fig. 10 presents the signals from the fibre Bragg sensor subsystem. The impact with the pier was also observed. Note that our system uses the method for pre-installed sensor deformation, which is described in [13].

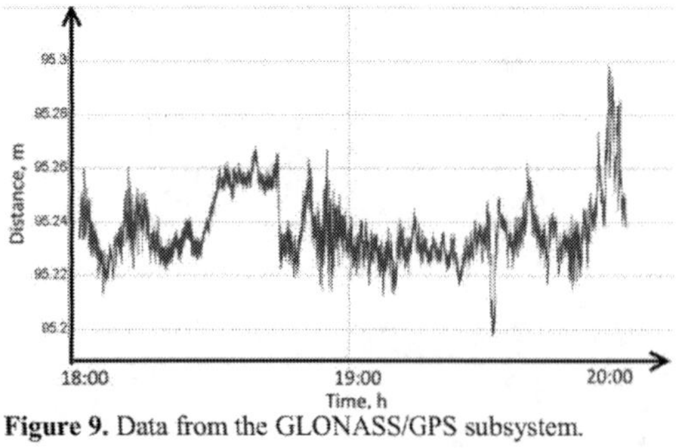


**Figure 9.** Data from the GLONASS/GPS subsystem.

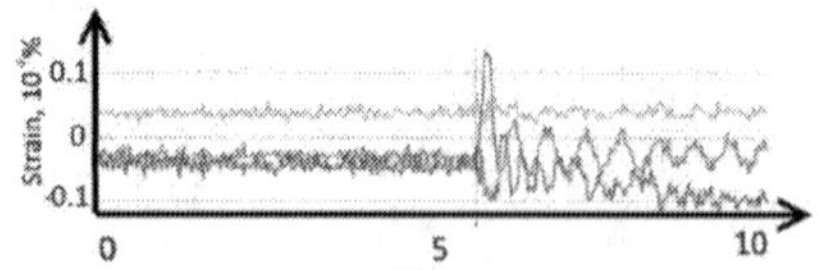


**Figure 10.** Data from a fibre Bragg grating sensor.

The data arriving at the processing server were recorded on data carriers. Based on the collected data, the behaviour of the object during the expedition can be traced. At this time, additional attention should be paid to the moment when the vessel struck the pier.

## 6 Conclusions

A current trend in ocean engineering is the installation of structural health monitoring systems on ships. These prevent accidents, which substantially reduces the cost of technical support.

This paper presents the results of a project on implementing a complete and practical multipurpose monitoring system for ships. Basic operating principles of the developed monitoring system are described along with the results of a 3D finite element analysis of the ship model. The benchmarks and test results of the system during a two-day expedition on *Akademik Tryoshnikov* are presented. The GLONASS/GPS subsystem for large-scale deformation measurement and fibre Bragg subsystem using the compensation algorithm for local deformation measurement were successfully tested. Gyroscopes and accelerometers were useful to identify linear and angular accelerations of different ship parts. Shafting and steering monitoring subsystem measurements were used for vessel motion parameter calculations in different ice loads.

In June and July of 2016 is planned more significant tests of monitoring system during Arctic expedition. Russian fleet has many ice-class ships and platforms, so successful benchmarks may be promising for the massive usage of our system.

## References

1. Bao, X., Chen, L. "Recent progress in distributed fiber optic sensors." Sensors 12 (2012): 8601-8639.
2. Sielski, R.A. "Ship structural health monitoring research at the Office of Naval Research." JOM 64 (2012): 823-827.
3. Zaghloul, M.S. "Online ship control system using Supervisory Control and Data Acquisition (SCADA)." International Journal of Computer Science and Application 3 (2014): 6-10.
4. Xiong, X.-Y., et al. "Vibration monitoring system of ships using wireless sensor networks." IEEE International Conference on Mechatronics and Automation (ICMA), 90 2014.
5. Decò, A., Frangopol, D.M. "Real-time risk of ship structures integrating structural health monitoring data: Application to multi-objective optimal ship routing." Ocean Engineering 96 (2015): 312-329.
6. Structural Monitoring System http://www.scimar.com/
7. SMARTSTRESS http://www.bmtseatech.co.uk/
8. SENSFIB http://www.lightstructures.no/
9. Hull Monitoring System http://www.mcaco.com/
10. Hullmos http://www.rrouvari.fi/
11. Naviscan http://www.sst21c.com/
12. StressAlert II http://www.strainstall.com/
13. Fedorov, A.K., et al. "Structural monitoring system with fiber Bragg grating sensors: implementation and software solution." Journal of Physics: Conference Series 594 (2015): 012049.

DOI: 10.1051/matecconf/20167504006

# The study of the dynamics of the spacecraft landing on a celestial body with microgravity under different fixing conditions

Chen Danhe and Vsevolod V. Koryanov

*Bauman Moscow State Technical University, Dynamics and flight control of motion of rockets and spacecraft Department, 105005, Moscow, Russian Federation*

**Abstract.** The features of angular motion of the spacecraft when landing on the surface of a small celestial body are studied. Different variants of the spacecraft landing using clamping engines and cable are proposed and the definition of the parameters of the landing system works using a fixing system under different conditions. Optimal conditions are recommended for securing the space lander on the surface of a small celestial body.

## 1 Introduction

Exploration of the small celestial body (asteroids, comets, act.) is one of the important space activities in recent years and it is carries out by launching the spacecraft to the small celestial bodies. In previous works of authors has been proposed the method determining the longitudinal motion parameters of spacecraft during landing on the surface of a small celestial body [2], and the method of calculation of dynamics of the angular movement while spacecraft landing on the surface of the celestial body with small gravitational field [3, 4].These methods are developed on the base of tripod lander with influence of clamping engines and cable tension, in order to secure the space lander on the surface of small celestial body [6, 7]. The structure of lander and vector of forces acting on the lander are shown in picture 1, 2. Where, the OXYZ is the surface coordinate system (UCS) - X, Y, Z, is associated with the surface of the celestial body (X and Z axes lie in the plane of the landing surface, the X axis lies in the plane of the first gear).

## 2 Modelling

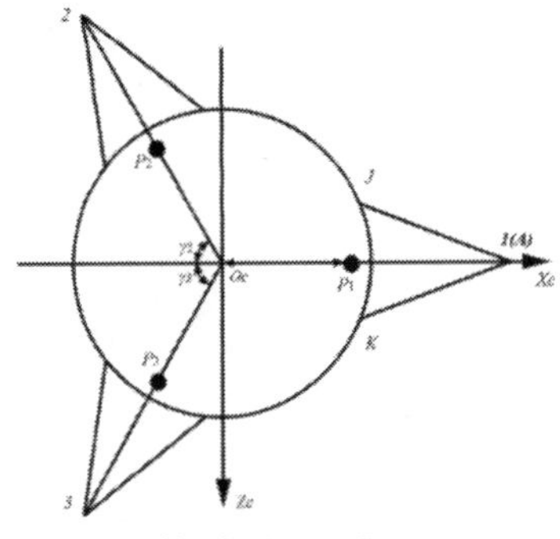

**Figure 1.** The structure of lander in *OXcZc*

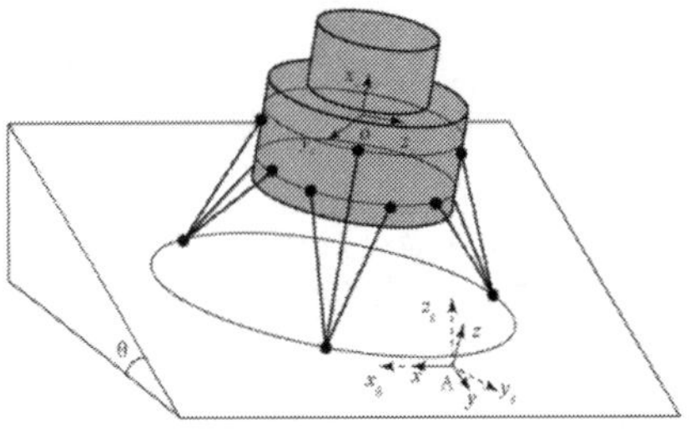

**Figure 2.** View of lander on the surface of celestial body

In Fig. 1: P1, P2, P3 – where fixed clamping engines; $\theta$-slope surface. The differential equation of the spatial movement of lander while landing on the surface of planet is described as below:

$$
\begin{aligned}
& m\frac{dV_x}{dt} = \sum_{j=1}^{n} F_{Txj} + T_x + P_x + mg_x \\
& m\frac{dV_y}{dt} = \sum_{j=1}^{n} F_{Nj} + T_y + P_y + mg_y \\
& m\frac{dV_z}{dt} = \sum_{j=1}^{n} F_{Tzj} + T_z + P_z + mg_z \\
& \frac{dx}{dt} = V_x, \frac{dy}{dt} = V_y, \frac{dz}{dt} = V_z \\
& J_x\frac{d\omega_x}{dt} = \sum_{j=1}^{n} M_{xcj} + M_{Tcx} + (J_y - J_z)\omega_y\omega_z \\
& J_y\frac{d\omega_y}{dt} = \sum_{j=1}^{n} M_{ycj} + M_{Tcy} + (J_z - J_x)\omega_x\omega_z \\
& J_z\frac{d\omega_z}{dt} = \sum_{j=1}^{n} M_{zcj} + M_{Tcz} + (J_x - J_y)\omega_y\omega_x
\end{aligned} \tag{1}
$$

$$\dot{\theta} = \sin\gamma\omega_y + \cos\gamma\omega_z$$

$$\dot{\varphi} = \frac{\cos\gamma\omega_y - \sin\gamma\omega_z}{\cos\varphi}$$

$$\dot{\gamma} = \omega_x - \frac{\cos\gamma\sin\varphi\omega_y - \sin\gamma\sin\varphi\omega_z}{\cos\varphi}$$

where, $F_{Tx}, F_N, F_{Tz}$ -impact on the gears when contacting with the surface; $T_x, T_y, T_z$ - cable tension; $P_x, P_y, P_z$ - clamping force of the engines; $mg_x, mg_y, mg_z$ - projections of gravity on OXYZ axis; $V_x, V_y, V_z$ - the projections of the velocity vector of the CPA on *OXYZ* axis; $x, y, z$ - coordinates of the cm CPA; $\vartheta, \psi, \gamma$ - Euler angles of lander.

In the equations (1) considered two modes of fixing for lander while landing on the surface of celestial body, cable with specific anchor-penetrator, and clamping engines. Equation (1) can be solved by the following initial landing conditions for analyzing the characteristic of dynamical movement of space lander and the influences of fixing on it.

## 2.1 Analysis of influence of clamping engines

Consider different cases using clamping engines that are installed on the space lander. Clamping engines are switched under different conditions. Note that the number NV- embodiment when power is pressing, NV = 1: the motors are switched on when any one leg of lander touches the surface; NV = 2: the motors are turned on when any two legs touch the surface; NV = 3: engines activated when the rate of the space lander on the OY axis is positive. The traction of clamping engines is P = 200N and cable tension force T = 0N, the results of the calculations are shown in the following figures.

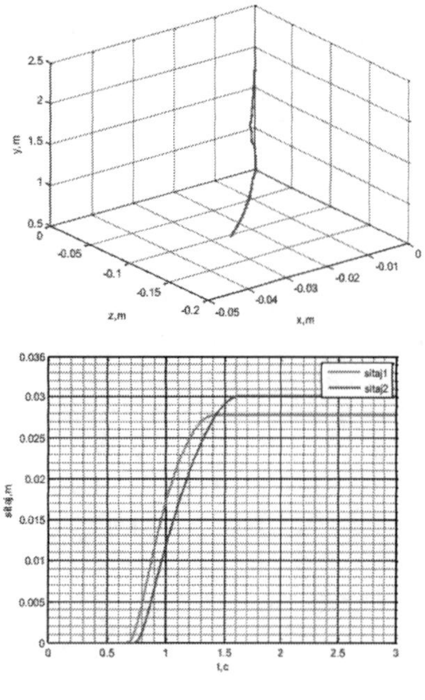


**Figure 3.** Engines operate under option NV=1

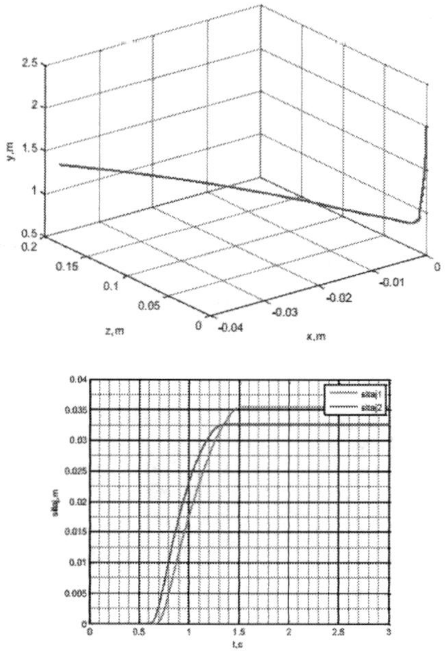


**Figure 4.** Engines operate under option NV=2

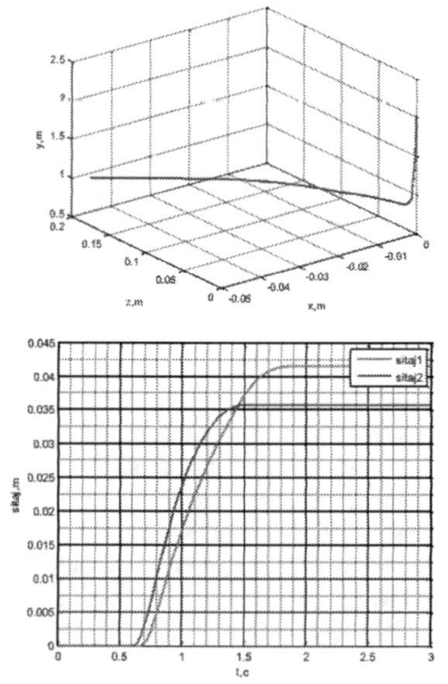


**Figure 5.** Engines operate under option NV=3

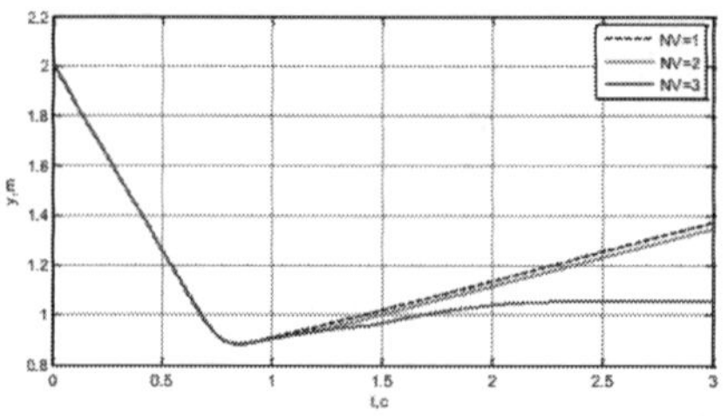


**Figure 6.** Comparison of the characteristics of the vertical position of space lander

DOI: 10.1051/matecconf/20167504006

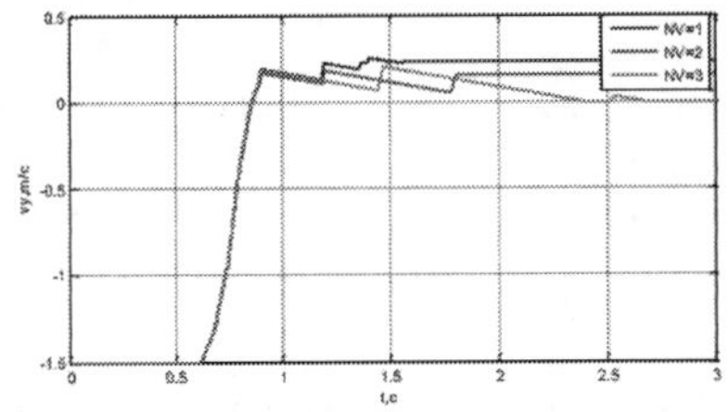

**Figure 7.** Comparison of the vertical velocity characteristics of space lander

The results of calculation results show that landing with the use of clamping engines with a certain tractive force could not afford to provide a secure fixation. Among the three modes clamping engines, which switched on if the space lander obtain positive vertical velocity, have greater stability to ensure a successful landing, compared with other options.

## 2.2 Analysis of influence of cable tension

Landing on the surface of a small celestial body using the clamping engines always contains limitation. At the same time, this study shows the use of special device - Anchor-penetrator, which shoot into the ground surface of a celestial body on landing and is given constant force tension down.

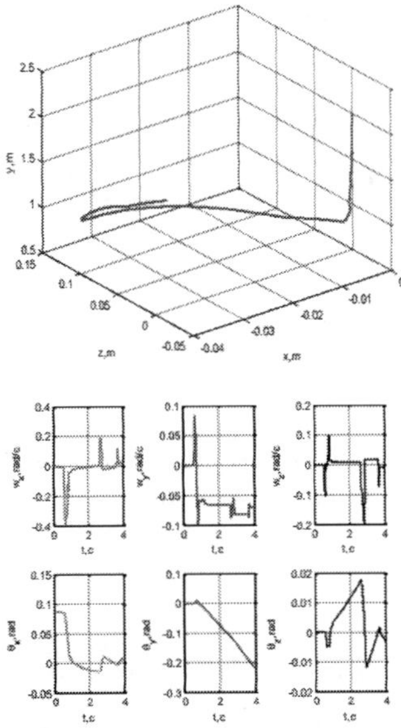

**Figure 8.** The cable works for all landing time, T = 200N, P = 200N

The value of the power cable tension is given by the calculation program, which is here 200N. The work time of the power cable tension is determined by calculating two options. In this paper we propose two options:

The strength of the tension arises immediately after the shot penetrator and is valid for the time of the landing process.

The strength of the cable tension arises under the condition of increasing the length of the rope, the rope starts to pull the unit at a height of one meter before the end of the time of landing.

In such cases, the operating mode of the clamping engines NV = 3 is used. Now consider the case of landing in the presence of the cable tension force for all the time of landing. The results are shown in the following figures:

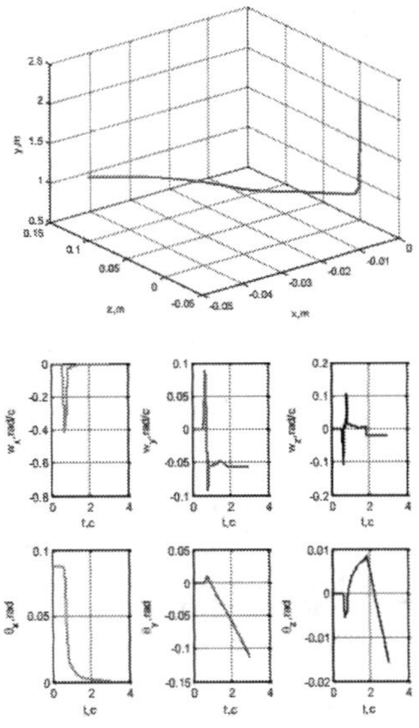

**Figure 9.** The cable operates at a height of one meter, T = 200N, P = 200N

Figure 8 and Figure 9 showed that when the cable is valid for all the time of landing, it is possible to provide a more stable landing.

Now we will analyze the impact of the value of a cable tension on landing. Let the constant force of tension is T=300N and T=400N. In both cases the cable is valid for all the time of landing. The results are shown in the following figures:

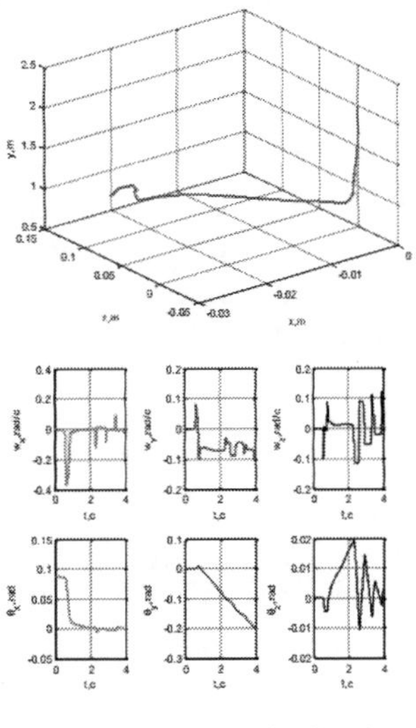

**Figure 10.** The cable works during all landing time, T=300N, P=200N

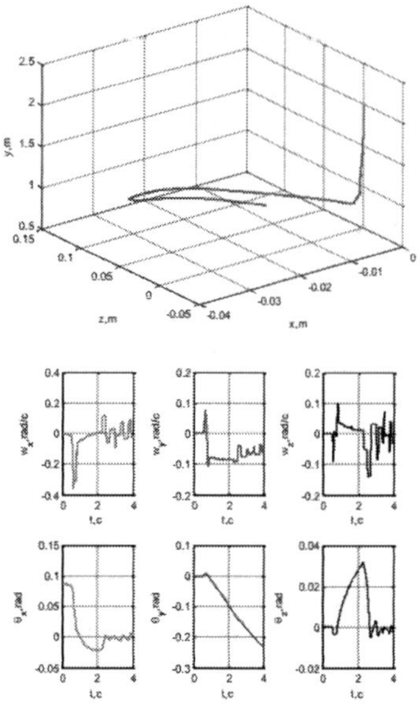

**Figure 11.** The cable works during all landing time, T = 400N, P = 200N

From above presented calculation results it can be concluded that the amount of force the cable tension affects the landing. If the tension relatively small, the cable will not be able to pull the space lander to the surface of the celestial body, in the presence of enough tension force could ensure a successful landing.

## 3 Conclusion

The Analysis of results indicates that the complexation system (clamping engines and cable system) securing the space lander provides a successful landing on the surface of a small celestial body.

Thus, the most recommended conditions for landing on the surface of a small celestial body are such initial parameters specified as the action of a cable system for all landing time with a tension force T = 400N, getting started clamping engine when the speed of the Y-axis is positive. This work is important to find the optimal conditions of landing the space lander implementation and its future challenges. The successful landing of the space lander allows further, not only to study the dynamics of flight spacecraft, but also to develop the space program on the whole.

## References

1. *"Designing descent unmanned spacecraft"* edited by Corresponding member of the Academy of Sciences of the USSR Kovtunenko V.M. M.: "Engineering" (1985)
2. V.P. Kazakoztsev, Chen Danhe, V.V. Koryanov, Method for calculation spacecraft longitudinal motion parameters during landing on the surface of a small celestial body. Herald of the Bauman Moscow State Technical University, series: Mechanical Engineering. 1 (2014) (94). p. 26-35
3. V.P. Kazakovtsev, V.V. Koryanov, Chen Danhe. Dynamics of landing a spacecraft on a celestial body with small field of gravity // 2015 5th international workshop on computer science and engineering: information processing and control engineering, WCSE 2015-IPCE (Information Processing and Control Engineering, Programme (2015) p. 191-195
4. V.P. Kazakovtsev, V.V. Koryanov, Chen Danhe. Investigation of the dynamics of the angular motion of the spacecraft when landing on a celestial body with a small field of gravitation. // Flight.10 (2014) p. 47-53
5. M. Hilchenbach, O. Kuechemann, H. Rosenbauer, Impact on a comet: Rosetta Lander simulations. Planetary and Space Science 48 (2000) p.361±369
6. D Steltzner Adam, K Nasif Annette. Anchoring Technology for In Situ Exploration of Small Bodies[C]//IEEE Aerospace Conference. Big Sky. Montana, USA, (2000) p.507-518
7. Zhijun Zhao, Research on the asteroid landing mechanism and its penetrating and anchoring. Dissertation for the Doctoral Degree in Engineering. Harbin Institute of Technology. (2014) p.2-25
8. E. Robert Lavender. Touchdown Dynamics Analysis of Spacecraft for Soft Lunar Landing. NASA Technical Note D-2001 (1964)
9. V.V. Koryanov, Method of calculating the descent of the spacecraft in the atmosphere using technology adaptation landing in different environmental conditions // Proceedings of the International Astronautical Congress, IAC 8 (2014) p.5628-5632
10. V.V. Koryanov, V.P. Kazakovtsev, Method for calculating the motion parameters of spacecraft. Natural and Technical Sciences. 9-10 (77) (2014) p. 179-184

DOI: 10.1051/matecconf/20167504007

# One-way FSI simulation of the phase and the geometric parameters of the model of compressor blades on the oscillating gas-dynamic processes

Butymova Liliia Nikolayevna and Modorskii Vladimir Yakovlevich

*Federal State-Funded Education Institution of Higher Professional Education Perm National Research Polytechnic University, Center of high-performance computing systems, Perm, Russia (29 Komsomolsky prospekt, Perm – GSP, Perm krai, Russia, 614990)*

**Abstract.** The paper presents the results of numerical simulations to assess the impact of the phase shifts and the geometric design parameters on oscillations in inter-blade channel model compressor gas pumping unit. The blades are considered to be cantilevered plate. At low flow rates and at different angles of phase shift between oscillations of neighboring blades investigated gasdynamic forces acting on the structure. It is shown that the phase shift between the change in the corresponding gas-dynamic forces acting on the blades and vanes oscillations of depends on the phase shift between the oscillations of of the blades.

## 1 Introduction

Analysis of vibrating processes in compressors of gas pumping unit (GPU) shows that the complex vibrations occur comprising a "blade" and "rotor" and the other frequencies.

Investigation of mechanisms of formation of oscillations and the need to develop measures to prevent their extinction or requires computational tools to evaluate not only the frequency and amplitude of the oscillations, but also their phase. The latter is difficult, as it implies the study fast flowing multidimensional, multidisciplinary process, as in the cavity gas-dynamic, and in complex structures, taking into account the rotation.

In present article we study the effect of the phase model of compressor blades oscillations and geometric parameters of the structure on the amplitude, frequency and phase, acting on the blades of gas-dynamic forces. At the same time to facilitate the analysis does not take into account the vibration with the "rotary" frequencies for this selected low flow velocity along the blade, imitating the rotation at low speeds.

## 2 Discussed problems

The problem of occurrence and vibration reduction is closely associated with the problems of ensuring efficiency, noise reduction and meet the requirements of International Civil Aviation Organization (ICAO). There are a number of papers [1]-[13] associated with reduced vibration and noise, the problem is with the actual. This is due to the need to improve existing methods of designing and in-depth study of the mechanisms of occurrence of oscillatory processes.

## 3 Physical model

Physical statement of the problem should be formulated:

- processes are considered to be three-dimensional dynamic; there in the air, which is regarded as compressible as the working fluid; calculation scheme includes two fixed bodies that simulate two adjacent compressor blades; movable bodies are cantilevered; applied to the blades vibration load varying harmonically.

It is assumed that the blade can be independently the same oscillation amplitude and frequency but differing in phase. In this case the blades of gas-dynamic force is applied distributed. It is assumed that if the resultant gas-dynamic forces of the in-phase component of the disturbing vibration, the vibrations may increase. If these oscillations occur with a phase shift, depending on its magnitude may both amplification and attenuation of vibrations. Compressor operation is modeled flow at a low speed oncoming. The phase shift between the oscillations of neighboring blades in the calculations was assumed to be fixed as follows: 0, $\pi/4$, $\pi/2$, $3\pi/4$, $\pi$. Calculation scheme is shown in Fig. 1.

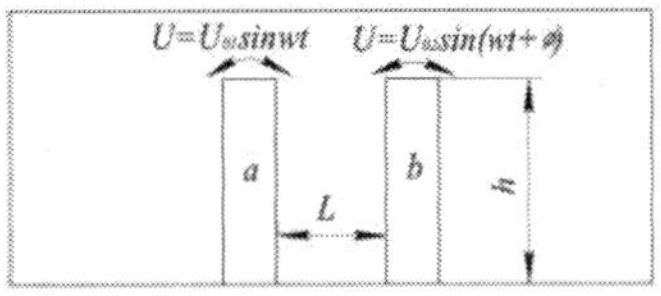


**Figure 1.** Calculation scheme

## 4 Mathematical model

The mathematical model includes the following relations:

- law of conservation of mass

$$\frac{\partial \rho}{\partial t}+\nabla(\rho \vec{V})=0, \quad (1)$$

- law of conservation of momentum

$$\frac{\partial \rho \vec{V}}{\partial t}+\nabla(\rho \vec{V}\otimes\vec{V})= =-\nabla P+\nabla\left((\mu+\mu_t)\left(\nabla\vec{V}+(\nabla\vec{V})^T\right)\right), \quad (2)$$

- law of energy conservation

$$\frac{\partial(\rho H)}{\partial t}+\nabla(\rho\vec{V}H)=\frac{\partial P}{\partial t}+\nabla\left(\left(\frac{\lambda}{c_p}+\frac{\mu_t}{\mathrm{Pr}_t}\right)\nabla H\right), \quad (3)$$

where $t$ - time; $\rho$ - density; $V$ - velocity; $P$ – pressure; $\mu$ - dynamic viscosity; $\mu_t$ - turbulent dynamic viscosity; $H$ - enthalpy; $c_p$ - heat capacity; $\lambda$ - coefficient of thermal conductivity; $\mathrm{Pr}_t$ - turbulent Prandtl number.

It uses well-known $k-\varepsilon$ model of turbulence. We introduce the following relations:

$$\frac{\partial(\rho k)}{\partial t}+\nabla(\rho\vec{V}k)=\nabla\left(\left(\mu+\frac{\mu_t}{\sigma_k}\right)\nabla k\right)+\mu_t G-\rho\varepsilon, \quad (4)$$

$$\frac{\partial(\rho\varepsilon)}{\partial t}+\nabla(\rho\vec{V}\varepsilon)= =\nabla\left(\left(\mu+\frac{\mu_t}{\sigma_\varepsilon}\right)\nabla\varepsilon\right)+C_1\frac{\varepsilon}{k}\mu_t G-C_2 f_1\rho\frac{\varepsilon^2}{k}$$

где $G=D_{ij}\frac{\partial V_i}{\partial x_i}$, $D_{ij}=S_{ij}-\frac{2}{3}\left(\nabla\cdot\vec{V}+\frac{\rho k}{\mu_t}\right)\delta_{ij}$,

$S_{ij}=\frac{\partial V_i}{\partial x_j}+\frac{\partial V_j}{\partial x_i}$, $\sigma_k$ = 1, $\sigma_\varepsilon$ = 1,3, $C_1$ =1,44, $C_2$ =1,92, $f_1$ =1, $k$ - turbulent energy, $\varepsilon$ - turbulent energy dissipation rate, $G$ - determines the rate of generation of turbulent energy, $S_{ij}$ - double rate of deformation tensor.

The equation of turbulent viscosity is calculated using the formula of Kolmogorov-Prandtl:

$$\mu_t=C_\mu\rho\frac{k^2}{\varepsilon}, \quad (5)$$

where $C_\mu$=0,09 – constant.

- equation of state

$$P=\rho RT, \quad (6)$$

where $R$ - gas constant.

Mathematical model closes initial and boundary conditions. The initial values of temperature and pressure as follows: $T_0$ = 20°С , $P_0$=0.1МПа , $V_0$=0,1 м/с .

# 5 Algorithm of numerical modelling

Method of calculation was developed for the calculation of the gas-dynamic interblade channel, comprising the steps of:

1) Development calculated scheme, which is a tube of rectangular cross section, arranged in it two cantilevered blades.

2) Design of gas-dynamic cavity is the space between the blades, as well as the inlet and outlet pipes of the model.

3) Construction of the gas-dynamic calculation grid cavity is a box in one cell width.

4) It is necessary to assign to the lateral surface of the blade model compressor boundary condition «Wall», appoint perpendicular to the shaft axis of the plane on the surface bounding Gasdynamic cavity boundary condition «Symmetry», assign to the side surfaces of the blade condition «Wall», with a special option to Mesh Motion grid " Specified Displacement ».

5) Set properties of the gas flowing in the gas cavity. To do this, choose from a library of materials ANSYS CFX software package, you need a working body.

6) Define settings for solving the problem, convergence parameters and monitoring the progress of calculation in the software package ANSYS CFX, appointing Solver Control and Output Control option.

7) Identify the necessary resources to solve the problem. Running high-performance computing solutions to complex of Perm National Research Polytechnic University allows an assessment of the necessary cluster resources.

8) Processing gas-dynamic calculation of the results, the conclusion of necessary variables as fields fills, graphs, tables, vectors, streamlines.

# 6 Computational examples and analysis

If we consider the process of gas-dynamic flow between two adjacent blades of the rotor rim, it is possible in addition to the specification of "rotary" and "blade" of frequencies, to clarify the mechanism of occurrence of vibrations, namely in the part related to the spread of the gas-dynamic flow between the shoulder blades. As a first approximation, the analysis is reduced to the search for a model similar to the problem of motion of the gas-dynamic waves in an open pipe or on the motion of the gas-dynamic waves in the open at one end of the tube. Significant factors in the analysis of this process is the interscapular distance and the height of the blades. In some cases, one factor dominates, in others - the other.

Tables 1, 2, 3 shows the relationship between the predetermined phase shift and structure oscillations obtained in the course of the numerical solution of the phase shift between gas-dynamic forces acting on the blades, and oscillations phase shift between the blade and the gas-dynamic force acting on it.

**Table 1.** Phase shifts in oscillations in the gas and construction at L = 50 mm and a cantilevered

| C onstr uction | Gas dynamics | Gasdynamics-construction |
|---|---|---|
| | | *The phase oscillations between the blade and the gas-dynamic force acting on it* |

| φ c *The phase shift of neighboring blades oscillations* | φ gas *The phase oscillations of gas-dynamic forces acting on the adjacent blades* | φ left *left blade* | $F_{left}$, H *Gas-dynamic force* | φ right *right blade* | $F_{right}$, H *Gas-dynamic force* |
|---|---|---|---|---|---|
| 1 | 2 | 3 | 4 | 5 | 6 |
| 0 | 0 | 0, -π/2 | 0,02 | 0, -π/2 | 0,03 |
| π/4 | 0 | π/4 | 0,6 | 0 | 0,69 |
| π/2 | π | - π/2 | -0,8 | - π | 0,7 |
| 3π/4 | - π/2 | - π/4 | 0,6 | 0 | 0,63 |
| π | π/4 | - π | 0,01 | 0 | 0,01 |

**Table 2.** Phase shifts in oscillations in the gas and construction at L = 100 mm and a cantilevered

| C onstr uction | Gas dynamics | Gasdynamics-construction *The phase oscillations between the blade and the gas-dynamic force acting on it* | | | |
|---|---|---|---|---|---|
| φ c *The phase shift of neighboring blades oscillations* | φ gas *The phase oscillations of gas-dynamic forces acting on the adjacent blades* | φ left *left blade* | $F_{left}$, H *Gas-dynamic force* | φ right *right blade* | $F_{right}$, H *Gas-dynamic force* |
| 1 | 2 | 3 | 4 | 5 | 6 |
| 0 | 0 | 0 | 0,1 | 0 | 0,1 |
| π/4 | - π/2 | - 3π/4 | 0 | π/2, 0 | 0,44 |
| π/2 | π | - π/2 | 1 | - π | -1,75 |
| 3π/4 | - π/2 | - π/2 | 0,09 | 0 | 0,5 |
| π | - π/2 | - π/4 | 0,01 | 0 | 0,01 |

**Table 3.** Phase shifts in oscillations in the gas and construction at L = 139 mm and a cantilevered

| C onstr uction | Gas dynamics | Gasdynamics-construction *The phase oscillations between the blade and the gas-dynamic force acting on it* | | | |
|---|---|---|---|---|---|
| φ c *The phase shift of neighboring blades oscillations* | φ gas *The phase oscillations of gas-dynamic forces acting on the adjacent blades* | φ left *left blade* | φ c *The phase shift of neighboring blades oscillations* | φ gas *The phase oscillations of gas-dynamic forces acting on the* | φ left *left blade* |
| 1 | 2 | 3 | 4 | 5 | 6 |
| 0 | 0 | 0 | 0,09 | 0 | 0,06 |
| π/4 | - π/2 | π/4 | 0,5 | π/4 | 0,08 |
| π/2 | π | π/2 | -1 | π/2 | 1,49 |
| 3π/4 | - π/2 | 3π/4 | 0,58 | 3π/4 | 0,18 |
| π | - π/2 | π | 0,01 | π | 0,01 |

**Table 4.** By the analysis of oscillatory processes

| L, m | $h_b$, m | a, m/s | $f_L$, Hz | $f_h$, Hz | $f_L \backslash f_h$ |
|---|---|---|---|---|---|
| 0,05 | 0,15 | 330 | 1320 | 2200 | 0,6 |
| 0,1 | | | 6600 | | 3 |
| 0,139 | | | 4700 | | 2,1 |

Tables 1, 2, 3, F> 0 corresponds to a positive working gas dynamic forces, for F <0 - negative. Analysis tables allowed to obtain dependence gas dynamic forces of inter-blade distance (F=f(L)) for different shifts phases of neighboring blades(φ ). Example dependences obtained is shown in Fig. 2.

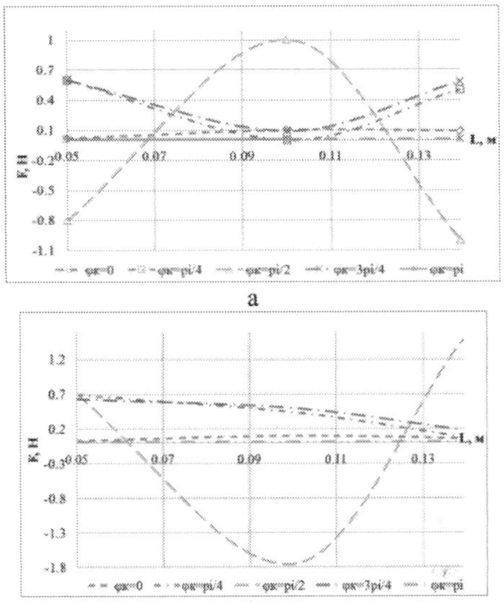


**Figure 2.** Dependence of gas dynamic forces from the inter-blade distance:
a – left blade; b – right blade

If inter-blade distance L is much greater than the height of the blade $h_n$ , ($L>>h_n$ ) and φ =0, then:

- with oscillating movements of the blades in interblade space compression waves and expansion waves. They move toward each other and meet in the inter-blade space. Accordingly, real time, which is necessary for a travel is reduced by almost half ($L>>h_n$ ). The natural frequency in this case is calculated as f=a/L.

If the blade height $h_n$ much larger inter-blade distance dimensions L, namely $h_n >> L$:

- gas-dynamic flow is distributed in the inter-blade space as in a tube closed at one end. This changes the oscillation frequency of the gas-dynamic force, which is calculated according to the formula f=a/2L. In this case, the height of the blade determines the lower the natural frequency of the gas cavity between the shoulder blades ($h_n >> L$).

As a result of numerical experiment results were as follows.

In Table 4 $f_L$ – lower natural frequency inter-blade space in the tangential direction, L – the distance between blades, a – velocity of sound in the environment, $h_b$ – blade height, $f_h$ – lower natural frequency inter-blade space radially.

Table 4 shows that the inter-blade distance of 100 mm, the resonance interaction occurs, since in this case there is a multiplicity of eigenfrequencies.

In deriving the dependency of the amplitude oscillations of the gas-dynamic forces interscapular distances for different phase shifts between the oscillations of the displacement of neighboring blades ($\varphi$ ), the results were:

If L = 50 mm:

- on the left blade, when the gas-dynamic work force is greater than zero ($A_F$>0) there is a maximum amplitude in the gas-dynamic forces at $\varphi = 3\pi/4$. When the gas-dynamic work force is less than zero ($A_F$<0), maximum gas dynamic force amplitude observed at $\varphi = \pi/2$, minimum – at $\varphi = \pi$ (Fig. 2a);
- on the right blade, when the gas-dynamic work force is greater than zero $A_F$>0 maximum gas dynamic force amplitude observed at $\varphi = \pi/2$, minimum – at $\varphi = \pi$ (Fig. 2b);

If L = 100 mm:

- on the left blade, when the gas-dynamic work force is greater than zero $A_F$>0 maximum gas dynamic force amplitude observed at $\varphi = \pi/2$, minimum – at $\varphi = \pi$ (Fig. 2a);
- on the right blade, when the gas-dynamic work force is greater than zero $A_F$>0 maximum gas dynamic force amplitude observed at $\varphi = 3\pi/4$. When $A_F$<0 maximum gas dynamic force amplitude observed at $\varphi = \pi/2$, minimum – at $\varphi = \pi$ (Fig. 2b);

If L = 139 mm:

- on the left blade, when the gas-dynamic work force is greater than zero $A_F$>0 maximum gas dynamic force amplitude observed at $\varphi = 3\pi/4$. When $A_F$<0 maximum gas dynamic force amplitude observed at $\varphi = \pi/2$, minimum – at $\varphi = \pi$ (Fig. 2a);
- on the right blade, when the gas-dynamic work force is greater than zero $A_F$>0 maximum gas dynamic force amplitude observed at $\varphi = \pi/2$, minimum – at $\varphi = \pi$ (Fig. 2b);

## 7 Conclusions

1) These studies make it possible to use them as a base for carrying out complicated calculations and analysis of the results, taking into account: real geometry, rotation at different speeds, two-way FSI - interaction.
2) The study results allow to specify the "blade" of the frequency and the factors influencing them, and thus to clarify the issues related to the resonant interaction in the system "gas - construction" during the vibrations
3) Installing bandage on the shoulder can afford to change the oscillation frequency, and tune out the resonant interaction in the system "gas - construction."

## Acknowledgements

The research work was supported by Russian Science Foundation under Grant No. 14-19-00877.

## References

1. A. Shmakov, V. Modorskii, Modeling of oscillating processes in elements construction SRM, Scientific and Technical Volga region Bulletin, **6**, 368-371 (2013).
2. E. Mekhonoshina, V. Modorskii, On a phase-shift of waves at the medium interface, Computer Optics, **39**(3), 385 (2015).
3. G D Ujade and M B Bhambere, Review of Structural and Thermal Analysis of Gas Turbine Blade, International Journal of Mechanical Engineering and Robotics Research, **3** (2), 347-352, (2014).
4. A. Shmakov, V. Modorskii. Energy Conservation in Cooling Systems at Metallurgical Plants, Metallurgist, **59**(9), 882-886 (2016)
5. A. Shmakov, V. Modorskii. Numerical Modeling of Gasdynamic Processes and Processes of Deformation in Compressor of Model Test Bench of the Gas-Distributing Unit, AMM, 799-800, 865-869 (2015)
6. M. Dimentberga, S. Kolesnikov, Vibrations in thechnics. Vibrations machines, structures and their elements (Moscow, Mashinostroyeniye, 1980)
7. V. Modorskii and other, The solution of engineering problem on the high-performance computer system of the Perm national research polytechnical university: monograph. (Perm national research polytechnical university, 2014)
8. Butymova L.N., Modorskii V.Ya., Petrov V.U., Numerical simulation of model compressor blade oscillation influence on gas-structure system, Scientific and Technical Volga region Bulletin, **6**, 161–163 (2015).
9. Modorskii V.Ya., Sokolkin U.V., "Gasoelastic processes in power plants: monograph. (FIZMATLIT, 2007).
10. Balakrishnan, A. V., Aeroelastic flutter analysis in viscous flow: continuum models, **16** (3), 221 (2009)
11. Mekhonoshina E.V., Modorskii V.Ya., Petrov V.Yu, Numeric simulation of the interaction between subsonic. (Source of the Document CEUR Workshop Proceedings, Conference «Information Technology and Nanotechnology, (ITNT-2015), Mathematical Modeling) 211-218 (2015)
12. Ponik A. N., Modorskii V. Ya., Kozlova A. V., Sokolkin Yu. V., Numerical modeling of flow mixing and cooling processes in a perforated exhaust duct, Russian Aeronautics (Iz VUZ), **57**(2), 181-186 (2014)
13. Modorskii V.Y., Petrov V.Y., Aeroelastic Deformation of a Solid-Propellant Rocket Engine During Supercharging, Russian Aeronautics (Iz VUZ), **54** (3), 272-275 (2011)

# An Increase in Estimation Accuracy Position Determination of Inertial Measurement Units

Ladislav Beran[a], Pavel Chmelar and Lubos Rejfek

*Department of Electrical Engineering, Faculty of Electrical Engineering and Informatics, University of Pardubice,Pardubice, Czech Republic*

**Abstract.** This paper deals with an increase in measurement accuracy of the Inertial Measurement Units (IMU). In the Inertial Navigation Systems (INS) a fusion of gyroscopes, accelerometers and in some cases magnetometers are typically used. The typical problem of cheap IMU is non-stationary offset and high level of noise. The next problem of IMU is a problem with a bumpy floor. For this case it is necessary to a have high quality chassis to eliminate additional noise. Also, it is possible to eliminate this noise by using some algorithm, but results are still poor. These properties lead to the inaccurate position estimation in the integration process. Even a small offset error leads to a big mistake in position determination and grows quickly with a time. This research is focused on the elimination of these poor properties and increase of accuracy of position estimation using Kalman Filtration.

## 1 Introduction

Precise navigation of mobile robots in unknown areas is common issue in much research. For the robot navigation in exterior the Global Navigation Satellite System (GNSS)is commonly used. For better results in outdoors is also possible to use the Satellite Based Augmentation Systems (SBAS). In Europe the European Geostationary Navigation Overlay Service (EGNOS)is usually used, while in US Wide Area Augmentation System (WAAS)is used. The benefit of using GNSS navigation systems is the achieved accuracy, typically better than 15 meters. Our research is focused on the autonomous navigation of a robotics platform in indoor spaces like ruins, mines, etc. The GNSS navigation system is not suitable for indoor areas because in the buildings navigation does not have a signal received from satellites or base stations. For these cases, it is necessary to develop an alternative navigation system.

The navigation of a robotics platform in indoor spaces, a modern navigation system called visual odometry can be used. This system is known from the eighties, but massive penetration is connected with fast Digital Signal Processors (DSP). This system is based on capturing a scene by the camera. The algorithm calculates interesting points from an actual and previous captured scene. The algorithm searches pairs of interesting points and afterwards calculate a distances between these pairs, called pixel distance. Also it is necessary to use some type of transformation to transform pixel distances to real coordinates. The distance estimation is performed using the common elevation angle of a sensing camera, the size of a sensing chip and focal length. This problem is described in detail in papers [1], [2], [3] and [4].

An alternative navigation system is called Inertial Navigation System (INS). This system uses accelerometers and gyroscopes. High level of noise and non-stationary offset is typical for cheap measurement units. These properties lead to the inaccurate acceleration estimation in the integration process. This is typical especially for low-cost accelerometers. Even a small offset error leads to a big mistake in position determination and grows quickly with time.Normally, the high level of noise is possible to eliminate using advanced digital filters [5], but in this case we need a special filter like Kalman filter. In our research we are focused on the eliminating a wrong position estimation using a Direct Cosine Matrix, Odometrymeasurement and Kalman filter [6]. The final device will be used as fusion of a navigation of robotics platform with an optical measurement device [7], [8], [9].

## 2 Rotation formalism

### 2.1 Direct cosine matrix (DCM)

The Direct Cosine Matrix provides transformation vectors in Euclidean space around an origin by an angle $\theta$. In practical use it is necessary to have some rotation formalism to provide a rotation around three independent axes called *Yaw*, *Pitch*, *Roll* (in some cases called $\alpha$, $\beta$ and $\gamma$). These axes are shown in Figure 1.

Basic clockwise rotation around an x-axis by an angle $\gamma$ in three dimension space can be written in a matrix form (1). Similar equations (2), (3) exists for an y-axis and an z-axis.

[a] Corresponding author: ladislav.beran@student.upce.cz

$$R_x(\gamma) = \begin{bmatrix} 1 & 0 & 0 \\ 0 & \cos\gamma & -\sin\gamma \\ 0 & \sin\gamma & \cos\gamma \end{bmatrix} \quad (1)$$

$$R_y(\beta) = \begin{bmatrix} \cos\beta & 0 & \sin\beta \\ 0 & 1 & 0 \\ -\sin\beta & 0 & \cos\beta \end{bmatrix} \quad (2)$$

$$R_z(\alpha) = \begin{bmatrix} \cos\alpha & -\sin\alpha & 0 \\ \sin\alpha & \cos\alpha & 0 \\ 0 & 0 & 1 \end{bmatrix} \quad (3)$$

where $R_n$ is the rotation matrix defined on an axis and $\alpha$, $\beta$ and $\gamma$ are angles of clockwise rotation.

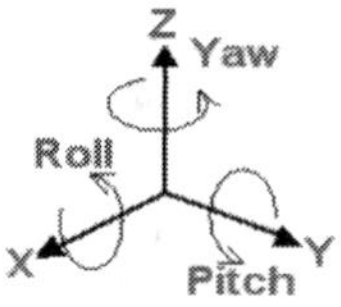


**Figure 1.** Euler angles.

A final three-dimension rotation matrix is created by a simple matrix multiplication (4).

$$R = R_x(\gamma)R_y(\beta)R_z(\alpha) \quad (4)$$

### 2.2 Position estimation correction using DCM

The correction of position estimation using DCM by simple matrix multiplication is provided using (4).

The corrected results using DCM transformation was presented in [1]. Precision of results was poor and it is necessary to increase precision to use this system for fusion with a higher precision navigation system based on visual odometry. In this time, we are testing a suitability of increasing anestimation accuracy position determination of IMU only on direct path. The DCM correction will be implemented to the final version of robotics platform navigation – navigation of robotics platform using a fusion of visual odometry, Inertial Measurement Unit and optical odometers at every wheel.

### 2.3 Position estimation correction using moving average filters

The moving average filters are usually used to eliminate the extreme deviations in a data series. The diagram of fifth order Moving Average Filter (MVA) is shown inFigure 2.

In our research we tried to use this type of filtration to eliminate extreme deviations in our measured data using IMU. The results of using this type of filtration are shown inFigure 6.

$$y_k = \sum_{i=-\frac{M-1}{2}}^{\frac{M-1}{2}} (A_i x_{k+i}) \quad (5)$$

where $M$ is an order of filter, $Ai$ are coefficients of filter, $x_{k+i}$ are input data, $y_k$ are filtered output data

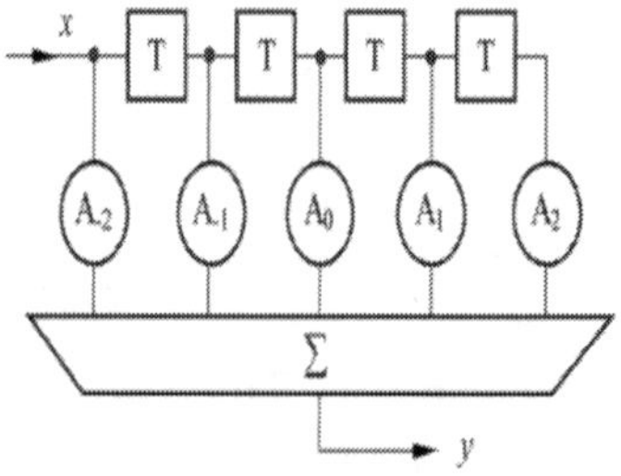


**Figure 2.** The moving average filter for fifth order [7].

## 3 Kalman filter and it´s Implementation to our system

The Kalman filter is a modern, effective and robust method for solving problems of dynamic systems with random signals. The implementation and using of Kalman Filter is described in papers [10], [11]. The Kalman filter is established using equation (6) and (7).

$$\boldsymbol{x}(t+1) = \boldsymbol{A}\cdot\boldsymbol{x}(t) + \boldsymbol{B}\cdot\boldsymbol{u}(t) + \boldsymbol{v}(t) \quad (6)$$

$$\boldsymbol{y}(t) = \boldsymbol{C}\cdot\boldsymbol{x}(t) + \boldsymbol{D}\cdot\boldsymbol{u}(t) + \boldsymbol{e}(t) \quad (7)$$

where $\boldsymbol{x}(t)$ is state vector in time $t$, $\boldsymbol{y}(t)$ is a vector with measurements in time $t$, $\boldsymbol{u}(t)$is thedrive state vector, $\boldsymbol{v}(t)$ is process noise vector, $\boldsymbol{e}(t)$ is the measurement noise vector, ***A,B,C,D*** are matrixes,which providesbehaviour of system.

In our case, a model of traveled distance we can establish using equation (8).

$$x(t) = x_0 + v\cdot t + \frac{a\cdot t^2}{2} \quad (8)$$

where $x(t)$ is a distance in time, $x_0$ is an initial distance, $v$ is a speed, $a$ is an acceleration, $t$ is a time.

Using equation (9) we can establish a time-dependent system of equations.

$$\begin{bmatrix} x_{m1} \\ x_{m2} \\ \vdots \\ x_{mN} \end{bmatrix} = \begin{bmatrix} x_0 \\ x_0 \\ \vdots \\ x_0 \end{bmatrix} + \begin{bmatrix} v \\ v \\ \vdots \\ v \end{bmatrix} \cdot \begin{bmatrix} t_1 \\ t_2 \\ \vdots \\ t_N \end{bmatrix} + \frac{a}{2} \cdot \begin{bmatrix} t_1^2 \\ t_2^2 \\ \vdots \\ t_N^2 \end{bmatrix} \quad (9)$$

where $\boldsymbol{x}_m$ is a vector of measured results, $\boldsymbol{x}_0$ is a vector containing the initial distance (before measurement),$\boldsymbol{v}$ is a vectorcontaining theinitialspeed (before measurement)

The vector $x_m$ is a vector which contains measured distances according to time and it is reference to our system.

$$y_e(t) = y_e(t-1) + \frac{a(t)}{2}t^2 \quad (10)$$

where $y_e(t)$isan actual estimated path using acceleration from accelerometers, $y_e(t-1)$is a previous value of estimated path, a(t) is time-depend acceleration acquired from accelerometers.

In our case we can rewrite an equation (6) and (7) to easier form, because theinformation about the influence of the control signalshave not been assumed.

$$\boldsymbol{x}(t+1) = \boldsymbol{A}\cdot\boldsymbol{x}(t) + \boldsymbol{v}(t) \quad (11)$$

$$\boldsymbol{y}(t) = \boldsymbol{C} \cdot \boldsymbol{x}(t) + \boldsymbol{e}(t) \quad (12)$$

The behaviour of a system is defined by the simple state vector. This vector is created using measured distance and calculated speed.

$$\mathrm{x(t)} = \begin{bmatrix} \mathrm{x(t)} \\ \mathrm{s(t)} \end{bmatrix} \quad (13)$$

where *s(t)* is calculated speed - derivation a distance by time.

$$\mathrm{x(t+1)} = \begin{bmatrix} \mathrm{x(t+1)} \\ \mathrm{s(t+1)} \end{bmatrix} = \begin{bmatrix} \mathrm{x(t) + s(t) \cdot T} \\ \mathrm{s(t)} \end{bmatrix} =$$

$$= \begin{bmatrix} 1 & \mathrm{T} \\ 0 & 1 \end{bmatrix} \begin{bmatrix} \mathrm{x(t)} \\ \mathrm{s(t)} \end{bmatrix} = \mathrm{A \cdot x(t)} \quad (14)$$

$$\mathrm{y(t)} = [1 \quad 0] \cdot \begin{bmatrix} \mathrm{x(t)} \\ \mathrm{s(t)} \end{bmatrix} = \mathrm{C \cdot x(t)} \quad (15)$$

The last variables are $\boldsymbol{v}(t)$ and $\boldsymbol{e}(t)$. The noise process vector $\boldsymbol{v}(t)$ have two elements because the state vector $\boldsymbol{x}(t)$ also have two elements (measured distance *x(t)* and it´s derivation–speed *s(t)*).

$$\mathrm{v(t)} = \begin{bmatrix} \mathrm{v_x(t)} \\ \mathrm{v_s(t)} \end{bmatrix} \quad (16)$$

$$\mathrm{e(t)} = \begin{bmatrix} \mathrm{e_y(t)} \end{bmatrix} \quad (17)$$

The Kalman filter usingVarianceMatrix of noise of process signed Q and Variance Matrix of noise of measurement signed R. The both matrixes are possible to establish using equations (16), (17), (18) and (19)

$$\mathrm{Q} = \mathrm{var}\big(\mathrm{v(t)}\big) = \begin{bmatrix} \mathrm{var(v_x)} & \mathrm{cov(v_x, v_S)} \\ \mathrm{cov(v_S, v_x)} & \mathrm{var(v_S)} \end{bmatrix} \quad (18)$$

$$\mathrm{R} = \mathrm{var}\big(\mathrm{e(t)}\big) = \big[\mathrm{var(e_y)}\big] \quad (19)$$

The Kalman filter minimalizes the mean value of trace of the correlation matrix.

$$\mathrm{P(t)} = \mathrm{E}\Big[\big(\mathrm{x(t)} - \hat{\mathrm{x}}(\mathrm{t})\big) \cdot \big(\mathrm{x(t)} - \hat{\mathrm{x}}(\mathrm{t})\big)^{\mathrm{T}}\Big] \rightarrow \mathrm{MIN} \quad (20)$$

where *E[...]* is mean value, $\hat{x}(t)$ is an optimal result (the best estimation value of *x(t)*)

In the Kalman filter is recalculation step and predictive step. In the recalculation part, the estimations of the state vector and a correlation matrix of error are increasing to the time *t-1*.

$$\mathrm{K} = \mathrm{P(t-1) \cdot C^T \cdot (C \cdot P(t-1) \cdot C^T + R)^{-1}} \quad (21)$$

$$\mathrm{P(t)} = \mathrm{P(t-1) - K \cdot C \cdot P(t-1)} \quad (22)$$

$$\hat{\mathrm{x}}(\mathrm{t}) = \hat{\mathrm{x}}(\mathrm{t}-1) - \mathrm{K} \cdot (\mathrm{y(t)} - \mathrm{C} \cdot \hat{\mathrm{x}}(\mathrm{t}-1)) \quad (23)$$

$$\boldsymbol{P}(t+1) = \boldsymbol{A} \cdot \boldsymbol{P}(t) \cdot \boldsymbol{A}^T + \boldsymbol{Q} \quad (24)$$

$$\hat{\boldsymbol{x}}(t+1) = A \cdot \hat{x}(t) \quad (25)$$

Equations (21) to (25) provide a recursive solution of a model which is described by the equations (11) and (12).

The last step of The Kalman filter design is setting of initial values. The best estimated value $\hat{\mathrm{x}}(\mathrm{t})$ is equal to zero, because the initial speed and distance is also equal to zero.

$$\hat{\mathrm{x}}(\mathrm{t}=0) = \begin{bmatrix} \mathrm{x(0)} \\ \mathrm{s(0)} \end{bmatrix} = \begin{bmatrix} 0 \\ 0 \end{bmatrix} \quad (26)$$

## 4 Propossed approach

In our research we are using a wireless-controlled robotics platform. On this platform is fitted a cheap Inertial Measurement CHR-6Dand user-made optical odometer. The parameters of the optical odometer are shown at Table 1. In Table 2 the chassis parameters are shown.

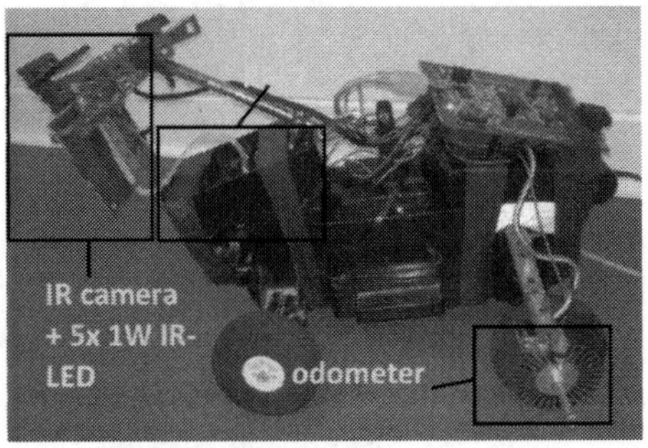


**Figure 3.** Our robotics platform.

**Table 1.** Optical encoder wheel parameters.

| **Optical encoder wheel parameters** | |
|---|---|
| **Parameter** | **Value** |
| Outside Radius | 60mm |
| Inner Radius | 45mm |
| Pixel length (at outside radius) | 2,94mm |
| no. of steps | 64 steps |
| bits | 1 bit |

**Table 2.** Chassis parameters of our robotics platform.

| **Robotics chassis configuration** | |
|---|---|
| **Parameter** | **value** |
| Nr. of Wheels | 4 |
| Wheel diameter | 58 mm |
| Wheel circumference | 182,21 mm |
| Chassis type | Automobile type - controllable front part |

The IMU is based on accelerometer ADXL335 and gyroscopes LP510AL for roll and pitch axes and LY510ALH for yaw axis. All parameters of IMU are written in Table 3.The odometer is based on an infrared light barrier and a 64 step encoder wheel. The parameters of optical wheel are written inTable 1. The wheel rotation is captured by a STM32F4Discovery development kit. The measured data from the odometer is transferred to the Raspberry PI 2 via a serial link every 0.25 seconds. In this time, data from odometer and IMU are transferred via a wired connection, but in near future this connection will be upgraded to the Bluetooth connection. In this time, the Raspberry Pi 2 is used for recording data stream. Then, the captured data is processed in Matlab software. In the future, all algorithms will be implemented at this platform. In Figure 3 our robotics platform is shown.

**Table 3.** Inertial measurement unit parameters.

| IMU POLOLU CHR-6D | |
|---|---|
| **Parameter** | **value** |
| Gyroscopes LP510AL and LY510 ALH | 300°/full scale |
| ADXL335 | 16bit, ±3g |
| | 300mV/g |

## 5 Position estimation

The position estimation of the path using IMU is calculated by equations (27) and (28).

$$v(t) = v(t-1) + a(t) \cdot Ts \qquad (27)$$

$$s_a(t) = s_a(t-1) + v(t) \cdot Ts \qquad (28)$$

where $S_a(t)$ is the calculated distance using data from accelerometers, $v(t)$ is the calculated speed, $a(t)$ is measured acceleration, $Ts$ is sampling frequency.

The position estimation of path using optical odometer is calculated by equation (28).

$$s_o(t) = s(t-1) + c(t) \cdot \frac{W_c}{N} \qquad (29)$$

where $so(t)$ is calculated distance using data from the odometer, $c(t)$ is count of impulses measured by odometer in defined time, $Wc$ is a wheel circumference (in our case 182,21 mm), $N$ is a number of impulses(steps) in one rotation of a wheel (in our case 64 steps).

The maximal error caused by resolution of odometer wheel can be calculated using equation (30) and (31).

$$\alpha_l = \frac{360°}{N} \qquad (30)$$

$$e_o = \pm\frac{\alpha_l}{2} \cdot \frac{W_c}{360} \qquad (31)$$

where $e_o$ is maximal error caused by the resolution of the odometer wheel; $\alpha_l$ is an angle covered in one step; $Wc$ is wheel circumference, $N$ is a number of steps on the odometer wheel (in our case 64 steps).

## 6 Results and discussion

At present, in our solution we are using post-processing calculations in Matlab software, but in future all calculations will be implemented into a powerful microcontroller, like Raspberry Pi 2 or equal.The captured data is calculated in pair in Matlab software. The first data array contains a number of counted impulses from the odometer which depends on time; the second data array contains acceleration in depends on time; the third data array contains angular velocity also in depends on time. In this time we are using only a first and second data array for position estimation. The third data array (measured angular velocities) will be used in the following research.

At the Figure 4, the original measured data is shown. This measured data contains a lot of noise caused by the bumpy floor. For this case it is necessary to use some type of filtration. The result using a moving average filter of the tenth order is shown in Figure 6. Using of digital filters is described in detail in [7].

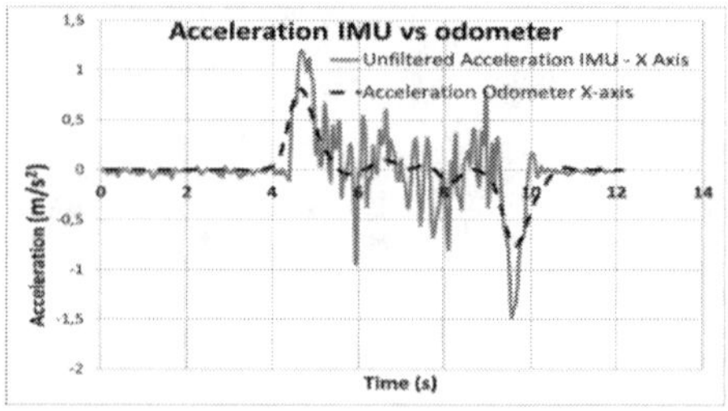


**Figure 4.** Original acceleration measured using inertial measurement unit vs calculated from odometer.

In Table 4 and Table 5, estimated path length using an odometer and an IMU is shown. In the table results calculated from filtered data using a moving average filter is shown. For better readability, the first five measurements from fifteen are shown in the tables. Totally, each path had fifteen repetitions. The statistical values are calculated from all measurements.

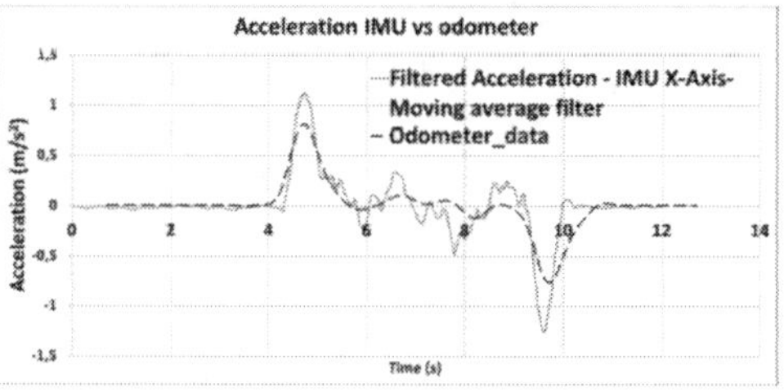


**Figure 5.** Filtered acceleration measured by IMU using Moving average filter.

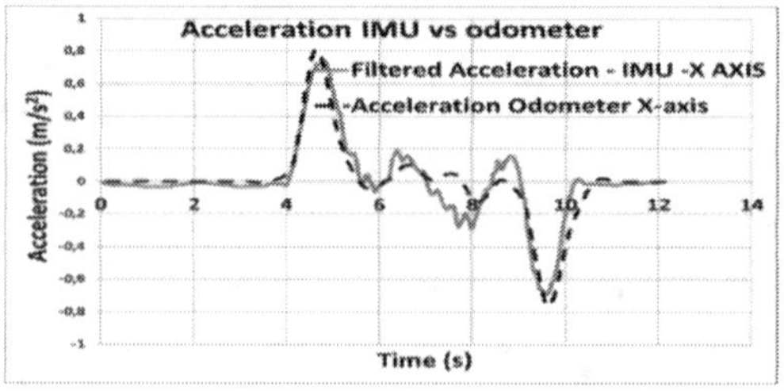


**Figure 6.** Filtered acceleration measured by IMU using Kalman filter.

At the Figure 6, the filtered data using Kalman filter is shown.

In Table 6 the filtered data using MVA is shown. InTable 7, the filtered data using the Kalman Filter is shown. From the tables it is possible to find that, the Kalman filter rapidly increase a precision compared with unfiltered data. The results using MVA has better precision than unfiltered data but worse than parameters in case of using the Kalman Filter.

**Table 4.** Estimated path length using odometer.

| Real Path length (m) | Estimated Path Length Using Odometer | | | | | | |
|---|---|---|---|---|---|---|---|
| | Measure number | | | | | Aver. (m) | STD. DEV (m) |
| | 1 | 2 | 3 | 4 | 5 | | |
| 3 | 2,99 | 3,02 | 2,98 | 3,01 | 3,00 | 3,005 | 0,013 |
| 4 | 4,08 | 3,96 | 3,98 | 4,01 | 4,02 | 3,992 | 0,041 |
| 5 | 5,09 | 5,02 | 5,03 | 4,94 | 5,02 | 5,070 | 0,061 |
| 10 | 10,01 | 9,93 | 10,15 | 9,98 | 10,10 | 10,041 | 0,080 |

**Table 5.** Estimated path length using IMU.

| Real Path length (m) | Estimated Path Length Using IMU | | | | | | |
|---|---|---|---|---|---|---|---|
| | Measure number | | | | | Aver. (m) | STD. DEV (m) |
| | 1 | 2 | 3 | 4 | 5 | | |
| 3 | 3,25 | 3,15 | 2,91 | 3,25 | 3,09 | 3,1268 | 0,108 |
| 4 | 4,28 | 3,95 | 3,95 | 4,01 | 3,90 | 4,135 | 0,118 |
| 5 | 5,19 | 5,27 | 5,16 | 4,91 | 4,92 | 5,219 | 0,165 |
| 10 | 10,58 | 10,25 | 10,36 | 9,51 | 10,13 | 10,391 | 0,476 |

From Table 4 and Table 5 it is possible to recognize that, the odometer has more precise measurement than IMU.

**Table 6.** An estimated path using IMU – MVA.

| Real Path length (m) | Estimated Path Length Using IMU – MVA | | | | | | |
|---|---|---|---|---|---|---|---|
| | Measure number | | | | | Aver. (m) | STD. DEV (m) |
| | 1 | 2 | 3 | 4 | 5 | | |
| 3 | 3,12 | 3,11 | 2,95 | 3,14 | 3,09 | 3,0722 | 0,078 |
| 4 | 4,17 | 3,94 | 3,93 | 3,95 | 3,90 | 4,085 | 0,082 |
| 5 | 5,12 | 5,18 | 5,07 | 4,94 | 4,94 | 5,180 | 0,127 |
| 10 | 10,31 | 10,17 | 10,28 | 9,79 | 10,10 | 10,173 | 0,131 |

Table 7. An estimated path using IMU - Kalman filter.

| Real Path length (m) | Estimated Path Length Using IMU – Kalman Filter | | | | | | |
|---|---|---|---|---|---|---|---|
| | Measure number | | | | | Aver. (m) | STD. DEV (m) |
| | 1 | 2 | 3 | 4 | 5 | | |
| 3 | 3,07 | 3,09 | 2,97 | 3,11 | 3,05 | 3,051 | 0,053 |
| 4 | 4,1 | 3,99 | 3,97 | 4,00 | 3,97 | 4,078 | 0,098 |
| 5 | 5,09 | 5,16 | 5,12 | 4,95 | 4,94 | 5,091 | 0,117 |
| 10 | 10,27 | 10,17 | 10,20 | 9,81 | 10,09 | 10,191 | 0,228 |

## 7 Conclusion

In this paper an increasing estimation accuracy position determination of Inertial Measurement Units was described. This paper is focused on the implementation of the Kalman Filter to our robotics platform. Firstly this research used an alternative to Kalman filtration – Average Moving Filter (AMF). Results were better compared with unfiltered data, but precision was still poor. For this case it is necessary to implement predictive filtration -Kalman filter.

The results recorded using the Kalman filter has very well precision. The main problem of the Kalman filter is that, it is necessary to have a precision reference signal – for this case we are using an Odometer.

The odometer has the best precision.The disadvantage of the odometer is that, measuring only path length (impulses) and does not care about curves etc. For this case, it is necessary to have gyroscopes for recording a rotation of robotics platform. Also a big disadvantage of the odometer is wheel slip. In case of a slippery floor, the odometer counts impulses, but the position of robotics platform is the same. The disadvantage of IMU is a big noise in measured signal and lower precision than the odometer. The IMU have problems with additional noise and have a lower precision than the odometer.

Also big disadvantageof IMU is a problem with bumpy floor. In case of bumpy floor, it is impossible to measure acceleration and rotation using IMU without high quality chassis. To eliminating this problem we are using a fusion of Odometer and IMU.

In this time, we are testing suitability of our theory only on a direct path. The DCM correction will be implemented to final version of a robotics platform – navigation of robotics platform using a fusion of visual odometry, Inertial Measurement Unit and mechanical odometers at every wheel. A very big advantage of our visual odometry is precision of position estimation [12]. The final device will be used as a fusion of navigation of the robotics platformwith an optical measurement device to create a 3D space map of dangerous spaces like a ruins, abandoned mines etc. [7], [8], [9]

## Acknowledgment

The research was supported by the Internal Grant Agency of University of Pardubice, the project No. SGSFEI_2016_022.

## References

1. L. Beran, P. Chmelar, and M. Dobrovolny, Navigation of robotic platform with using Inertial Measurement Unit and Direct Cosine Matrix, 56thInternation-al Symposium Electronics in Marine(Elmar), **56**, 87-90 (2014)
2. L. Beran, P. Chmelar, and L. Rejfek, Navigation of Robotics Platform using Monocular Visual Odometry, 25th International Conference Radioelektronika, **25**, 213-216 (2015)

3. J. Campbell et.al, A Robust Visual Odometry and Precipice Detection System Using Consumer-grade Monocular Vision, Robotics and Automation. ICRA 2005, 3421-3427, (2005)
4. CH. Yang, M. Maimone, and L. Matthies, "Visual Odometry on the Mars Exploration Rovers," Systems, Man and Cybernetics, **1**, 903 (2005)
5. L. Rejfek,Z. Mosna, D. Kouba, J. Boska, and D. Buresova, Application of digital filters to check quality of the automatically scaled ionograms, Journal of Electrical Engineering, ISSN: 1335-3632, **66**, 3 164-168 (2015)
6. J. Borenstein and L. Feng , Measurement and Correction of Systematic Odometry Errors in Mobile Robots, IEEE Transactions on Robotics and Automation, **12**, 6 869-880, (1996)
7. P. Chmelar, L. Beran, N. Kudriavtseva , The Laser Color Detection for 3D Range Scanning Using Gaussian Mixture Model, 25th International Conference Radioelektronika 2015, **25**, 248 253 (2015)
8. P. ChmelarAnd M. Dobrovolny, The Fusion of Ultrasonic and Optical Measurement Devices for Autonomous Mapping. 2013, 23rd International Conference Radioelektronika, **23**, 292-296 (2013)
9. P. Chmelar, L. Beran, N. Kudriavtseva., "Projection of point cloud for basic object detection," 56th International Symposium (ELMAR 2014), **56**, 11-14 (2014)
10. B. Gersdorf and U. Frese, A Kalman Filter for Odometry using a Wheel Mounted Inertial Sensor. 10th International Conference on Informatics in Control, Automation and Robotics, **10**, (2013)
11. J.Z. Sasiadek and P. Hartana, Sensor data fusion using Kalman filter, in, FUSION 2000. Proceedings of the Third International Conference, **3**, (2000)
12. L. Beran, P. Chmelar, and L. Rejfek, Navigation of Robotics Platform Using Advanced Image Processing Navigation Methods, V. Eccomas Thematic Conference on Computional Vision and Medical Image Processing(VIPIMAGE), **5**, 341-346 (2015)

MATEC Web of Conferences **75**, 05002 (2016) DOI: 10.1051/matecconf/20167505002
*ICMIE 2016*

# A Decision-Making Method for Landing Routes of Aircraft on the Carrier

Yu Wu, Wenqian Tan, Liguo Sun and Xiangju Qu

*School of Aeronautic Science and Engineering, Beihang University, Beijing 100191, China*

**Abstract.**The landing safety of aircraft is of great significance for the normal operation of an aircraft carrier. This paper presents an optimal method for landing routes of aircraft. Owing to the fuzziness of environmental information and human judgment, a decision-making method is introduced to reduce the workload of landing console operators (LCO). Specifically, a route option method for aircraft landing is proposed based on Fuzzy Multi-attribute Group Decision Making (FMAGDM). Firstly, the route option problem is described. Then the essential elements of route option are described. A group decision-making method is proposed, and the most reasonable landing route in current environment is obtained considering the knowledge composition and the weight of each decision maker. Finally, simulations under different environments are conducted. The results indicate that the decision-making method is able to adapt to the environment changes, and determine the most reasonable route for each validation case.

## 1 Introduction

Landing aircraft on an aircraft carrier is a task laden with inherent risks [1]. There exist numerous uncertainties during the fight before the aircraft landing, e.g. the flight condition and the maritime environment [2], [3]. As a result, it increases the uncertainty of landing. The landing console operator (LCO) is in charge of determination of landing route [4], [5]. He/she makes decisions for the guidance of aircraft within a short time according to his knowledge and the information displayed on the instruments [6]. However, reasonable judgment is usually not able to be made quickly and accurately due to the fuzziness of the information. Artificial intelligence (AI) is suitable for dealing with fuzziness problem and is introduced to the route decision-making system of aircraft in order to reduce the workload of LCO.

Unlike the path planning problem [7], [8], the route modes for landing of carrier aircraft are fixed. The landing route is chosen for aircraft before they reach the aircraft carrier. It is the first step in the landing guidance support system. This issue is obviously essential to the automatic landing system and thus worthwhile to study in depth.

The aircraft usually adopt different routes to land depending on its performance, the traffic conditions, a variety of weather conditions and visibility [9].Some biases or even errors may occur if we only rely on the judgments of human. Therefore, a decision-making system for route option is essential to reduce the workload of staffs and improve the safety level of landing.

The main contribution of this paper is to point out the necessity and importance of researching the route option problem for aircraft landing, and have a comprehensive description of the problem. The establishment of mathematical model and the proposed method provide a valuable reference for assistant decision of the pilot and LCO. This group decision-making method is crucial to realize the intelligent decision-making.

## 2 Problem description and essential elements analysis

### 2.1 Problem description

The air control system sends return command to an aircraft in flight. Specifically, Tactical Air Navigation (TACAN) provides the position and distance information. The airborne early warning (AEW) provides the overall air traffic situation. The sea weather is measured by the observatory set up on the aircraft carrier, and the weather measurements are transmitted to a pilot. The route option problem is further depicted in Fig. 1.

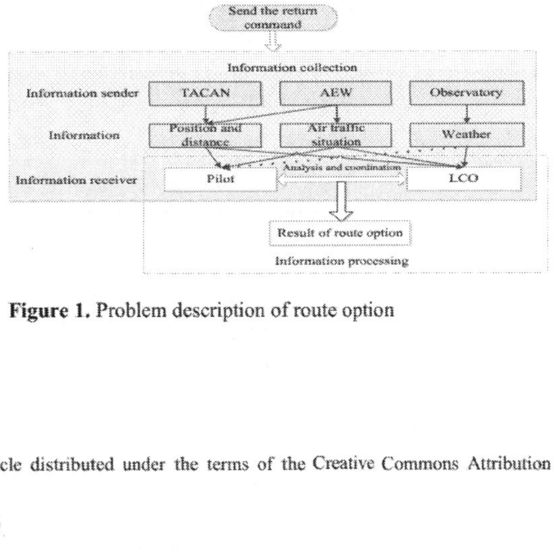


**Figure 1.** Problem description of route option

The dotted arrow denotes the receiver will only view the information as a reference to make a judgment. The pilot and the LCO are both decision makers when solving the route option problem. However, the LCO is chosen as the major decision maker in order to better master the air traffic situation and enhance the safety level of landing.

### 2.2 Analysis of essential elements

There are several essential elements in a route option problem. They will be discussed in this section.

#### *2.2.1 The alternative routes*

There are four modes of landing route according to the differences in atmospheric environment, performance of an aircraft and the air traffic situation.

① $l_1$ : The weather, visibility, performance of an aircraft and air traffic situation are all good when landing is permitted.The pilot lands the aircraft by vision under such conditions.

②$l_2$:The weather and visibility are at a medium level and the performance of the aircraft and air traffic are in good conditions.A landing mode with higher safety standard will be adopted when clouds become heavy. It becomes dangerous to land the aircraft by vision.

③ $l_3$ : The weather and visibility are bad, the performance of the aircraft and air traffic are in good conditions.It is also unsafe to land the aircraft by vision in such conditions.

④$l_4$: The performance of aircraft is poor or the air traffic is in a bad condition.No proper conditions exist for aircraft landing on the carrier. In this case, the pilot has to get in touch with the land base for emergency landing.

#### *2.2.3 The contributing factors:*

The contributing factors for a route option problem are concludedas follows:

① Weather

It is usually difficult to predict the weather around a large area accurately. Only the nearby weather condition can be mastered continuously.

② The height of clouds

The vision of a pilot will be affected when the clouds are low or cover a wide range of the sky. The heavy cloud may even lead to a bump of an aircraft during the flight.

③ Visibility

The level of visibility plays an important role in affecting the visual observation and the judgment of a pilot.

④ Performance of aircraft

The aircraft is usually in good performance during flight. However it may get damaged when accidents occur.

⑤ Air traffic situation

The aircraft under consideration refers to the status of other aircraft which are near to the landing routes.

## 3 FMAGDM (Fuzzy multi-attribute group decision making) based method of route option

The FMAGDM method has a good description of the current environmental information and the weights of decision-makers [10], [11]. According to the developed fuzzy TOPSIS (Technique for Order Preference by Similarity to an Ideal Solution) approach [12], the triangular membership function is used to describe the performance ratings of contributing factors (short for PRCF for convenience in the rest of this paper). The PRCF is the knowledge base of the model which can be evaluated in advance. Similarly, the current environmental vector which is the input of the route option model can be also described by the triangular membership function. Subsequently a mathematical model is established to measure each alternative route. Based on this, the fuzzy judgment method for group decision-making is introduced to get the most reasonable one for the current environment.

### 3.1. Description of the PRCF and the current environmental vector

The PRCF or the current environmental vector needs to be described using fuzzy linguistic variables. The triangular membership function is introduced to define the value of each fuzzy linguistic variable. The properties of the triangular membership function are presented.

***Definition 1***: The mean value of a triangular membership function $a$=[$a^L$,$a^M$,$a^U$] is defined as

$$s(a) = (a^L + 2a^M + a^U)/4 \tag{1}$$

The mean valuecan be used to sort the triangular membership functions. In view of the triangular membership functions $a$ and $b$, there exists a definition as follows: $a \geq b$if $s(a) \geq s(b)$ and $a < b$if$s(a) < s(b)$.

***Definition 3***: Assume $U$ is a universe of discourse and define the$\alpha$-cut set$a_\alpha$ as

$$a_\alpha = \{x \in U \mid \mu_{\tilde{A}}(x) \geq \alpha\}, \alpha \in [0,1] \tag{2}$$

where $a_\alpha$can be expressed as$a_\alpha$=[$a^L(\alpha)$,$a^R(\alpha)$], as shown in Fig. 2.

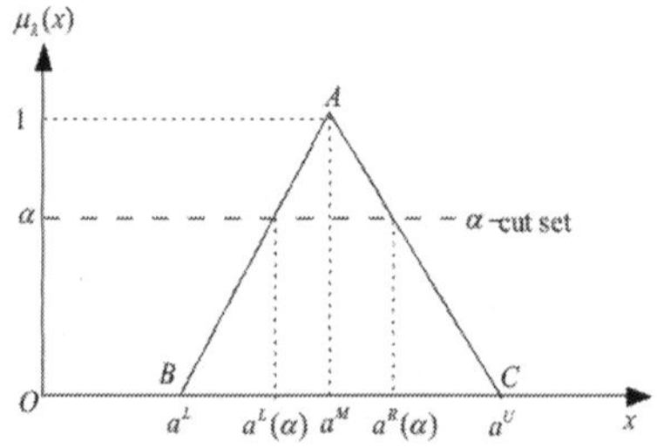


**Figure2.** Definition of $\alpha$ -cut set

***Definition 4***:Assume $a$ and$b$ are twotriangular membership functionsand $a \geq b$, the fuzzy distance between $a$ and $b$ is defined as follows.

 DOI: 10.1051/matecconf/20167505002

$$d(a,b)=[\max\{\int_0^1 d^L(\alpha)d\alpha,0\},d^L(1),\int_0^1 d^R(\alpha)d\alpha] \quad (3)$$

where $d^L(\alpha)$ and $d^R(\alpha)$ are defined as below:

$$\begin{cases} d^L(\alpha)=a^L(\alpha)-b^R(\alpha) \\ d^R(\alpha)=a^R(\alpha)-b^L(\alpha) \end{cases} \quad (4)$$

### 3.2 Decision-making with the developed fuzzy TOPSIS approach

5 contributing factors determine the result of route option. They (weather, clouds height, visibility, performance of aircraft and air traffic situation) are written as $f_1$, $f_2$, $f_3$, $f_4$ and $f_5$ respectively. The performance ratings of contributing factors are shown in Table 1.

**Table 1.** Description of performance ratings

| | $f_1$ | $f_2$ | $f_3$ | $f_4$ | $f_5$ |
|---|---|---|---|---|---|
| $l_1$ | Good | >3000 feet | >5 km | Medium | Good |
| $l_2$ | Bad | 1000-3000 feet | >5 km | Good | Good |
| $l_3$ | Terrible | < 1000 feet | < 5km | Good | Good |
| $l_4$ | Indifferent | Indifferent | Indifferent | Bad | Bad |

Thedescriptions mentioned above need to be unified with the same linguistic variables as shown in Table 2.

**Table 2.** Unified linguistic variables of performance ratings.

| Linguistic variable | No | Low | Medium | High | Very high |
|---|---|---|---|---|---|
| Expression | NO | LO | ME | HI | VH |

All linguisticvariables in Table 2 will be described by triangular membership functions in later usage. Suppose that the performance rating of $l_j$($j$=1,2,3,4)with respect to $f_i$($i$=1,2,3,4,5)is written as $e_{ij}$=[$e_{ij}^L$,$e_{ij}^M$,$e_{ij}^U$] (the value of $e_{ij}$ can be chosen from Table 2), then the fuzzy decision matrix for the route optionproblem can be established as follows:

$$D=\begin{bmatrix} e_{11} & e_{12} & e_{13} & e_{14} \\ e_{21} & e_{22} & e_{23} & e_{24} \\ \vdots & \vdots & \vdots & \vdots \\ e_{51} & e_{52} & \cdots & e_{54} \end{bmatrix} \quad (5)$$

where matrix $D$ is the knowledge base which has been mastered in advance. The matrix $D$ can assist a pilot or LCO to make optimal decision.

The detailed procedure of the method will be provided.

***Step 1*: Normalize of the fuzzy decision matrix**

Different contributing factors use different measurement units when collecting raw data. Therefore, normalization is needed to transform $e_{ij}$ into the interval of [0,1] in order to eliminate anomalies. In this paper, a linear transformation is adopted to get the normalized fuzzy decision matrix $X$:

$$X=\begin{bmatrix} x_{11} & x_{12} & x_{13} & x_{14} \\ x_{21} & x_{22} & x_{23} & x_{24} \\ \vdots & \vdots & \vdots & \vdots \\ x_{51} & x_{52} & \cdots & x_{54} \end{bmatrix} \quad (6)$$

Since the PRCF denote efficiency in route option, $x_{ij}$ can be normalized using the largest value, i.e., efficiency, of this contributing factor:

$$x_{ij}=[e_{ij}^L/e_i^{U\max},e_{ij}^M/e_i^{U\max},e_{ij}^U/e_i^{U\max}] \quad (7)$$

with $e_i^{U\max}=\max_{1\le j\le 4}\{e_{ij}^U\}$ ($i$=1,2,...,5 ;$j$=1,2,3,4). Using Eq. (7), all linguistic variables are transformed into the interval of [0,1].

We write the pilot and the LCO as decision maker 1 and 2. The normalized fuzzy decision matrix $X_1$ and $X_2$ can be established according to the knowledge and experience of decision makers.

$$X_1=\begin{bmatrix} HI & ME & LO & NO \\ VH & HI & LO & NO \\ VH & VH & LO & LO \\ ME & HI & HI & LO \\ ME & HI & HI & LO \end{bmatrix} X_2=\begin{bmatrix} HI & ME & NO & NO \\ HI & ME & LO & NO \\ HI & HI & LO & NO \\ ME & HI & HI & ME \\ HI & VH & VH & LO \end{bmatrix}$$

***Step 2*: Integrate the current environment vector reported by a decision maker into the normalized fuzzy decision matrix**

The environmental vector can also be described using linguisticvariable, as shown in Table 3.

**Table 3.** Linguistic variable of environmental vector.

| Linguistic variable | Too poor | A little poor | Normal | Good | Very good |
|---|---|---|---|---|---|
| Expression | TP | LP | NO | GO | VG |

We denote the environmental evaluation of each contributing factor as $\omega_i$=[$\omega^L$,$\omega^M$,$\omega^U$], which compose the environmental vector ($i$=1,2,…,5, the value of $\omega_i$can be chosenfrom Table 3). The normalized fuzzy decision matrix integrating the environmental vector is calculated:

$$Z=(z_{ij})_{5\times4}=(\omega_i x_{ij})_{5\times4}=\begin{bmatrix} z_{11} & z_{12} & z_{13} & z_{14} \\ z_{21} & z_{22} & z_{23} & z_{24} \\ \vdots & \vdots & \vdots & \vdots \\ z_{51} & z_{52} & \cdots & z_{54} \end{bmatrix} \quad (8)$$

***Step 3*:Determine the positive and negative ideal solutions**

The positive ideal solution is obtained when each PRCF chooses its highest value from the fuzzy decision matrix $D$. On the contrary, the negative solution is obtained by choosing the lowest value for each performance rating.

The fuzzy positive ideal reference point (FPIRP) and negative point (FNIRP) are written as $P$ and $N$in the remainder of this paper for convenience. The FPIRP and FNIRP can be defined as $P=(p_1,p_2,...,p_5)$ and $N=(n_1,n_2,...,n_5)$with $p_i=\max_{1\le j\le 4}\{z_{ij}\}$ and $n_i=\min_{1\le j\le 4}\{z_{ij}\}$.

***Step 4*: Calculate the fuzzy distance between $l_j$($j$=1,2,3,4)and P (or N)**

According to Eq. (3), the fuzzy distance between$l_j$($j$=1,2,3,4)and P (or N) are defined as follows:

$$d(l_j,P)=\sum_{i=1}^{5}d(z_{ij},p_i);d(l_j,N)=\sum_{i=1}^{5}d(z_{ij},n_i) \qquad (9)$$

***Step 5*:Obtain the closeness coefficient**

The larger the closeness coefficient is, the closer the alternative route is to the fuzzy ideal solution. Eq. (10) is adopted to calculate the closeness coefficient and distinguish the importance between FPIRP and FNIRP.

$$C(l_j)=\mu\frac{d(l_j,P)}{d(l_j,P)+d(l_j,N)}+(1-\mu)\frac{d(l_j,N)}{d(l_j,P)+d(l_j,N)} \qquad (10)$$

where $\mu$(0<$\mu$<1)is a weight for different distances from FPIRP and FNIRP, it is determined by the subjective attitude of a decision maker.

### 3.3 Group decision-making process

The main idea of solving the group decision problem is to regard the pilot and the LCO as two contributing factors, and view the group decision-making problem as a single person decision-making problem. The group fuzzy decision matrix is constructed as follows:

$$O=\begin{bmatrix} O_{11} & O_{12} & \cdots & O_{14} \\ O_{21} & O_{22} & \cdots & O_{24} \end{bmatrix} \qquad (11)$$

where $O_{st}$ ($s$=1,2 ;$t$=1,2,3,4) is the closeness coefficient$C(l_j)$, which is calculated using Eq. (10). The group solution is depicted by the following steps:

***Step 1*:** The weights of pilot and LCO are assumed as$\lambda_1$=[$\lambda_1^L$,$\lambda_1^M$,$\lambda_1^U$]and $\lambda_2$=[$\lambda_2^L$,$\lambda_2^M$,$\lambda_2^U$] respectively. The weighted normalized group fuzzy decision matrix is written as $V$=$(v_{kj})_{2\times4}$($k$=1,2 ;$j$=1,2,3,4).

***Step 2*:** Write the fuzzy positive group ideal reference point (FPGIRP) and negative point (FNGIRP) as *GP* and *GN* respectively. The FPGIRP and FNGIRP can be formulated as $GP=(gp_1,gp_2)$ and $GN=(gn_1,gn_2)$ where $gp_k=\max_{1\le j\le4}\{v_{kj}\}$ and $gn_k=\min_{1\le j\le4}\{v_{kj}\}$.

***Step 3*:** Calculate the fuzzy distance between $l_j$($j$=1,2,...,4)and the FPGIRP (or FNGIRP), and calculate the closeness coefficient.

$$d(l_j,GP)=\sum_{k=1}^{2}d(v_{kj},gp_k);d(l_j,GN)=\sum_{k=1}^{2}d(v_{kj},gn_k) \qquad (12)$$

$$GC(l_j)=\eta\frac{d(l_j,GP)}{d(l_j,GP)+d(l_j,GN)}+(1-\eta)\frac{d(l_j,GN)}{d(l_j,GP)+d(l_j,GN)} \qquad (13)$$

where$\eta$(0< $\eta$<1) denotes the attitude of the group. It reflects the different importance of FPGIRP and FNGIRP.

***Step 4*:** Calculate the mean values$(GC(l_j))$ according to Eq. (1).The route corresponding to $\max_{1\le j\le4}\{s(GC(l_j))\}$ is the most reasonable one in the current environment.

## 4 Results and analyses of route option in different environments

Real cases are presented to illustrate the validity of the proposed method. Two different environmental conditions are used as test scenarios. On the basis of a-priori knowledge, experience, current environment evaluations and weights of decision makers, the decision-making method is expected to choose the most reasonable route for the current environment. Further analysis of the simulation results implies that the proposed method has a bright future for real application in a practical project.

### 4.1 initialization and their fuzzy descriptions

The expressions of performance ratings and environment evaluations are shown in Table 4.

**Table 4.** Performance ratings and environment evaluations

| | Performance ratings | Environment evaluations | Membership function |
|---|---|---|---|
| Linguistic variables and expressions | NO | TP | [0,0,0.1] |
| | LO | LP | [0.1,0.3,0.5] |
| | ME | NO | [0.3,0.5,0.7] |
| | HI | GO | [0.5,0.7,0.9] |
| | VH | VG | [0.9,1,1] |

In order to compare the results in different environments, 2 pairs of simulations are carried out independently. The environment settings and evaluations of decision makers are listed in Table 5 and Table 6.

**Table 5.** Environment set to simulate

| Serial number of simulation | Environments |
|---|---|
| i | The weather, clouds height and visibility are good while the performance of aircraft and air traffic situation are normal. |
| ii | The weather, clouds height and visibility are bad while the performance of aircraft and air traffic situation are good. |

**Table 6.** Environment evaluations of decision makers

| Contributing factors | Simulation i | | Simulation ii | | Weight |
|---|---|---|---|---|---|
| | Pilot | LCO | Pilot | LCO | |
| $f_1$ | VG | GO | TP | TP | Pilot :$\lambda_1$= |
| $f_2$ | VG | GO | TP | TP | [0.4,0,5,0.6] |
| $f_3$ | VG | GO | TP | TP | LCO :$\lambda_2$= |
| $f_4$ | NO | GO | GO | GO | [0.8,0.9,1] |
| $f_5$ | NO | GO | GO | VG | |

### 4.2 Results and analyses

The proposed decision-making method is applied to determine the optimal route for different environment conditions. The results are given in Table 7 and Table 8.

**Table 7.** Results of simulation i

| Alternative route | Closeness coefficient | Mean value | Route option result |
|---|---|---|---|
| $l_1$ | [0,0.82,161.06] | 40.67 | |
| $l_2$ | [0,0.80,159.28] | 40.22 | |
| $l_3$ | [0,0.51,107.62] | 27.16 | $l_1$ |
| $l_4$ | [0,0.20,55.49] | 13.97 | |

**Table 8.** Results of simulation ii

| Alternative route | Closeness coefficient | Mean value | Route option result |
|---|---|---|---|
| $l_1$ | [0,0.51,5.24] | 1.57 | |
| $l_2$ | [0,0.80,5.89] | 1.87 | |
| $l_3$ | [0,0.80,6.41] | 2.00 | $l_3$ |
| $l_4$ | [0,0.20,2.84] | 0.81 | |

From Simulation i to Simulation ii, the environment condition set for each validation case is degrading gradually. With the changes of the environment setups, the decision-making method is able to adapt to the environment changes, and determine the most reasonable route for each validation case. The results indicate that the proposed method has the capability of dealing with the fuzziness in description successfully and can solve the troubles introduced by artificial decision-making. Above all, the proposed decision making approach is able to assist the pilot and the LCO to guide the aircraft to land safely and automatically. The proposed method can be used for air traffic control in an aircraft carrier. It can assist the LCO to make the landing route scheduling decision, and can provide the security information for the returning aircraft after having finished a combat or patrol mission.

## 5 Conclusion

In this paper, the route option problem of aircraft landing on an aircraft carrier is solved by proposing a new decision-making method. Firstly, the route option problem and theessential elements are described. Then using the improved fuzzy TOPSIS approach, fuzziness in description and judgment can be solved by introducing linguistic variables. Subsequently, a mathematical model, which can be later applied to measure the rationality of each alternative route for the current environment, is established. Finally, the group decision-making approach is introduced to take into account the judgments of both the LCO and the human pilot. Validation results for different environment setups indicate that the proposed decision-making method can adapt to the changes of environment and can make a quick and realistic decision.

The decision-making method proposed in this paper can reduce the workload of the pilot and the LCO by pointing out the optimal route option for the returning aircraft. The work of this paper provides a solution for landing route scheduling, which should be addressed at the first order to realize the automatic landing of carrier aircraft. Automatic flight controller design should be viewed as the future work under the framework of realizing automatic landing of aircraft.

## Acknowledgement

This research work is financially supported by the National Natural Science Foundation of China (NSFC) with the project reference number of 11502008.

## References

1. S. Newman, Shipborne aviation: an introduction. *Proceedings of the Institution of Mechanical Engineers, Part G: Journal of Aerospace Engineering* 219.5 (2005): 363-368.
2. M.B. Subrahmanyam, H-infinity design of F/A-18A automatic Carrier Landing System. *Journal of guidance, control, and dynamics* **17**.1 (1994): 187-191.
3. M.G. Petovello, et al. Measuring aircraft carrier flexure in support of autonomous aircraft landings. *Aerospace and Electronic Systems, IEEE Transactions on* **45**.2 (2009): 523-535.
4. K. L. Gold, and Alison K. Brown. A hybrid integrity solution for precision landing and guidance. *Position Location and Navigation Symposium, 2004. PLANS 2004*. IEEE, 2004.
5. N. Polmar, *The Naval Institute guide to the ships and aircraft of the US fleet*. Naval Institute Press, 2005.
6. K. M. Adams and T. J. Meyers. The US Navy carrier strike group as a system of systems. *International Journal of System of Systems Engineering* **2**.2-3 (2011): 91-97.
7. A. Shinkafi and C. Lawson, Development of a tool to study aircraft trajectory optimisation in the presence of icing conditions. *Proceedings of the Institution of Mechanical Engineers, Part G: Journal of Aerospace Engineering* **229**.8 (2015): 1464-1484.
8. Y. Wu and X.-J. Qu. Obstacle avoidance and path planning for carrier aircraft launching. *Chinese Journal of Aeronautics* **28**.3 (2015) : 695-703.
9. A. G. Shem, Thomas Mazzuchi, and Shahram Sarkani. Addressing uncertainty in UAV navigation decision-making. *Aerospace and Electronic Systems, IEEE Transactions on* **44**.1 (2008): 295-313.
10. P. Liu, A weighted aggregation operators multi-attribute group decision-making method based on interval-valued trapezoidal fuzzy numbers. *Expert Systems with Applications* **38**.1 (2011): 1053-1060.
11. B. Vahdani, et al. A new compromise solution method for fuzzy group decision-making problems with an application to the contractor

selection.*Engineering Applications of Artificial Intelligence* **26**.2 (2013): 779-788.
12. R. K. Singh and L. Benyoucef. A fuzzy TOPSIS based approach for e-sourcing. *Engineering Applications of Artificial Intelligence,* **24** 3 (2011): 437-448.

DOI: 10.1051/matecconf/20167505003

# Segmentation of the Optic Disc and Optic Cup Using Histogram Feature-Based Adaptive Threshold for Cup to Disk Ratio

Gibran Satya Nugraha and Indah Soesanti

*Department of Electrical Engineering and Information Technology, Faculty of Engineering, Universitas Gadjah Mada, Yogyakarta, Indonesia*

**Abstract.** Glaucoma is a condition of increased intraocular pressure within the eyes. Such increase then causes the damage on optic nerves as the organ bringing information to be processed in brain. One of the parameters to detect the glaucoma is the ratio between the optic cup and optic disc that can be identified through an examination towards the retinal fundus image of the patient. The ratio is obtained by firstly calculating the width of the area of the optic cup and the optic disc. This research was aimed to propose a method of the segmentation of the optic cup and optic disc with the adaptive threshold. The value of the adaptive threshold was obtained once calculating the mean value and standard deviation on the retinal fundus image of the patient. Before conducting the segmentation, the red component of the image would firstly be extracted followed by doing the contrast stretching. The last one was to perform the morphological operation such as closing and opening to remove the blood vessel to make the ratio calculation more accurate. This method has been tested in a number of retinal fundus images coming from DRISTHI-GS and RIM-ONE.

## 1 Introduction

Glaucoma is an eye disease that can cause the loss of sight if not early treated. In common, its emergence is related to the genetic factor and its symptom is unseen until certain time. Glaucoma is the second main factor of the loss of sight after the cataract in the world and in Indonesia. Different from the cataract in which the loss of sight can be healed after the cataract operation, the blindness (loss of sight) due to glaucoma is permanent. Glaucoma has caused the damage in the optic nerve due to the increase of intraocular pressure within eye [1]. The optic nerve itself is a nerve tissue that acts to continue the stimuli of light coming from the retina to the brain. The optic nerve brings all information that will be processed in the brain. Thus, human are able to see an object or a thing. Based upon the cause of the increase of the intraocular pressure on the eye, then the glaucoma can be divided into two: open angle glaucoma and angle closure glaucoma. Open Angle Glaucoma (OAG) is a type of glaucoma that is mostly occurred by 90% of total cases of glaucoma [2]. It develops gradually and sometimes is present without any signs of sight disorders for couple years. The main cause of OAG is related to the aqueous humor flowing through trabecular meshwork within the eye experiencing a blockage in its canal [3]. Meanwhile, Angle Closure Glaucoma (ACG) is a type of glaucoma taken place when the aqueous humor flowing through the trabecular meshwork is blocked due to the formation of the angle between iris and cornea [4].

The early detection of the glaucoma can prevent a total loss of sight. The use of Optical Coherence Tomography (OCT) and Heidelberg Retinal Tomography (HRT) mostly used by the psychiatrists nowadays still has a weakness in terms of the high cost compared to the detection of glaucoma using the image from fundus camera [5]. Fundus camera is a low-level microscope located inside the camera and used to document the retina [6]. One of parameters to detect the glaucoma possible to be detected by processing the image from the fundus camera is the cup to disk ratio or the comparison between optic cup and optic disc. Optic disc is a part of retina in which the axon from the optic nerve comes from and leaves the eye, optic disc and optic cup in the eye as seen in Fig. 1. Using the use of the image processing technique, the optic disc and optic cup can be segmented or separated prior to be calculated for each of their ratio. Ratio between optic disc and optic cup within the normal eyes was at 0.3 for the glaucoma of the ration ratio above 0.3 [7]. In this research, a segmenting method of optic disc and optic cup used the adaptive threshold based upon the histogram feature from the image of the camera fundus such as the mean values and standard deviation. After successfully segmenting the optic disc and optic cup, they would further be calculated for the ratio between the optic cup and optic disc.

## 2 Methodology

### 2.1 Retinal fundus image database

Retinal fundus image from DRISTHI-GS will be used to test and develop the proposed method. DRISTHI-GS used PNG format at the resolution of 2049x1750 pixels. The example of the retinal fundus image of DRISTHI-GS is presented in Fig. 1 [8]. DRISTHI-GS image database also provides the result of the segmentation of optic disc and optic cup in ground truth version. In addition, the measurement result of the cup to disk ratio was based on the measurement of four experts provided. Furthermore, retinal fundus image from RIM-ONE is also used, but it has smaller resolution of 503x268 pixels in JPG format. RIM-ONE database also provides a ground truth version, but only using two experts for validate the calculation [9].

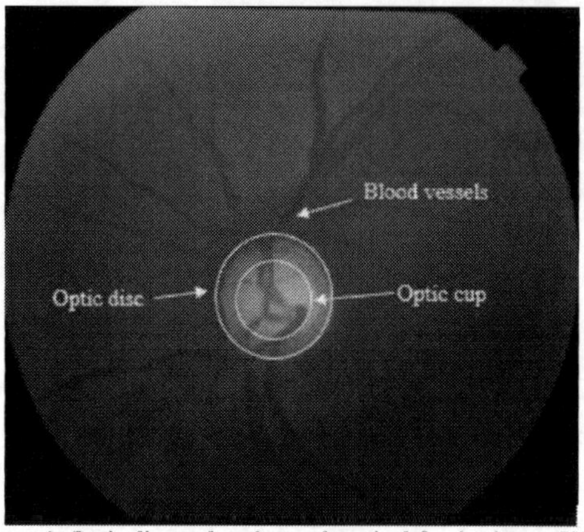


**Figure 1**. Optic disc and optic cup in retinal fundus image [8]

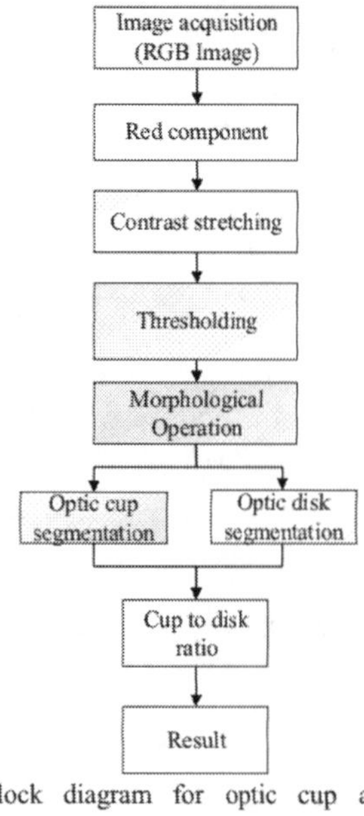


**Figure 2**. Block diagram for optic cup and optic disc segmentation

## 2.2 Optic disc and optic cup segmentation

One of parameters to detect the glaucoma is the ratio between the optic cup and optic disc that is above 0.3 for the eyes suffering from glaucoma. If it is less than 0.3, it means that the eyes are normal [7]. To identify the ratio of both, each of the optic cup and optic disc must be segmented or separated from their retinal image to make the measurement of the ratio more accurate. The diagram of the block from the proposed method to segment the optic cup and optic disc is presented in Fig. 2.

The retinal fundus image with a poor contrast will affect the process of the segmentation of optic disc and optic cup. Noted here, image with the low contrast frequently occurs due to the poor or unequal lighting. This can be caused by the sensors of image capturing that are not linear [10]. To change the distribution of the pixel intensity, a contrast stretching is necessary to be conducted using the formula below [11].

$$g(y,x) = \alpha f(y,x) \tag{1}$$

Based upon the formula above, $f(y, x)$ states the value of a pixel on the $f$ image and $g(y, x)$ states the pixel from the processing result of $f(y, x)$. Contrast will increase if the value of $\alpha > 1$ and contrast will decrease if $\alpha < 1$.

To segment the optic disc and optic cup this paper propose the threshold method. The segmentation with threshold refers to the simplest segmentation method with the value lower than the value of the threshold treated as the first and larger area compared to the values of the threshold categorized as the second area. In this case, one of the areas is positioned as the background. The segmentation with threshold method can be conducted using the formula below [12].

$$g(y,x) = \begin{cases} 1, \text{ if } f(x,y) > T \\ 0, \text{ if } f(x,y) \le T \end{cases} \tag{2}$$

In the formula, $T$ refers to the threshold, $f(x, y)$ is the value of the image of $f$ and $g(x, y)$ is the pixel of the result of the segmentation of the image of $f(x, y)$. In practice, the value 1 or 0 in equation can be exchanged. The threshold value will later be adaptive dependent upon the value of the histogram feature such as the mean and the standard deviation on image. Mean refers to the rate value of the image histogram; while the standard deviation is a size describing the level of data spread from the mean values. To calculate the mean value, it can be applied using the following equation [11].

$$m = \sum_{i=0}^{L-1} i.p(i) \tag{3}$$

In this case, $m$ refers to mean value, $I$ is the grey border in the image of $f$, and $p(i)$ states the probability of the appearance of $i$, and $L$ states the highest value of the grey border. The formula above will result if the average of the object brightness. Meanwhile, the formula for the standard deviation is presented as follows [11].

$$\sigma = \sqrt{\sum_{i=1}^{L-1} (i-m)^2 p(i)} \tag{4}$$

In this case, $\sigma^2$ called as the variant or the normalized two-order moment as $p(i)$ states the probability of the appearance $i$ and as the function of opportunity. $L$ states the highest border values and $m$ refers to the mean value. This feature provides the size of the contrast.

 DOI: 10.1051/matecconf/20167505003

The mean value and standard deviation can be observed from the distribution of the histogram of an image. The histogram of image is a diagram illustrating the frequency of each intensity value emerging in all pixels of an image. A large intensity value shows the white color while the small one shows the black one. The large frequency value states that the pixels have very large number of intensities [11]. An even distribution of histogram after the process of contrast stretching will help to determine an accurate threshold value. Fig. 3(a) shows an image histogram before given stretching and Fig. 3(b) is the image histogram after stretching. If looking at the Fig. 3(b), it is found the intensity value of pixel at 250 with the frequency of about 0.1. It can be stated that the optic disc and optic cup are located in the intensity and the frequency value. It is because the optic disc has the highest level of brightness compared to the other parts of the retinal fundus image. Hence, the threshold value for the segmentation of the optic disc and optic cup can be obtained from the intensity.

Having conducted an in-depth analysis to obtain the threshold value when doing the segmentation of the optic disc based upon the highest intensity, the mean value and the standard deviation from the retinal fundus image can be used through the following formula.

$$Tdisc = mean + 2 * std \tag{5}$$

Meanwhile for the segmentation of the optic cup, it can be applied using the following formula.

$$Tcup = mean + 4 * std \tag{6}$$

The value of 2 and 4 at Formula 5 and 6 is a constanta in which used to obtain the threshold value when doing the segmentation of optic disc and optic cup By setting the constanta value of 2, then multiplied it with the value of standard deviation and also mean value is added, it can be obtained a threshold value approaches the highest intensity value of the image, so the segmentation process will be more accurate. While the optic cup has a higher brightness level compared to the optic disc, therefore the intensity value of optic cup is also bigger, so it requires the bigger constanta value, in this case at 4.

To remove the blood vessels, a number of the techniques of morphological operation have been used such as opening and closing. The operation of the opening tends to remove the bright but small parts. The closing operation meanwhile is to maintain the small and bright objects. Both opening and closing operation in their process are involved in the dilation and erosion operation. Dilation process refers to a process of operation used to obtain the effect of broadening towards the pixel valued 1; meanwhile, the erosion process is a process that has an effect to reduce the image structure [11]. The opening operation can be conducted using the following formula [13].

$$A \circ B = (A \ominus B) \oplus B \tag{7}$$

where the opening process $A$ by $B$ refers to the erosion process $A$ by $B$ and then the results of the erosion is followed with the dilation process by $B$. Meanwhile, the closing process can be conducted using the following formula [13].

$$A \bullet B = (A \oplus B) \ominus B \tag{8}$$

where the closing process $A$ by $B$ refers to the dilation process $A$ by $B$ and the result of the dilation is followed with the erosion process by $B$.

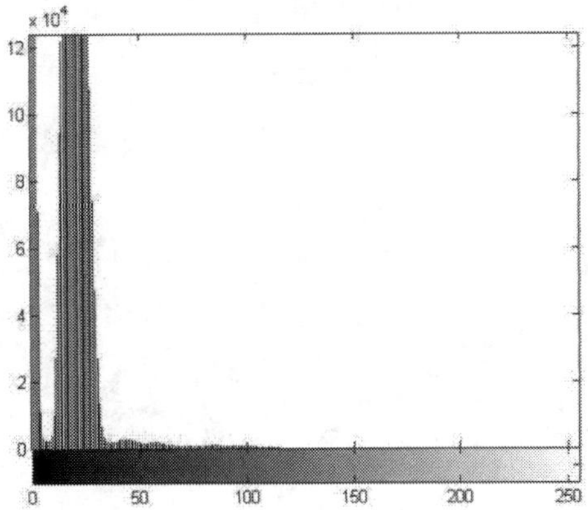

(a)

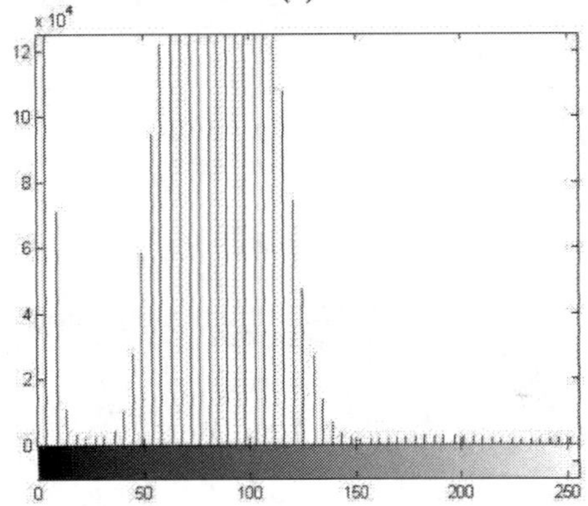

(b)

**Figure 3**. Image histogram, (a) Before stretching process, (b) After stretching process

Once the morphological operation has been done, each of optic disc and optic cup that has been segmented were counted for the width and the number of the white pixels in each of the optic cup and the optic disc. To calculate the ratio between the optic disc and cup to the disc ratio (CDR) was conducted by dividing the number of the white pixels in optic cup with the number of white pixels in the optic disc. The formula used is as follows.

$$CDR = \frac{white\ pixels\ of\ optic\ cup}{white\ pixels\ of\ optic\ disc} \tag{9}$$

## 3 Result and discussion

In this research, a total of 100 retinal fundus image will be used to test the proposed method, each of DRISTHI-GS and RIM-ONE provide 50 images [8], [9]. The image

DOI: 10.1051/matecconf/20167505003

consisted of three color components: red, green and blue (commonly abbreviated as RGB). The red component was used for having the best quality of the brightness in comparison to the other components. This comparison is illustrated in Fig. 4.

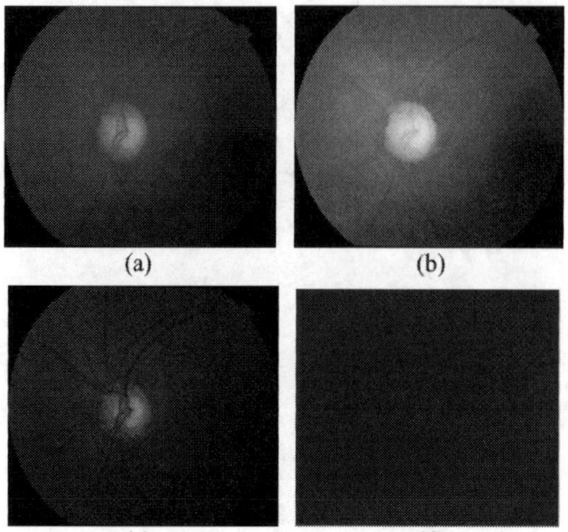


(c) (d)

**Figure 4**. Retinal fundus image in RGB format, (a) Original image, (b) Red component, (c) Green component, (d) Blue component

The contrast stretching was conducted to make the distribution of the pixel intensity even. This can facilitate the process of determining the threshold value. Fig. 5 shows the image with the red component before and after the contrast stretching.

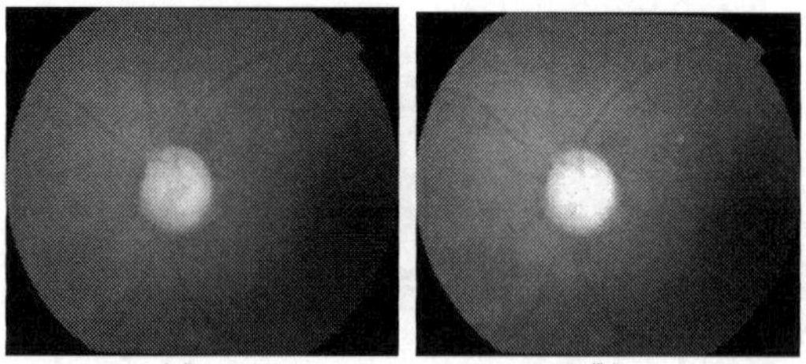

(a) (b)

**Figure 5.** Contrast stretching in red component, (a) Before given stretching, (b) After given stretching

After passing through the process of the contrast stretching, the optic disc and optic cup were segmented with the image of the background using the threshold value obtained from Equation 5 and Equation 6. The threshold value was obtained through the calculation of the mean and the standard deviation of the image. Furthermore, the morphological process was used to remove the blood vessel to make the calculation of the width of the optic disc and optic cup more accurate. The result of this segmentation later on would be in the form of binary image as seen in Fig. 6. To identify the accuracy of the result of the segmentation from the proposed method, it can be compared with the version of ground truth.

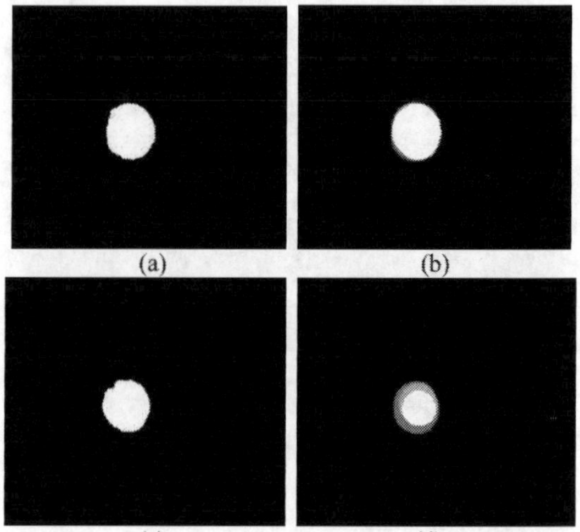


(c) (d)

**Figure 6**. Image segmentation, (a) Optic disc using proposed method, (b) Optic disc in ground truth version, (c) Optic cup using proposed method, (d) Optic cup in ground truth version

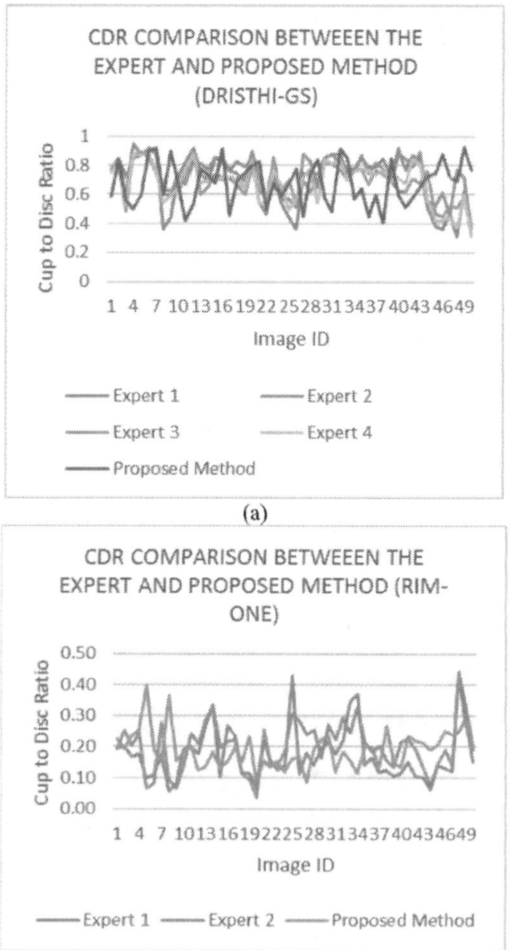


(b)

**Figure 7**. Comparison graph between the expert and proposed method, (a) DRISTHI-GS, (b) RIM-ONE

Fig. 7(a) and (b) show a graph comparison between cup to disk ratio using proposed method and also by the experts, four experts from DRISTHI-GS and two experts from RIM-ONE. DRISTHI-GS provides retinal fundus image which affected by glaucoma, it can be seen from CDR calculation result by the expert or proposed method that exceed 0.3, while RIM-ONE provides a healthy retinal fundus image. It can be seen that the result of CDR with proposed method was not much different from the result of the measurement of the experts.

The proposed method is quite suitable to do the segmentation of the optic disc and the optic cup for the CDR, both for the eye which affected by glaucoma or not, it can be seen in Fig. 7(a) and (b). This cannot be apart from the accurate determination of the threshold value based upon the feature of the histogram of mean and standard deviation. The process of the calculation involving both features can make it simple to obtain the accurate threshold value for the segmentation of the optic disc and optic cup.

The significant different result in the calculation between proposed method and the expert has caused a lot by the image acquisition process which is not good, for example because the lighting is too bright which make the optic disc area has the same pixel intensity with another area of the retina, so that the segmentation process becomes more difficult.

## 4 Conclusion

The segmentation method with adaptive threshold has been developed in this research. Such method began by extracting an image to take its red component before conducting the contrast stretching to make the intensity distribution even. Furthermore, it was to conduct the segmentation with the adaptive threshold method in which the threshold value was obtained from the measurement of the mean and standard deviation. The last process was the morphological process in the form of closing and opening to remove the blood vessels. The method was tested in the data of the retinal fundus image for diagnosing the glaucoma disease that is by segmenting the optic cup and optic disc from their background image. After this, the ratio was calculated between optic cup and optic disc in which the ratio above 0.3, then the eyes are suffering from the glaucoma and if the ratio is less than 0.3, then the eyes is categorized normal. The results of the calculation of the ratio between the optic cup and optic disc used the segmentation method proposed showed insignificant difference compared to the ones conducted by the experts. This method can be a media for the ophthalmologists to ascertain more their diagnosis when performing an examination to the patients. In addition, it is expected that this method can be used for the image with poor quality.

## References

1. R. Gayathri, R. Dr.P..V, and S.Aruna, Int. Conf. Adv. Comput. Commun. Informatics, p2808–2812 (2014).
2. T. Khalil, Sci. Inf. Conf., p438–442 (2014).
3. PRIMARY OPEN-ANGLE GLAUCOMA (POAG). Online. Available: https://www.glaucomafoundation.org/Primary_Open-Angle_Glaucoma.htm.
4. WHAT IS GLAUCOMA? Online.. Available: http://www.huntingtoneyecare.com/what-is-glaucoma/.
5. S. Kavitha, S. Karthikeyan, and K. Duraiswamy, Second Int. Conf. Comput. Commun. Netw. Technol., no. 1, p2–6 (2010).
6. P. J. Saine and M. E. Tyler, *Ophthalmic Photography: Retinal Photography, Angiography, and Electronic Imaging, 2nd Edition*, (Butterworth-Heinemann Medical, 2002).
7. H. Ahmad, A. Yamin, A. Shakeel, S. O. Gillani, and U. Ansari, 2014 Int. Conf. Robot. Emerg. Allied Technol. Eng. iCREATE 2014 - Proc., p321–324 (2014).
8. J. Sivaswamy, S. R. Krishnadas, and A. Chakravarty, JSM Biomed Imaging Data Pap 2(1) 1004, **2**, p1–7, (2015).
9. C. Pena-Betancor, M. Gonzalez-Hernandez, F. Fumero-Batista, J. Sigut, E. Mesa, S. Alayon, and M. De la Rosa, IOVS, pIOVS–14–15592 (2015).
10. A. K. Jain, *Fundamentals of Digital Image Processing*, (Prentice Hall, 1989).
11. A. Kadir and A. Susanto, *Pengolahan Citra Teori dan Aplikasi*, (ANDI, Yogyakarta, 2012).
12. S. S. Al-amri, N. V Kalyankar, and S. D. Khamitkar, J. Comput., **2**, no. 5, p83–86 (2010).
13. G. V Tcheslavski, *Morphological Image Processing : Basic Concepts*, (Spring, 2009).

MATEC Web of Conferences **75**, 05004 (2016) DOI: 10.1051/matecconf/20167505004
*ICMIE 2016*

# Smooth Path Planning by Fusion of Artificial Potential Field Method and Collision Cone Approach

Yong Hwi Kim[1], Wan-Sic Son[2], Jin Bae Park[1], and Tae Sung Yoon[2]

[1] *Yonsei University, Department of Electrical and Electronic Engineering, Seoul, Korea*
[2] *Changwon National University, Department of Electrical Engineering, Changwon, Kyongnam, Korea*

**Abstract.** A variety of path planning methods have been developed for different purposes, such as minimum path length and minimum processing time and smoothness of the path. This paper presents a novel path planning method by fusion of artificial potential filed (APF) and collision cone approach. The proposed algorithm provides smooth path and avoids local minima problem of the existing APF. And the smoothness of the path produced from the proposed algorithm is estimated by calculating the accumulated force which occurs while a robot passes a whole path. The proposed smooth path planning also can be helpful for considering vehicle kinematics and the safety of vehicle passengers. The simulation results demonstrate the performance of the proposed method.

## 1 Introduction

The obstacle avoidance problem is an important component of path planning. In the past few decades, a lot of methods have been studied and modified to deal with obstacle avoidance and path planning.

One of the most commonly used methods is the artificial potential field (APF) [1]. The APF method has the advantages of simple processing, small burden, and real time realization. The basic idea of the APF method is as following. First, goal generates attractive force and obstacles makes repulsive force. Then, a robot moves through the sum of those forces. However, there are some drawbacks such as local minima and oscillation situations. To avoid the problems, a hybrid approach, which integrates a priori knowledge of circumstance with local perceptions for executing the assigned mission [2]. And APF is modified by utilizing quantum particle swarm optimization which modifies the parameters of APF for adapting variety environments [3]. Other improved APF methods for dealing with local minima problem are proposed in [4-6]. In addition, the relative distance between robot and target into repulsive force to provide the global minimum is introduced in [7]

Another familiar method for path planning is the collision cone approach. The collision cone is the potential criterion that will result in a collision between a robot and an obstacle at some moment in time. If the relative velocity between the robot and the obstacle maintains a velocity outside the collision cone, collisions are guaranteed not to occur [8]. This algorithm has been improved for multi obstacles and real-time application.

In this paper, we present a smooth path planning method by applying a collision cone approach to the APF algorithm. The proposed algorithm is able to overcome local minima and provides smooth path. With the APF method, since the repulsive force depends on the distances from a robot and the obstacles, the path bends sharply at nearby obstacles. The force which avoids the collision cone is introduced to smoothen the path, and it can also escape a local minima problem.

The smooth path gives safeness and stability for the robot or robot's passengers. In this paper, the smoothness of the produced path is estimated by the accumulated force when the robot moves through the path. Two kinds of forces can be customarily treated when a robot moves. One is accelerating force and another one is centrifugal force.

The remainder of this paper is organized as follows: The next section presents related works and background of path panning methods. The traditional APF algorithm and modified APF algorithm are described, and the collision cone approach is also introduced. Next, in Section 3, our novel path planning is presented which provides smooth path and avoids the local minima problem. To verify the proposed method, we performed some simulations and analyzed the results in Section 4. Finally, conclusions and future work are given in Section 5.

## 2 Related works

### 2.1. Artificial potential field

The traditional APF is composed of the attractive field of goal and the repulsive filed of obstacles.

$$U(q) = U_{att}(q) + U_{rep}(q) \quad (1)$$

Here, $q = (x, y)^T$ is the position vector of robot's Cartesian coordinates. And artificial force is defined by the negative gradient of APF.

$$F(q) = -\nabla U(q) = F_{att}(q) + F_{rep}(q) \quad (2)$$

where $F_{att}(q)$ is the attractive force which is generated by goal, and $F_{rep}(q)$ is the repulsive force which is generated by obstacles. And they are described as following.

$$F_{att}(q) = -k(q - q_g) \quad (3)$$

$$F_{rep}(q) = \begin{cases} 0 & , \|q - q_o\| \geq d_{rep} \\ \eta\left(\frac{1}{\|q - q_o\|} - \frac{1}{d_{rep}}\right)\left(\frac{1}{(\|q - q_o\|)^2}\right)\frac{(q - q_g)}{\|q - q_g\|} & , \|q - q_o\| < d_{rep} \end{cases} \quad (4)$$

where $k$ and $\eta$ are scaling factors, $q_g$ is the position vector of goal, $q_o$ is the position of an obstacle, and $d_{rep}$ is the threshold for the repulsive force.

In a particular environment, there is a risk of collision with obstacles. To avoid the risk, the attractive force is modified as (5) by introducing the threshold for the attractive force $d_{att}$, in [4].

$$F_{att}(q) = \begin{cases} -k(q - q_g) & , \|q - q_g\| \leq d_{att} \\ d_{att}k\frac{(q - q_g)}{\|q - q_g\|} & , \|q - q_g\| > d_{att} \end{cases} \quad (5)$$

### 2.2 Collision cone approach

The collision cone approach provides collision avoidance strategy, so it can be used to predict the possibility of collision between a robot and an obstacle.

To make a collision cone, it is assumed that the robot and obstacle move with a constant velocities $v_t$ and $v_o$, respectively. Moreover, the shapes of them are assumed as a circle. If the relative velocity vector $v_t - v_o$ is contained in a collision cone, a potential collision occurs. Finally, it is assumed that the robot is a point and the size of the robot is added to that of the obstacle. The concept of collision cone is shown in Figure 1.

## 3 Proposed path planning method

APF method has some disadvantages such as local minima and unstable oscillation. In this paper, we proposed the new path planning method to deal with the disadvantages and provide a smoother path by fusion of APF method and collision cone approach.

### 3.1. Artificial force from a collision cone

In order to avoid collision effectively, an additive artificial force is presented from a collision cone approach

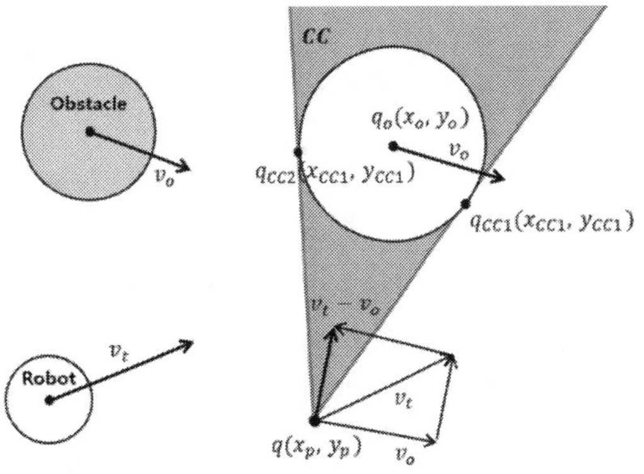


**Figure 1.** Concept of collision cone between a robot and an obstacle.

$$F_{CC}(q) = \begin{cases} 0 & , v_t - v_o \notin CC \\ \mu\frac{(q_o - q_{CC})}{\|q_o - q_{CC}\|} & , v_t - v_o \in CC \end{cases} \quad (6)$$

where $\mu$ is a scaling factor, $q_{CC}$ is a closer point between $q_{CC1}$ and $q_{CC2}$ from the relative velocity vector $v_t - v_o$ as shown in Figure 1. For example, in that case the relative velocity vector $v_t - v_o$ is closer to $q_{CC1}$ than $q_{CC2}$, $q_{CC1}$ becomes $q_{CC}$.

If there is a potential risk of a collision ($v_t - v_o \in CC$), the proposed artificial force $F_{CC}$ is added to the existing attractive and repulsive forces. Therefore, the entire artificial force is redefined as following.

$$F(q) = F_{att}(q) + F_{rep}(q) + F_{CC}(q) \quad (7)$$

## 4 Simulation results

In this section, we analyze the proposed path planning method with two aspects, local minima problem and smoothness. The simulation results about local minima problem are shown in Figure 2 and 3. The results about smoothness are presented in Figure 4 and 5.

The simulation conditions are arranged as following. The robot starts from (0, -11) to the goal which is set (0, 11). A static obstacle is located at (0, 1) in a simulation of local minima problem, a static obstacle is placed at (-2, 1) for a simulation of smoothness. In order to use the collision cone approach, the sum of the sizes of the robot and obstacle is set as a radius of 4. In addition, the parameters are set as follows: $k$ =0.5, $\eta$ =1, $d_{att}$ =10, $d_{rep}$ =10, $\mu$ =1.

DOI: 10.1051/matecconf/20167505004

### 4.1 Local minima problem

Local minima problem generally occurs when the attractive and repulsive forces are equal as shown in Figure 2. And there are some solutions for local minima problem such as changing the direction of repulsive field which is caused by circle obstacles.

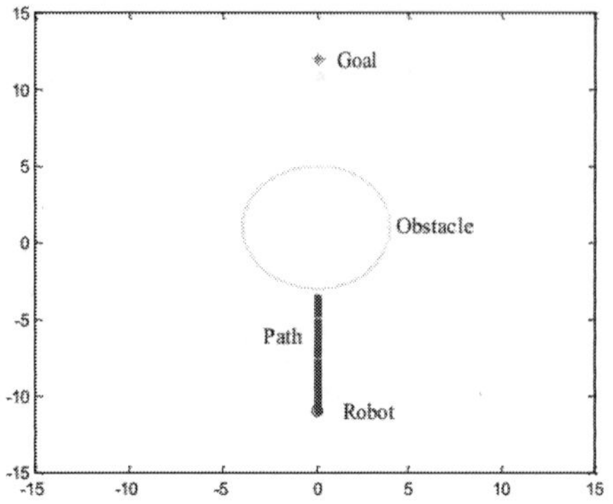


**Figure 2.** Local minima problem with APF.

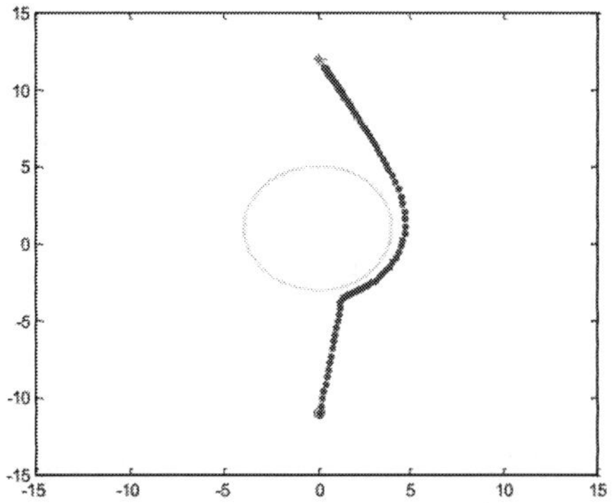

**Figure 3.** Local minima avoidance with the propose method.

The proposed method can solve the local minima problem easily by introducing an additive artificial force to avoid collision cone. The simulation result about local minima avoidance with the proposed method is displayed in Figure 3.

### 4.2 Smoothness of the path

The smoothness of paths can be evaluated by using some physical variables. The path curvature and the derivative of path curvature are adopted for the calculating the smoothness [9]. And the angular velocity and the angle difference between the nodes of path are used for considering the smoothness [10].

In this paper, we test the smoothness of the path by calculating the forces applied to the robot. Generally, two kinds of forces can be occurred when the robot moves. One is the accelerating force $f_a$ and another one is the centrifugal force $f_c$.

$$f = f_a + f_c = ma + mR\omega^2 \tag{8}$$

The robot drives through the whole path, and the accumulated force (energy) $E$ applied to the robot can be represented as follows:

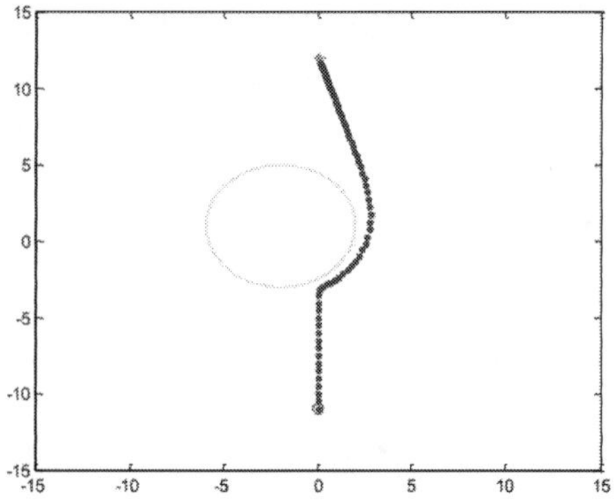

**Figure 4.** Path planning with APF.

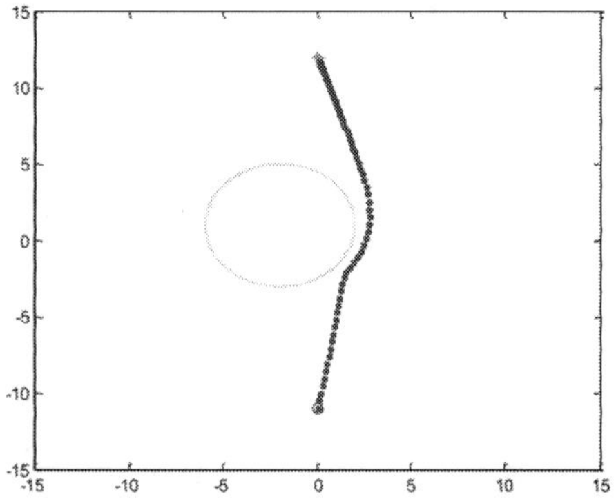

**Figure 5.** Path planning with the proposed method.

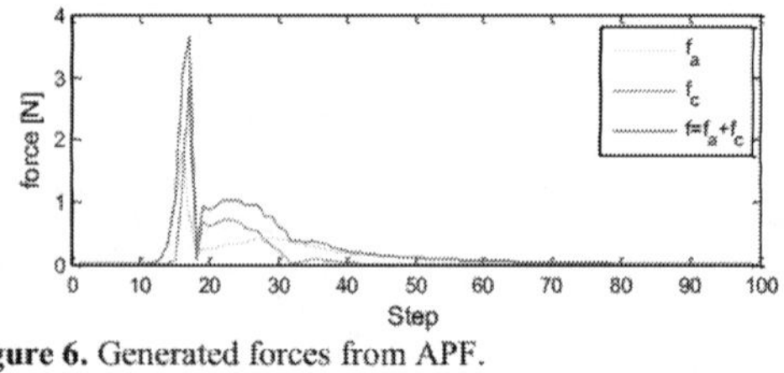


**Figure 6.** Generated forces from APF.

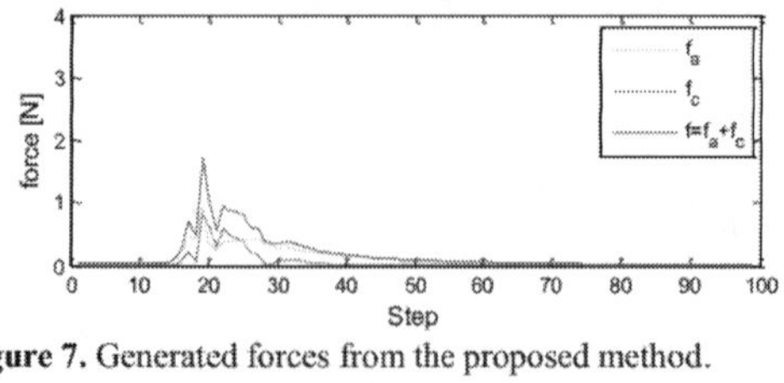


**Figure 7.** Generated forces from the proposed method.

$$E = \int_{t=t_0}^{t=t_f} ma + mR\omega^2 \qquad (9)$$

The simulation results of path planning with APF and the proposed method are shown in Figure 4 and 5. And Figure 6 and 7 display the generated forces from APF the proposed method. The accumulated force by using APF ( $E$ =26.4324) is much bigger than that by using the proposed method ( $E$ =16.6892), when $m$ and sampling time are assumed to be 1 and 0.1, respectively.

## 5 Conclusions and future works

In this paper, we present a novel path planning method by applying a collision cone approach to APF algorithm. The proposed algorithm avoids the local minima problem of the existing APF, and it is verified in simulation results. The smoothness of the path from the proposed algorithm is compared with that of the path from the previous algorithm. Although the smoothness is easily noticed by the paths' shapes, we considered the forces which occur when a robot passes a whole path. As a result of comparison, the superiority of the proposed algorithm is proved. And if the parameters for path planning are tuned well, the performance will be better. In addition, the smooth path planning is also able to assist to consider vehicle kinematics, and be helpful for the safety of vehicle passengers if the algorithm is adapted to the field such as an intelligent wheelchair.

In the future works, we attempt to improve the smoothness of the planning path by tuning the parameters, and consider multi obstacles complicated environment, and real-time application.

## Acknowledgment

This research was supported by Basic Science Research Program through the National Research Foundation of Korea (NRF) funded by the Ministry of Education, Science and Technology (No. 2013013607)

## References

1. O. Khatib, "Real-Time Obstacle Avoidance for Manipulators and Mobile Robots," *International Journal of Robotics Research*, Vol. 5, No. 1, pp. 90-99, 1986.
2. A. Sgorbissa and R. Zaccaria, "Planning and Obstacle Avoidance in Mobile Robotics," *Robotics and Autonomous Systems*, Vol.60, pp.628-638, 2012.
3. H. Zhang, Y. Liu, Z. Guo, and Y. Cao, "The Dynamic Path Planning Research for Mobile Robot Based on Artificial Potential Field," *Proceedings of 2011 International Conference on Consumer Electronics, Communications and Networks (CECNet2011)*, pp. 2736-2739, 2011.
4. G. Li., A. Yamashita, H. Asama, and Y. Tamura, "An Efficient Improved Artificial Potential Field Based Regression Search Method for Robot Path Planning," *International Conference on Mechatronics and Automation (ICMA)*, pp. 1227-1232, 2012.
5. W. Shi, X. Huang, and W. Zhou, "Path Planning of Mobile Robot Based on Improved Artificial Potential Field," *Journal of Computer Application*, Vol.30, No.8, pp. 2021-2023, 2010.
6. B. Zhang, W. Chen, and M. Fei, "An Optimized Method for Path Planning Based on Artificial Potential Field," *Proceedings of the Sixth International Conference on Intelligent System Design and Applications (ISDA2006)*, pp. 226-231, 2006.
7. B. he, G. Liu, J. Gao and H. Wang, "A Route Planning Method Based on Improved Artificial Potential Field Algorithm", *Proceedings of 2011 IEEE 3rd International Conference on Communication Software and Networks (ICCSN2011)*, pp.550-554, 2011.
8. A. Chakravarthy and G. Debasish, "Obstacle Avoidance in A Dynamic Environment: A Collision Cone Approach," *IEEE Transactions on Systems, Man and Cybernetics, Part A: Systems and Humans*, Vol. 28, No.5, pp. 562-574, 1998.
9. Y. Kanayama and B. Hartman, "Smooth Local Path Planning for Autonomous Vehicles," *Proceedings of IEEE International Conference on Robotics and Automation*, Vol. 3, pp. 1265-1270, 1989.
10. X. Gao, Y. Ou, Y. Fu, Z. Li, D. Dai, L. Cui, Y. Zhou, and W. Guo, "A Novel Local Path Planning Method Considering both Robot Posture and Path Smoothness," *Proceedings of IEEE International Conference on Robotics and Biomimetics*, pp. 172 - 178, 2013.

MATEC Web of Conferences **75**, 05005 (2016) DOI: 10.1051/matecconf/20167505005
*ICMIE 2016*

# Control Algorithm Trajectory Planning for Dual Cooperative Manipulators with Experimental Verification

Ahmed Omar, Pan Ri and Yajun Zhang

*College of Mechanical and Electrical Engineering, Beijing University of Chemical Technology, Beijing 100029, China*

**Abstract.** In order to achieve cooperation of multi-robots simultaneously, this paper presents an approach of Trajectory planning for two decoupled robots with six degrees of freedom (DOF) each to execute packing task. The Denavit-Hartenberg (D-H) representation is used to model robot links and solve the transformation matrices of each joint. The inverse kinematics solution using for a 6- DOF Robotic arm is presented, for given initial and final configurations of the robots, the robots are placed so that their workspaces overlap. An algorithm successfully finds a set of configurations to ensure, collision free transition from start to end configuration. Simulation and experiments based on 6-dof robot are carried out and the results verified the effectiveness of the proposed algorithm.

## 1 Introduction

Dual arm manipulators demand has increased with the public expecting robots that mimic human's behavior. therefore, studying dual-arm manipulators is a natural avenue of researches [1]. variety of robot platforms used for dual arm manipulation, while some researches are carried out on systems built by simply placing two single arm manipulators for sharing same workspace [2].

The difficulty in mechanical analysis for dual arms comes mainly from the closed chain system, which is formed during the manipulation of the objects, and which is much more kinematically and dynamically complex than the typical serial manipulator configuration [3].

In general, control of multiple robot cooperation systems, the master-slave method is still the state of art for industrial robots [4]. The Master-Slave control strategy has to identify one robot as Master, the other as Slave.

A number of research works have been presented relating to the modelling of kinematics. Shi et al. have proposed the solution for forward kinematic problem of 6 DOF robot [5]. comprehensive survey can be found in [6].

In [7] analyzed the problem of path planning for two cooperating arm robots carrying an object between two given configurations, in a setup which includes obstacles. the method proposed extends the randomized potential field planning techniques introduced in [8].

The two keys contributions of this work are: (1) Designing control algorithm of the robots to conduct the prescribed task. (2) Cooperative robot manipulators system performing the assembly task. Suggested system is including two arm robots. Target of these robots is cooperating for Drug Packing.

This paper is structured as follows: firstly the proposed robotic platform and developed kinematic models introduced. Trajectory planning approach was discussed, then explains the design details of the electronic hardware. Finally Experimental results for two Cooperated robots are presented and concluded the work.

## 2 Coordinate frames assignment

The proposed system is containing two 6-DOF robots manipulators. This Robot has five rotational joints and a moving gripper. Figure 1 illustrates the graphical view of all the joints of the presented robot. Joint 1 represents the base and its axis of motion $z_1$. This joint provides a rotational $\theta_1$ angular motion around $z_1$ axis in $x_1y_1$ plane. Joint 2 is identified as the Upper Arm and its axis is perpendicular to Joint 1 axis. It provides a rotational $\theta_2$ angular motion around $z_2$ axis in $x_2y_2$ plane. $z_3$ axes of Joint 3 and Joint 4 (Wrist) are parallel to Joint 2 z-axis; they provide $\theta_3$ and $\theta_4$ angular motions in $x_3y_3$ and $x_4y_4$ planes respectively.

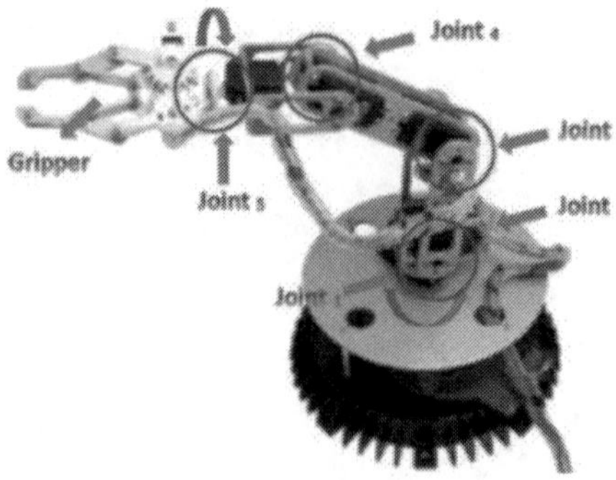

DOI: 10.1051/matecconf/20167505005

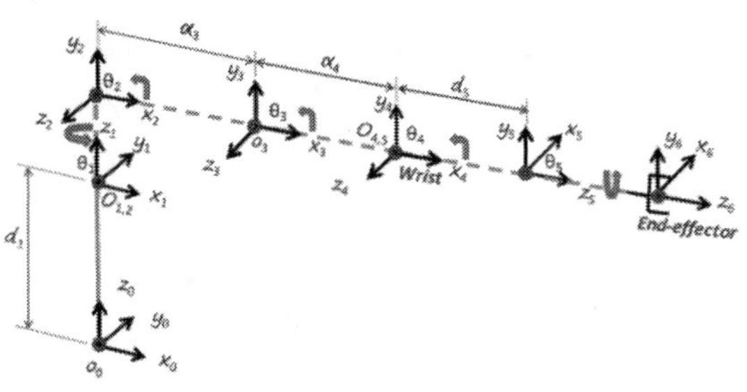


**Figure 1.** Coordinate robot frame.

### 2.1 Denavit-Hartenberg (D-H) parameters

The Denavit-Hartenberg analysis is one of the most used. depending on each mechanism. Based on the analysis, once one coordinate transformation between two frames is defined, D-H parameters for this robot defined for the assigned frames in Table 1.

**Table 1.** Link parameters of slave manipulators (D-H parameters)

| No | $\alpha_{i-1/(°)}$ | $a_{i-1/(mm)}$ | $d_{i/(mm)}$ | $\theta_{i-1/(°)}$ |
|---|---|---|---|---|
| *1* | *0* | *0* | *0* | $\theta_1$ |
| *2* | *90* | *0* | *93* | $\theta_2$ |
| *3* | *0* | *80* | *0* | $\theta_3$ |
| *4* | *0* | *81* | *0* | $\theta_4$ |
| *5* | *-90* | *0* | *22* | $\theta_5$ |
| *6* | *0* | *0* | *0* | *Gripper* |

Based on these parameters, the transformation matrix includes the overall rotation and translation of tool frame {B} with respect to base frame {A}.

Assume that, $a_i$ is the length of Manipulator's link i, $\theta_i$ is the angle of joint i, $\alpha_i$ is the angle between link i and link i-1, $d_i$ the distance between link i and link i-1 , where i =1,2,...,6. $a_i$ , $\alpha_i$ , $d_i$ are all known and $\theta_i$ is proposed to be obtained In this paper, the link parameters of robot are given in Table 1 above.

According to calculated ${}^{O_2}_{P_2}\boldsymbol{T}$ , suppose that:

$$\theta_1 = arc\tan 2(x, y) \tag{1}$$

$$\theta_2 = arc\tan 2(s_2, c_2) \tag{2}$$

$$\theta_3 = arc\tan 2(s_3, c_3) \tag{3}$$

$$\theta_{234} = arc\tan 2(s_{234}, c_{234}) \tag{4}$$

$$\theta_4 = arc\tan 2(s_4, c_4) \tag{5}$$

$$\theta_5 = arc\tan 2(s_5, c_5) \tag{6}$$

## 3 Analysis and calculation of dual robots

In space geometry, using a unique 4×4 homogeneous transformation matrix ${}^{B}_{A}\boldsymbol{P}$ can describe the motion from coordinate system $\{B\}$ of Point B to coordinate system $\{A\}$ of Point A, i.e:

$${}^{B}_{A}\boldsymbol{P} = \begin{bmatrix} {}^{B}_{A}\boldsymbol{R} & {}^{B}_{A}\boldsymbol{T} \\ \boldsymbol{0} & 1 \end{bmatrix} \tag{7}$$

where ${}^{B}_{A}\boldsymbol{R} \in \mathbf{R}^{3\times3}$ is rotation matrix, ${}^{B}_{A}\boldsymbol{T} \in \mathbf{R}^{3\times3}$ is translation matrix. If a series of motions appear, composite transformation will be a good choice to describe.

Assume that, $\{O_1\}$ ( $\{O_2\}$ ) is the base coordinate system of master (slave) manipulator and $\{P_1\}$ ( $\{P_2\}$ ) is the end coordinate system of master (slave) manipulator as shown in Fig. 2. Then ${}^{O_1}_{P_1}\boldsymbol{T}$ ( ${}^{O_2}_{P_2}\boldsymbol{T}$ )is a transformation matrix from the end coordinate system to the base coordinate system of master (slave) manipulator.

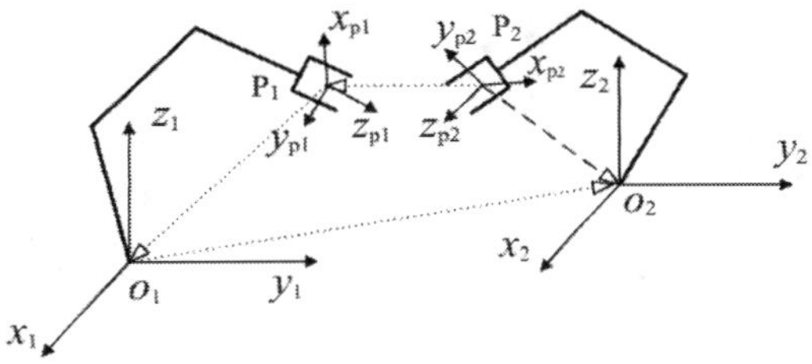

**Figure 2.** The simplified model of dual manipulators cooperating at $t_0$ .

${}^{O_2}_{P_2}\boldsymbol{T}$ could be calculated through composite transformation. It is a transformation from $\{P_2\}$ to $\{P_1\}$ , then from $\{P_1\}$ to $\{O_1\}$ , and from $\{O_1\}$ to $\{O_2\}$ . Complying with the rules of the rotation matrix multiplication, we obtain ${}^{O_2}_{P_2}\boldsymbol{T}$ as:

$${}^{O_2}_{P_2}\boldsymbol{T} = {}^{O_2}_{O_1}\boldsymbol{T} \cdot {}^{O_1}_{P_1}\boldsymbol{T} \cdot {}^{P_1}_{P_2}\boldsymbol{T} \tag{8}$$

${}^{O_2}_{O_1}\boldsymbol{T}$ is known because two bases of manipulators must be fixed before they work. After the master manipulator teaching, ${}^{O_1}_{P_1}\boldsymbol{T}$ is obtained. According to given tasks, we can obtain ${}^{P_1}_{P_2}\boldsymbol{T}$ . So the transformation ${}^{O_2}_{P_2}\boldsymbol{T}$ can be obtained in the case where ${}^{O_2}_{O_1}\boldsymbol{T}$ , ${}^{O_1}_{P_1}\boldsymbol{T}$ , ${}^{P_1}_{P_2}\boldsymbol{T}$ are known. Therefore, the transformation matrix ${}^{O_2}_{P_2}\boldsymbol{T}$ at any time $t$ can be written as:

$${}^{O_2}_{P_2}\boldsymbol{T} = {}^{O_2}_{O_1}\boldsymbol{T} \cdot {}^{O_1}_{P_1}\boldsymbol{T}_{(t)} \cdot {}^{P_1}_{P_2}\boldsymbol{T}_{(t)} \tag{9}$$

If the dual-manipulator collaborative movement is coupled motion, ${}^{P_1}_{P_2}\boldsymbol{T}_{(t)}$ is a constant matrix because the

DOI: 10.1051/matecconf/20167505005

relative relation of the extremity pose between master manipulator and slave manipulator during their movements is constant. It is particular case of dual manipulators collaborative movements and can be described as:

$$ {}^{O_2}_{P_2}\boldsymbol{T} = {}^{O_2}_{O_1}\boldsymbol{T} \cdot {}^{O_1}_{P_1}\boldsymbol{T}_{(t)} \cdot {}^{P_1}_{P_2}\boldsymbol{T} \quad (10) $$

## 4 Trajectory planning

Trajectory planning for robotic manipulators is the process of creating trajectories free of collisions allowing the manipulators to perform the required task. the trajectory calculation formulas of Dual Manipulators under two different collaborative movements were put forward in literature (10). trajectory planning of Dual Manipulators is transformed into the teaching trajectory of single Manipulator through calculation. However, the algorithm given in the literature (10) is complicated. Different formulas are bound to be used in different circumstances.

### 4.1Transformation from initial to final point

In this experiment work Master Manipulator was required to move from initial position $A_1$ [0,0,0] to a given position $B_1$ [0, 290, 20]to pick up the Bottle, then to position $C_1$ [0, 250, 170] to position $D_1$ [250, 0, 170] in the Cooperated area and finally to initial position again $A_1$ [0, 0, 0] in 12 Seconds. And the Slave Manipulator was required to move from initial position $A_2$[0,0,0] to a given position $B_2$ [0, 290, 20] to pick up the Box, then to position $C_2$ [0, 250, 100] to position $D_2$ [-250, 0, 100] in the Cooperated area then to position $E_2$ [-210, 210, 100] to locate it in the Delivery area and finally to initial position again $A_2$ [0, 0, 0] in 14 Seconds. Figure 3 illustrates the top view of packing process task.

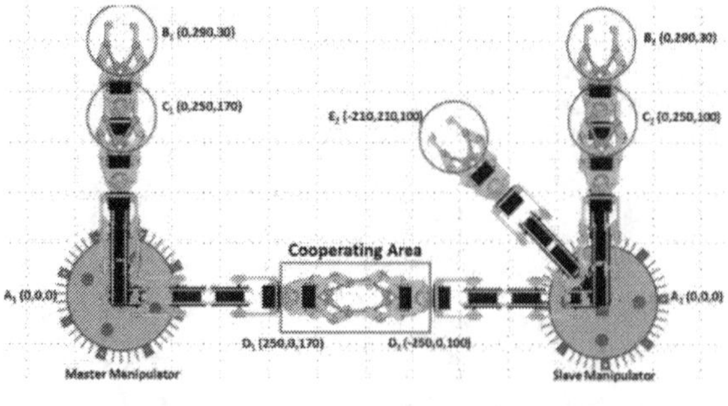


**Figure 3.** Cooperation process (master & slave) manipulators.

The motion was separated to four segments for Master Manipulator and five segments for Slave Manipulator, defined by the time period .Figure 4-5 illustrate the angle Values of the Joints of Master and Slave Manipulator in whole motion segments, the initial position (0,0,0) will be used as a reference to all joint angles of the robots.

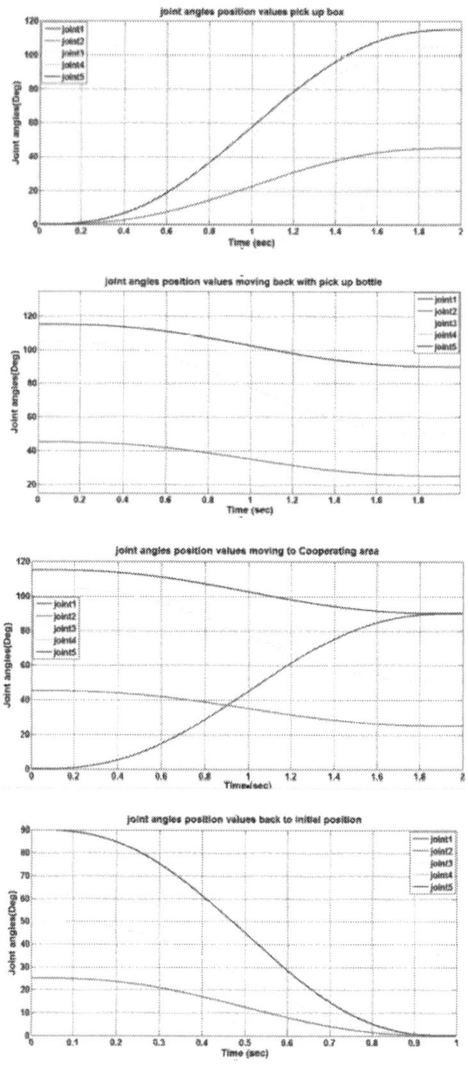


**Figure 4.** Angle values of master manipulator joints.

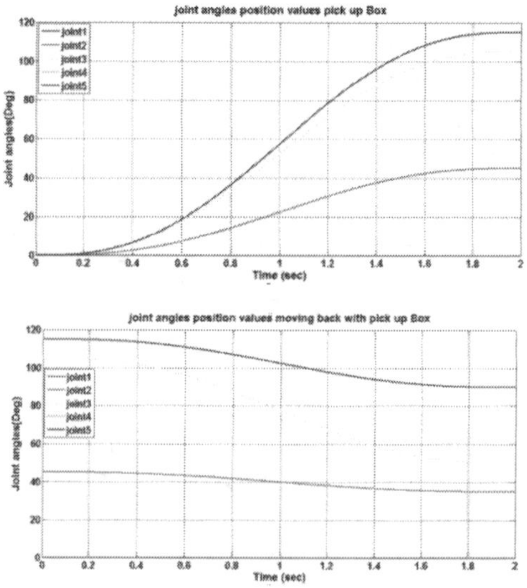

DOI: 10.1051/matecconf/20167505005

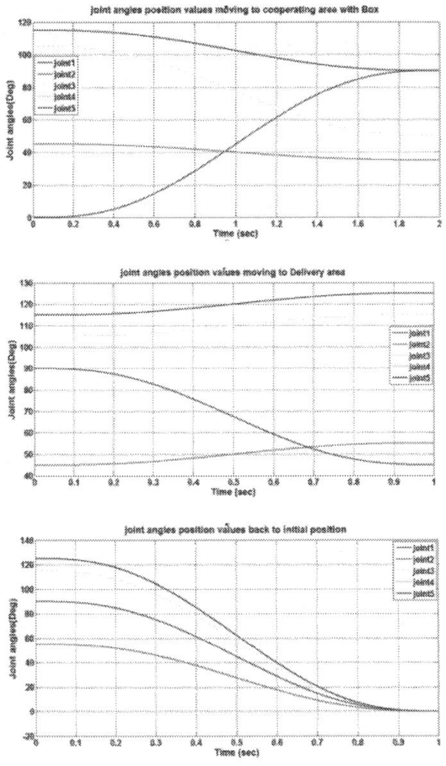


**Figure 5.** Angle values of slave manipulator joints.

## 5 Experimental platform

Double Manipulators, Arduino controller board, PC, 2 infrared distance sensors, 2 external power supplies (5V/1A & 6V/3A) are needed in this experiment. Arduino controller board with Atmegal 168.The sensors GP2Y0A21 can measure the distance of less than 80 cm.

## 6 Experimental results and discussion

Experiments of Dual arm Robots in Cooperating task motion are carried out. As shown in Figure 6, Master robot picks a bottle while Slave carries a Drug box. The slave robot tracks the trajectories on the box while the master transports the bottle. sensory feedback used to add an advantage of closed loop controller for the control system.

Execute the program and perform operation, the whole process will be as follows:

Master and slave Manipulators were moved from initial to final positions as discussed previously. whole real sequence process of Dual Manipulators cooperation for drug Packing shown in Figure 6.

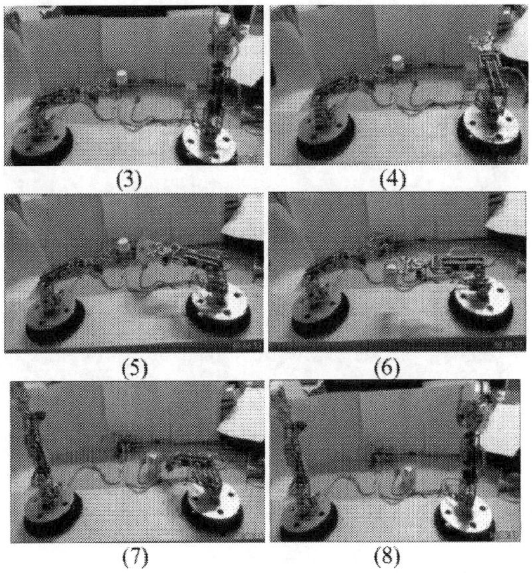


**Figure 6.** Cooperation process task of dual manipulators.

In this experiment, the control system was set up and the task of dual manipulators cooperating for drug packing was performed using the method put forward before. By the above figures, it is clear that the two robots can move in a Cooperative motion Therefore, the results verified the effectiveness of the proposed algorithm.

## 7 Conclusion

The aims of this paper were to establish the theoretical foundation for automated cooperative trajectory planning of Dual arm robots. Motion constraints between cooperative master and slave robots are the basis for this method. *Denavit-Hartenberg* (D-H) representation is used to model robot links and solve the transformation matrices of each joint. The inverse kinematics solution for a 6- DOF Robotic arm was presented. the proposed algorithm was successfully implemented on the packing Cooperating task. this work was motivated bearing in mind that the two arm robots could be controlled by one controller simultaneously in centralized fashion that would reduce the cost of production effectively.

## Acknowledgement

This work was financially supported by the Fundamental Research funds for the Central Universities (YS1403, ZY1521).

## References

1. Ch. Smith, Y. Karayiannidis, L. Nalpantidis, X. Gratal, P. Qi, D. V. Dimarogonas, D. Kragic, Dual arm manipulation – A survey, Robotics and Autonomous.

2. F. Caccavale, P. Chiacchio, A. Marino, L.Villani,impedance control of dual-arm cooperative manipulators, IEEE/ASME Transactions on Mechatronics **13**, 5 576 586 (2008)
3. S. Al-Yahmadi, J. Abdo and T. C. Hsia, Modeling and control of two manipulators handling a flexible object, Journal of the Franklin Institute, **344** 349-361 (2007)
4. KUKA Roboter GmbH. KUKA.CR Issued: 21 Aug 2007.Motion Cooperation 2.1 For KUKA System Software (KSS) 5.3/5.4/5.5, Version: 00.
5. X. Shi and N. G. Fenton, A complete and general solution to the forward kinematics problem of platform-type robotic manipulators, IEEE International Conference on Robotics and Automation Proceedings, 3055-3062 (1994)
6. Latombe, J. C. Robot Motion Planning. Kluwer Academic Publishers, Boston (1991)
7. Y. Koga, J.-C. Latombe, Experiments in dual-arm manipulation planning, in: IEEE International Conference on Robotics and Automation, **3**, 2238–2245 (May 1992)
8. J. Barraqu and J. C. Latombe, Robot motion planning: a distributed representation approach, Stanford University, (May 1989)

# Analysis of the reasonableness of the current level of required accuracy of satellite navigation of consumer systems civil mobile means

Lev N. Lysenko, Vsevolod V. Koryanov, and Alexey G. Toporkov

*Bauman Moscow State Technical University, 105005 2-ya Baumanskaya ul. 5, Moscow, Russian Federation*

**Abstract.** Analysed the paper presents the analysis of the essential accuracy level expected from the modern fundamental coordinate and time data of both existing and prospective satellite navigation and telecommunication systems, while solving the real-world problems of ballistic and navigational support. The paper shows that the respective analysis of the effect of the navigational coordinate and time data uncertainty should be performed based on the estimates of the required ballistic and navigational support of the satellite systems, taking into account the differences between the theoretically attainable level and the practically feasible level of the accuracy for the fundamental coordinate and time data.

## 1 Introduction

Determining essential elements of the modern coordinate-time and navigation support (CTNS) is created and successfully operated in the Russian Federation global navigation satellite system (GLONASS).

As a consequence, modernity has emerged self-sufficient integrated system that provides users the ability to create an independent specialized consumer system (CS), able to solve the problem of CTNS in the specific conditions of operation with a given level of reliability and accuracy.

Basic determinant of the achievement of CS limit capabilities is the level of accuracy of the fundamental coordinate and time support (FCTS), without which it is impossible to create a satisfying private demands CS [1].

The interaction between FCTS and CTNS is clear: the level of CTNS is uniquely determined by the achieved level of development of FCTS, on the other hand highly uncertain due to the lack of correctly and reliably limits the accuracy of demand characteristics and parameters of FCTS and especially the impossibility of strict accounting, the level of uncertainty in the ballistic and navigation support (BNS) satellite navigation systems (SNS).

In this regard, note that the key performance indicators of CTNS CS, integrally defined locating accuracy, availability, integrity and continuity of navigation field, may not be directly linked to numerical precision the values of the parameters of FCTS.

At the same time, it is clear that the new level of requirements to CTNS not possible without further research in the field of FCTS. These studies are very diverse and extremely complex and unique measuring tools [2]. The exploration and development of the corresponding field of knowledge is a time-consuming, great effort and considerable material resources, which makes it relevant to the assessment of justification of requirements to the level of required accuracy of navigation definitions on the basis of radio-navigation satellite systems (RNSS).

## 2 The concept and indicators of the accuracy of navigation satellite definitions

Ensuring operational ephemeris (OE) navigation satellites (NS) includes the determination and prediction of orbits in the inertial parameters of the CS, the calculation of the Earth rotation parameters (ERP) and the recalculation of parameters of motion of the satellite with the use of ERP in the non-inertial coordinate system (CS) current version (characterized by the existing level of uncertainty of the parameters of FCTS), update as often as necessary, these data on Board NS for frames of digital information in the navigation signals, including parameters of the almanac.

Calculation of ephemerides and their accuracy factor is based on the results of measurements of the distance call-outs and/or pseudo-range of no-request systems, and, in some cases, data of Very Long Baseline Interferometry (VLBI).

The orbits are specified on the daily plan for each stage, and measuring the interval refinement of ephemeris corresponds to the circuit of the satellite tracks overlap. In systems of high-accuracy determination of ephemeris and time corrections in the interests of ephemeris-time support processing only measurements of no-request systems global network. They also calculate long-term (90 days) forecast of the state almanac and operational frames of digital information for up to 10 days to maintain the assisting technologies.

Usually considered that the error actually determine the ephemeris from the standard accuracy is about 2-3 times lower than the errors caused by errors in assigning frequency-time parameters.

Operational frequency-time support implements the synchronization of on board clock relative to the time scale of the system by calculating and predicting the on-Board time scale relatively time scale of the system update on Board Time-Frequency Corrections (TFS) placed in the navigation frame.

To predict Board time scale 8...12 hours is used the two-day array of results, which is approximated by a linear polynomial on the interval of extrapolation.

When you upgrade the TFS at every turn of the standard deviation (SD) errors in the mutual synchronization of any two Board time scale of navigation satellite is approximately from 5 to 8 ns, which introduces error in the determination of the location of the consumer at the rate of 2.0 to 2.8 m (2 SD), and to ensure the TFS equal precision data for all satellites of the system in principle is not possible. At the present level differences in accuracy between the TFS "best" and "worst" navigation satellite can exceed 5-6 times.

It would seem that the above considerations can serve as a quite distinct grounds for unequivocal determination of the precise characteristics of the domestic GLONASS and American GPS (Global Positioning System) and map them.

Usually given in the literature data should be considered as rather arbitrary and approximate. The case is as follows.

The accuracy of navigation definitions has significant impact on the location of the satellite in the constellation relative to the user. For planning navigation support in different parts of the Earth it is necessary to have cards of equal accuracy, which should be marked with lines of equal values of Position Dilution Of Precision (PDOP). Therefore, for global satellite navigation system can only talk about the potential accuracy of navigation definitions, which together form "field precision of the satellite navigation system ". When this field precision varies with time due to the motion of the satellite relative to the Earth's surface and variations of the configuration of the selected constellation.

If the structure of the satellite navigation system regular, i.e. it repeats itself after a certain period of time, the precision can be allocated to the elementary structural zone, which is repeated in time and space. For the prototype GPS system "Navstar", for example, the size of such a zone area was 1.5 hours time and 60 degrees in longitude. In the period under consideration (early–mid 80-ies of the last century) standard deviation determine the location of the object using said system, respectively 7…25 m, height 10…43 m, the maximum measurement error of radial velocity was 0.06 m/s.

A comparison of the level of navigation accuracy for different satellite systems is possible only when there is at least similar prerequisites for achieving the same goal, which is not always taken into account. As a consequence, we compare the accuracy characteristics of GLONASS with GPS in the structure of the global system of WAAS (Wide Area Augmentation System) differential navigation (GDGPS), utilizing up to 30 stations in redundant form, or at least 12 stations that are guaranteed to be optimally located throughout the globe.

With all the conventions of existing assessments though objectively it is possible to consider [3] that at low angles of shading (at the angle designated signals up to 5°) orbital group (OG) GLONASS-24, providing 100 percent availability of navigation except for small areas at latitudes ±28°…± 30°, the accuracy is inferior to GPS on almost 20%; at high angles of shading (elevation ≥ 25°, which corresponds to the urban and mountainous conditions) OG GLONASS-24 1.5–2 times lower than on the global availability of not only GPS but also GALILEO satellite navigation system-27 (European Union) and COMPASS (China).

Leaving aside the problems of accessibility, let us state the assumption that the existing backlog on navigation accuracy from GPS (USA) was the reason for the language requirements of the current stage of development of GLONASS aimed at achieving a precision navigation definitions up to 0.1...0.6 m (SD), which significantly exceeds the guaranteed accuracy of operating system.

## 3 Assessment of demand indicators of present level of accuracy of satellite navigation.

The achievement of a previously specified level of accuracy of navigation definitions put forward, according to numerous statements of representatives of the Institute of applied astronomy of Russian Academy of Sciences (RAS), the accuracy requirement is fundamental segment KBO at the level of about 3 cm in determining the position of the center of the geoid (the Geocenter) with an error less than 1 cm, ground CS – $5 \cdot 10^{-10}$ (3 mm coordinates), and in the sky CS – location accuracy should characterize the level of 0.02 ms for the arc. of the standard ICRF and 0.1ms of the arc. for geodetic sources.

To understand the reachability of such requirements, we note that they cannot be guaranteed without the installation of beacons for VLBI observations at the poles of the moon, determining the ephemerides of Solar system bodies at the level of $10^{-11}$, the daily non-stability frequency standards and time of $10^{-16}$ to $5 \cdot 10^{-17}$ with finally, the daily forecast of the orbits of GLONASS satellites with a precision better than 1 mm. About the unfeasibility, at least, the latter condition is, in principle, cannot speak.

But if so, the question naturally arises about the practical validity of the formulated requirements to guarantee the level of accuracy of FCTS.

Moreover, we note that the analysis of the impact of uncertainty, knowledge of the parameters of CTNS for RNSS on the basis of the accuracy of their required BNS need to be implemented taking into account the differences between the theoretically achievable level of accuracy of parameters of FCTS and their practical level of precision determined by the requirements of specific consumers.

So, there are two aspects to answer the question posed.

1. Are there any alternatives to improving the accuracy of FCTS to achieve the desired level of accuracy of navigation definitions of consumers?

2. What is the intrinsic demand of modern level of accuracy of satellite navigation, at least from the major consumer of civilian systems?

For a better understanding of the substance of the first part of the matter under discussion will limit ourselves to the analysis of problems BNS geostationary satellites. No matter how accurately we have not defined the parameters of FCTS, the absolute accuracy of their knowledge is impossible to achieve and therefore almost impossible to enforce the actual conditions of stationarity.

Objectively, uncertainties knowledge of factors such as off-center gravity of Earth, the inaccuracy of the knowledge of attraction of the Sun, moon, Venus, Jupiter and the other inner planets of the Solar system, direct and reflected radiation pressure, tidal deformation of the Earth, the ocean tides, the corrections General relativity and other factors, will inevitably lead to a shift (drift) of the satellite relative to the earth.

Despite the fact that the acceleration experienced from, for example, a direct light pressure (level $10^{-7}$ m/s$^2$) and tidal deformations of the Earth ($10^{-10}$ m/s$^2$) are, at first glance, almost intangible value, after 30 days the resulting maximum deviation from the nominal setups will be respectively about 2.4 and 8.5 m.

As a consequence, suggest the following possible solutions to the problem: either to increase the accuracy and reliability of knowledge of the parameters of FCTS as possible, and technically to implement precision BNS, either consciously (taking into account the existing possibilities of technical systems) limit the accuracy of BNS to the level of allowed deviation caused by drift, keeping the parameters of the orbit closest to the ideal stationary values in the neighborhood of a given point of standing through the management of the movement of the center of mass of geostationary satellites, which consists in the periodic correction of its orbit.

The first generation displayed on geostationary Earth orbit (GSO) satellites is adjusted, as a rule, only the parameters that define the offset of the satellite in longitude to the satellite is guaranteed to remain within "geostationary belt" ("Clarke belt").

Since that time much has changed. Used in past classical methods for determining the ephemeris and, as a consequence, astronomical constants of the moon and planets were based mainly on optical observations. Revolutionary was the widespread use of radio and especially laser astronomical measurements. Thanks to this data situation, at least for the inner planets has drastically changed. Accumulation of measurement results obtained using the AMC (mainly Viking, Pathfinder, MGS and Odyssey), had developed high-precision theory of the ephemerides of the inner planets with a precision of milliseconds of arc. The standard deviation of the observations about estimated values for mercury is 1.4 km for Venus and Mars – 0,7 km, for AMC Viking – 8.8 m, Pathfinder – 5.1 m, MGS and Odyssey – 1.4 m.

As a result, managed to create a dynamic model of high-precision ephemerides of planets and the moon (EPM). The best of existing foreign models of DE405/LE405 in this case exceeds, for example, the previously used model DE200/LE200 by the accuracy of assigning astronomical unit (a.u.) length (a.u. = 149597870696,3; 0,1) is several orders of magnitude. Developed the Institute of Applied Astronomy RAS model are comparable in accuracy with the specified DE405/LE405. Domestic dynamic model EPM2004 (Pitjeva, 2004) takes into account mutual perturbations of the major planets and the moon, compression of the Sun, the effects of physical libration of the moon, perturbations from the 301 biggest asteroids and the massive asteroid ring located in the plane of the Ecliptic. Ephemeris converted to the ICRF system and agreed with the resolution - mi XXIII and XXIV General assemblies of the International astronomical Union.

In relation to the refinement of geodetic parameters of the Earth to Refine the resonant harmonics of the gravity of Earth, primarily in the interests of GPS, in the middle of 1980 in the U.S. program was implemented COGOES (international program of optical observations of geo - stationary satellites).

Note, incidentally, that in the USSR at that period there were no conditions for the implementation of geodetic and geophysical softwares required level, due to the lack of national special spacecraft, similar to that used in the USA to improve the accuracy of ephemeris support navigation satellite systems such as GPS.

This, in turn, has improved the level of accuracy BNS excretion geostationary satellites to officially declare a point of standing in GSO, as well as to find an acceptable compromise on the size of the dead zone of permissible deviations of the satellite from the nominal value, defining the value of energy losses to the correction of the orbit on the one hand, and, on the other, restrictions on the minimum allowable angular distance between adjacent satellites in GSO, set by international agreements, reasons of electromagnetic compatibility of satellite radio. The relevant regulating document serves as "the radio Regulations" / Edition of 2004. No. 119-04-REV-04, imposing the restriction of ±0.1° in the recent past, adjusted up to ±0.05° currently, for the deviation of the satellite in longitude relative to the nominal setup [1].

In order to perform the required corrective maneuvers hold the satellite at a given longitude and latitude with minimal energy consumption and with a sufficient precision, needed to have the most accurate information about the parameters of the real orbits. The navigation process is based on Measurements of Current Navigation Parameters (MCNP) using special measurement tools and statistical processing of results of measurements using the most rational of mathematical models and computational schemes that are part of BNS utilization.

While for most practical applications geostationary satellites, in principle, it is sufficient to account for the influence of the following perturbation:

• the force components of the gravity of Earth with the deduction of harmonics of the spherical functions of the geopotential, not above the 8th order (with the fame para - meters 36 harmonic model WG84 and 70 for model EGM96);

• the gravitational attraction of the Sun and moon as point masses;

• direct light pressure;

• implementation errors of the thrust vector of the propulsion unit (module and orientation).

Data from the generalized requirements of most CS CTNS follows that none of the consumer systems does not imply the need reaches the level of the average values of SD centimeter accuracy for solving problems of navigation definitions of satellite navigation systems.

As an exception can be noted only marginal (maximum) accuracy levels the accuracy of determining when approaching and landing aircraft according of the International Civil Aviation Organization categories (max 0.2 m, min 2 m) and prospective levels of precision maneuvering and conducting the specialized works in the water area of bodies of water and ports (max 0.1 m min 3.5 m). We emphasize again that we are talking about extreme levels of precision, which in most cases is not common to use in design development.

Individual attention requirements of substation surveying software and high-precision monitoring of displacements of engineering structures difficult.

Note that for the solution of applied problems of geodesy measurements performed on points of the geodetic control network using the methods of relative definitions. State geodetic network in this case is based not only on satellite geodetic network of 1 class, but also on a fundamental astronomic-geodetic network, as well as high-precision geodetic network. In this system besides written long existing and widely used network triangulation and polygonometry different classes. Everything else for reading surface, in particular in the CS-95, adopted not the true surface of the Earth, and the surface reference ellipsoid of Krasovsky.

In this setting the output centimeter level of accuracy of astronomical and geophysical networks in combination with new high precision points of the satellite network of permanent differential stations, of course, is not an unsolvable problem. Moreover it should be considered real and millimeter level, relevant to the solution of some problems of geodynamics (earthquake prediction in seismically active areas). As declared millimeter level accuracy for cadastral surveying and fixing the boundaries of land use, such claims cannot be taken seriously. Here, at least, desirable (but not substantiated) is valid.

This gives grounds to assume that from the point of view of the practical demands of the consumers of satellite navigation, at least for civil purposes, the existing requirements for the level of accuracy of the means for FCTS is overstated [1].

As to the provision of the required level of accuracy of even the "precision" of space objects, as you know, updated currently means Ground Computer-Aided Control Complex (GCACC) allow or will allow in the near future to carry out the measurement of the orbit parameters of spacecraft in the altitude range 200...40 000 km with an accuracy (maximum error) in the range of 1...30 m and the radial velocity of 0.001...0,003 m/s. Such accuracy is quite sufficient for high-precision navigation definitions most of the existing types of spacecraft, in particular, for example, to keep the type of communications satellites "Express-A" in the permissible range is ±0.05°, implying the need to define the parameters of their orbits at: 0,9 arc. minute – latitude of sub-satellite point; 0,3 arc. minute – longitude of sub-satellite point; 0.15 – period of satellite revolution.

The satisfaction of increased demands BNS advanced space funds can only be secured through the involvement of satellite navigation equipment.

## 4 Conclusions

1. The level of requirements for navigation software from various categories of consumers, gives reason to conclude that none of the existing substation CS CTNS not imply the need reaches the level of the average values of SD centimeter accuracy for solving problems of navigation definitions of satellite navigation systems.

In this regard, requirements to the complex of FCTS on the level of knowledge of the position of the Geocenter with error not exceeding 1 cm, ground CS – $5 \cdot 10^{-10}$ (3 mm for coordinates), the celestial CS – 0.02 ms for the arc of the standard ICRF and 0.1 ms of the arc. for geodetic radio sources, is overstated.

2. Given that the achievement of the stated values focus on the need to ensure in principle unattainable daily predict the orbits of GLONASS satellites with a precision better than 1 mm, these requirements appear to be superfluous not only for the state of the art, but also in relation to the long term.

3. Navigation errors are not only caused by errors of determination of orbit parameters according to the processing results of MCNP, but also the forecast errors due to inaccurate descriptions of the forces at work, including errors in the execution of maneuvers.

Not accounting for this circumstance is caused first of all by those considerations that the maximum interval of forecast ephemeris and TFS used in the ephemeris-time support system is always less than a predetermined interval of "aging", which, as is generally assumed, can always be reduced if it is used, the value begins to have a noticeable impact on the precision technical characteristics of ephemeris-time support system.

Currently, the interval of "aging" ephemeris-time information in the GLONASS system for normal functioning period of the system is 12...14 hours. Thus, it is common to assume that the semidiurnal forecast ephemeris and TFS for almost MCNP should not introduce errors, or they are so small that they can be neglected. In fact, everything is determined by the required level of accuracy of navigation definitions. For today's "meter-level" errors semidiurnal forecast can be neglected, but the exit to "the centimeter level" has radically changed the situation. Reducing the interval of "aging", which is widely accepted to refer to as an

DOI: 10.1051/matecconf/20167505006

"elementary solutions" is not unlimited and cannot serve as a guarantee for the solution.

4. In order to provide a predetermined sufficiently small value of the range of aging ephemeris-time information is necessary to measure network keys, which help to clarify the ephemeris and TFS, would not break, required large intervals of "aging" and that the topology of filling of the measuring stations the GCACC would ensure the bookmarks with the corresponding period. The current network of filling GCACC allows you to increase the frequency of bookmarks by no more than 1-2 hours, which is insufficient for achieving a promising level of accuracy of satellite navigation.

## References

1. Lysenko L.N., Koryanov V.V., Toporkov A.G. Evaluation of satellite navigation accuracy requirements based on the analysis of the current state of commercial coordinate and time navigation support systems, Vestnik BMSTU, Mech. Eng, 47 (2015).
2. Finkelstein A.M., Ipatov A.V., Skurikhina E.A. The fundamental coordinate and time support for the GLONASS system means RSDB-network Quasar-KVO. Proc. of IAA RAS, **77**, 608 (2007).
3. Stupak G.G., Lysenko L.N., Betanov V.V. Status and Perspective for Improving Orbital Structures of Navigation Satellite Systems, Vestnik BMSTU, Priborostr, 3 (2014).
4. Lysenko L.N., Betanov V.V., Zvyagin F.V. *Theoretical Foundations of ballistics Navigation Support Space Missions*w, 518 (2014).
5. Urlichich Yu.M., Yezhov S.A., Zhodzishsky A.I., Kruglov A.V., Makhnenko Yu.Yu, Modern technology navigation geostationary satellites, 271 (2006).

# Mutual Coupling Reduction between Very Closely Spaced Microstrip Antennas Using CPW Structure

Xiao-zun Ma[1,2], Ming-yang Zhao[2], De-feng Zhang[2] and Peng-jun Liu[2]

[1]*Dept. of Electronic Engineering,Beijing University of Aeronautics and Astronautics, Beijing,China*
[2]*Luoyang Electronic Equipment Test Center of China, Luoyang, Henan, China*

**Abstract.** An efficient H-plane mutual coupling reduction method is introduced for very closely placed dual-element microstrip antennas operating at 2.4GHz. A simple coplanar waveguide (CPW) structure is introduced in the ground, energy from one element is guided to another element, artificially generating new kind of mutual coupling current. By optimizing size of CPW structure to achieve purpose that new mutual coupling current cancel out with original existing mutual coupling. More than 36dB isolation ($|S_{21}|$) is achieved with dual-element broadside array that distance is less than 0.024 $\lambda_0$ ( $\lambda_0$ is free space wavelength).

## 1 Introduction

Reduction mutual coupling has currently attracted considerable attention, antenna performance can be improved if mutual coupling is suppressed , such as inner-antenna spacing reduction and channel-capacity increment in MIMO application, side-lobe level (SLL) control and direction of arrival (DOA) estimation .

Coupling between antenna elements is mainly caused by surface waves and near field [1]-[3]. Near field coupling arises when one antenna is placed in near field zone of another antenna. Near field coupling becomes dominant over surface wave coupling when space between antenna elements decreases.

To degrade mutual coupling, various types of electromagnetic band-gap (EBG) have been proposed [4]-[7]. Because size of the cell structure of EBG is not so small, these methods are mainly applicable when substrate is thick, dielectric constant is relatively high and distance between antenna elements is not so close. Besides, EBG often involves via or two dielectric layers which increase the complexity and cost. Recently, simple and efficient methods are proposed in [8] and [9], but both of them are E-plane mutual coupling reduction.

In this letter, a novel method that etches a simple CPW in the ground is proposed for more degradation of H-plane mutual coupling and distance between array elements. The CPW can introduce a new coupling from the one element to another element, new mutual coupling current cancel out with original existing mutual coupling. The dual-element broadside array that distance is only 0.024 $\lambda_0$ and the isolation is more than 36 dB.

## 2 Procedure for CPW

### 2.1 Antenna structure

The geometry of array along with proposed CPW etched in the ground is illustrated in Fig. 1. The front of proposed antenna is in Fig. 1(a) and the back is in Fig. 1(b). This array comprises two microstrip antennas operating at 2.4GHz. The dielectric board is $F_4Bm$-2 with relative permittivity of 2.2, loss tangent of 0.001 and thickness of 2 mm.

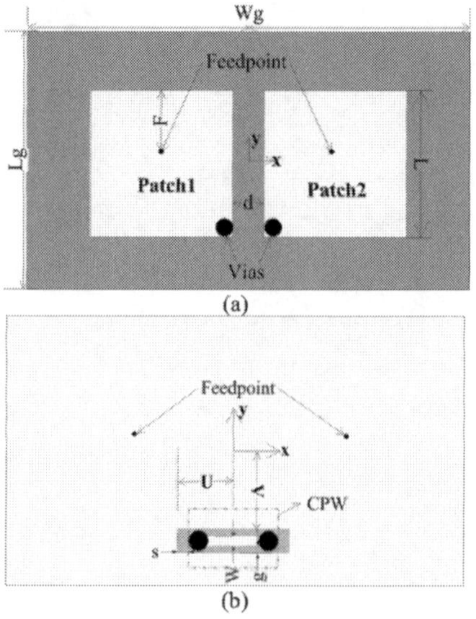


**Figure 1.** Geometry of the proposed patch antenna array: (a) front; (b) back.

In Fig. 1(a), there are two square patches which are on common ground with inter-patch spacing (edge to edge) of d that is about $0.024\lambda_0$. The side length of the patches is L that is about $0.5\lambda_g$. *F* denotes distance between feedpoint and edge of patch. The ground size is depicted by $Wg \times Lg$. In Fig. 1(b), there is a CPW on the ground which is indicated by dotted bordered rectangle. In order to avoid the center line of the CPW connecting to the ground, a slot is opened at each end of the CPW. Through the via, CPW is connected to patches. *U* and *V* denote respectively the distance between the CPW and the *y*-axis or *x*-axis. *W* and *g* denote the width of center line and gaps of the CPW respectively. The width of the slot is s.

### 2.2 Parametric studies

CST Microwave Studio was used to study parameters of the proposed structure. Through parameters sweeps and optimization, *U and V* were found important in this design. The results of parametric studies are depicted in Fig. 2 and Fig. 3.

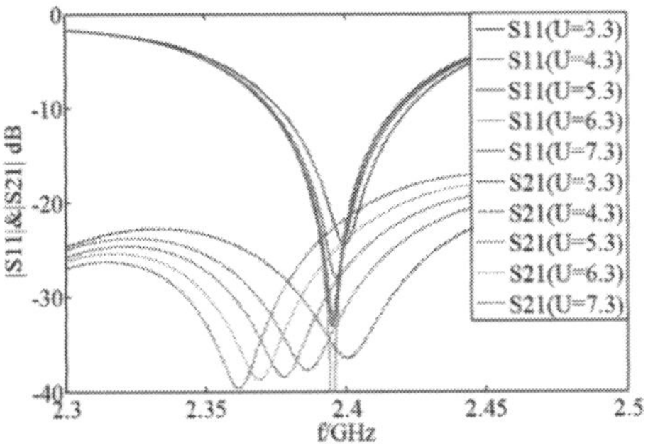


**Figure 2.** S-parameters varying with U (unit of U: mm).

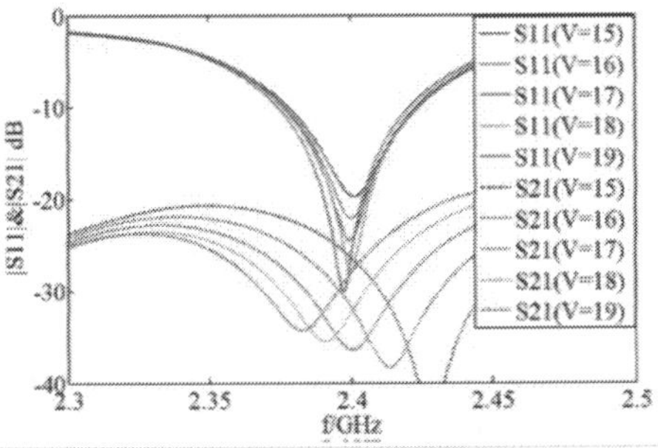


**Figure 3.** S-parameters varying with V(unit of V: mm).

Fig. 2 shows the S-parameters along with the parameter *U*. when the other parameters are set to *d*=3 mm, *V*=17 mm, *F*= 12 mm, *W*=0.8 mm, *g*=0.4 mm, *Lg*=73.5 mm, *Wg*=116 mm, *s*=0.8 mm, L=40mm. The solid line shows the $S_{11}$, while the dotted line shows the $S_{21}$ (the same below).As shown, the operating frequency changes very small along with *U*, while the frequency of the minimum value of $S_{21}$ moves to lower frequency when *U* increased. This is because that the field distribution is different at different place of patches, which affects the new coupling field that CPW introduces.

Fig. 3 shows the S-parameters along with the parameter *V*. The parameter *V* has almost the same effect of *U*, but the minimum value of $S_{21}$ is increased when *V* is increased.

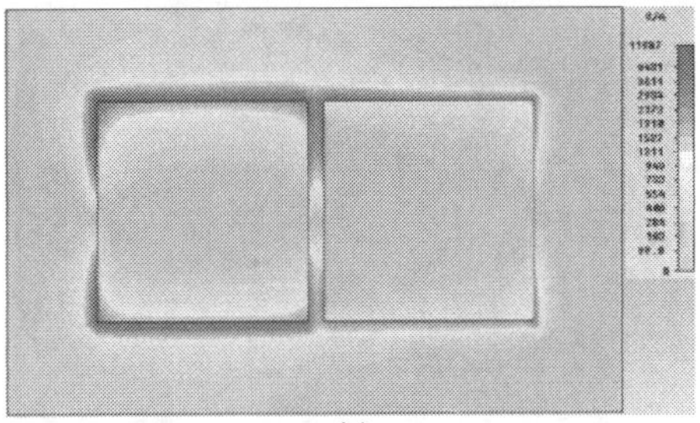

(a)

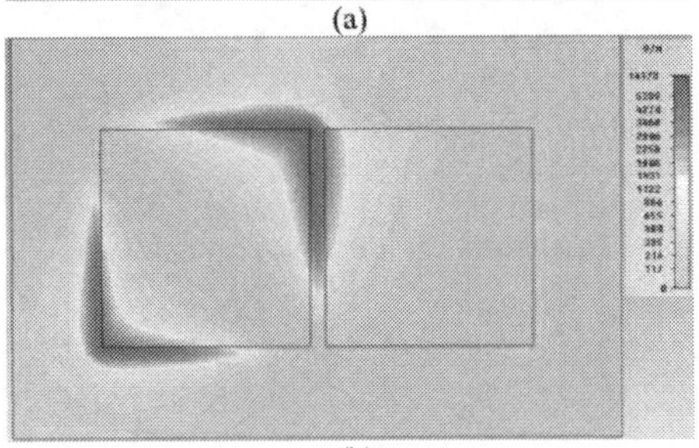

(b)

**Figure 4.** Absolute electric-field distribution:(a) the normal; (b)the proposed.

## 3 Result and discussion

In order to understand the mechanism of the proposed method, the electric-field distribution comparisons between the proposed antennas and normal microstrip antennas are shown in Fig. 4(Patch 1 is excited while the Patch 2 is terminated with a 50-Ω impedance). The mutual coupling for normal antenna elements is serious with respect to the proposed antennas. In Fig.4 (b), the electric-field distribution along the *x*-axis direction is uneven. At the position and the symmetric position of vias, there is no electric-field distribution. In the proposed structure, the vias worked as a probe, which is connected to the CPW. The energy from Patch 1 is guided to Patch 2 through the vias and the CPW, which offsets the original existing coupling. To Patch 2, it is that the conduction current (introduced by CPW) and the displacement current (mainly caused by original existing near field coupling) offset each other.

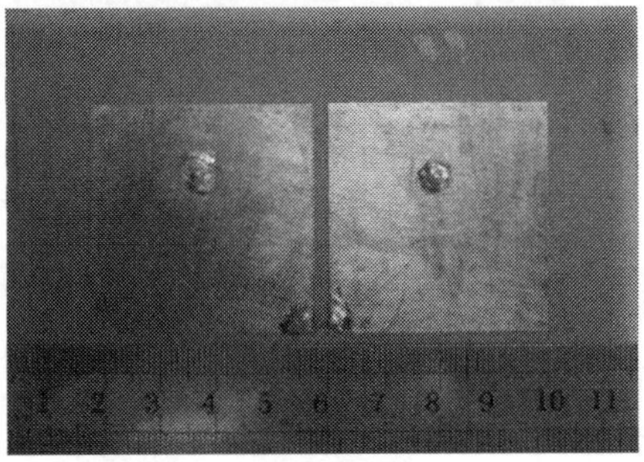

(a)

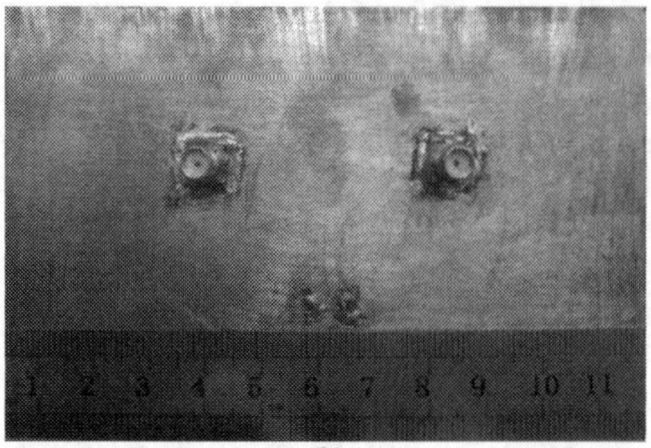

(b)

**Figure 5.** Photograph of the two fabricated prototypes:(a) front; (b) back.

Finally, proposed designs were fabricated and measured using Agilent vector network analyzer-8510B. Photographs of the fabricated prototypes are shown in Fig. 5. Taking into account processing convenience, through-vias are replaced by soldered conductor pillars. In fact, the CPW is so small that is difficult to be found in the photograph.

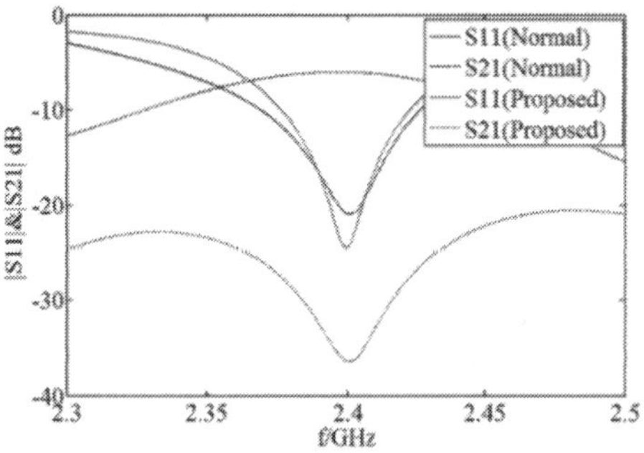


**Figure 6.** S-parameters comparison of normal and proposed microstrips.

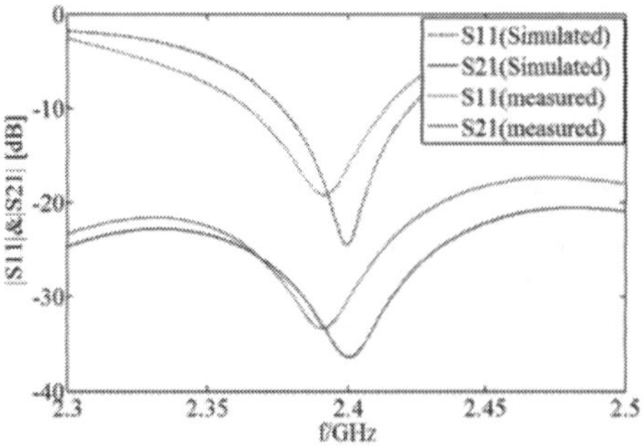


**Figure 7.** Simulated and measured S-parameters.

Results of S-parameters of the proposed structure in comparison to the normal structure are presented in Fig. 6. As shown, the proposed structure provides a significant improvement of isolation about 30 dB compared to conventional normal microstrip antennas. It should be noted that patch dimensions and feed-point position have been done a slight adjustment. Fig. 7 shows comparisons between simulation and measured results. Compared to the simulation, the measurement results are slightly worse and there is a little frequency offset, which are mainly caused by the machining error.

Radiation patterns in H-plane and E-plane of proposed structure in comparison to single patch antenna and normal patch antennas at 2.4GHz are presented in Fig. 8. For the H-plane, the main lobe direction of array element of proposed antennas is slightly different from the single, while main lobe direction of array element of normal antennas is significantly different. The former may be due to the asymmetry of the ground plane ; the later is because that radiation of coupling patch offsets or enhances original radiation. For E-plane, these simulation results don't show any significant differences, but directivity of normal antennas is smaller than others.

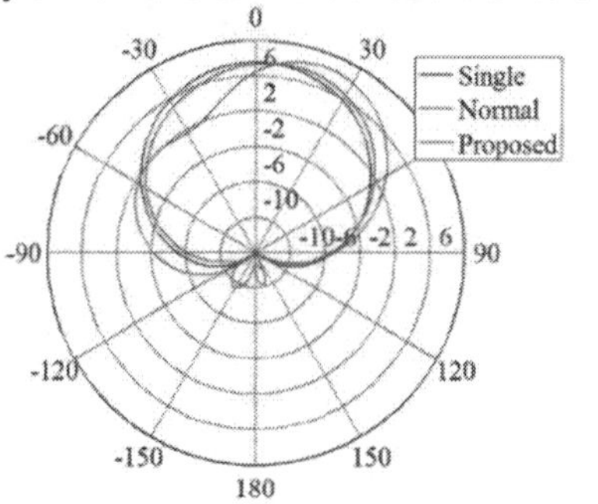


(a)

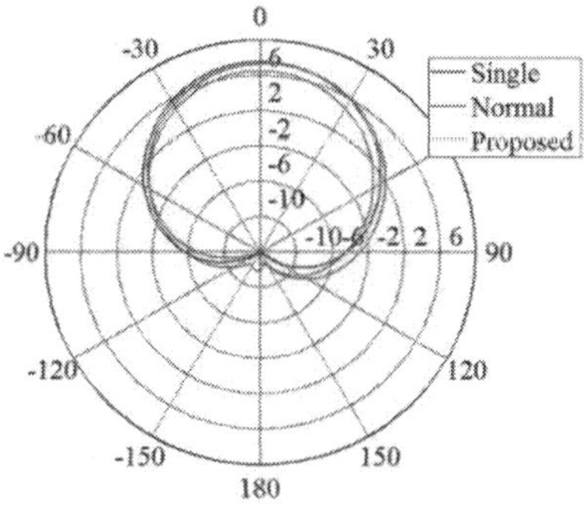


(b)

**Figure 8.** Radiation patterns of single, normal and proposed antennas at 2.4GHz:(a) H-plane (xz-plane). (b)E-plane (yz-plane).

Finally, to further validate the radiation effects of the proposed structure, simulated values of the directivity, main lobe direction and side lobe lever (SLL) of the designs compared to the single patch antenna and normal antennas are summarized in Table 1. The table shows that the proposed structure, compared to normal structure, improves significantly the side lobe lever and main lobe direction. At the same time, the directivity of the proposed structure is close to the single patch.

**Table 1.** Comparison of radiation.

| Design | Directivity(dBi) | | Main lobe direction(°) | | SLL(dB) | |
|---|---|---|---|---|---|---|
| | xz | yz | xz | yz | xz | yz |
| Single | 7.5 | 7.5 | 0 | 0 | -20.6 | -20.6 |
| Normal | 8.4 | 6.5 | 21 | 0 | -6.4 | -18.3 |
| Propo | 7.5 | 7.3 | 10 | 3 | -16.9 | -20.0 |

## 4 Summary

In this letter, an efficient but simple technique is proposed for more reduction of the mutual coupling and distance between the antenna array elements. Through a simple CPW on the ground plane, more than 36dB isolation is achieved with inter-antenna spacing of less than 0.024 $\lambda_0$. The technique is suitable for MIMO antennas and compact arrays.

## References

1. A. Bhattacharyya, Characteristics of space and surface waves in a multilayered structure, IEEE Trans. Antennas Propag., **38**, 8, pp.1231–1238, (1990).
2. D. Jackson, J. Williams, A. Bhattacharyya, R. Smith, S. Buchheit, and S. Long, Microstrip patch designs that do not excite surface waves, IEEE Trans. Antennas Propag., **41**, 8, pp. 1026–1037, (1993).
3. M. M. Bait-Suwailam, O. F. Siddiqui and O. M. Ramahi, Mutual Coupling Reduction Between Microstrip Patch Antennas Using Slotted-Complementary Split-Ring Resonators, IEEE Antennas Wireless Propag. Lett., **9**, pp.876-878, (2010).
4. M. Coulombe, S. F. Koodiani and C. Caloz, Compact Elongated Mushroom (EM)-EBG Structure for Enhancement of Patch Antenna Array, IEEE Trans. Antennas Propag., **58**, 4, pp. 1076-1086, (2010).
5. F. Yang and Y. Rahmat-Samii, Microstrip Antennas Integrated With Electromagnetic Band-Gap (EBG) Structures: A Low Mutual Coupling Design for Array Applications, IEEE Trans. Antennas Propag., **51**, 10, pp. 2936-2946, (2003).
6. E. Rajo-Iglesias, Ó. Quevedo-Teruel and L. Inclán-Sánchez, Mutual Coupling Reduction in Patch Antenna Arrays by Using a Planar EBG Structure and a Multilayer Dielectric Substrate, IEEE Trans. Antennas Propag., **56**, 6, pp. 1648-1655, (2008).
7. H. S. Farahani, M. Veysi, M. Kamyab and A. Tadjalli, Mutual Coupling Reduction in Patch Antenna Arrays Using a UC-EBG Superstrate, IEEE Antennas Wireless Propag. Lett., **9**, pp.57-58, (2010).
8. A. Habashi, J. Nourinia and C. Ghobadi, et al. Mutual Coupling Reduction Between Very Closely Spaced Patch Antennas Using Low-Profile Folded Split-Ring Resonators (FSRRs), IEEE Antennas Wireless Propag. Lett., **10**, pp.862-865, (2011).
9. J. O. Yang, F. Yang and Z.M. Wang, et al. Reducing Mutual Coupling of Closely Spaced Microstrip MIMO Antennas for WLAN Application, IEEE Antennas Wireless Propag. Lett., **10**, pp.310-313, (2011).

# BER Performance of QAM and QPSK Modulation Technique for DCT Based Channel Estimation of STBC MIMO OFDM

Suzi Seroja Sarnin, Siti Maisurah Sulong and Nurhuda Abu Hasan Asa'Ari

*Department of Communication Engineering,Faculty of Electrical Engineering, Universiti Teknologi MARA, Shah Alam, Malaysia*

**Abstract.** Consumer demand for high speed data communication system is increasing day by day. However the major challenges for wireless communication system are Inter Symbol Interference (ISI) and Inter Carrier Interference (ICI) that occur during data transmission. This interference will distort signal and increase error rate in the system. Combination of Orthogonal Frequency Division Multiplexing (OFDM) and Multiple Input Multiple Output (MIMO) give a reliable high data rate. Discrete Cosine Transform (DCT)-based channel estimation is used in STBC MIMO-OFDM and simulated using MATLAB software. The system performance is compared with multiple antenna configuration and different modulation technique. Multiple antenna configurations that involve during simulation are 2x2, 4x4 and 6x6 while modulation techniques are Quadrature Amplitude Modulation (QAM) and Quadrature Phase Shift Keying (QPSK). This simulation will prove that the higher number of transmit and receive antenna, the lower BER for the system.

## 1 Introduction

Nowadays, wireless broadband communication system is well known throughout the world. Demanding for wireless network technologies were increasing years by years and it is proven by over past few years, data rates for wireless network increasing approximately from 10Kbps to 10Mbps and beyond. It is full responsibility to achieved higher capacity of wireless communication system with better performance. Wireless broadband communication system is a system where signal been transmitted from transmitter to receiver passing through space. Throughout this system there should be degradation in terms of system performance that needs to be enhanced.

MIMO technology had become the advance technology in communication system where using multiple antennas at both transmitter and receiver in order to send many data at one time and increase the capacity of data transmit. MIMO-OFDM been introduced because of [1] its ability to enable high data rate transmission over multipath and frequency selective fading channel. OFDM is a method used as multi-carrier digital modulation technique [2]. This technique where dividing the high-speed information signal into multiple narrowband subcarrier channel. The multiple subcarrier channel transmit simultaneously at different frequency in parallel and waveforms are orthogonal to each other where it overlapping and not interfere with adjacent subcarriers. OFDM is robust against narrowband interference [3], has high spectral efficiency and reduce ICI (Inter Carrier Interference). Combination of MIMO and OFDM system can give high data rates, speeds and high spectral efficiency [1] which needed for future broadband communication system. Channel estimation is important in order to maintain and recover the signal at receiver from noisy channel by using accurate channel estimator. Previous researcher [4] also use DCT instead of DFT. FFT introduced by Joseph Fourier shows FFT is a simple algorithm that allows conversion back and forth between the frequency domain and the time domain. In [5], shows that DCT is enhance of the DFT and FFT. This paper will analyze the performance of the DCT-based channel estimator in MIMO-OFDM wireless communication system. QAM and QPSK is used in MIMO OFDM as a modulation scheme and compare it which modulation scheme works the best for MIMO-OFDM using DCT-based channel estimation.

## 2 Methodology

### 2.1 MIMO system

Multiple Input Multiple Output (MIMO) had become advance technology in communication system. A radio antenna technology that uses multiple antennas at the transmitter and receiver enables to send many signal paths as antennas carry the data [6]. MIMO had increase the system capacity and user data rates without using additional power or bandwidth. Spatial multiplexing is one of the fancy versions of MIMO where different data is fed to the multiple antennas and sends the information in parallel. At the receiver it is able to separate these data out and combine it back together.

DOI: 10.1051/matecconf/20167506002

In MIMO system, a transmitter sends multiple streams by multiple transmit antennas. The transmit streams go through matrix channel which consist of all $N_tN_r$ path between the $N_t$ transmit antennas at the transmitter and $N_r$ receive antennas at the receiver. Then, the receiver gets the receive signal vectors by the multiple receive antennas and decode the receive signal vectors into the original information [7].
MIMO system can be modeled as

$$Y = Hx + n \tag{1}$$

which $x$ is the transmitted symbol from each antenna and $Y$ is the received signal at the receiver. $H$ is the channel matrix for multiple antennas. $H$ channel can be express by matrix according to the number of transmit and receive antennas and n were the channel noise, AWGN channel.

$$H = \begin{bmatrix} h_{1,1} & h_{1,2} & \cdots & h_{1,N_t} \\ h_{2,1} & h_{2,2} & \cdots & h_{2,N_t} \\ \vdots & \vdots & \cdots & \vdots \\ h_{N_r,1} & h_{N_r,2} & \cdots & h_{N_r,N_t} \end{bmatrix}$$

H channel indicates the number of transmit and receive antennas, as the number of antenna increases for transmit and receive antenna, increases the transmit stream of MIMO channel.

## 2.2 Space time block code

Diversity is one of technique in multipath antennas, that used to decrease the error of the performance due to wireless fading channels. Space time frequency and spatial diversity are various type of diversity implemented in order to decrease error performance. Space Time Code (STC) system illustrate a concept of space-time diversity that employs in multiple transmit antenna. STC make used of the spatial diversity technique to encode the information into space-time code word and transmitted over symbol times decode back at the receiver. STC can be classifies into Space time block code (STBC) and space-time trellis code (STTC).

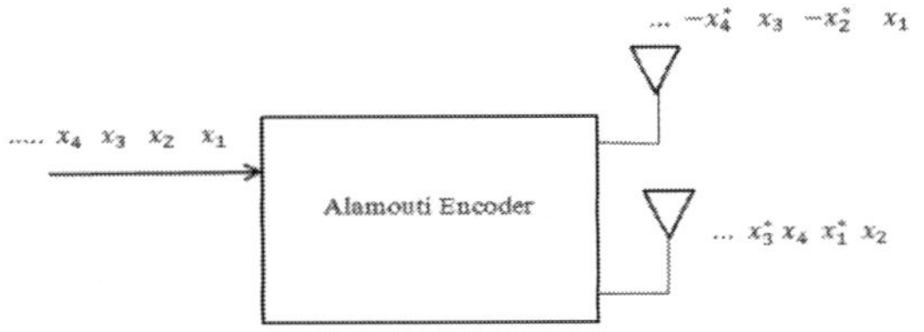


**Figure 1**. Alamouti code of STBC.

Alamouti code is the very first STBC which is a complex orthogonal space-time code specialized for two transmit antennas. In Alamouti, transmitted symbol are encoded within two symbol period and it will take two time-slots to transmit two symbols. Fig. 1 shows during the first period, two symbols $x_1$ and $x_2$ were transmitted from each of antenna. During the second period, the symbol are copied and transmitted again with $-x^*_2$ and $x^*_1$ from first and second transmit antennas. This is the orthogonally between symbol after receiving process at the receiver.

The diversity scheme example for 2x2 antenna configuration, it can be seen four copies of $x_1$ and four copies of $x_2$ at the receiver regarding to the $H$ channel. Since the code is not interfering with each other, the code is orthogonal. Hence, the diversity order for 2x2 antenna configuration is four.

## 2.3 OFDM

Orthogonal Frequency Division Multiplexing is dividing the total channel bandwidth into many small bandwidth channels. The individual frequencies are carefully chosen such that they are orthogonal to each other, although the carrier can overlap [6]. Individual frequency is known as subcarriers and each of the subcarriers is modulated with a conventional scheme such ASK, PSK and QAM.

In transmission process, a data stream is converted into a parallel from high rate data stream to low rate data stream. The number of data streams corresponding to the number of available OFDM sub-carriers. Then, each of these low data rate stream is used to modulate each of the subcarrier frequencies [6] like example in the Fig. 2.

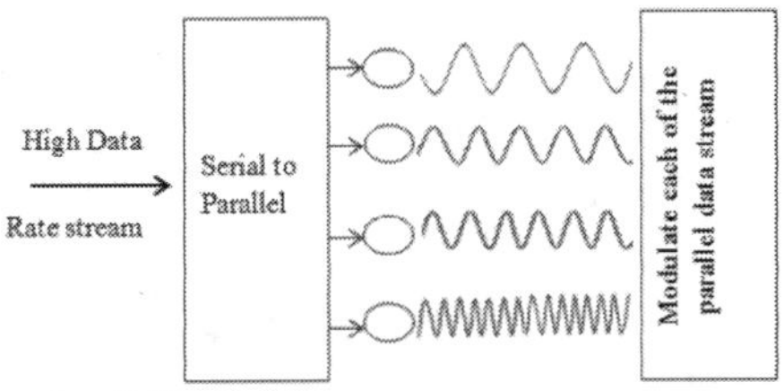


**Figure 2**. OFDM system

## 2.4 MIMO-OFDM system

MIMO OFDM system been widely used in wireless communication system such IEEE 802.11n for WLAN where the standard for wireless broadband communication and been used throughout this project. MIMO OFDM been proved [2], [8] can improve the system capacity and increase speed and high data rate. Previous researcher stated that MIMO OFDM fulfill the requirement because of its unique properties such give high data rates, high spectral efficiency and resistance towards multipath propagation [2].

In order to enhance MIMO-OFDM system performance, channel estimation been added to MIMO OFDM [2]. In this paper, DCT-based channel estimation with STBC in MIMO OFDM system is used in order to increase the performance of the system. Besides, the system also tested on the MIMO system where multiple antennas of 2x2, 4x4 and 6x6 are used in this project with different digital modulation technique.

## 2.5 Discrete cosine transform (DCT)

Discrete Cosine Transform (DCT) or inverse DCT (IDCT) can be expressed as a sequence of finitely many data points in terms of a sum of cosine functions oscillating at different frequencies [3]. DCT is proved to be more efficient than DFT as mention by previous researcher[5], [7], [9], [10]. DCT are mostly used in science and engineering application such as lossy compression of audio and images (eg. JPEG). In [3], they said that DCT is a Fourier-related transform similar to the discrete Fourier transform (DFT), but using only real numbers.

Using DCT, a complex exponential function set can be used as orthogonal basis to implement multi carrier scheme [9]. The real time high-frequency components of DFT can be reduced by DCT. DCT employs mirror extension of N-point data sequence to 2N-point data sequence, which removes the discontinuous edge [9].

The output signal of IDCT/DCT-based OFDM system written as:

i. Using IDCT-based at transmitter:

$$x(n) = IDCT[y(k)] \tag{2}$$

$$x(n) = \sum_{k=1}^{N} w(k)y(k)\cos\left(\frac{\pi(2n-1)(k-1)}{2N}\right) \tag{3}$$

$k = 1,2 \ldots, N$

where

$$w(k) = \begin{cases} \frac{1}{\sqrt{N}} & k = 1 \\ \sqrt{\frac{2}{N}} & 2 \le k \le N \end{cases}$$

ii. Using DCT-based at receiver:

$$Y(k) = DCT[x(n)] \tag{4}$$

$$Y(k) = w(k)\sum_{n=1}^{N} x(n)\cos\left(\frac{\pi(2n-1)(k-1)}{2N}\right) \tag{5}$$

$k = 1,2 \ldots, N$

where;

$$w(k) = \begin{cases} \frac{1}{\sqrt{N}} & k = 1 \\ \sqrt{\frac{2}{N}} & 2 \le k \le N \end{cases}$$

N refers to the length of data symbol obtain from modulation constellation QAM and QPSK while w(k) is the coefficient of DCT that reconstruct sequence very accurately and it is useful property for applications requiring data reduction [5]. IDCT/DCT is a Fourier-transform where transform from frequency to time domain and vice versa.

### 2.6 Quadrature phase shift keying (QPSK)

Phase Shift Keying is shifting the phase of carrier by modulating signal with phase measured relative to the previous bit interval. In M-ary modulation where two or more bits been grouped together to form a symbol and a signal, s(t). This project used 4-PSK where $M = 2^n$, n=2 shows transmitting 2 bit per symbol to be mapped into phase [10].

QPSK can be express as:

$$s(t) = A\cos(2\pi f_c t + \theta^\circ) \tag{6}$$

where $f_c$ is the frequency carrier and $\theta^\circ = 0^\circ, 90^\circ, 180^\circ, 270^\circ$ is a modulating signal phase shift by $90^\circ$ as in constellation diagram Fig. 4.

### 2.7 Quadrature amplitude modulation (QAM)

QAM is a modulation technique that combines with Amplitude Shift Keying (ASK) and Phase Shift keying (PSK). QAM consist of two amplitudes of wave and called quadrature when $90^\circ$ amplitude are out-of-phase with each other. Bit per symbol for M-QAM level used in this project is 4-QAM. QAM can be express as:

$$s(t) = A_I\cos(2\pi f_c t) + A_Q\sin(2\pi f_c t) \tag{7}$$

$A_I$ and $A_Q$ were phase (I) and quadrature (Q) where the data bit stream split into I and Q are grouped together using QAM modulator and mapped into complex symbol [11]. The complex symbol been transmitted and splits up at the receiver using QAM demodulator in order to get the output data.

### 2.8 DCT-OFDM system

Fig. 3 shows the MIMO-OFDM system using DCT-based channel estimation. The system works with input binary data modulated by QAM and QPSK modulation scheme. The idea of converting from serial to parallel stream and modulate each of the subcarrier frequencies has been explained in OFDM system section.

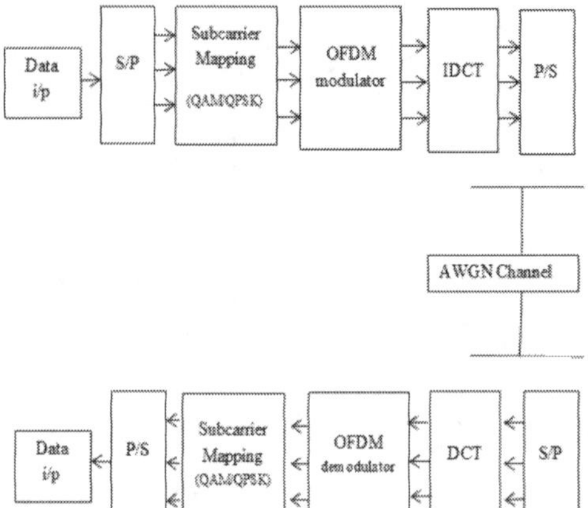


**Figure 3.** MIMO-OFDM system using DCT-based channel estimation.

Before the signal had been transmitted, the IDCT-based channel estimation is placed at the transmitter and the sets of low data rate channels carrying the information in parallel being transmitted with multiple antennas through AWGN channel. At the receiver, the signal is change to frequency-domain by inserting DCT-based and demodulated by the OFDM demodulator. Lastly, In order to achieve original data from data stream, the QAM or QPSK demodulator is used.

## 3 Simulation results

Table 1 shows the parameter used for MIMO-OFDM system based on the WLAN (IEEE 802.11n).

**Table 1.** Parameter of MIMO-OFDM system

| Parameter | Value |
|---|---|
| Antenna configuration | 2x2, 4x4, 6x6 |
| Data mapping | 4-QAM/ QPSK |
| Channel estimation | IDCT/DCT |
| IDCT and DCT size | 52 |
| No of subcarrier | 52 |
| Channel used | AWGN |
| Frame length | 100 |
| Cyclic prefix | 16 |

### 3.1. QPSK modulation

Fig. 4 shows graph of BER versus Eb/No, for QAM modulation scheme. At Eb/No=5dB, BER for 6x6 antenna configuration is equal to $1.257\times10^{-2}$ while 4x4 is $5.399\times10^{-2}$ and 2x2 antenna is $9.689\times10^{-2}$. MIMO OFDM system of 6x6 antenna configuration has the lowest BER when using QPSK modulation technique. The performance has proven that high number of antenna can make the system become more efficient because transmitted data will reach receiver without being spread to other. Besides that, it shows that the system has also achieved high diversity order. Bits per symbol transmit for M-QAM level can be determined by $M=2^N$ where M is the QAM level and N is the bits per symbol. Higher bits per symbol sent will increase the probability of bits error.

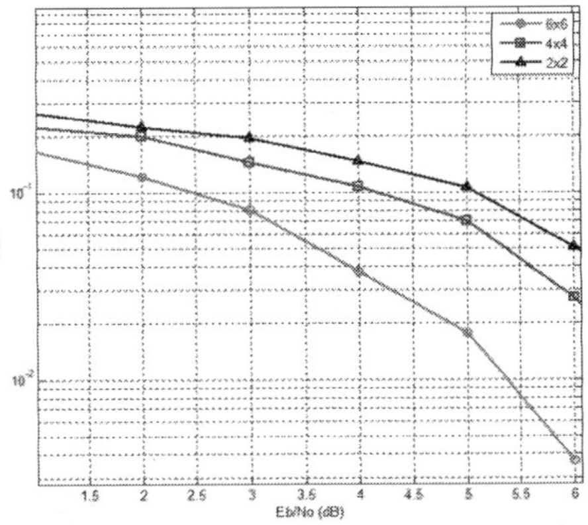


**Figure 4** . BER vs. Eb/No MIMO-OFDM system using DCT-based using QAM modulation.

Fig. 5 shows graph of BER versus Eb/No for QPSK modulation scheme. At Eb/No=5 dB, BER for 6x6 antenna configuration is equal to $1.775\times10^{-2}$ while 4x4 is $7.027\times10^{-2}$ and 2x2 antenna is $1.058\times10^{-1}$. From the result, it shows that when the number of the antenna is increasing, the BER will decrease and successfully reduce interference in the wireless system.

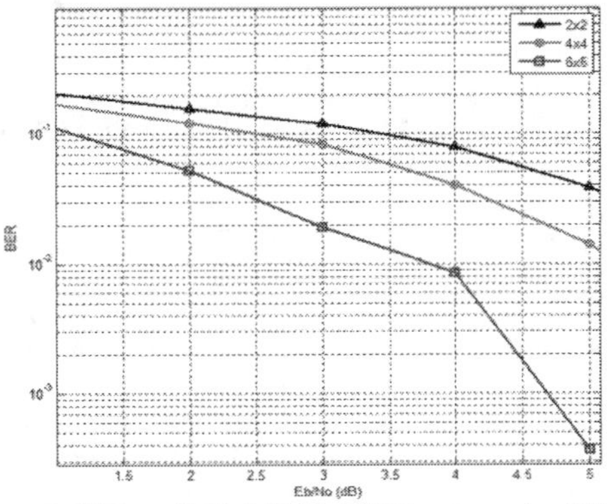


**Figure 5.** BER vs. Eb/No MIMO-OFDM system using DCT-based using QPSK modulation.

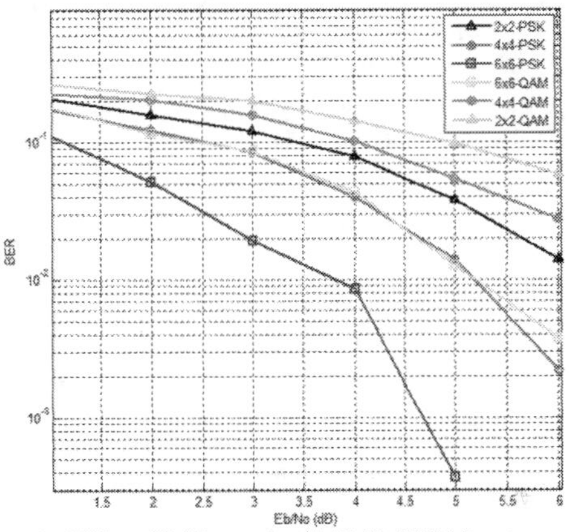


**Figure 6.** BER vs Eb/No graph of MIMO-OFDM system using DCT-based channel estimation with 2x2, 4x4 and 6x6 antenna configuration for QAM and QPSK modulation scheme.

Based on the Fig. 6 above, we can see the difference between performances of modulation. With 6x6 antenna configurations, QPSK has the lowest BER while 2x2 QAM has highest BER during simulation. QAM utilise both amplitude and phase variations. Although QAM is widely used in wireless communication system in terms of data carrying capacity, but QAM is more susceptible to noise compare to QPSK. This is because QAM has larger distance between constellation point compare to QPSK. Even though QAM is seen to have the same spectrum and bandwidth efficiency as QPSK but still it is difficult to demodulate in the presence of noise [12]. Addition of STBC in the system also becomes a big help for the antennas to achieve maximum diversity.

## References

1. A. Kakadiya, M.M. Solanki, Analysis of adaptive channel estimation techniques in MIMO OFDM

system, International Journal of Advancement in Research & Technology, **2**, 4, April (2013).
2. P. Venkateswarlu, Channel estimation techniques in MIMO-OFDM LTE System, International Journal Engineering Research and Applications, **4**, 7, pp.157-161, July (2014).
3. S.M. Sajid, C.R. Lakshmi, DCT based improved OFDM communication, International Journal of Engineering and Innovative Technology (IJEIT), **2**, 12, June (2013).
4. V. Khare, M. Ahmed, Precoded DCT-OFDM system for Baseband and Wireless Communication using QAM modulation over AWGN and Multipath Rayleigh fading channel, Internation Journal of Research in Engineering and Scientific Applications (IJRESA), **1**, 1, Jul-Aug (2014).
5. L. Patidar, A. Parikh, BER comparison of DCT-based OFDM and FFT-based OFDM using BPSK modulation over AWGN and Multipath Rayleigh Fading channel, International Journal of Computer Applications, **31**, 10, October (2011).
6. Bridging Technologies, understanding Long Term Evolution, *Building Communication Channels (Orbitage)*, Version 3.1, March (2013).
7. P. Bhatnagar, M. Tiwari, Enhancement of OFDM system Performance with MIMO technique, International Journal of computer Technology and Electronics Engineering (IJCTEE), **1**, 3.
8. Shazeeda, T.M. Shashidhar, B.G .Anupama, FPGA based design 1D DCT/IDCT for MIMO OFDM channel estimation, European Scientific Journal, **10**, 30, October (2014).
9. S. Saleem, Q.-ul-Islam, On Comparison of DFT-based and DCT-based Channel Estimation for OFDM System, International Journal of Computer Science Issues(IJCSI), **8**, 3, May (2011).
10. Kharagpur, *Data Communication Fundamentals*, Version 2 CSE, IIT
11. C. Poongodi, P. Ramya, and A. Shanmugam, BER Analysis of MIMO OFDM System Using M-QAM over Rayleigh fading channel, in Communication and Compitational Intelligence (INCOCCI), International Conference on 2010, pp.284-288, (2010).
12. D. R. Smith, *Digital Transmission System*, Washington: Kluwer Academic Publisher, (1999).

# Performance Evaluation of Modified SLM Technique in OFDM System Using Selected Codeword Shift

Ezmin Abdullah[1], Azlina Idris[1] and Azilah Saparon[2]

[1]Wireless Communication Technology Group (WiCoT), Faculty Electrical Engineering, UniversitiTeknologi Mara,40450, Shah Alam, Selangor DE, Malaysia
[2] Faculty Electrical Engineering, UniversitiTeknologi Mara,40450, Shah Alam, Selangor DE, Malaysia

**Abstract.**Selective Mapping (SLM) is a well-known technique used in OFDM System to reduce high Peak-Average Power Ratio (PAPR). High PAPR occurred during signal processing which multiple data are converted into electromagnetic signal via Inverse Fast Fourier transform (IFFT) at the transmitter. Even though SLM technique has a potential to reduce high PAPR, it disrupts the overall performance of bit error rate (BER) at receiver part. In this paper, a modified SLM has been proposed using circulant shift and embedded bit side information to overcome the BER performance degradation at the receiver part as well as to further improve the PAPR performance at the transmitter part. The results have shown an encouraging improvement of PAPR and BER to the OFDM system performance.

## 1 Introduction

The demands of mobile communications have progressively led to global energy consumption around the world. Power amplification process contributes to about 50% of power consumption distribution in radio access technology. In this level, the main component to reduce power consumption relies on the High-Power Amplifier (HPA) efficiency [1].

The efficiency of HPA in multicarrier modulation systems such as OFDM system is associated to the Peak-Average Power Ratio (PAPR) which occurred during the modulation process via Inverse Fast Fourier Transform (IFFT). High PAPR causes power deficiency in the RF amplifier, signal distortion and energy spilling to adjacent channel. The drawbacks will lead to the cost of extending higher range of RF amplifier [2], [3] and computational complexity [4].

Selective Mapping (SLM) technique is widely used in OFDM system in order to reduce high PAPR because of its advantage of avoiding signal distortion [5]. However, SLM technique requires a side information (SI) transmitted together with the signal to allow recovery of original data. In many cases, the erroneous of the SI will cause a confusion of in detecting the correct SI. This problem could affect BER performance [6], [7]. The BER problem can be resolved by adding forward error check (FEC) encoder and decoder at the transmitter and receiver respectively as proposed by Wen [8] et al. However, this application will increase the complexity of the system as well as its cost.

In this paper, a modified SLM technique has been proposed to overcome the BER drawback in SLM as well to further improve the PAPR performance. This technique applied a circulant shift onto parallel bit data and embedded bit SI before digital modulation process in transmitter. The results shows the comparison of PAPR and BER performance between modified SLM, Method 1 of Data Position Permutation (DPP)[8] and Selective Codeword Shift (SCS).

## 2 Selective mapping (SLM) technique

In the conventional Selective Mapping (SLM) technique as seen in Fig. 1, after undergone digital modulation, $u$number of OFDM symbol candidates are organized in parallel. Every candidate hold a same OFDM symbol sequences, $X = [X(0), X(1), \dots X(N-1))]^T$ .The multiplication of every candidate with phase factor, $\mathrm{P}^u = [\mathrm{P}^0, \mathrm{P}^1, \dots, \mathrm{P}^{U-1}], 0 \leq u \leq U-1$ will produce$u$ number of alternative OFDM symbol sequences. The alternative symbol sequences can be described as

$$\mathrm{X}^u = \prod_{n=1}^{N} X_n \otimes \mathrm{P}^u, \qquad 0 \leq u \leq U-1 \qquad (1)$$

where $\otimes$ denotes the component-wise multiplication of two vectors. The phase sequence, $\mathrm{P}^u$ is generated by using a complex number magnitude, $\mathrm{P}^u = e^{j\varphi_u}$ where $\varphi_u \in [0,2\pi]$. These alternative input sequences, $\mathrm{X}^u$ are then undergone Inverse Fourier Transform (IFFT) process to generate OFDM signal. The signal in the time domain is written as

$$x^u(t) = \frac{1}{\sqrt{N}} \sum_{k=0}^{N-1} X^u \cdot e^{j2\pi f_k t}, \quad 0 < t < NT \quad (2)$$

where $e^{j2\pi f_k t}$ is the orthogonal series of sinusoids performed by IFFT. Finally, the minimum Peak-Average Power Ratio (PAPR) OFDM signalscandidate will be selected for transmission. The transmitted signals are obtained by

$$s^u(t) = \underset{0 \le u \le U-1}{\operatorname{argmin}} \left\{ \frac{\max|x^u(t)|^2}{E|x^u(t)|^2} \right\} \quad (3)$$

Side information (SI) is embedded together with the transmitted signal in order to decode the received signal at the receiver. The erroneous of the received SI cause the confusion to detect the correct SI index that resulting BER performance degradation.

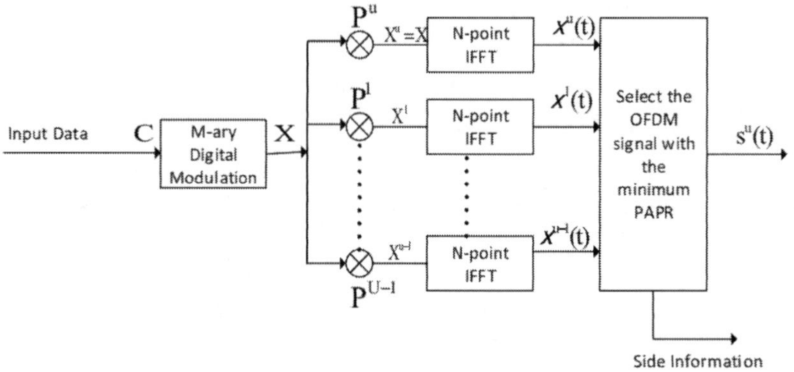


**Figure 1.** Block diagram of the conventional selective mapping (SLM) OFDM system

## 3 Modified selective mapping (SLM)

A modified Selective Mapping using circulant shift is proposed in this paper, and it shows to have an encouragingPeak-Average Power Ratio (PAPR) reduction performance as compared to conventional SLM.

Based on Eq. (2), Orthogonal Frequency Division Multiplexing (OFDM) signals are generated by multiplication of data symbol $X_n$ and orthogonal sinusoids IFFT, $e^{j2\pi f_k t}$. These two components, $X_n$ and $e^{j2\pi f_k t}$ are the important parts to determine PAPR as power symbols and calculated below

$$X^S_{pow} = \frac{|X^S|^2}{2} \quad (4)$$

Data symbol sequences$X_n$ are the complex form of data modulated by the digital modulation QAM to the codeword bits.In this paper, we proposed to manipulate codeword bits, C using circulant shift (Fig. 2). This technique is limited to the M-ary modulation scheme as the circulant shift will need two or more bits in a codeword to shift the bits in parallel.

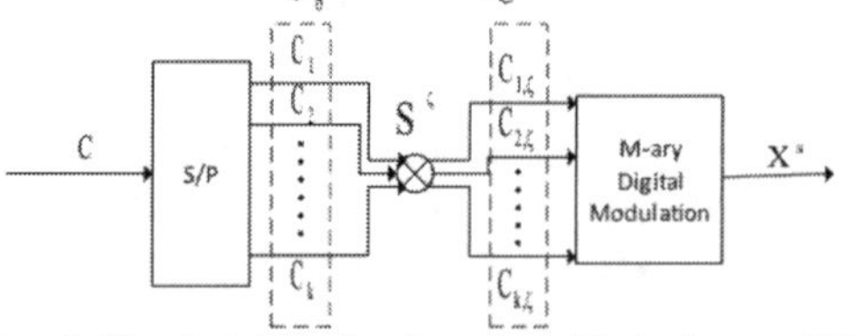


**Figure 2.** Circulant shift of codeword sub-blocks for one shift factor.

In order to achieve a large PAPR reduction, it is proposed for an increase in the probability of various alternative symbol sequences, $X_n$. This technique is taking place at the bits data before digital modulation process. As illustrated in Fig. 2, the bits data, C is indicated as $c_1, c_2, \ldots, c_d]$ and divided into sub-blocks, $C_b = [C_1, C_2, \ldots . C_K]$, where $= \frac{d}{m}$, is the number of sub-block and $m = \log_2$(M-ary QAM), is the number of bits per symbol. Codeword C in sub-block is represented as

$$C_b = \prod_{k=1}^{K} C_k, \quad 1 \le k \le K \quad (5)$$

where $C_1 = (c_1, \ldots, c_m)$ , $C_2 = (c_{m+1}, \ldots, c_{2m})$ and so on until $C_K$ are arranged in parallel.

The modification of SLM is started when circulant shift (one shift to the right) is applied to every sub-block of the parallel bits data. Thus, the alternative codewords are represented as

$$C^\zeta = \prod_{k=1}^{K} C_k \otimes S^\zeta \quad 0 \le \zeta \le m-1 \quad (6)$$

Codeword shift sequences become $C^\zeta = [C_1 \otimes S^\zeta, C_2 \otimes S^\zeta, \ldots, C_K \otimes S^\zeta] = [C_{1,\zeta}, C_{2,\zeta}, \ldots, C_{K,\zeta}]$ . The codeword shift sequences are then modulated by M-ary digital modulation. The outputs of the symbol sequences are $X^S = [X_{1,\zeta}, X_{2,\zeta}, \ldots, X_{K,\zeta}]$. The process afterward follows the conventional SLM which is illustrated in Fig.

3. The new alternatives OFDM symbol for modified SLM is given by

$$X^{S,u} = \prod_{k=1}^{K} X_k^S \otimes P^u, \qquad 0 \le u \le U-1 \tag{7}$$

Therefore, the alternative OFDM signals in the time domain is represented as

$$x^{S,u}(t) = \frac{1}{\sqrt{N}} \sum_{k=0}^{N-1} X^{S,u} \cdot e^{j2\pi f_k t} \tag{8}$$

Finally, the minimum PAPR alternative OFDM signals are obtained by

$$s^{S,u}(t) = \underset{0 \le u \le U-1}{\operatorname{argmin}} \left\{ \frac{\max|x^{S,u}(t))|^2}{E|x^{S,u}(t)|^2} \right\} \tag{9}$$

In this paper, the number of $U$ for the phase factor in SLM part is limited where it is based on the number of bit per symbol, $m$. It can be noted that side information of $s^{S,u}$ index is included together in the transmitted signals in order to properly retrieve the data at the receiver.

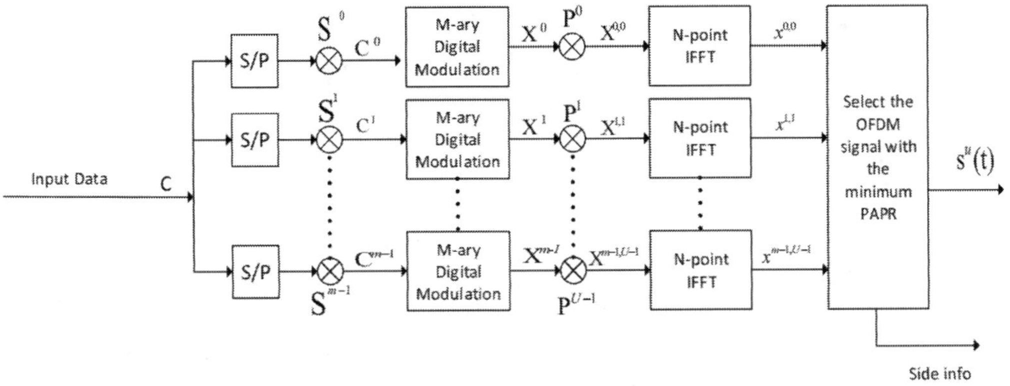


**Figure 3.** Modified selective mapping (SLM) OFDM system

### 3.1. Embedded Bit Side Information (SI)

Side information (SI) is an important index which is transmitted together with the transmission signal in order to retrieve the original data at the receiver. The index indicates which candidate has been selected for transmission. At the receiver, the data is retrieved based on the information from the SI index.

An embedded bits SI has been developed in this paper. The SI is embedded in the form of binary bits. The positions of $m$ bits of binary SI are situated at the head and at the middle of the codeword block as shown in Fig. 4.

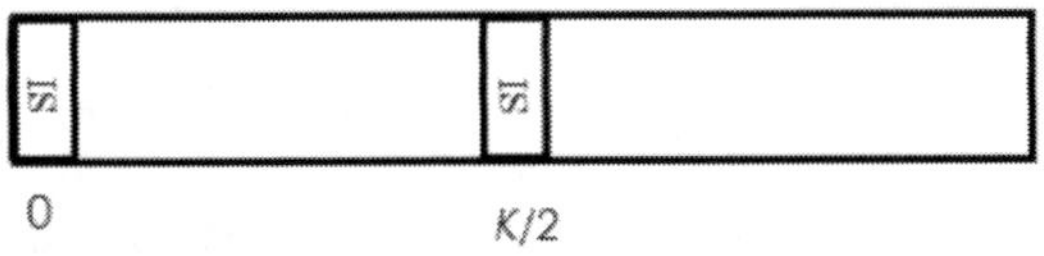


**Figure 4.** Codeword blocks format with embedded bits side information

The embedded SI can be illustrated as

$C_1$=[SI($m$ bits), $c_1$, $c_2$, $c_3$, $c_4$, $c_5$, $c_6$, $c_7$, $c_8$, $c_9$, SI($m$ bits), $c_{10}$,

$C_2$=[SI($m$ bits), $c_6$, $c_1$, $c_2$, $c_3$, $c_4$, $c_5$, $c_{12}$, $c_7$, $c_8$, SI($m$ bits), $c_9$,

$C_3$=[SI($m$ bits), $c_5$, $c_6$, $c_1$, $c_2$, $c_3$, $c_4$, $c_{11}$, $c_{12}$, $c_7$,SI($m$ bits), $c_8$,

$C_4$=[SI($m$ bits), $c_4$, $c_5$,$c_6$, $c_1$, $c_2$, $c_3$, $c_{10}$,$c_{11}$, $c_{12}$,SI($m$ bits), $c_7$,

$C_5$=[SI($m$ bits), $c_3$, $c_4$, $c_5$, $c_6$, $c_1$, $c_2$, $c_9$, $c_{10}$, $c_{11}$, SI($m$ bits), $c_1$

$C_6$=[SI($m$ bits), $c_2$, $c_3$, $c_4$, $c_5$, $c_6$, $c_1$, $c_8$, $c_9$, $c_{10}$, SI($m$ bits), $c_{11}$

## 4 Results and discussions

The performance evaluation of proposed technique has been validated and compared by computer simulations using Matlab.The OFDM system parameters used in the simulation are based on 3rd Generation Partnership Project Long Term Evolution (3GPP-LTE) System [9] standard as shown in Table 1. Therewere $10^4$ of OFDM signals have been considered to calculate the PAPR complementary cumulative distribution functions (CCDF). The performance evaluation consists of modified SLM, conventional SLM and method 1 of DPP are shown in PAPR CCDF and BER graph.To make the computational complexity to be equal, the number of IFFT used in SLM is set to be the same as the number of IFFT used in the modified SLM.

As shown in Fig. 5, the PAPR performance of the modified SLM is effectively reduced the PAPR by about 2.9 dB from the original OFDM. The modified SLM outperformed the conventional SLM, DPP and SCS with 27.6% improvement as compared to the SLM, DPP and SCS, 8.6%, 12.4% and 19.0% respectively. The number of IFFTs block was set to be equal in every reduction technique by following the proposed method minimum system which is $m$=6. Table 2 summarizes the results in Fig. 5.

**Table 1.** Simulation parameters for the 3rd Generation partnership project long term evolution (3GPP-LTE) system [9].

| Parameter | Value |
|---|---|
| Bandwidth (BW) | 1.25 MHz |
| Sampling frequency | 1.92 MHz |
| Sampling time | $5.208 \times 10^{-7}$ sec |
| IFFT size | 128 |
| Used subcarrier | 76 |
| Modulation technique | 64 QAM |
| Guard interval | ¼ |
| Chanel model | Rayleigh Fading |

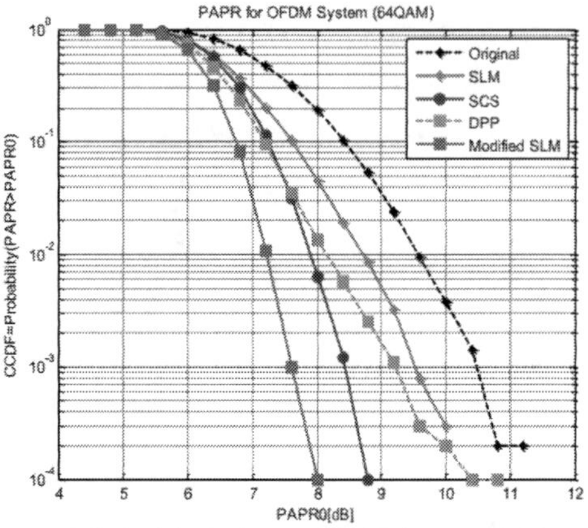


**Figure 5.** CCDF of the PAPR Modified Selective Mapping (SLM) as compared to original OFDM signal, conventional Selective Mapping, Selected Codeword Shift (SCS) and Data Position Permutation (DPP).

**Table 2**. PAPR analysis for various PAPR reduction techniques at clip rate $10^{-3}$

| OFDM System (64 QAM, 128IFFT) | PAPR in dB | Improvement (%) |
|---|---|---|
| Original OFDM | 10.5 | - |
| SLM | 9.6 | 8.6 |
| DPP | 9.2 | 12.4 |
| SCS | 8.5 | 19.0 |
| Modified SLM | 7.6 | 27.6 |

The BER performance of the system is very important in order to evaluate its robustness. Taking that into account, the investigation of BER performance was conducted in the Rayleigh fading channel and the result are shown in Fig. 6. The BER result compares the Modified SLM with original OFDM, conventional SLM, DPP and SCS. Conventional SLM and DPP shown a worse BER performance when compared to the original OFDM. In many approaches, they were suggested to apply Forward Error Check (FEC) encoder and receiver at the transmitter and receiver respectively to overcome the BER performance degradation. However, in this paper, the application of FEC is not considered because it will increase the complexity of the system as well as leading to cost increment. Meanwhile, the modified SLM shown a better improvement when compared to the conventional one even though it slightly exceeded the original OFDM value.

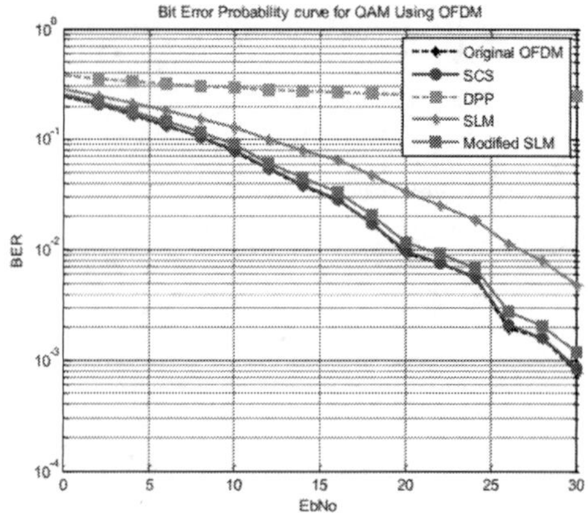


**Figure 6.** Comparison between BER of the Modified SLM, original OFDM, conventionalSLM, DPP and SCS.

The relationship between side information (SI) to the performance of BER is investigated by comparing the Modified SLM using bit SI and symbol SI. The bit SI which in the form of binary bits are embedded into the codeword sequence before the digital modulation process. Meanwhile, the symbol SI which in the form of complex number are embedded into the symbol sequence before signal transmissions. The BER performance of bit SI and symbol SI are shown in Fig. 7. As shown in Fig. 7, the symbol SI exceeded the original OFDM while bit SI indicates a similar performance as original OFDM. This finding explains that the symbol SI are tend to be interfered.

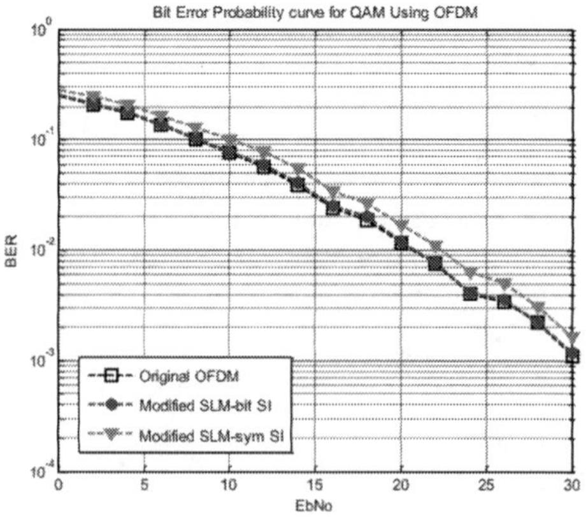


**Figure 7.** Comparison between BER of Modified Selective Mapping (SLM) using embedded bits SI and symbol SI.

## 5 Conclusions

Selective Mapping is widely used to reduce high PAPR in OFDM system in the cost of BER performance degradation. In this paper, a modified Selective Mapping technique based on circulant shift technique has been proposed to overcome the BER degradation that occurred using SLM and further reduce the high PAPR. Furthermore, theproposed embedded bits SIoffer a better performance of BERwhen to compare to symbol SI that had been used in conventional SLM.

## Acknowledgement

We are grateful to the Universiti Teknologi Mara, UiTM for providing scholarship to Author 1. This paper is part of research work that supported by the Faculty of Electrical Engineering, Universiti Teknologi Mara, Shah Alam, Selangor Malaysia.

## References

1. L. M. Correia, D. Zeller, O. Blume, D. Ferling, Y. Jading, I. Godor, G. Auer, and L. Van der Perre, Challenges and Enabling Technologies for Energy Aware Mobile Radio Networks, *IEEE Signal Process. Mag.*, **48**, 11, pp. 66–72 (2010).
2. S. H. Wang, J. C. Sie, C. P. Li, and Y. F. Chen, A Low-Complexity PAPR Reduction Scheme for OFDMA Uplink Systems, *IEEE Trans. Wirel. Commun.*, **10**, 4, pp. 1242–1251, Apr (2011).
3. M. Chauhan, Different Techniques to Reduce the PAPR in OFDM System, *Int. J. Eng. Res. Appl.*, **2**, 3, pp. 1292–1294 (2012).
4. H. B. Jeon, J. S. No, and D. J. Shin, A Low-Complexity SLM Scheme Using Additive mapping Sequences for PAPR Reduction of OFDM Signals, *IEEE Trans. Broadcast.*, **57**, 4, pp. 866–875 (2011).
5. L. Ning, M. Yang, Z. Wang, and Q. Guo, A novel SLM method for PAPR reduction of OFDM system, in *IEEE Vehicular Technology Conference* (2012).
6. S. Y. Le Goff, S. S. Al-Samahi, B. K. Khoo, C. C. Tsimenidis, and B. S. Sharif, Selected mapping without side information for PAPR reduction in OFDM, *IEEE Trans. Wirel. Commun.*, **8**, 7, pp. 3320–3325 (2009).
7. A. Vallavaraj, B. G. Stewart, D. K. Harrison, and F. G. McIntosh, Reducing the PAPR of OFDM using a simplified scrambling SLM technique with no explicit side information, in *Proceedings of the International Conference on Parallel and Distributed Systems - ICPADS*, pp. 902–907(2008).
8. J. Wen, S. Lee, and C. Kung, SLM-Based Data Position Permutation Method for PAPR Reduction in OFDM Systems, *Wirel. Commun. Mob. Comput.*, **13**, 11, pp. 985–997 (2015).
9. F. Shamshad, U. Javed, and S. Saleem, Physical Layer Aspects of 3GPP s Long Term Evolution ( LTE ), *Adv. Comput. Sci. its Appl.*, **2**, 1, pp. 287–294 (2012).

# Dynamic Multi Threshold Priority Packet Scheduling Algorithms

Tamer Dag and Sezer Uzungenc

*Computer Engineering Department, Kadir Has University, 34083Istanbul, Turkey*

**Abstract.**Packet scheduling algorithms are developed in order to use shared transmission resources efficiently. Various application packets such as real and non-real time packets might have different QoS requirements and traditional scheduling algorithms might be insufficient to respond to the applications needs. In this paper, two packet scheduling algorithms are proposed to overcome this problem: Dynamic multi threshold priority packet scheduling (DMTPS) and dynamic multi threshold priority with urgency packet scheduling (DMTPUS). The proposed algorithms aim to provide a better QoS level with a decrease in delay time and loss ratio for the low priority packets while still maintaining acceptable fairness towards high priority packets. To evaluate the performance of DMTPS and DMTPUS algorithms, they are compared with the commonly used scheduling algorithms such as first come first served (FCFS) and fixed priority. Simulation results illustrate that the dynamic multi threshold priority packet scheduling algorithms can provide a better QoS for low priority packets without decreasing the QoS levels of high priority packets.

## 1 Introduction

A Wireless Sensor Network (WSN) consists of tiny sensor nodes which are low-cost and low-energy and with sensing, data processing and communicating components. A WSN can be realized with many sensor nodes that are deployed closely. The feature of sensor networks is the cooperative effort of the sensor nodes. Sensor nodes are fitted with an onboard processor. Instead of sending the raw data to the nodes responsible for the fusion, they use their processing abilities to locally carry out simple computations and transmit only the required and partially processed data [1].

WSNs are used in various areas such as military, health and home. For military applications, the rapid deployment, self- organization, and fault tolerance characteristics of WSNs make them a very promising sensing technique for military command, control, communications, computing, intelligence, surveillance, reconnaissance, and targeting systems. In health, they are used for monitoring patients and assisting disabled patients with deployment of sensor nodes. Managing inventory, monitoring product quality, and monitoring disaster areas are some other applications where WSNs are used [1].

In WSNs, sensor nodes can be mobile and heterogeneous. They are scalable to large scale of deployment. Design of a WSN is cross-layer and it is becoming an important study area for wireless communications. Cross layer design is used for improving the transmission performance which is energy efficiency, data rate and Quality of Service (QoS).

WSNs may be deployed in various environments and energy determines the lifetime of a WSN. Communication in sensor nodes is a key component and protocols and algorithms must satisfy the issues such as increasing of lifespan, self-configuration, robustness and fault tolerance. WSNs have a problem with energy consumption. Energy is the most important resource of WSN nodes [2], [3].

Scheduling is the strategy by which the system decides which task should be executed at any given time. Generally, it is used for load balance and system resources to be served effectively and with desired quality [4]. Scheduling algorithms are classified as preemptive and non preemptive. In preemptive scheduling, arrival of high priority data suspends current lower priority data. In non preemptive scheduling, the current lower priority data is processed until it is completed. Preemptive schedulers provide better performance for system utilization. Non preemptive schedulers encounter the starvation problem [5]. Non preemptive schedulers may give better performance and robustness on reduced systems [6].

The major scheduling algorithms for WSNs are FCFS and fixed priority non-preemptive scheduling. In FCFS algorithm, each priority level of data's loss ratio and delay time results are assumed to be the same since priority is not criteria in this algorithm [7]. In fixed priority non-preemptive scheduling algorithm, lower priority level data can get lost and delayed excessively and the higher priority level data are processed more quickly and there is a definite unfairness between priority levels [8].

In this work, we develop a new scheduling algorithm for WSNs which can provide a better QoS and resolve the disadvantages of the current algorithms. The proposed Dynamic Multi Threshold Priority packet scheduling algorithms try to maximize the level of QoS provided to different priority packets by reducing the loss ratio and waiting times. Also, the proposed scheduling algorithms are non preemptive due to structure of reduced system load of WSNs.

The rest of the paper is organized as follows: Section 2 describes some of the most commonly used Queue Scheduling Algorithms for WSNs, Section 3 introduces the proposed dynamic multi threshold priority packet scheduling algorithms. Simulation results of the algorithms are shown in Section 4 and finally Section 5 concludes this paper.

## 2 Packet scheduling algorithms

In this section, the commonly used scheduling algorithms FCFS and fixed priority non-preemptive scheduling are described briefly. By comparing these algorithms against proposed DMPTS and DMPTUS algorithms, the effectiveness of the proposed algorithms is evaluated.

FCFS algorithm has a disadvantage of leading very high loss ratios and large delay times for higher priority real-time data. FCFS treats all packets the same way since data packets are ordered and transmitted according totheir arrival times only. The priority of a packet is not a criterion for the ordering process [7]. As a result, large delay times for higher priority data are experienced and delay time guarantees might not be supplied for some cases. Similarly, the loss ratios for any type of packets will be similar. Providing same level of QoS to different types of applications is the main disadvantage of FCFS.

In fixed priority non-preemptive scheduling algorithm, the packets are scheduled according to their arrival times and their priorities [8]. When a new packet arrives, it is ordered according to its priority level in the queue. However, this algorithm favors only high priority packets. Lower priority packets can be dropped excessively as they will be placed towards the end of the queue. With fixed priority non-preemptive scheduling algorithm, the QoS levels provided for high priority packets will be very high. On the other hand, lower priority packets will be in danger of starvation [9]. A steady arrival of large priority packets would result in never transmitting low priority packets, as they will always be waiting for the completion of transmission of high priority packets. Thus, the algorithm might be very unfair for some application types. Thus, the main disadvantage of fixed priority non-preemptive scheduling is unfairness.

## 3 Dynamic multi threshold priority packet scheduling algorithms

In this section, the proposed dynamic multi threshold priority packet scheduling (DMTPS) and dynamic multi threshold priority with urgency packet scheduling (DMTPUS) algorithms are described.

### 3.1 Dynamic multi threshold priority packet scheduling algorithm (DMTPS)

The first proposed algorithm is DMTPS. With DMTPS, packets are placed into the waiting queue according to their priorities and each priority level has a distinct threshold value. The threshold levels with the occupancy of the queue is used to determine the ordering of the packets. The basic threshold is shown in Fig. 1. When a new packet arrives, the priority level change decision is made based on if the queue occupancy is higher than the threshold value.

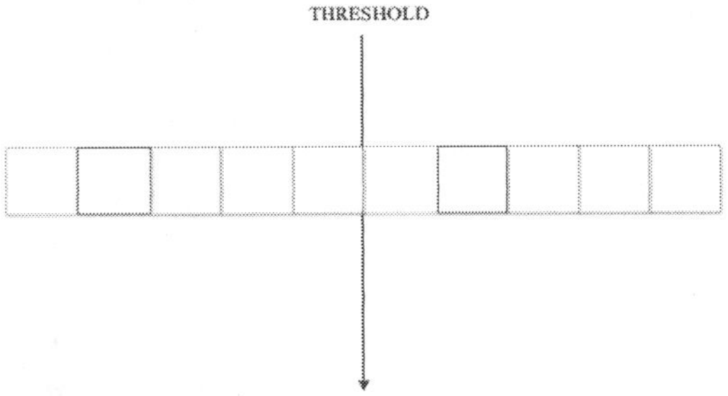


**Figure 1.** Basic threshold structure at the waiting queue.

Three levels of priorities are defined for the DMTPS algorithm. High, medium and low priority. The number of priority levels may be adjusted to a desired level, based on the real system but for the rest of this paper, three levels of priorities will be used.

The threshold values for each priority level are specified with different values:

For priority 1 or high priority packets, the threshold is equivalent to %$k$ of the total capacity in the waiting queue.

For priority 2 or medium priority packets, the threshold is equivalent to %$m$ of the total capacity in the waiting queue.

And for priority 3 or low priority packets, the threshold is equivalent to %$n$ of the total capacity in the waiting queue.

In order to be fair towards low priority packetsbut at the same time satisfy the QoS requirements of high priority packets, the threshold values are ordered as:

$n > m > k$.

These values are also dynamic, and can be changed depending on the application types or a congestion in the network.

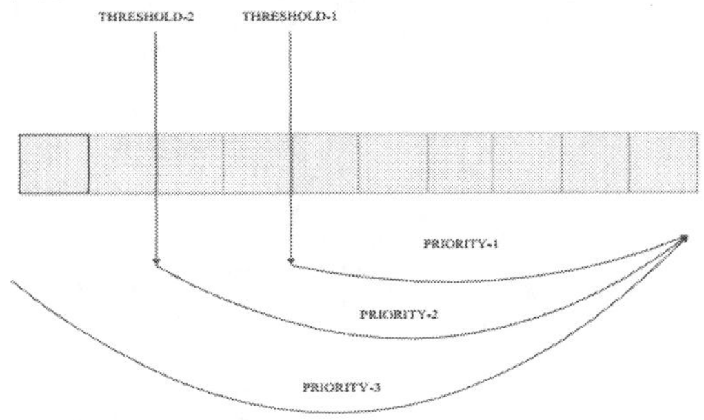


**Figure 2.** Three levels of priority with their corresponding threshold values.

DOI: 10.1051/matecconf/20167506004

Fig. 2 shows an example of the representation of initial values of threshold values. For this figure, it is assumed that for low priority packets, the threshold level is %100.

The basic working principle of DMTPS algorithm is as follows:

When a packet arrives and if the queue occupancy is less than the threshold value for that packet's priority level, the packet is placed at the end of the queue just like in FCFS. The newly arriving packet will not be placed in front of lower priority packets as it is still possible to satisfy its QoS requirements. This case allows, low priority packets to be transmitted without starvation.

When a packet arrives and if the queue occupancy is greater than the threshold value for that packet's priority level, the new packet needs to be placed somewhere in the queue between its threshold and the head of the queue to satisfy its QoS requirements. If a lower priority packet is available between the corresponding threshold value and the head of the queue, it is replaced with the newly arriving packet. If the queue is full and a lower priority packet is available when a new packet arrives, the newly arriving packet will take the place of the lower priority packet, but in this case, lower priority packet will need to be dropped.

When a packet arrives and if the queue occupancy is greater than the threshold value for that packet's priority level or if the queue is full, and there is no lower priority packet between its threshold and the head of the queue, the new arrival will be treated as in FCFS.

### 3.2 Dynamic multi threshold priority with urgency packet scheduling algorithm (DMTPUS)

The main objective of the DMTPUS algorithm is similar to the DMTPS algorithm. The multi threshold structure and packet ordering is the same as DMTPS. However, the major difference is the urgent status of the packets. An urgent status packet needs to be processed as quickly as possible with DMTPUS.

DMTPUS is an efficient technique for scheduling of packets which are urgent and should be processed immediately. Urgent status is defined as a field ($u$) in the packet and if a packet is marked as urgent it is upgraded. Table 1 shows the priority levels of packets with or without urgent status.

**Table 1.** Urgent status

| Priority Level | Urgent | Not Urgent |
|---|---|---|
| Priority 1 | Marked as the Most Valuable Packet (The highest priority level and higher than Priority 1) | Priority 1 |
| Priority 2 | Priority 1 | Priority 2 |
| Priority 3 | Priority 2 | Priority 3 |

Packets which have urgent statuses are evaluated as follows:

A priority 1 or the highpriority level packet with urgent status is considered to be the most urgent packet in the system. Such a packet needs to be delivered immediately. A packet in this case is directly added to head of the waiting queue because it should be processed immediately. All remaining packets in the queue will shift one step back with the arrival of such a packet. Fig. 3 shows an example of the behavior of the system when a high priority level packet with urgent status arrives.

A priority 2 or the medium priority level packet with urgent status is upgraded and marked as a priority 1 packet. This packet afterwards is treated as a priority 1 packet in the DMTPS algorithm.

A priority 3 or the low priority level packet with urgent status is upgraded and marked as a priority 2 packet. This packet afterwards is treated as a priority 2 packet in the DMTPS algorithm.

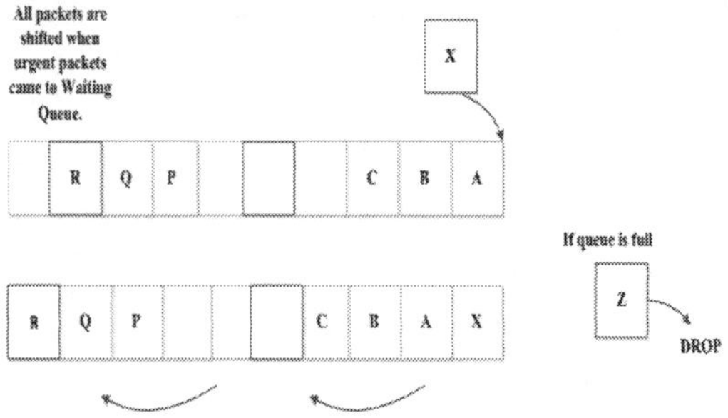


**Figure 3.**Arrival a packet X, priority level 1 with urgent status

## 4 Simulation results

In this section, the performance of the proposed DMTPS and DMTPUS algorithms are analyzed through simulation studies. The simulations are performed using discrete event system simulation technique in Java. For the simulations, two important QoS metrics are evaluated: The loss ratio and the average delay. These metrics for the proposed algorithms are compared with FCFS and fixed priority non-preemptive scheduling algorithms.

The utilization of the queue is a factor that has impact on the value of the loss ratio and the delay time results. The value of both loss ratio and the delay time increases with increased utilization. The capacity of the waiting queue is also an important factor that affects the QoS metrics. The loss ratio seen in the system decreases with the increase of capacity.

For the simulations, in order to observe the operation of the algorithms under heavy traffic, the utilization of the queue is assigned as 0.99. Capacity of the waiting queue is assigned with values between 20 and 100.

The threshold level values described in Section 3.1 are chosen as 60, 80 and 100 for the k, $m$ and $n$ values respectively. The urgent status described in Section 3.2 for the DMTPUS algorithm is applied randomly to %20 of the packets of any priority level.

For DMTPS, DMTPUS, FCFS and priority scheduling algorithms, high, medium, and low priority results are plotted separately.

DOI: 10.1051/matecconf/20167506004

### 4.1 Loss ratio

Loss ratio results of all algorithms within each priority levels are shown in Fig. 4.

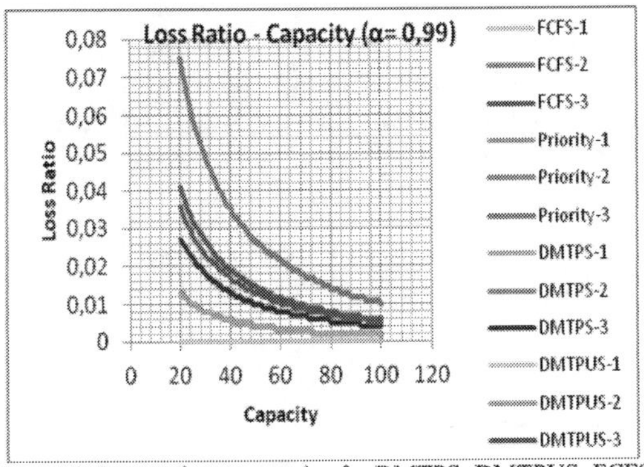


**Figure 4.** Loss ratio vs. capacity for DMTPS, DMTPUS, FCFS and priority scheduling.

In FCFS algorithm, all loss ratio results of priority levels are similar with each other. Because, priority value of packet is not important for FCFS algorithm. In fixed priority non-preemptive scheduling algorithm, the higher priority packets whose priorities are 1 and 2 level's loss ratio are less than packets which have the lowest priority level. Because priority is important at the fixed priority non-preemptive scheduling algorithm and order is specified according to packet's priority. However, the lowest priority packets loss is very high.

Dynamic multi threshold priority packet scheduling algorithms are given higher QoSlevels for the lowest priority when compared with fixed priority non-preemptive scheduling algorithm.

DMTPS and DMTPUS algorithms have similar loss ratio results. Loss ratio results of packets which are higher priority level (Priority-1 and Priority-2) are similar to the fixed priority non-preemptive scheduling algorithm and the lowest priority level packets loss ratio results are similar with the FCFS algorithm. Threshold algorithms improve loss ratio results and these algorithms provide efficiency to all priority levels. The significant alteration is observed at the lowest priority level packets loss ratio results.

### 4.2 Average delay time

Average delay time results of all algorithms within each priority levels are shown at Fig. 5.

In FCFS algorithm, total delay time and average delay time results are similar and equal to each other for all priority levels.

In fixed priority non-preemptive scheduling algorithm, total delay time level and average delay time of each priority is associated with loss ratio results. Delay time results of threshold algorithms are similar. Comparing threshold algorithms with fixed priority non-preemptive scheduling algorithm and FCFS algorithm have different results at average delay time analysis. The lowest priority level packets are delayed more than threshold algorithms, but the higher priority level packets are delayed less than threshold algorithms. FCFS algorithm has consistent results on all priority levels.

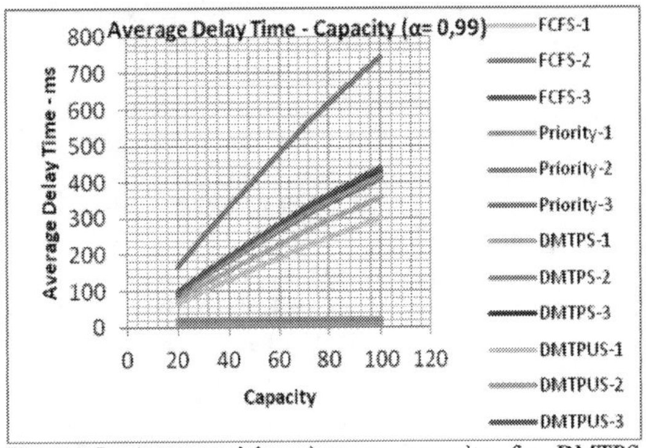


**Figure 5.** Average delay time vs. capacity for DMTPS, DMTPUS, FCFS and priority scheduling.

Threshold algorithms aim at the lowest priority packets delay less than fixed priority non-preemptive scheduling algorithm and the higher priority packets delay less than FCFS algorithm. The efficiency on the average delay time is reached with dynamic threshold algorithms. DMTPS and DMTPUS algorithms provide improvement on all priority level's loss ratio and delay time results.

## 5 Conclusions

All algorithms are experimented and analyzed for improving various packet scheduling algorithm's QoS. In this paper, we propose dynamic multi threshold priority packet scheduling algorithms for WSNs. These algorithms ensure a decrease in delay time and loss ratio for the lower priority levels with acceptable fairness towards higher priority data. Dynamic multi threshold priority packet scheduling algorithms have better performance than the existing FCFS and fixed priority non-preemptive scheduling algorithm in terms of the average delay time and loss ratio. All algorithms are analyzed according to contribution of QoS. The results show that proposed algorithms improve the QoS of selected attributes. DMTPS algorithm provides higher QoS than DMTPUS algorithm which algorithm also has a good QoS in decreasing of average delay time. Urgency and threshold optimization factors are another factor that can be improved for future works.

## References

1. I. F. Akyildiz, W. Su, Y.Sankarasubramaniam, E. Cayirci, A Survey on Sensor Networks, Georgia Institute of Technology, IEEE Communications Magazine, August (2002).
2. M. Bernard T. and Fouchal H., A low energy consumption MAC protocol for WSN, IEEE Communication Conference, ICC, (2012).
3. S. Kim, Y. Kim, An energy efficient priority-based MAC protocol for wireless sensor networks,

International Conference ICT Convergence ICTC, (2012).
4. K. Ramamritham, J.A. Stankovic, Scheduling Algorithms and Operating Systems Support for Real-Time Systems, University of Massachusetts, Amherst, (1994).
5. A. Iqbal, A. Zafar, B. Siddique, Dynamic Queue Deadline First Scheduling Algorithm for Soft Real Time System, IEEE International Conference on Emerging Technologies, (2005).
6. Aloysius K. Mok, Wing-Chi Poon, Non-Preemptive Robustness under Reduced System Load, Proceedings of the 26th IEEE International Real-Time Systems Symposium, (2005).
7. L. XinYan, L. PingPing, Research and Improvement of Real-Time Queue Scheduling Algorithm, International Forum on Information Technology and Applications, (2010).
8. C. Taddia, G. Mazzini, On the Jitter Performance of FIFO and Priority Queues mixture, The 17th Annual IEEE International Symposium on Personal, Indoor and Mobile Radio Communications, University of Ferrara, Italy, (2006).
9. N. Nasser, L. Karim, T. Taleb, Dynamic Multilevel Priority Packet Scheduling Scheme for Wireless Sensor Network, IEEE Transactions on Wireless Communications, **12**, No.4, (2013).

# Titan TTCN-3 Based Test Framework for Resource Constrained Systems

Artem Yushev [a], Manuel Schappacher and Axel Sikora

*Institute of Reliable Communication and Electronics of Offenburg University of Applied Sciences, Offenburg, 77652 Germany*

**Abstract.** Wireless communication systems more and more become part of our daily live. Especially with the Internet of Things (IoT) the overall connectivity increases rapidly since everyday objects become part of the global network. For this purpose several new wireless protocols have arisen, whereas 6LoWPAN (IPv6 over Low power Wireless Personal Area Networks) can be seen as one of the most important protocols within this sector. Originally designed on top of the IEEE802.15.4 standard it is a subject to various adaptions that will allow to use 6LoWPAN over different technologies; e.g. DECT Ultra Low Energy (ULE). Although this high connectivity offers a lot of new possibilities, there are several requirements and pitfalls coming along with such new systems. With an increasing number of connected devices the interoperability between different providers is one of the biggest challenges, which makes it necessary to verify the functionality and stability of the devices and the network. Therefore testing becomes one of the key components that decides on success or failure of such a system. Although there are several protocol implementations commonly available; e.g., for IoT based systems, there is still a lack of according tools and environments as well as for functional and conformance testing. This article describes the architecture and functioning of the proposed test framework based on Testing and Test Control Notation Version 3 (TTCN-3) for 6LoWPAN over ULE networks.

## 1 Introduction

With the upcoming Internet of Things (IoT), wireless communication systems and its devices more and more become part of our daily live and the overall connectivity increases rapidly since everyday objects become part of the global network. Several new wireless protocols have been developed for this purpose whereas 6LoWPAN (IPv6 over Low power Wireless Personal Area Networks) has emerged as one of the most important protocols.

The main idea of 6LoWPAN is to allow the communication using the widely spread IPV6 protocol even on resource constrained devices. In its original version it uses the IEEE802.15.4 standard as its lower transport medium with 127 bytes of payload per frame where it specifies; e.g. a compression layer [1] or fragmentation mechanism of the link-layer frames to overcome IPv6 protocol restrictions such as the Maximum Transmission Unit (MTU). However, new approaches exist to use different technologies as a basis for 6LoWPAN. One of the recent adaptions is the 6LoWPAN communication over DECT Ultra Low Energy (ULE) networks which is currently driven by the ULE Alliance.

This rapid and large development leads to the question of how such systems can be verified in general and how to guarantee conformance amongst them. Although there are lots of software implementations available (even the majority of them being available as open-source) there is still a lack of tools and environments to test such systems. This holds true for the functional testing as well as for interoperability testing. Nevertheless with the Testing and Test Control Notation version 3 (TTCN-3) a scripting language and specification exists that is commonly known and used in conformance testing of communicating systems [2]. With the help of so-called TTCN-3 compilers such as Titan TTCN-3, the target system's performance can be evaluated and, moreover, the resulting test system can verify the system under test's conformity with respective to standards or the interoperability with other implementations. Taking recent efforts of the ULE Alliance to specify and to develop 6LoDECT as a reason, the authors started to concentrate on the development of a TTCN-3 based, flexible and generic test framework to address the current lack of testing tools for wireless communication systems mainly in the field of IoT. The main goal of the testing framework is to give the possibility to create functional and compliance tests without depending on a specific communication protocol but still being flexible enough to address common challenges of different systems. The test framework was designed and specified by the authors and a reference implementation has been created with the purpose to test and verify 6LoDECT implementation recently developed by the ULE Alliance.

This paper is structured as following: Section 2 provides a brief overview of IPv6 for DECT ULE networks, as well as displays the need to perform tests. Section 3 outlines the selection of a programming

[a] Corresponding author: artem.yushev@hs-offenburg.de

language and a tool for test cases. The description of the Test System and all respective components follows in Sections 4 and 5. Finally, Section 6 gives an example of a typical test case execution routine.

## 2 IPv6 for DECT ULE networks

The official process of adopting the IPv6 related family of standards for IEEE 802.15.4 technologies originates in *6lowpan*, an IETF working group concluded several years ago. The results of this activity became a push for the successor's WG *6lo* to apply accumulated knowledge and define new standards for a broad range of resource constrained networks. Current 6lo's contribution extends the original scope of the *6lowpan* group to communication technologies like BACnet, Bluetooth Low Energy (BLE), ITU-T G.9959 and DECT Ultra Low Energy (ULE). The subject of the authors' work was to verify that the ULE specific *6lo* adoption fulfils the latest standard draft [3].

Usage of IPv6 functionality for any communication system obliges to comply with several requirements not directly related to the core standard, such as fragmentation/reassembly of link-layer frames and multi-hop frames delivery [4]. Due to the fact that several network technologies provide proprietary support for required features some of them can be excluded from a formal specification. This holds true for the DECT ULE standard draft which defines the usage of the underlying fragmentation/reassembly support for frames as well as a rule of thumb to restrict large Application Layer payload length. Moreover the native RPL protocol for routing in 6LoWOPAN networks is excluded from the standard proposal as not required due to the star topology of DECT ULE networks. Also a fully functional 6lo-enabled DECT ULE network should have a full spectrum of commissioning protocols like the support of multiple prefixes. In other words there is a difference in composition of generic standard 6LoWPAN implementations and the target system, thus bringing the requirement to pose a reliable and repeatable testing framework for 6lo-enabled DECT ULE systems.

To make use of existing test frameworks the Testing and Test Control Notation Version 3 (TTCN3) was selected because of its long history of usage for telecommunication protocols [2] and as a tool officially selected for 6LoWPAN plugtests driven by the European Telecommunications Standards Institute (ETSI).

## 3 Testing and test control notation version 3

TTCN is a scripting language with strong typing used in conformance and interoperability testing. This language—developed and maintained by ETSI—does not define specific communication interface to the System Under Test (SUT), though it provides an abstraction of the communication port towards the SUT which in turn should be implemented in another programming language. Also TTCN-3 has native support for ASN.1, IDL and XML type definitions. Figure 1 depicts a simple test case in TTCN3 notation to resolve an ETSI domain name where the Domain Name System (DNS) acts as SUT. While the current TTCN version is three (TTCN-3), the previous Tree and Tabular Combined Notation (TTCN-2) was essentially a different language sharing some common properties, nevertheless it has been a major testing language for telecommunication protocols such as UMTS, 2G, 3G, SIP, etc.

```
testcase TC_resolveEtsiWww() runs on DnsClient
{
    timer t_ack;
    srvPort.send(m_dnsQuestion("www.etsi.org"));
    t_ack.start(1.0);
    alt {
        []srvPort.receive(mw_dnsAnswer("172.26.1.17"))
{
            setverdict (pass);
        }
        []srvPort.receive { // any other message
            setverdict(fail);
        }
        [] t_ack.timeout {
            setverdict(inconc);
        }
    }
    t_ack.stop;
}
```

**Figure 1.** ETSI domain name resolver test case example in TTCN-3

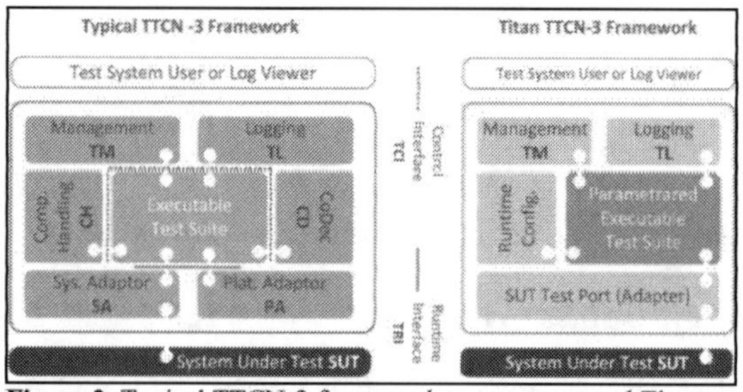

**Figure 2.** Typical TTCN-3 framework components and Titan TTCN-3 framework components

As with other programming languages the corresponding organization standardizes only syntax and abstract usage guide thus giving flexibility for 3rd parties to provide a compiler to generate an executable file, basically translating TTCN syntax into another programming language; e.g. C++ or Java.

Despite that ETSI formalized the TTCN-3 language syntax and made it publically available more than a decade ago there are only a few open source compilers which can offer continuous maintenance, development and supplementary tools. Probably one of the best among them is very well known within Ericsson Titan TTCN project, which a team of developers recently published on github under Eclipse Public License (EPL).

Titan TTCN uses C++ code as a supplementary language on which users should implement their communication port adapters. Moreover the Titan compiler generates C++ code from TTCN scripts and compiles afterwards the whole testing system using the preferred C++ compiler.

Figure 2 depicts the difference in structure of a typical TTCN-3 compatible compiler/framework and the Titan TTCN toolset. To be more specific, the generic TTCN approach defines the usage of two interfaces between layers. Namely, Test Control and Test Runtime interfaces,

DOI: 10.1051/matecconf/20167506005

TCI and TRI respectively, static API of which a developer can use for a test system. In contrast the Titan TTCN-3 framework specifies the usage of so-called test ports or adaptors implemented in C++. The Titan TTCN-3 code is generic: the interfaces between the test system and the system under test are specified at the level of the exchanged abstract data messages and signals. Setting up and maintaining the transport connections as well as sending/receiving "real" messages and signals are the tasks of interface adaptors. Titan's main purpose is functional testing (conformance, fuzzing, function, integration verification, end-to-end and network integration testing), and can also be used for performance testing. Titan proposes to use a C++ API for adaptors that would complete the test with the connectivity layer(s) between the test system and the system under test. However, in the authors' opinion this proven and elegant approach requires several modifications and customization for distributed test systems.

Based on the Titan TTCN toolset together with some improvements the authors have developed a test system for one specific 6lo Implementation Under Test, although the overall facility is not designed specifically for one System Under Test, but follows a more generic and abstracted approach.

## 4 Test system overview

### 4.1 System requirements

The test framework should comply with the following requirements: 1) It should be flexible enough to operate with virtual devices and with real devices without changes. Depending on the source code of a SUT, sometimes final images for different systems don't behave identically thus bringing the requirement to perform test cases on a virtual setup during the development phase and on a real setup with real devices; 2) The test system should have an option to control SUTs remotely. Sometimes it runs test cases without being physically on the same machine with the SUTs. In this case the authors consider the test framework a distributed system; 3) The test system should have an option to deliver control information and messages to the SUT without a direct usage of a specific communication interface. Due to the distributed character of the testing framework it might not be a rare case when SUT being a device have only a low data rate, serial connection as the sole communication interface. In this case, if a fully automated test system launches test cases from a remote server, it shouldn't have a direct connection with each SUT instead utilize one channel with high data rate to some multiplexing application which manages all the connections with the SUTs [5].

An inspiration for the overall design is [6]. Despite an extensive description of a Generic SUT adaptor it cannot be directly applied to the Titan based test system as the original proposal deploys TRI and TCI API written in Java.

Therefore, the authors propose a modular Communication Interface for Titan TTCN-3 [7]. Figure 4 demonstrates the generic view on the Test Framework with all possible scenarios, whereas Figure 3 depicts a more detailed comparison between different communication setups, namely LCM inter-process communication (IPC) for virtual setup and serial interface for setup with real devices. The test system architecture shares some features with the GNU Debugger (GDB) toolset architecture. GDB uses a client-server model of communication where a GDB-server connects directly to a target via the hardware debugger/programmer and the GDB-client can be located anywhere and send instructions to the IP address of the machine on which the GDB server runs.

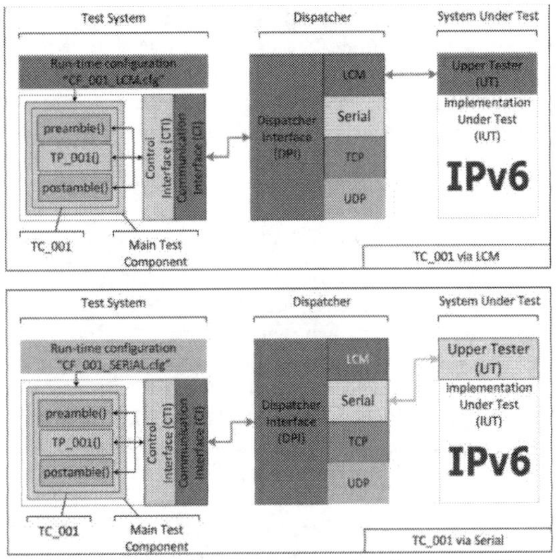


**Figure 3.** Test system architecture overview for 6lo over ULE networks implementation. Test System on the left side, Dispatcher in the middle and System Under Test on the right side

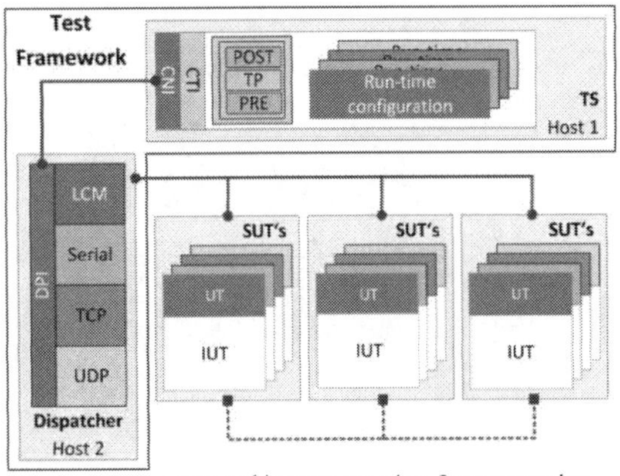


**Figure 4.** Test system architecture overview. Interconnection overview

In the current implementation the Test System acts in the same role as a GDB client located on the left in Figure 3, in the middle, the Dispatcher (DPI) acts as a GDB server which is located nearby the test field and finally a System Under Test is on the right side, where this SUT is a target of debugging.

### 4.2 General system overview

Generally *Host 1* (cf. Figure 4) controls the test cases' execution and manages the verdicts, as well as provides logging. A TCP/IP based communication link interconnects *Host 1* and *Host 2*. *Host 2* in its turn on another end—they can be separate devices or the same with different opened sockets—waiting for incoming messages. Moreover *Host 2* possesses all possible communication interfaces to the SUTs whether it be a company's proprietary or generic protocol. More precisely there are three entities in the testing framework:

1. The Test System (TS) which controls the execution of test cases.
2. The Dispatcher (DPI) which routes requests and responses from TS to SUT and backward.
3. The System Under Test combines the Upper Tester (UT), which maps commands from TS to direct API calls, and the Implementation Under Test (IUT).

### 4.3 Test system

The Titan compiler builds an executable TS (ETS) from the TTCN-3 test suite, called abstract test suite (ATS), adapter code and the Titan runtime library. The ETS can't be launched without a configuration file as Titan provides a very flexible runtime parametrization of the test cases; e.g. it is possible to select the exact type and properties of the connection between DPI and SUT, settings for the test cases etc.; the values of runtime parameters need not be defined in development time—though default values can be specified—but they can be provided just before the test execution session Therefore flexible test scenarios can be launched with different configurations for different setups (virtual, serial, TCP/IP or proprietary) without rebuilding the ETS.

The Titan TTCN-3 tool set has a runtime library which makes available the runtime control with the Main Controller (MC), where the latter has several tasks:

- It distributes a runtime configuration to all test components.
- It controls the execution of all test cases and generates verdicts.
- It has a failsafe engine which allows to catch runtime "errors" of a specific test case, to clean allocated resources and to proceed to the next test case.
- It logs all performed action on the system using the logging level predefined in the runtime configuration.

Given that the MC launches the Main Test Component (MTC) which controls other components including the test steps procedure, the standard routine in each test case can be divided on three phases:

1. The pre amble phase sets a remote IUT in initial and defined state, ready for a test purpose (subject of the test case) execution, also referred to as Stimulus;
2. The test purpose phase checks the behavior and message flow between nodes as well as sends extra requests if required;
3. The post amble phase brings the system in the final state possibly cleaning tracks of activity on SUT.

### 4.4 Dispatcher

When a test case transmits control data or a message via the CTI interface to the Dispatcher entity, the latter parses received packet, executes commands if required and dispatches data to the requested SUT. In the current version of the testing framework the Dispatcher entity is written in C++ language, as on the one hand this language follows the OOP paradigm which offers code modularity with flexibility and on the other hand, because of its high performance it minimizes a possible processing delay of data between TS and SUT.

The Dispatcher accepts the following requests from the TS:

- *Open*, to open a connection with a SUT, assign a SUT Interface Identifier (ID) and return it to the TS;
- *Close*, to close a specific connection with a SUT Interface using the SUT Interface ID received as a parameter of the command;
- *Control*, to perform managing actions with a SUT Interface as well as to forward control payloads towards a SUT using SUT Interface ID received as a parameter of the command;
- *DataOut*, to forward directly the payload to the selected SUT Interface.

Also, the Dispatcher generates responses about SUT interfaces as well as forwards data to the CTI of the TS. Possible responses from the Dispatcher are the following:

- *Opened*, to notify the TS about the open status of the SUT Interface and also to transmit a SUT Interface Identifier, which can be used for all further requests;
- *Closed*, to notify the TS about the closed connection status with the SUT Interface;
- *Status*, to transmit the response to a control request from a SUT;
- *DataIn*, to transmit a payload from a SUT to the TS.

### 4.5 System under test

As was indicated previously the SUT is composed of the Implementation Under Test and the Upper Tester entities. The SUT requires the UT in order to translate specific configuration commands sent from the TS into direct API calls. Here are some examples of the usage of UT:

- To reset, initialize or de-initialize the IUT.
- To read temporary characteristics of the IUT.
- To process data from the inner side to the outer or vice versa.
- To emulate user actions.

However, it is also possible to have a SUT without an Upper Tester component, although in this case a distinct component should persist nearby which allows to inject data into a common network.

## 5 Communication interfaces

There are two types of communication interfaces on the TS (cf. Figure 3):

- The Connection interface (CNI) links together Test System and Dispatcher interface. It establishes a stable bidirectional LCM-based connection [8].

- The Control interface (CTI) manages the SUT Interface and provides and API to the TS. Also the CTI hides the exact SUT Interface selection from the user using a simple communication protocol with fixed format and 8 types of messages: OPEN; OPENED; CLOSE; CLOSED; CONTROL; STATUS; DATAOUT; DATAIN, described previously in Sections 4.3 and 4.4.

These two pairs of interfaces play two distinct roles in the test scenario. When the control interface has the abstract static API available for the Test System by which the latter can control the SUT Interface driver running on the Dispatcher unit, the connection interface provides a possibility to deliver all the commands and data to a remote system and backward via a stable and reliable connection.

## 6 Test execution

Figure 6 shows a specific test setup configuration for 6lo over ULE networks. For the purpose of only 6lo implementation conformance testing, the provided library has a full network stack excluding DLL and PHY layers, instead the software establishes a UDP connection, thereby emulating the properties of a connection with other nodes of the network.

The test setup consists of the Test System with the Main Test Component (MTC), the Dispatcher and two Systems Under Test which represent a router entity and a local node. Two SUTs have an Upper Tester component to link the Test system with the Implementation Under Test, each of the UTs have a library to parse commands and translate them into direct API calls. The authors propose to have a modular approach for the UT's software architecture in order to provide developers with a library with a flexible substitution of Command or Transport engines.

Each performed test case for this setup requires at least several components: 1) The test case code in TTCN-3 language; 2) The CI (CTI and CNI) adapter implementation in C++ which will deliver the test case commands' responses as well as raw binary frames from the SUTs; 3) The Codec/Decodec (CoDec) entity which is generated automatically by the Titan TTCN-3 tool set from TTCN-3 code (cf. Figure 5).

```
/* @desc    6lowpan Frame (RFC 4944/6282) */
type union SixLo_frame
{
    SixLoUCIpv6      ucipv6,     // Uncompressed IPv6
    SixLoHC6282      hc,         // Compressed IPv6
    SixLoMesh        mesh,       // Mesh Header
    SixLoBroadcast broadcast,// Broadcast Header
    SixLoFrag1       frag1,      // Fragment Header 1
    SixLoFragN       fragN       // Fragment Header N
} with { variant "TAG(
        ucipv6,      dispatch = '01000001'B;
        hc,          dispatch = '011'B;
        mesh,        dispatch = '10'B;
        broadcast,  dispatch = '01010000'B;
        frag1,       dispatch = '11000'B;
        fragN,       dispatch = '11100'B )"
}
```

**Figure 5.** An example of 6lo frame definition in TTCN-3 with Titan automatic coder/decoder generation

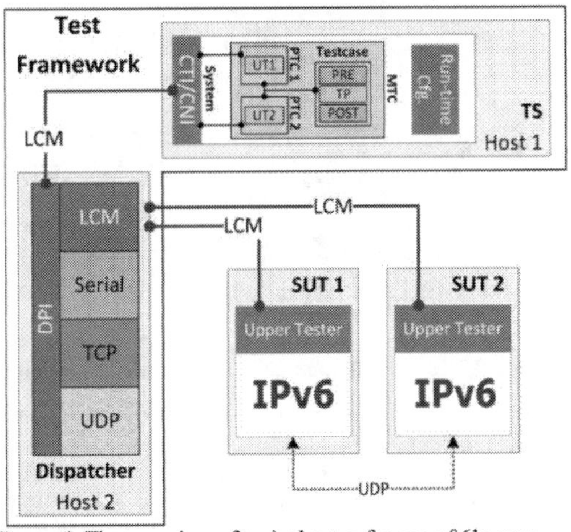

**Figure 6.** The overview of typical setup for one of 6lo over ULE networks test cases

A Titan test case works with abstract data; e.g, like a 6LoWPAN link-layer frame definition listed in Figure 5. That is why the CoDec entity deserialize binary data into a TTCN-3 data type; e.g. if an incoming frame starts with a *011* binary sequence the CoDec creates the *SixLo_frame* object with only one internal field, *hc*, inside which the *dispatch* field comes first, followed by the remaining payload.

## References

1. Z. Shelby, and C. Bormann, "6LoWPAN: The wireless embedded Internet", Vol. **43**, John Wiley & Sons, (2011)
2. J. Grabowski, et al. "An introduction to the testing and test control notation (TTCN-3)." Comp. Net. **42**.3 (2003)
3. Transmission of IPv6 Packets over DECT Ultra Low Energy, draft-ietf-6lo-dect-ule-03, (2015)
4. M. Schappacher, et al. "A flexible, modular, open-source implementation of 6LoWPAN." IDAACS-2015, 2015 IEEE 8th Int. Conf. on, Vol. **2**, (2015)
5. A. Yushev, et al. "Extended performance measurements of scalable 6LoWPAN networks in an Automated Physical Testbed." I2MTC-215, 2015 IEEE Inter. Conf., IEEE, (2015)
6. A. Hyrkkänen, "General purpose SUT adapter for TTCN-3." MS thesis, Tampere Univ. of Tech., (2005)
7. J. Z. Szabó, and T. Csöndes. "TITAN, TTCN-3 test execution environment." Infocomm. J. **62**.1 (2007): 27-31
8. A. S. Huang, et al. "LCM: Lightweight communications and marshalling." Intelligent robots and systems (IROS), IEEE/RSJ Intl. Conf., (2010)

# Automated Physical Testbeds for Emulation of Wireless Networks

Axel Sikora, Jubin Sebastian E, Artem Yushev, Edgar Schmitt and Manuel Schappacher

*Institute of Reliable Embedded Systems and Communication Electronics (ivESK), Offenburg University of Applied Sciences, 77652 Offenburg, Germany*

**Abstract.** Institute of Reliable Embedded Systems and Communication Electronics, Offenburg University of Applied Sciences, Germany has developed an automated testing environment, Automated Physical TestBeds (APTB), for analyzing the performance of wireless systems and its supporting protocols. Wireless physical networking nodes can connect to this APTB and the antenna output of this attaches with the RF waveguides. To model the RF environment this RF waveguides then establish wired connection among RF elements like splitters, attenuators and switches. In such kind of set up it's well possible to vary the path characteristics by altering the attenuators and switches. The major advantage of using APTB is the possibility of isolated, well controlled, repeatable test environment in various conditions to run statistical analysis and even to execute regression tests. This paper provides an overview of the design and implementation of APTB, demonstrates its ability to automate test cases, and its efficiency.

## 1 Introduction

Over the last decade, wireless networks have been deployed at almost all the areas. The exciting capabilities made possible by the inexpensive ubiquity of wireless networking technology, has led to a massive amount of research activity to improve the performance of existing wireless networks, and develop new applications based on this technology such as wireless sensor networks and internet of things. Even though many wireless standards are available, to meet the specific requirements in terms of low energy, scalability etc, we need to develop optimum solutions. Conducting such type of wireless network research is, none the less, a challenging effort due to the distributed and short lived nature of wireless signal propagation. There are many challenges in wireless network research which are nicely listed in [1], [18].The major approaches used for performance evaluation of wireless networks are simulation, virtualization, outdoor field test (test beds), and emulation.

Simulation is used to test the behaviour of the protocol and its implementation, i.e. the firmware on hardware abstraction layer level. The advantage of simulations is its scalability, but simulation is not helpful in debugging the underlying hardware and is totally missing the effects that hardware has on RF characteristics [1].

Virtualization of networks also used to analyse the protocol stack in a virtual environment of intended network. Network virtualization has the advantage that enables to emulate connection between applications, services, dependencies, and end users in a test environment without the need to test in real hardware. But, distinctive properties of wireless environment such as time variant channels, mobility, broadcast, attenuation, and specific access technologies makes convergence, sharing and abstraction difficult to achieve [10].

Outdoor field tests play a vital role for testing the capabilities and reliabilities of wireless systems, but the major disadvantages are irreproducible tests due to the uncontrollable nature of the outdoor environment. Also, outdoor field tests are time consuming and therefore expensive [1].

Emulation is the combination of simulation and testbed. As in the simulation approach emulators can generate predefined emulated network conditions and traffic dynamics for different use cases. Hardware-based emulation clearly achieves the most physical layer realism; practical considerations such as ease of development, control, and experimental repeatability have made emulation the dominant experimental technique [1].The requirements of the development of an emulator can be seen in [2].

Authors have developed first generation of our automated RF emulator for distributed wireless sensor networks for analysing EnOcean Radio Protocol (ERP) [2], and the second generation developed to test Ko-TAG subsystem [1]. In this paper we describe the architecture and implementation details our third generation fully automated emulator for wireless networks. This includes the physical networking nodes, RF elements to allow the controlling of RF environment between the participating nodes, and automated control software. We call this set up as Automated Physical Testbeds (APTB) for emulation of wireless networks.

APTB is an emulator to perform reproducible tests with static and dynamic attenuation between the participating wireless nodes as real laboratory set up. The nodes are connected over their antenna outputs to RF-waveguides. It is also possible to cut or switch the connections of the network, in this way many different topologies can be set up. APTB promises to achieve much of the flexibility of wireless emulators while maintaining great realism of real wireless networks [4].

This contribution is structured as follows: Ch. 2 gives a review on the previously developed emulator platforms by other researchers and the previous generation of emulator by author's Ch. 3 presents the APTB architecture and implementation details, whereas Ch. 4 gives an overview of experience with APTB, Ch. 5 concludes the paper.

## 2 State of the art

An overview of the previously proposed emulators by other work groups and authors team is given in this chapter.

### 2.1 Review of emulators

In the field of wireless network emulation we observed different approaches like a). Central control based wireless emulators, in which the wireless nodes connect to central server as seen in ONE [5], MIT's Click modular router [6], Dummynet [7] and NIST NET [8]. b). Simulation integrated wireless emulators, which combines the network emulation with existing network simulators which has rich models and libraries as VINT/ns [9]. c). Trace based wireless emulation, which collect the traces from the target network and then construct a wireless network model with collected traces, and then reproduces the traced network effect in a wired network and provide a reproducible environment [3].

Clearly these emulators provide a controllable and reproducible environment, but it generally lacks the support for network topology, mobility and an interactive interface for users. In some of the emulator setup the node mobility implemented using remotely controlled robots as in CMU [11], APE [12], Mobile Emulab [13], m-MiNT [14]. Associated with the mobility features a tracking system [14]-[16] is used to determine the position and orientation of each node. But such type of emulator platforms is not flexible and efficient to test the wireless system in the development stage. Using a wired connection of RF elements we overcome these drawbacks. EMWIN [3] is an IP based mobile wireless network emulation system; provide a unique feature of emulating multiple mobile hosts in one single emulator node, also to emulate the mobility using a wired network of computers.

To the best of our knowledge, we have not seen a system which uses wired connection of RF elements like attenuators, splitters, switches etc., to imitate the RF environment and manipulate the path characteristics. So, we decided to develop such type of emulator for wireless networks. Next section reviews the details of our previous generation of emulators.

### 2.2 Review of our previous emulators

The first generation consists of elements like network nodes, management nodes, a client computer, a deployment support network (DSN), with an embedded web server running web 2.0 technology for remote monitoring and control of its main elements. The emulator originated form the implementation of the EnOcean Radio protocol (ERP), the details of this emulator and experience are described in [2]. This presented an approach of management and sniffing tools for distributed wireless networks. The emulator was designed flexible and generic to extent this to other wireless protocols.

The second generation of our emulator developed for highly scalable IEEE 802.11p communication and localization subsystem in the Ko-FAS project [1]. This includes physical networking nodes, but models the RF environment using RF-waveguides. The RF emulator allows the controlling of path losses and connectivity between any of the nodes with the help of RF attenuator and programmable RF switches, while it's shielded against its surrounding RF environments in the lab. Details and experience of the system design can be seen [1].

These two versions of our emulator was designed for specific projects, we extend this development to easily reuse it for other wireless network research projects.

## 3 Automated physical testbeds (APTB)

We have developed a wireless emulator, APTB that enables both realistic and repeatable wireless experimentation by supporting accurate wireless signal transmission, propagation, and reception in an emulated physical space. APTB uses physical networking nodes, but where the RF environment is still modelled using RF-waveguides and RF elements like attenuators, splitters, switches, cables, terminators etc. In this environment, it is well possible e.g. to control the path loss between any of the nodes with the help of RF attenuators and to control the connectivity between any of the nodes with the help of programmable RF switches. APTB maintains the repeatability, configurability, the ability to modify wireless device behaviour, automated tests, support for a large number of nodes, manageability of simulation while retaining the support for real applications, and much of the realism of hardware testbeds. As a result, this APTB emulator can be a superior platform for wireless network experimentation [4].

### 3.1 Architecture

APTB consists of both custom and commodity hardware that must act in concert to enable users to accurately emulate arbitrary signal propagation environments and efficiently execute experiments. Developing hardware to achieve this has been a challenging task. Figure 1 describes the architecture that enables our physical layer APTB to achieve accurate emulation while enabling efficient experimentation. The hardware elements of the APTB are APTB nodes, RF elements and Controller units.

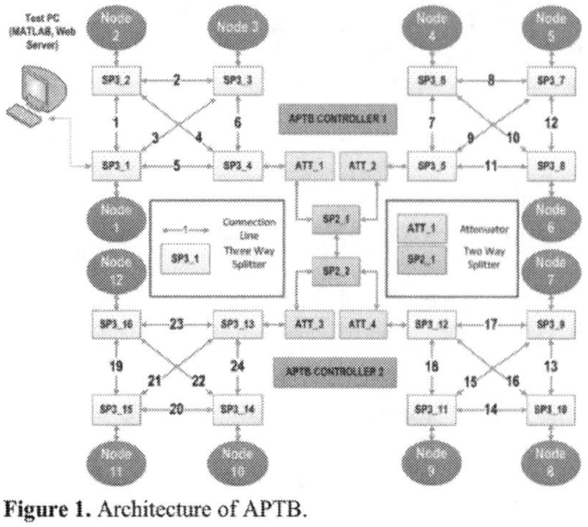

**Figure 1.** Architecture of APTB.

Any of the wireless modules with an antenna output used as APTB nodes. We encapsulate these APTB nodes in to shield boxes. The RF nodes are shielded from each other using RF shield boxes so that less communication occurs over air. Such types of 12 shield boxes are present in the APTB emulator as shown Figure 1. The RF elements used in the APTB are two way splitters/combiners (SP2), three way splitters/ combiners (SP3), variable and fixed attenuators (ATT), high isolation switches (SW), SMA cables, terminators and RF shielded test enclosure. All the communication between RF nodes occurs through the signal propagation environment modelled using these different RF elements within the APTB emulator. The controller units present in the APTB are two ARM Cortex-M3 based MCU to control the RF elements.

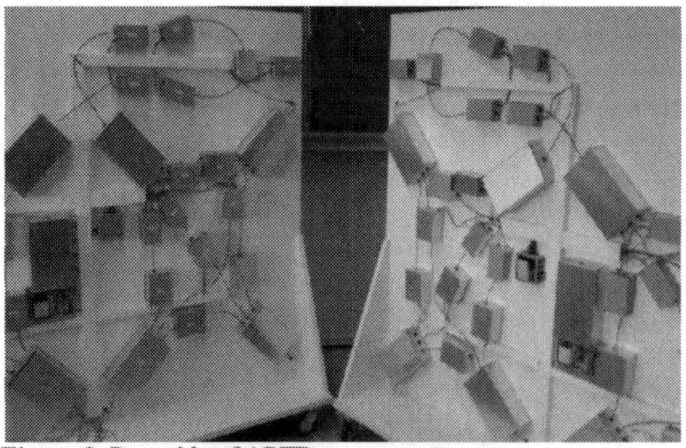
**Figure 2.** Rear side of APTB.

**Figure 3.** Front side of APTB.

In the author's Lab APTB is developed in two identical physical boards, the rear side of the testbed is shown in Figure 2 with splitters, attenuators, switches, SMA cables and APTB control unit. The front side of the APTB is shown in Figure 3 consist of shield boxes where we can encapsulate the APTB nodes.

## 3.2 APTB hardware

The hardware elements of the APTB are APTB nodes, RF elements and controller units.

- APTB Nodes

Any type of wireless nodes with antenna output can be used with APTB. For example, @ANY900-2 modules used as APTB nodes, which contains Atmel microcontroller ATmega 1281V (128 bytes of FLASH memory, 8K bytes SRAM, 4K bytes EEPROM), IEEE 802.15.4 radio transceiver AT86RF212 (868/915 MHz band) and 2Mbit serial data flash (AT25F2048), which can store the flash images of the used MCU [19]. We encapsulate these APTB nodes in to shield boxes. The RF nodes are shielded from each other so that no communication occurs over air. Such types of 12 shield boxes are present in the APTB emulator.

- RF elements

The RF elements used in the APTB are provided by "Mini Circuits" company follow: two way splitters/combiners (ZFRSC-183+), three way splitters/combiners (ZB3PD-63+), variable attenuators (ZX76-15R5-SP+), fixed attenuators (VAT-30+), high isolation switches (ZASWA-2-50DR+), SMA cables (141-6SM+), RF shield boxes (STE2200) and terminators (ANNE-50+). All the communication between physical nodes occurs through the signal propagation environment modelled using these different RF elements within the APTB emulator.

➢ Two way Splitters/combiners (ZFRSC-183+): are two-way, 0 degree, resistive (50Ω) splitter/combiner, which operates in wideband from DC to 18000MHz, and has coaxial connector. This provides low insertion loss and high isolation [20].

➢ Three way splitters/combiners (ZB3PD-63+): are three-way, 0 degree, in-phase splitter and combiner covers a wide frequency range (150-6000 MHz), making this splitter suitable for GPS, GSM applications, in addition to WiFi, Bluetooth and 802.11a uses. This model also features low insertion loss, amplitude and phase unbalance [21].

➢ Fixed attenuator (VAT-30+): is coaxial SMA fixed 30 dB attenuator of 0.5W, and has wide coverage from DC to 6000 MHz [22].

➢ Variable attenuators (ZX76-15R5-SP+): is a 50Ω RF digital step attenuator that offers an attenuation range up to 15.5 dB in 0.5 dB steps. The control is a 5-bit serial interface. The model operates on a single +3 volt supply [23].

➢ High isolation switches (ZASWA-2-50DR+): is 50Ω SPDT, coaxial high isolation switch, operate at DC to 5GHz, and has an integral TTL driver. It possesses high isolation of 82dB typical at 2 GHz [24].

➢ SMA cables (141-6SM+): low loss precision test, 141 series hand-flex coaxial cables are used for interconnection RF elements. It has wideband frequency

 DOI: 10.1051/matecconf/20167506006

coverage from DC to 18GHz and low Loss, 0.43dB, at 18GHz. Also it has excellent return loss, 23 dB at 18GHz [25].

➢ RF shield boxes (STE2200): is an RF shielded test enclosure [26] used to connect the APTB nodes and have a shield RF environment for test set up. In STE2200 heavy, rugged 0.090 and 0.125 aluminium is used throughout and double lip woven Monel flexible gaskets are used at all joint locations assuring a reliable RF tight closure. Oversized hinges and latches are used to provide a physically tight seal every time. RF absorbent foam lines the interior to provide a typical RF attenuation of -90dB @ 3GHz.

➢ Terminators (ANNE-50+): SMA Male, 50Ω terminator [27] used for test set up of APTB.

**Table 1.** RF elements overview.

| type | attenuation [dB] | frequency | No. used in APTB |
|---|---|---|---|
| two way splitter/ combiner | 6 | DC-18000MHz | 2 |
| three way splitter/ combiner | 5 | 150-6000MHz | 16 |
| fixed attenuator | 30 | DC-6000MHz | 12 |
| variable attenuator | 1-15 | DC-4000MHz | 22 |
| high isolation switch | 82 | DC-5GHz | 24 |
| terminator | - | DC-18GHz | 12 |
| RF shielded test enclosure | 90 | - | 12 |

All the communication between RF nodes occurs through the signal propagation environment modelled using these different RF elements within the APTB emulator.

- Controller Units

APTB Controller unit is an ARM Cortex-M3 based MCU (STM32F107VCT6*)* running together with a web interface on the PC to control paths and attenuation between different nodes. With these we can vary the topology and attenuation between the APTB nodes.

Based on the experience from earlier emulator setups [2] [1], all elements have additional shielding and are carefully mounted whereas links were fixed with a torque wrench. All elements and connections are characterized after they have been built-in with regard to attenuation, reflection, and phase shift (delay).

### 3.3 APTB software

The main software implementations for APTB are control software for RF elements and MATLAB scripts to automate, reproduce the tests. All the involved devices like APTB nodes, RF elements and controller units are automatically controlled by MATLAB program from the PC, integration with the APTB controller unit.

To control the RF elements of the APTB emulator uses web interface running together with APTB controller. In experiment setup the APTB consists of two parts, which helps on one side to control data traffic flow better and on other side to extend physical topology as much as it required for the test case. Two identical parts create a local network which is managed externally by an Ethernet based control interface. Used infrastructure with such devices helps to add different controlling extensions which may automatically perform tests and grab output data [1].The controlling devices are equipped with an embedded TCP/IP stack including an embedded webserver. This provides basic access mechanisms e.g. using an embedded website as shown in Figure 4. In APTB Low layer commands are used to directly access RF elements of the APTB. For example,

- APTB:RADELements:ATTenuator:SETAttenuation <id>,<val>
- APTB:RADELements:SWitch:SETEnable <id>,<val>
- APTB:NETements:NOde:SENDData <id>,<data>

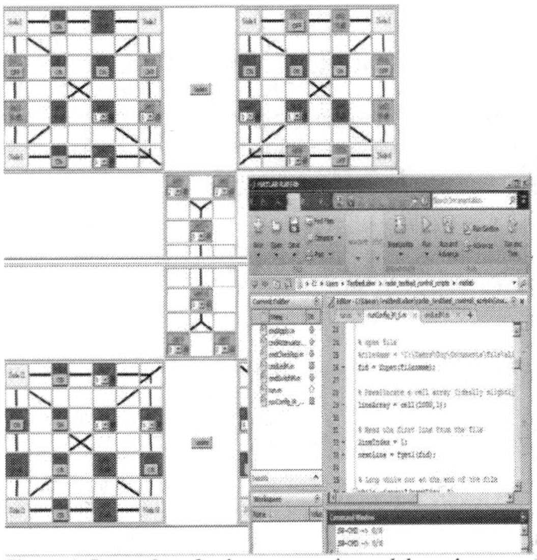

**Figure 4.** Screenshots for the programming and the web interface for APTB automated tests

These commands are interpreted and executed by the APTB Controller which includes an according "Standard Commands for Programmable Instruments" (SCPI) parser/handler. The RF Switch Control is fitted with the STZEDN's embedded TCP/IP protocol suite (emBetter) which is expanded by an SCPI module.

Furthermore the software of the controlling devices have been extended with a software module able to interpret the SCPI protocol commonly used to control and to monitor measurement devices. This makes it possible to control and to automate the network topology e.g. with MATLAB that is using Java-IO libraries to send the SCPI commands to the controlling devices. Self-defined configuration scripts can be used to create a time-controlled sequence of different topologies and therefore also provides the possibility to reproduce scenarios. MATLAB programs are used for high level user interface to describe the use cases and topologies. A sample MATLAB script for a scenario of link loss add reconnection after a specific amount of time can be described in a human readable manner (CSV, XML) as in the Table 2.

**Table 2.** MATLAB script for test setup.

| | |
|---|---|
| 0 ,SW,12,1*<LF>* | Enable Switch-ID 12 |
| 100,SW,12,0*<LF>* | Disable after 100ms |
| 200,SW,12,1*<LF>* | Reconnect after another 100ms |

Figure 4 shows the automated test setup together with webserver using MATLAB, the green colour between the nodes shows the enabled paths and red shows the disabled path.

## 4 Experiences with APTB

To analyse the accuracy of the emulator we have performed many tests like creating different topologies using automatic control of the attenuator and switches using MATLAB, and we have measured the attenuation between different nodes, compared this with the calculated attenuation. These measurement results confirmed the calibration accuracy of the APTB. We have extensively used APTB for testing and verification of the protocol stacks developed by author's team. These testing involve various steps: test case definition, configuration of APTB nodes, test execution and result analysis. The measurements of different test cases were performed for different topologies like chain, mesh and tree. Each test executed many times in a time span of 24 hours to one week. More details related to these experiments and results are seen in [4], [17].

## 5 Conclusion and outlook

In this paper we presented the state of art of our wireless network emulator APTB, which helps the wireless network researchers to set up different wireless time varying topologies and path characteristics automatically. The major advantage of using APTB against traditional mount and measure approach is concentrated in striving for isolated, well controlled, repeatable test environment which may help to prove a sustainable and reliable behaviour of wireless network protocols in various conditions. The authors would be happy to share more details about the implemented APTB, so that other work groups can build identical or extended testbeds, keeping compatibility especially on the software side.

## References

1. A. Sikora, M. Schappacher, L. Möllendorf, "Automated RF Emulator for a highly scalable IEEE802.11p communication and localization subsystem," in *6th Intl Workshop on Communication Technologies for Vehicles*, Offenburg, (2014)
2. L. Möllendorf, D. Schauenberg, N. Braun, D. Rahusen, A. Sikora, "A Distributed Embedded Web2.0 Based Automated Testbed for Wireless Mesh Networks," in *1st IEEE Int'l Workshop On AWN*, Leipzig, (2009)
3. Ni, Pei Zheng and Lionel M., "EMWIN: Emulating a Mobile Wireless Network using a Wired Network," in *In Proceedings of the 5th ACM international workshop on Wireless mobile multimedia*, (2002)
4. A. Yushev, P. Lehmann, A. Sikora, V.F. Groza, "Extended Performance Measurements of Scalable 6LoWPAN Networks in an Automated Physical Testbed," in *IEEE Int'l Conference $I^2$MTC, Pisa,* (2015)
5. M. Allman, A. Caldwell, and S. Ostermann, "ONE: The Ohio Network Emulator," Ohio University, *Tech Report TR-19972*, August (1997)
6. E. Kohler, R. Morris, B. Chen, J. Jannotti, and M. F. Kaashoek, "The Click Modular Router," *ACM Transactions on Computer Systems*, vol. **18**, pp. 263-297, (2000)
7. L. Rizzo, "Dummynet: a simple approach to the evaluation of network protocols," *ACM Computer Communication Review*, vol. **27**, (1997)
8. http://snad.ncsl.nist.gov/itg/nistnet
9. K. Fall, "Network Emulation in the VINT/NS Simulator", In *Proceedings of 4th IEEE Symposium on Computers and Communications*, (1999)
10. https://en.wikipedia.org/wiki/Network_virtualization
11. D. Maltz, J. Broch, and D. Johnson. "Experiences Designing and Building a Multi-Hop Wireless Ad-Hoc Network Testbed". *In CMU TR99-116,* (1999)
12. H. Lunndgren, D. Lundberg, J. Nielsen, E. Nordstrom, and C. Tscudin. "A Large-scale Testbed for Reproducible Ad Hoc Protocol Evaluations". *In Proc. of WCNC,* (2002)
13. D. Johnson, T. Stack, R. Fish, D. M. Flickinger, L. Stoller, R. Ricci, and J. Lepreau. "Mobile Emulab: A Robotic Wireless and Sensor Network Testbed" *In IEEE Infocom*, (2006)
14. http://www.ecsl.cs.sunysb.edu/mint/
15. L. Cremean, W. Dunbar, D. van Gogh, J. Hickey, E. Klavins, J. Meltzer, and R. M. Murray. "The Caltech Multi-Vehicle Wireless Testbed" *In Proc. of Conference on Decision and Control*, (2002)
16. D. Johnson, T. Stack, R. Fish, D. M. Flickinger, L. Stoller, R. Ricci, and J. Lepreau. "Mobile Emulab: A Robotic Wireless and Sensor Network Testbed" *In IEEE Infocom*, (2006)
17. A. Yushev, P. Lehmann, A. Sikora, "6LoWPAN with RPL Performance Measurements in an Automated Physical Testbed",$2^{nd}$ *IEEE IDAACS SWS, Offenburg, Germany*, pp. 29-33, (2014)
18. M. Ringwald, K. Römer, "Deployment of Sensor Networks: Problems and Passive Inspection", *Proc. of WISES '07, Madrid, Spain*, (2007)
19. http://www.ansolutions.de/documents/@ANY900_2_Module_Leaflet.pdf
20. http://194.75.38.69/pdfs/ZFRSC-183+.pdf
21. http://194.75.38.69/pdfs/ZB3PD-63+.pdf
22. http://194.75.38.69/pdfs/VAT-30.pdf
23. http://194.75.38.69/pdfs/ZX76-15R5-SP+.pdf
24. http://194.75.38.69/pdfs/ZASWA-2-50DR+.pdf
25. http://194.75.38.69/pdfs/141-6SM+.pdf
26. http://www.rfshieldbox.com/datasheets/STE2200.pdf
27. http://194.75.38.69/pdfs/ANNE-50+.pdf

# Speech Recognizing for Presentation Tool Navigation Using Back Propagation Artificial Neural Network

Nur Hasanah[a], Dessy Irmawati, Fatchul Arifin and Eko Marpanaji

*Electronics Engineering Education Department, Yogyakarta State University, Karangmalang, Yogyakarta 55281, Indonesia*

**Abstract.** Backpropagation Artificial Neural Network (ANN) is a well known branch of Artificial Intelligence and has been proven to solve various problems of complex speech recognizing in health [1], [2], education [4] and engineering [3]. Today, many kinds of presentation tools are used by society. One popular example is MsPowerpoint. The transition process between slides in presentation tools will be more easily done through speech, the sound emitted directly by the user during the presentation. This study uses research and development to create a simulation using Backpropagation ANN for speech recognition from number one to five to navigate slides of the presentation tool. The Backpropagation ANN consists of one input layer, one hidden layer with 100 neurons and one output layer. The simulation is built by using a Neural Network Toolbox Matlab R2014a. Speech samples were taken from five different people with wav format. This research shows that the Backpropagation ANN can be used as navigation through speech with 96% accuracy rate based on the network training result. Thesimulation can produce 63% accuracy based on 100 new speech samples from various sources.

## 1 Background

Artificial Neural Network (ANN) is a branch of Artificial Intelligence which has been widely applied to various applications and is proven to solve various complex problems. Backpropagation is one of the popular ANN architecture and is widely applied in research of speech recognizing, for example, speech recognition of a recently born baby [1], speech recognition of cardiac abnormalities [2], speech recognition to control robots [3] and speech recognition to learn foreign language [4].

Presentation tools usually consists of slides that can be displayed on the screen and the user can navigate using a keyboard or other controller devices such as mouse, wireless mouse and wireless pointer. The control device intended to facilitate users in giving presentations. But the use of a keyboard and mouse in the transition between slides still limit the movement of user because there is a maximum distance between the devices with a laptop or computer that is used. While wireless pointer provides only know next and previous orders and can not refer directly to a specific slide the user wants. The transition process between slides will be more easily done through speech, the sound emitted directly by the user during the presentation. The easier use of the device, the better support for the presentation process. Therefore, this research aims to create software for presentation tool navigation through speech, using Backpropagation ANN.

## 2 Theoretical basis

### 2.1 Fast fourier transform algorithm

FFT is a method for transforming the speech signal into a frequency signal, meaning that the speech recording is recorded in digital form in the form of wave-based speech frequency spectrum. The FFT algorithm is designed to perform complex multiplications and additions, even though the input data may be real valued [5].

### 2.2 Artificial neural network

An artificial neural network is an information-processing system that has certain performance characteristics in common with biological neural networks. Artificial neural networks have been developed as generalizations of mathematical models of human cognition or neural biology [6].

A neural net consists of a large number of simple processing elements called neurons, units, cells or nodes. Each neuron is connected to other neurons by means of directed communication links, each with an associated weight. The weights represent information being used by the net to solve a problem [6].

[a] Corresponding author: nurhasanah@uny.ac.id

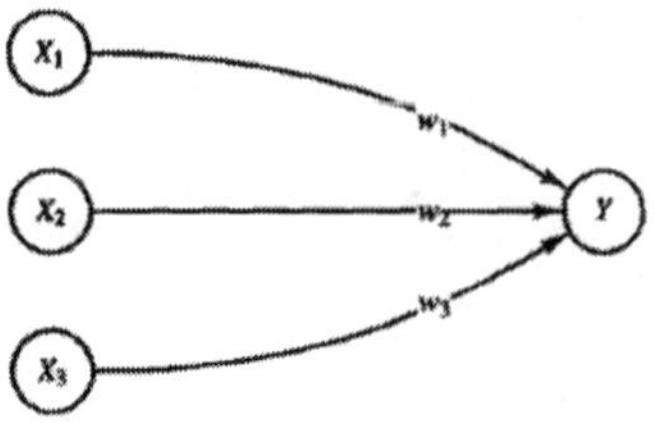


**Figure 1**. A simple (artificial) neuron.

Each neuron has an internal state, called its activation or activity level, which is a function of the inputs it has received. Typically, a neuron sends its activation as a signal to several other neurons. For example, consider a neuron Y, illustrated in Figure 1, that receives inputs from neurons $X_1$, $X_2$, and $X_3$. The activations (output signals) of these neurons are $x_1$, $x_2$, and $x_3$, respectively. The weights on the connections from $X_1$, $X_2$, and $X_3$ to neuron Y are $X_1$, $X_2$, and $w_1$, $w_2$ and $w_3$ respectively. The net input, y_in, to neuron Y is the sum of the weighted signals from neurons $X_1$, $X_2$, and $X_3$, i.e.,

$$y\_in = w_1 x_1 + w_2 x_2 + w_3 x_3 \tag{1}$$

The activation y of neuron Y is given by some function of its net input, y = f(y_in) [6].

### 2.3 Common activation functions

1. Identity function (Purelin)

$$f(x) = x \text{, for all } x \tag{2}$$

Single-layer nets often use a step function to convert the net input, which is a continuously valued variable, to an output unit that is a binary (1 or 0) or bipolar (1 or -1) signal [6].

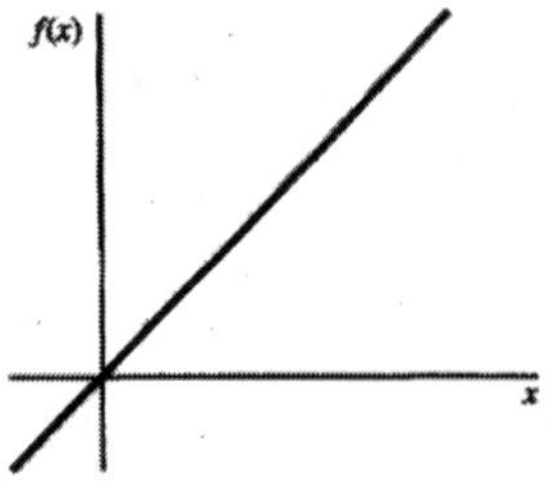


**Figure 2.** Identity function.

2. Binary Sigmoid function (Logsig)

$$f(x) = \frac{1}{1 + \exp(-\sigma x)}$$

$$f'(x) = \sigma f(x)[1 - f(x)] \tag{3}$$

During feedforward, each input unit ($X_i$) receives an input signal and broadcasts this signal to the each of the hidden units $Z_1, \ldots, Z_p$. Each hidden unit then computes its activation and sends its signal ($z_j$) to each output unit. Each output unit ($Y_k$) computes its activation ($y_k$) to form the response of the net for the given input pattern [6].

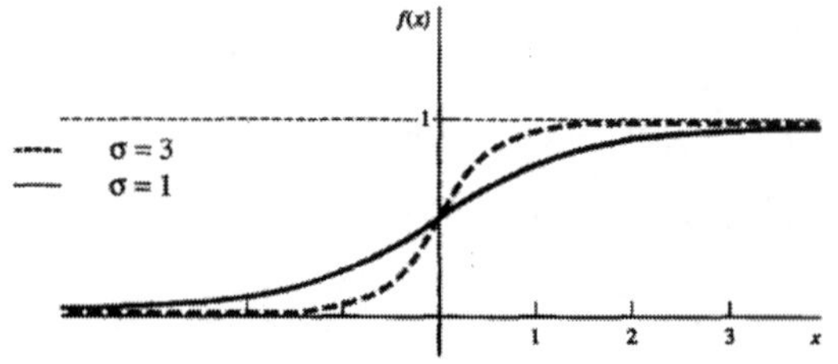


**Figure 3.** Binary Sigmoid. Stepness parameters a = 1 and a = 3

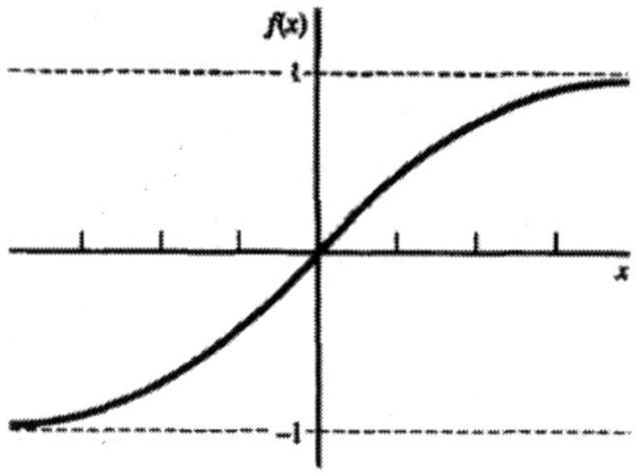


**Figure 4.** Bipolar Sigmoid.

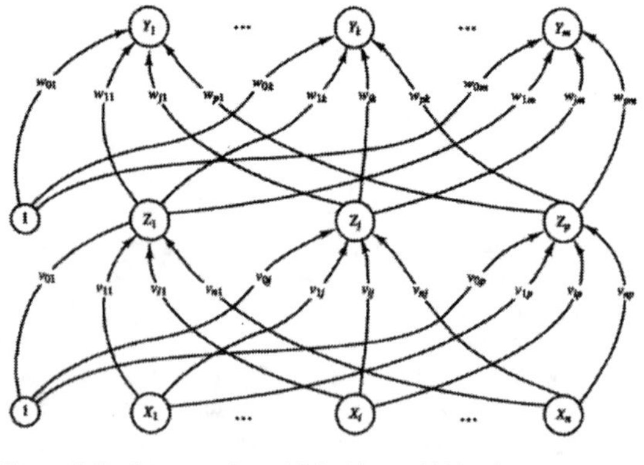


**Figure 5.** Backpropagation ANN with one hidden layer.

During training, each output unit compares its computed activation $y_k$ with its target value $t_k$ to determine the associated error for that pattern with that unit. Based on this error, the factor $\delta_k$ ( k = 1, ... m) is computed. $\delta_k$ is used to distribute the error at output unit $Y_k$ back to all units in the previous layer (the hidden units that are connected to $Y_k$). It also used (later) to update the weights between the output layer and the hidden layer.

After all of the δ factors have been determined, the weights for all layers are adjusted simultaneously. The adjustment to the weight $w_{jk}$ (from hidden unit $Z_j$ to output unit $Y_k$) is based on the factor $\delta_k$ and the

DOI: 10.1051/matecconf/20167507001

activation $z_j$ of the hidden unit $Z_j$. The adjustment to the weight $v_{ij}$ (from input unit $X_i$ to hidden unit $Z_j$) is based on the factor $\delta_j$ and the activation $x_i$ of the input unit [6].

Backpropagation learning has emerged as the standard algorithm for the training of multilayer perceptrons, against which other learning algorithms are often bench-marked [7].

## 3 Research Method

This study uses Research and Development to develop speech recognition simulation of number one, two, three, four and five. Speech samples were taken from five different people with wav format. The total of speech samples for network training is 75. Each number has 15 speech samples.

### 3.1. Feature extraction

In this stage, software for preprocessing is built using FFT, as shown in Figure 6.

### 3.2 Network training

The next stage is to design software and implement ANN Backpropagation algorithms to learn speech patterns. Backpropagation ANN is an autoassociative network type, input range that is processed into the same network with the range of output results. Backpropagation ANN development in general can be seen in Figure 7.

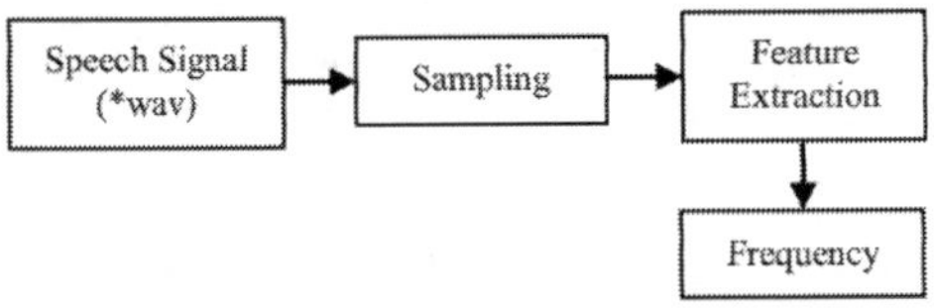


**Figure 6.** Preprocessing stage.

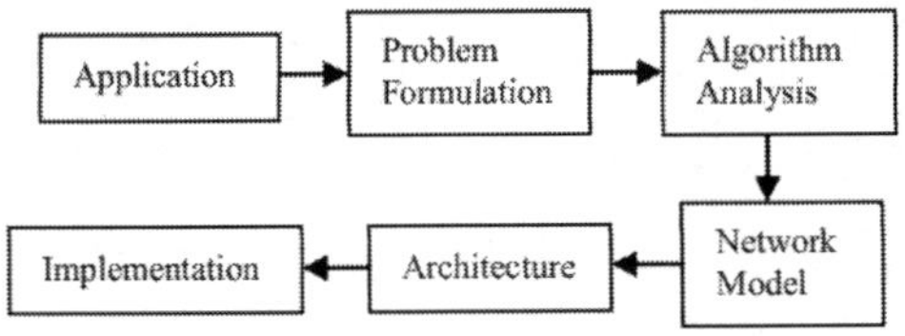


**Figure 7**. Development process of artificial neural networks.

Thus, the general system of this application follows a path as shown in Figure 8.

The Backpropagation ANN consists of one input layer, one hidden layer and one output layer. Activation function is used to connect the input layer to the hidden layer and also to connect the hidden layer to the output layer. Activation function used in the research is Tansig. A multilayer network learns much faster when the sigmoidal activation function is represented by a hyperbolic tangent [8].

## 4 Result and discussion

The Backpropagation ANN built has 1 hidden layer with 100 neurons. The activation function used is Tansig. There are two ways to determine the training process stops, by limiting the number of iterations or Mean Square Error (MSE).

The simulation is built by using a Neural Network Toolbox Matlab R2014a. Matlab is an interactive, matrix-based system for scientific and engineering numeric computation and visualization. Its strength lies in the fact that complex numerical problems can be solved easily and in a fraction of the time required with a programming language. The basic Matlab program is further enhanced by the availability of numerous toolboxes [9].

In the simulation, the number of iterations is limited to 1000 while MSE is limited to 1.00e-06. Network training results can be seen in Figure 9. The training process occurs until 23 iteration and training time during 0 second.

While Figure 10 shows the results of Confusion. The conclusion obtained is the level of accuracy in determining the output network is 96 %.

To measure the accuracy of Backpropagation ANN, it is necessary to test some new speech samples using a simulation program that can be seen in Figure 11.

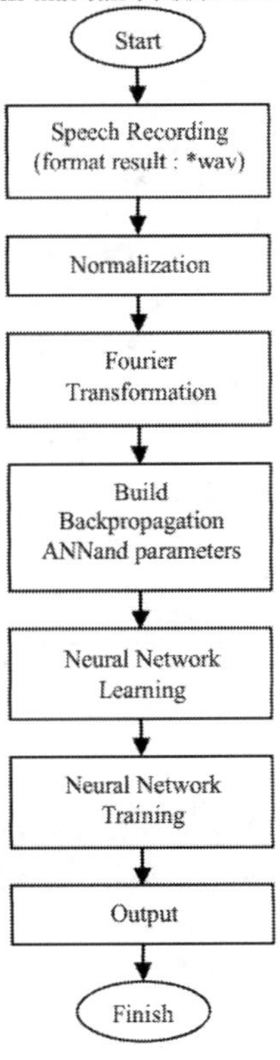


**Figure 8.** Flowchart of the simulation system.

DOI: 10.1051/matecconf/20167507001

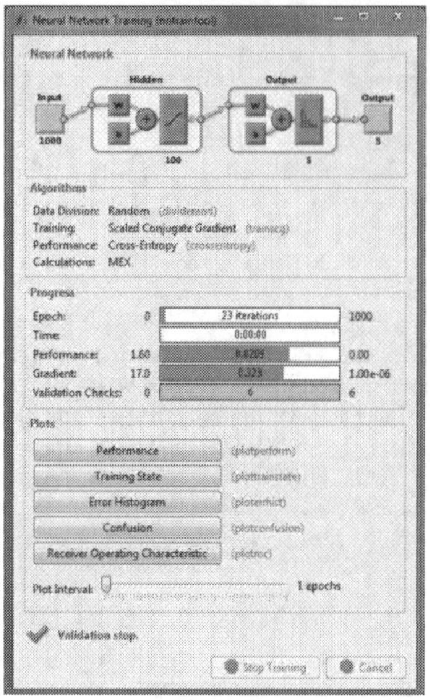


**Figure 9.** Result of network training.

This program provides menus to recordspeech, play speech, save speech and open a speech that have been saved. After a speech has beenrecorded, the program will display the results in terms of numbers as well as opening the intendedslide presentation tool. The speech will be recorded by the internal microphone on the PC / laptop.

Figure 12 shows part of codes from Matlab program that instructs to go to the slide based on the analysis result from the speech input.

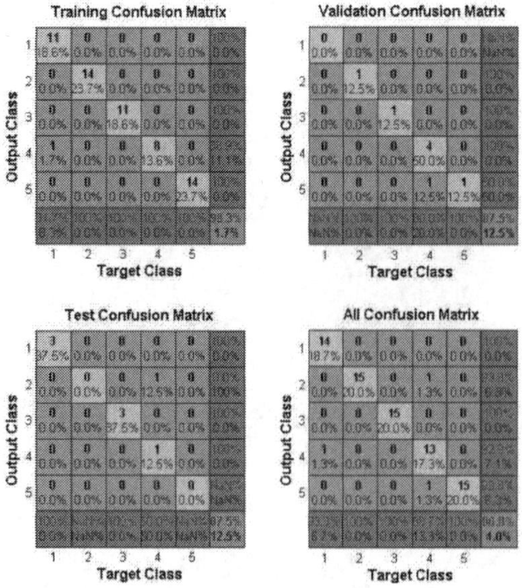


**Figure 10.** Result of confusion.

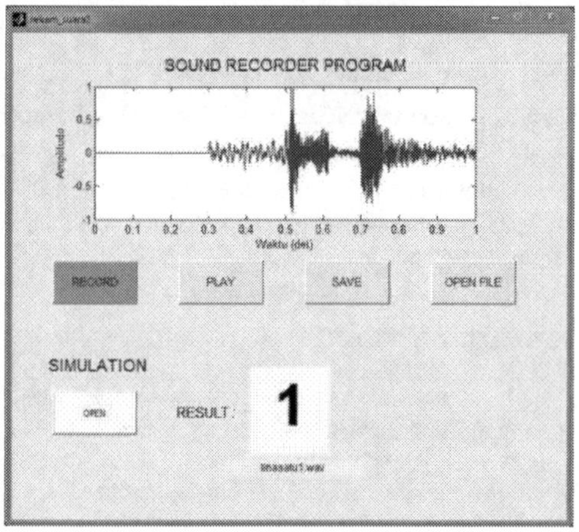


**Figure 11.** GUI of the simulation system.

```
hShow = s.SlideShowWindows.Item(1);

if y(1)==1
set(findobj('Tag','hasil'),'String','1');
set(findobj('Tag','nama'),'String',filename);
hShow.View.GotoSlide(1);

elseif y(2)==1
set(findobj('Tag','hasil'),'String','2');
set(findobj('Tag','nama'),'String',filename);
hShow.View.GotoSlide(2);
```

**Figure 12.** Matlab source codes for slide navigation.

In Figure 13, it can be seen that the program record a new speechsample then saved it as Sample_a2.wav. This speech instruction is to open the slide 2. It can be seen that the program succeeded in opening the slide 2.

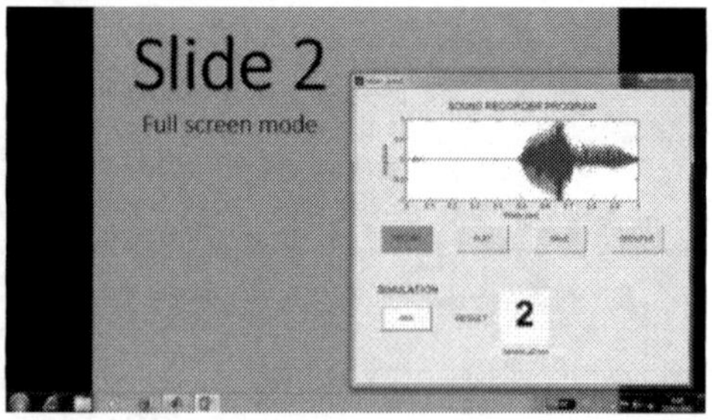


**Figure 13.** Testing new speech sample in opening slide 2.

While in Figure 14, it can be seen that the program record another new speechsample then saved it as Sample_a3.wav. This speech instruction to open the slide 3. This speech instruction is to open the slide 3. It can be seen that the program succeeded in opening the slide 3.

 DOI: 10.1051/matecconf/20167507001

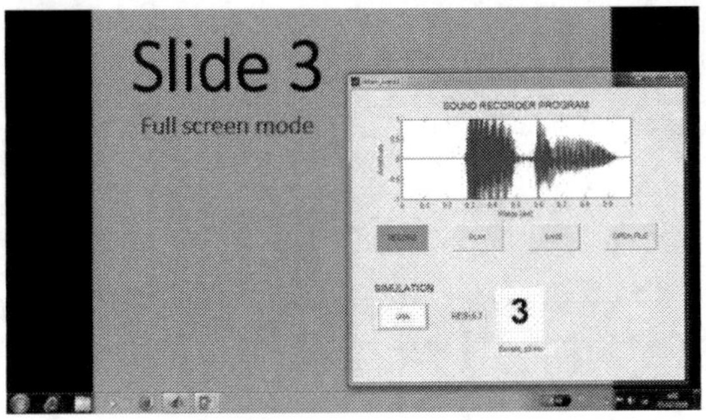


**Figure 14.** Testing new speech sample in opening slide 3.

The new speech samples tested were 100 speeches from different sources. Some results of the testing can be seen in Table 1.

The result showsthat the simulation can recognize63speeches correctly and can demonstrate the proper slide. While 37speeches can not be recognized properly by the system and showed wrong slides.

**Table 1.** Result of new speech samples

| No | File Name | Purposed Slide | Slide Result | Correct |
|---|---|---|---|---|
| 1 | Sample_a1 | One | One | Yes |
| 2 | Sample_a2 | Two | Two | Yes |
| 3 | Sample_a3 | Three | Five | Yes |
| 4 | Sample_a4 | Four | Four | Yes |
| 5 | Sample_a5 | Five | Five | Yes |
| 6 | Sample_b1 | One | Five | No |
| 7 | Sample_b2 | Two | One | No |
| 8 | Sample_b3 | Three | Three | Yes |
| 9 | Sample_b4 | Four | Five | No |
| 10 | Sample_b5 | Five | Five | Yes |

## 5 Conclusion

Based on test results and discussion, it can be concluded that the Backpropagation ANN can be used as a presentation tool navigation through speech with 96% accuracy rate based on the network training result. The simulation can produce 63% accuracy based on 100 new speech samples from various sources.

## References

1. R. Orion F. and R. Carlos Alberto, SPECOM 9th Conference Speech and Computer, A System for the Processing of Infant Cry to Recognize Pathologies in Recently Born Babies with Neural Network (2004)
2. Asian Network for Scientific Information, Classification of Heart Abnormalities Using Artificial Neural Network, 820–821 (2007)
3. W. Suryo. and Thiang, IPCSIT, Speech Recognition Using Linear Predictive Coding and Artificial Neural Network for Controlling Movement of Mobile Robot, **6** (2011)
4. V. Radha, Vimala C. and M. Krishnaveni, CSEIJ,Isolated Word Recognition System for Tamil Spoken Languange Using Backpropagation Neural Network Based on LPCC Features, **1** (2011)
5. P. John G. And M. Dimitris G, Digital Signal Processing Priciples, Algorithms, and Applications Third Edition. Prentice Hall International Edition (1996)
6. F. Lauren, Fundamental of Neural Network. Prentice Hall International Edition (1994)
7. H. Simon, Neural Network A Comprehensive Foundation. Prentice Hall International Edition (1999).
8. N. Michael, Artificial Intelligence: A Guide to Intelligent Systems, Pearson Education (2002)
9. K. I. Vinay and G.P. John, Digital Signal Processing Using Matlab V.4. PWS Publishing Company (1997)

 DOI: 10.1051/matecconf/20167507002

# Performance degradation estimation of robot in highly radioactive environment: Servo Motor Control

Shimeng Li[1], Hongfei Wang[1], Yuan F. Zheng[1], and Lei (Raymond) Cao[2]

[1]*Electrical and Computer Engineering, The Ohio State University, Columbus, Ohio 43210 USA*
[2]*Mechanical Engineering, The Ohio State University, Columbus, Ohio 43210 USA*

**Abstract.** For robot application in highly radioactive environment, the possible radiational damage to the components must be evaluated. Particularly, our work discussed in this paper chooses servo motor as the research object, considering motor control issues which is more realistic in practical robot application. We study the performance degradation of the robot servo motor by introducing the radiation effect on the specific motor constant in the motor model, establish a PID control scheme for motor speed control, and simulate the radiation-induced performance degradation.

## 1 Introduction and Review

Being aware of the high risk for humans working in the highly radioactive environment such as the nuclear power plants of Chernobyl and Fukushima Daiichi) for rescuing, repairing, and maintenance tasks, robots are adopted as good substitution for completing the complex missions in high-risk areas because of the intelligence, mobility and accuracy of the robots [1]. However, this idea has not yet been taken into practice since robot components can be damaged and loses functionality due to high dose radiation in the accident area [2,3]. Research concerns on radiation-hardened robot components and applications have been ever since 2007, when NASA carried on the Radiation Hardened Electronics for Space Environments (RHESE) project [4] for developing radiation hardened electronic components. In 2013, the DARPA grand robotic challenge was aiming at completing particular tasks that humanoid robots can do in the nuclear accident area [5].Eight tasks were established, simulating the situations that a rescuing robot may encounter in a nuclear power plant in accident.

One of the competing robots in the DARPA challenge, Hubo+, is shown in Figure 1. Hubo+ is an adult-sized humanoid robot platform with 38 Degrees of Freedoms (DOF), each of which consists of one servo motor and one harmonic drive as speed reducer. The servo motor is controlled by a joint motor control board, and all the boards receive command from an inside processor which serves as the main controller. The environmental feedback data retrieved from the sensors and cameras are also processed in the main controller. According to previous studies, the electronic components in the robot system, including the controllers, cameras, and sensors, have been proved vulnerable under radioactive environments and studied on how the performance will degrade after irradiation [6-8]. However, the radiation effect on the performance of the mechanical components, like the servo motor and speed reducer has never been well studied before.

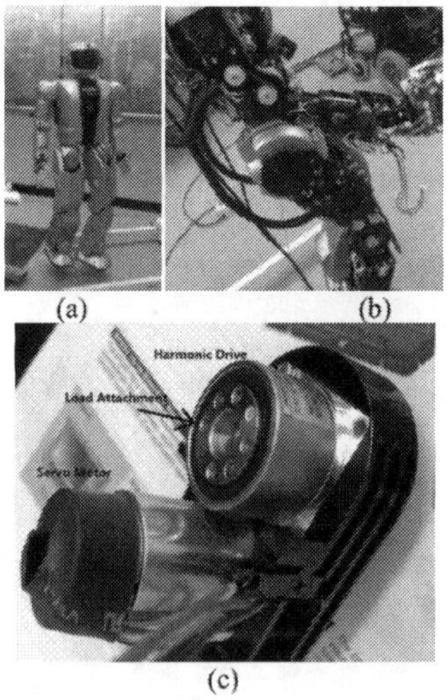

**Figure 1.** Hubo+ Humanoid Platform (a) close view (b) and joint view(c).

Since 2012, we have been working on the research of radiation sensitivity of robot components. Our research focuses on the harmonic drive and brushless DC motor (BLDC). The harmonic drive is an ideal component for robots which allows large speed reduction ratio in compact size and light weight taking advantage of flexible material; the BLDC motor, on the other hand, gets rid of brushes and can be controlled sensorlessly so that there is no brush commutation loss during operation, and noises are significant reduced. Our research on the harmonic drive [9-11] has proved that these effects are very little. As for the BLDC motor, since it is impossible to irradiate the whole component because of potential risks of radioactivation, we irradiated the motor magnet only to see the radiation influence, and simulated the motor performance variation based on an ideal BLDC motor model introducing the influence, obtaining the result that the BLDC will be degraded in power and possibly be out of work caused by the "magnetic trap" jerking effect [12].

From our previous studies, we can draw a conclusion that the motor could be affected to a much more significant extent than the harmonic drive. Therefore, the follow-up work we present in this paper will be primarily based on the radiation induced degradation on the servo motor. We will assume an appropriate working condition for a robot servo, build up a complete servo motor control scheme rather than the simple PWM input in [12], illustrate more detailed modeling work, and analyze the system response simulated in MATLAB.

## 2 System Modeling and Control

### 2.1 System configuration

The robot model we will discuss in this paper is shown in Figure 2. Comparing with the humanoid robot like Hubo+, a wheeled robot is much focused on the motor which drives the wheels. The configuration of the robot is defined in Table 1. Ignoring the motions of the robot arms, the model can be viewed as simple as a wheeled vehicle driven by one servo motor. Our paper will focus on the control of this single DOF.

For the convenience of our modelling work, we choose the U12M4T servo disc motor as our object of simulation. The parameters of the motor is shown in Table 2.

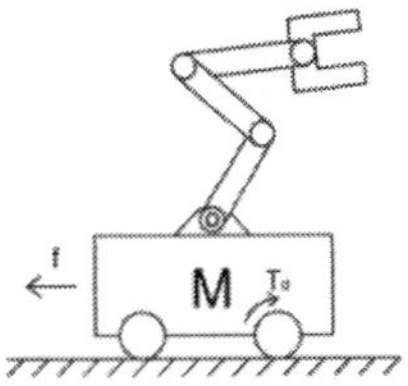


**Figure 2.** A wheeledmobile robot.

**Table 1.**Vehicle robot configuration.

| Parameter | Index | Value | Measurement |
|---|---|---|---|
| Weight | $M$ | 10 | kg |
| Max Speed | $v_{max}$ | 0.5 | m/s |
| Max Acceleration | $a_{max}$ | 0.2 | m/s$^2$ |
| Wheel Radius | $r_w$ | 0.1 | m |

**Table 2.** Servo motor parameters*.

| Parameter | Index | Value | Measurement |
|---|---|---|---|
| Peak Torque | $T_p$ | 852 | N*cm |
| Nominal Power | $P$ | 254 | W |
| Nominal Voltage | $U_n$ | 40.4 | V |
| Nominal current | $I_m$ | 8.71 | A |
| Idle Speed | $n_0$ | 6000 | rpm |
| Rotor Inertia | $I_R$ | 1.84 | kgcm$^2$ |
| Magnet Remainance | $B_r$ | 1.2 | T |
| Terminal resistance | $R$ | 0.75 | Ω |
| Terminal inductance | $L$ | <0.1 | mH |

* The parameters in this table can be found on http://www.kollmorgen.com/en-us/products/motors/brush-dc/servo-disc/_literature/servodisc_catalog_en-us_2003.pdf

### 2.2 Radiation demagnetization modeling

A regular motor dynamic model can be described as

$$\begin{cases} u = (R + \frac{d}{dt}L)i + e \\ \tau_e = ei/\omega \\ \frac{d\omega}{dt} = (\tau_e - \tau_{load} - b\omega_r)/j \end{cases} \quad (1)$$

in which$u$ is the applied input voltage, $i$ the current,$e$ the motor back-emf voltage, $j$ the moment of inertia, $b$ the damping coefficient, $\omega$ the motor angular velocity, $\tau_e$the output torque, and $\tau_{load}$ the equivalent torque load.

The electromagnetic equation of motor back-emfof one coil can be described as a function of motor rotation speed according to the law of induction [12]:

$$e = -\frac{d\Psi}{dt} = \omega \cdot \left(-\frac{d\Psi}{d\theta}\right) = \omega \cdot K_b$$

$$\text{and}\Psi = N_{turns} \sum_{i=1}^{m} \int_{\alpha_i}^{\alpha_i+\alpha_l} B(\theta, \alpha) d\alpha. (2)$$

where $N_{turns}$ is the number of turns in each armature, $m$ the number of armatures in a single phase, $\alpha_i$ the relative starting position of thecoil, and $\alpha_l$ the length of the coil, $B(\theta, \alpha)$ is a scalar function of rotor rotation $\theta$ and stator relative position $\alpha$ representing the B-field intensity on the stator surface.

Normally $d\Psi/d\theta$ is approximately periodic square wave function of $\theta$, and $K_b$ can be considered as a constant, which is called the motor back-emf constant or motor velocity constant. However for motor exposed to the radiation, the stator surface flux will be disturbed. The magnetic flux distribution can be modelled as a periodic function of angular position in the 1-D space [14]. In the electromagnetic model of BLDC motor, only the radial component is considered in the back-emf voltage calculation. The radiation causes a permanent change to the plant model, which is determined by the total absorbed dose by the magnet, the energy of radiation particle, and the distribution of radiation absorption of the magnet. The latter causes the reduction of the B-field intensity, i.e., demagnetization.Among those 3 factors, the first 2 has already been proved to have a considerable positive relationship with the total demagnetization [15]; the third factor, although has been observed in our previous experiment, has too many unpredictable and uncontrollable features, so we assume it to be an "inherent noise component" which is random in our model as [12]:

$$\delta = \frac{\Delta B}{B_0} = \frac{B_0 - B}{B_0} = k\tilde{d}(\theta)$$

$$\text{and } \tilde{d}(\theta) \sim \tilde{N}(\mu, \sigma_a^2) \quad (3)$$

where$k$ is the power coefficient determined by the power of radiation source, $\tilde{d}(\theta)$ is an random B-field intensity reduction ratio following the distribution $\tilde{N}(\mu, \sigma_a^2)$, in which $\mu_e$ is the average reduction ratio, $\sigma_a^2$ the variance of the same parameter.Our research in [16] has shown that the value of $d\Psi/d\theta$ always vary by a greater extent than the magnetic flux density $B$. Therefore, the motor back-emf constant $K_b$, will nolonger be a constant, but a number varying with the motor angular position.

Our study should be based on an obvious fact that the electronic driving circuits (if well protected from the radiation) will not know if the motor is irradiated and the magnet is demagnetized. So it will continue the original operation. As a result, we obtain different results if we give the same open-loop pulse-width-modulation (PWM) input to the "unirradiated motor plant model" and the"irradiated motor plant model". The latter will further show different performance whenthe radiation absorption by the motor is to different extent. However, for a servo motor on a robot, it is certain that we are able to take advantage of the sensory feedbacks, to realize control schemes like torque (current) control, speed control, and position control. For our study discussed in this paper, we still claim that the scheme and parameters of motor controller, and the sensory feedback data do not change with the state of motor plant "irradiated" or "unirradiated".

Particularly in this paper, we do not consider the possible commutation error as illustrated in [12], which is described as "magnetic trap" phenomenon, causing the motor jerk and quit working. This is because a servo motor is not necessarily a brushless one, and unlike the BLDC motor, the brushed DC motor does not have a mechanism of phase shifting control and the brushed commutation mechanism of motor armature currents cannot be influenced by radiation. Therefore, we focus on conditions where "magnetic trap" does not occur, and study how the change of motor velocity constant $K_b$, will affect the motor performance under properly controlled pattern of the radiation.

Normally the value of $K_b$is approximated by the mean value of $-d\Psi/d\theta$in the section of coil activation, as is shown in Figure 3. We take the data within the angular section between the two cursors as a sample. These data demonstrate little variation and can be considered as the constant $K_b$.

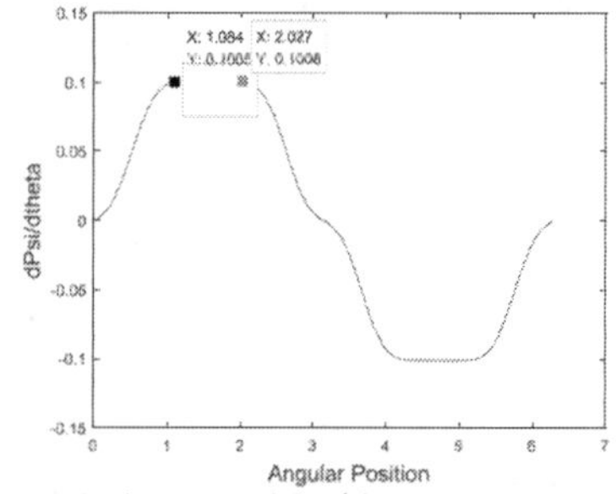

**Figure 3.** A typical pattern of $d\Psi/d\theta$ in one phase of BLDC motor. The constant portion is used for the value of $K_b$.

Then we set $k = 5$, $\mu = 0.03$, $\sigma_a = 0.05$ in (3), and calculate the $d\Psi/d\theta$. The result is shown in Figure 4. We can observe that $d\Psi/d\theta$ within the same angular section is varying in a random pattern around $K_b$in Figure 4 (a), and the probability of variationgenerally goes down while the extent of variation getting larger. It should be noted that the variation probability does not necessarily follow an exact Gaussian distribution, from a bunch of results simulated other than what is shown in Figure 4.

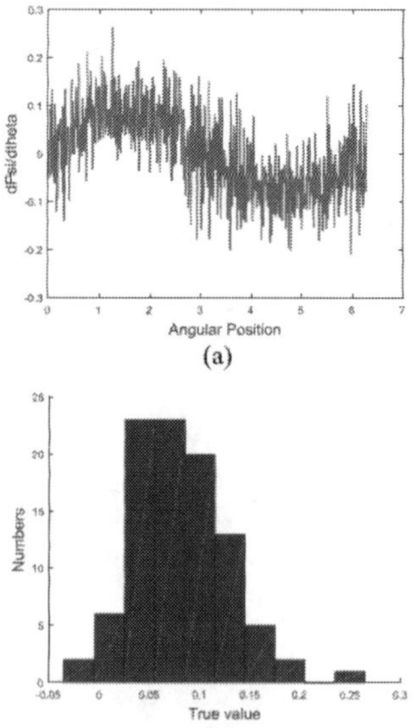

**Figure 4.** Irradiated pattern (a) and histogram (b) of $d\Psi/d\theta$ of the motor ($k = 5$, $\mu = 0.03$, $\sigma_a = 0.05$).

### 2.3 Control design

Generally, PID controller is used for many control problems because of its simple structure and easy implementation. For controlling a robot servo motor, the most usual application is position control and speed control. We also know that normally the motor torque control is achieved by maintaining the armature current knowing that the motor torque coefficient can be viewed as a constant. However motor behaviors will be changed after introducing radiation factors to the motor plant.

The position controller is designed as an outer loop of speed control, with a maximum speed defined; and the speed controller and the current controller are designed on two parallel loops, between which only one of the two can be activated at one time. As we are dealing with only on the wheel of the vehicle robot, we focus on the speed control of the servo motor.

## 3 Simulation and Analysis

The simulation works are performed in MATLAB. The system sampling frequency is set at 200 Hz. Since no practical experiment data is referred in this paper, we ignore the circuit delay, the environmental disturbance, and the sensor error. This allows us to have a more direct and clear view on the radiation impact on the motor servo system. Assume that the constant load torque is $f = 9.8$ N. The reference speed of the vehicle robot is set at 0.3 m/s.

As we have already illustrated in the previous section, although it is very difficult to grasp the pattern of $K_b$ variation after irradiation, it is possible for us to confirm a region of $K_b$ variation where the true value of $K_b$ is very probable to locate in. The distribution of $K_b$ can also follow the histogram in Figure 4 (b) since we are using real motor parameters as reference in MATLAB simulation.

The motor speed PID control functions in the time domain can be described as

$$\begin{cases} v(t) = u(t) * v_{nom} \\ \dot{\omega}(t) = \dfrac{\frac{K_b}{R}v(t) - \tau_{load}/n - (b + \frac{K_b K_\tau}{R})\omega(t)}{J} \end{cases}$$

and

$$u(t) = \begin{cases} k_p(\omega_d - \omega(t)) - k_d\dot{\omega}(t) + k_i\int_0^t \omega(t)dt = D & 0 < D \leq 1 \\ 0 & D \leq 0 \\ 1 & D > 1 \end{cases} \quad (4)$$

in which $v_{nom}$ is the nominal voltage of the motor, $n$ the speed reduction ratio, $\omega(t)$ the motor speed, $K_\tau = K_b$. since we are using SI units for all values, the control signal $u(t)$ is defined as PWM duty cycle varying from 0 to 1. $k_p, k_i,$ and $k_d$ are PID control gains, $\omega_d$ is the desired motor speed.

When the radiation influence has been involved, $K_b$ will no longer be constant but a periodic signal varying with the motor position. So the first part of (4) becomes

$$\begin{cases} v(t) = u(t) * v_{nom} \\ \dot{\omega}(t) = \dfrac{\frac{\tilde{K}_b(\theta)}{R}v(t) - \tau_{load}/n - (b + \frac{\tilde{K}_b^{2}(\theta)}{R})\omega(t)}{J} \end{cases}$$

and

$$\theta = \int_0^t \omega(t)\,dt - 2m\pi$$

$$m = \ldots, -2, -1, 0, 1, 2, \ldots \qquad 0 \leq \theta < 1 \qquad (5)$$

As we have illustrated before, the system itself does not know whether the motor is irradiated or not, so the same control reference are provided to the unirradiated and irradiated motors, respectively. We set reduction ratio $n = 100$, control gains $k_p = 0.0025$, $k_i = 0.0032$, and $k_d = 0.0004$, with the radiation parameters $k = 5$, $\mu = 0.03$, $\sigma = 0.05$, same as in Figure 4. Then we simulate the response of the speed control which is shown in Figure 5.

What we have observed is that: the speed control seems to be functioning normally on the irradiated motor; while the voltage/speed ratio of the motor has increased and the current/torque ratio has decreased, both by a significant scale. This variationis predictable because in Figure 4, we have seen from the histogram that the average $K_b$ has decreased significantly. The noises during operation does not exist in the simulation because the motor is rotating in a relatively high speed due to the existence of motor speed reducer.

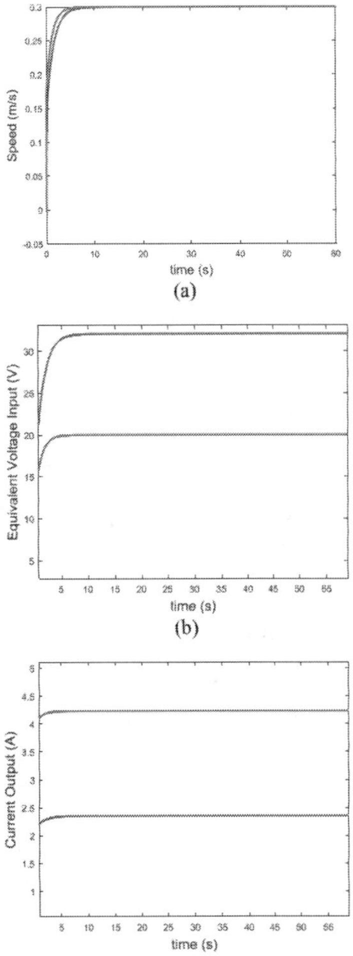


**Figure 5.** Simulation results of unirradiated (blue plot) and irradiated servo motor (red plot) ($k$=5).

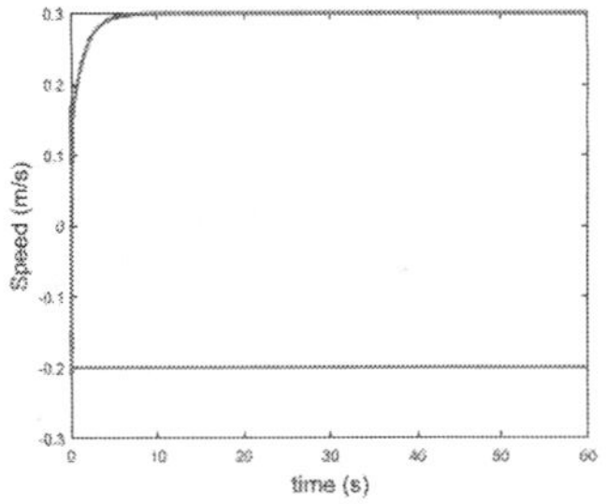


(a)

DOI: 10.1051/matecconf/20167507002

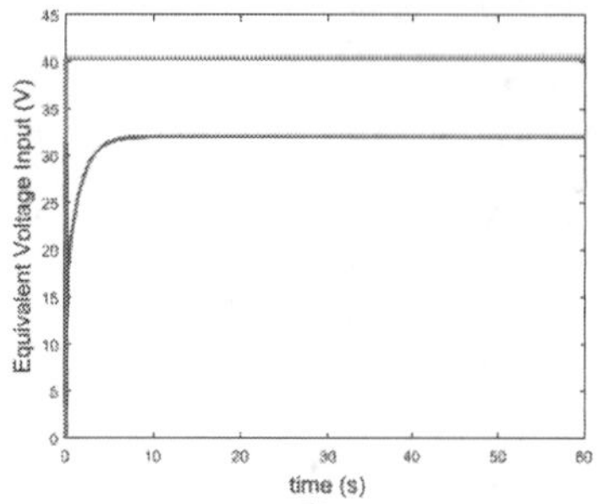

(b)

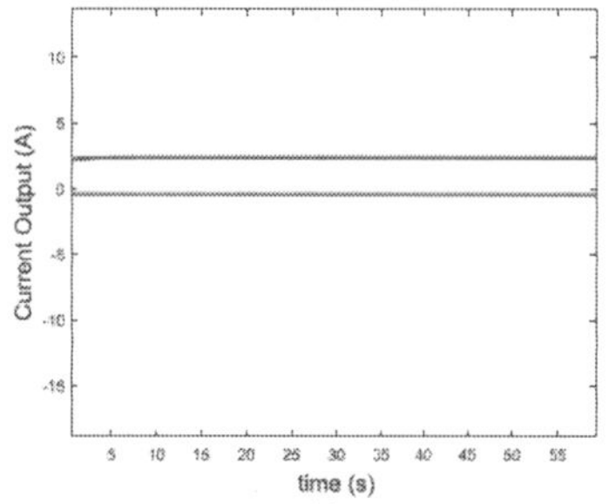

(c)

**Figure 6.** Simulation results of unirradiated (blue plot) and irradiated servo motor (red plot) ($k$=20).

The radiation parameters, $\mu$ and $\sigma$, are parameters related to the average particle energy and spacial distribution, respectively, which are difficult to control in reality. Therefore, we increase $k$, which has a positive relationship with the radiation dose, to see how the motor performance will change when the radiation dose increases. The simulation results are shown in Figure 6. Comparing to Figure 5, the motor speed control has failed, current dropped below zero, and the input voltage remains constantly maximum. And from the previous result, we can conclude that since $K_b$ has decreased in a greater extent, the motor itself cannot generate enough torque to overcome the load torque. In this condition, the motor is back driven by the constant load torque, and we can easily imagine that if a coulomb frictional load, of which the maximum is over the motor capability, is applied, the motor will be stalled with no rotation so there will be more risk of over-current damage to the motor.

## 4 Conclusion

In this paper, we have established a servo motor control scheme for the servo motor in a vehicle robot, and simulated the performance of speed control for the irradiated and unirradiated motor models. We model the radiation influence as a disturbance to the motor back-emf constant, and simulated the system response with PID control. The result shows that with PID motor speed control at high speed will eliminate the noises we have observed in previous research; on the other hand, since $K_b$ has decreased, the voltage/speed ratio of the motor has increased while the current/torque ratio has decreased, thus inducing greater risk that the motor could not generate enough torque, and when the motor is stalled, it is easier to get damaged by over-current.

## Acknowledgement

This work was supported by the U.S. Defense Threat Reduction Agency under Contract HDTRA1-13-1-0012.

## References

1. I. Karl, and S. Dubowsky, The 26th ASME Biennial Mechanisms and Robotics Conference, **2000**. (DETC 2000)
2. J. Wyss, D. Bisello, and D. Pantano, Nuclear Instruments and Methods in Physics Research Section A: Accelerators, Spectrometers, Detectors and Associated Equipment, **462**(3). (2001)
3. P.E. Dodd, and L.W. Massengill, IEEE Transactions on Nuclear Science, **50**(3). (2003)
4. A. Keys, J.H. Adams, and J.D. Cressler, et al., Technology, **4**. (2007)
5. P. Gill, and J. Manzo, Robotics & Automation Magazine, **20**(2). (IEEE 2013)
6. K. Nagatani, S. Kiribayashi, and Y. Okada, et al., International Symposium on Safety, Security, and Rescue Robotics, Kyoto, Japan, (IEEE 2011)
7. Y. Yao, F. Dai, and R.C. Jaeger, et al., IEEE Transactions on Industrial Electronics, **55**(7). (2008)
8. G. Piolain, F. Geffard, and A. Coudray, et al., 1st International Conference on Applied Robotics for the Power Industry, Montreal, Canada. (IEEE 2010)
9. S. Li, Y. F. Zheng , and A. J. Samin, et al., International Conference on Mechanical Design, Manufacture and Automation Engineering, Phuket, Thailand. (2014)
10. S. Li, Y. F. Zheng , and L. Cao, International Conference on Mechatronics and Automation, Tianjin, China. (IEEE 2014)
11. S. Li, Y. F. Zheng, and J. Qiu, et al., International Journal of Mechatronics and Automation (in press).
12. S. Li, Y. F. Zheng, and J. Qiu, et al., International Journal of Mechatronics and Automation (accepted).
13. P. Pillay, K. Ramu, IEEE Transactions on Industry Applications, **25**(2). (1989)
14. P. Kumar, P. Bauer, IEEE Transactions on Magnetics, **44**(10), (2008)
15. J. Alderman, P.K. Job, and R.C. Martin, et al., Nuclear Instruments and Methods in Physics Research Section A: Accelerators, Spectrometers, Detectors and Associated Equipment, **481**(1). (2002)
16. S. Li, Y. F. Zheng, and J. Qiu, et al., International Conference on Electro/Information Technology, Dekalb, IL, USA. (IEEE 2015)

MATEC Web of Conferences **75**, 07003 (2016) DOI: 10.1051/matecconf/20167507003
*ICMIE 2016*

# Robotic vision system for random bin picking with dual-arm robots

Sangseung Kang, Kyekyung Kim, Jaeyeon Lee and Joongbae Kim

*Electronics and Telecommunications Research Institute, 34129, Daejeon, Korea*

**Abstract.** Random bin picking is one of the most challenging industrial robotics applications available. It constitutes a complicated interaction between the vision system, robot, and control system. For a packaging operation requiring a pick-and-place task, the robot system utilized should be able to perform certain functions for recognizing the applicable target object from randomized objects in a bin. In this paper, we introduce a robotic vision system for bin picking using industrial dual-arm robots. The proposed system recognizes the best object from randomized target candidates based on stereo vision, and estimates the position and orientation of the object. It then sends the result to the robot control system. The system was developed for use in the packaging process of cell phone accessories using dual-arm robots.

## 1 Introduction

Automation has been most successfully used in manufacturing systems, and can be an effective solution in part-packaging processes at industrial sites because such processes are simple and repetitive tasks [1]. To achieve such automation, the solution applied must be able to provide reliable detection for the target parts during the packaging process. Further, it is important to operate the automated picking function of randomized parts stored in bins. Although an improvement in the accuracy of bin picking has already been extensively studied, estimating the position and pose of the target objects remains a difficult issue. Therefore, the development of applicable and reliable methods for solving the problems inherent to industrial bin picking is required.

There has been a large body of research on vision systems used for bin picking applications. Such vision systems use 2D or 3D feature information for object detection. Single-view based systems are affected by the environmental conditions of the handled objects, including the lighting reflections and an overlapping of the target objects. Rahardja and Kosaka [2] propose a vision-based bin picking architecture that utilizes simple visual cues as triggers toward recognition and pose estimation of target objects. Rodrigues et. al. [3] build a multi-light imaging system and develop a data-driven method for pose estimation using random ferns to map the patch appearance into pose hypotheses votes. Liu et. al. [4] use a multi-flash camera that extracts robust depth edges and a shape-matching algorithm called fast directional chamfer matching.

Most 3D vision systems exploit 3D object models for image interpretation. Such approaches include the use of 3D range sensors based on line laser scanners, time-of-flight cameras, or structured light-based area scanners. Boughorbel et al. [5] combined video and range images. They generate the 3D model of parts and objects in the bin and find the geometry of the bin contents. Fuchs et. al. [6] approach a bin picking system by combining an impedance controlled lightweight robot and a time-of-flight camera. They used the camera for fast modelling of the dynamic environment and for localizing the bin plus the objects therein. Scharstein and Szeliski [7] propose a method for acquiring high-complexity stereo image pairs with pixel-accurate correspondence information using structured light. Papazov et al. [8] present a 3D object recognition and pose estimation approach for grasping procedures using a Kinect sensor.

Model-based vision systems are expensive and require predefined models of the handling objects. Low-cost sensors such as a Kinect are attractive alternatives to expensive 3D sensors, but do not include reliable object detection owing to a poor quality of the input depth data [9].

In this paper, we present a robotic vision based bin picking system for industrial robotics applications. The proposed system is used in the packaging process for cell phone accessories using dual-arm robots. It recognizes the best picking object from randomized target objects based on stereo vision, and sends the position and orientation information of the object to the robot control system. The robot system conducts a packaging operation using a pick-and-place task.

The developed robotic vision system performs detection and recognition functions of the target object in multiple bins and provides the information for sequential executions of the pick-and-place operations to the robot control system. We set up a vision based random bin

DOI: 10.1051/matecconf/20167507003

picking test environment and performed experiments on the proposed method.

The remainder of this paper is organized as follows. Section 2 presents the method of object detection and recognition used by our robotic vision system for randomized bin picking based on stereo vision. Section 3 describes the selection of the candidate objects to be picked up by the proposed system and the related components. Section 4 presents the experimental results obtained from the position and pose estimation experiments. Finally, some concluding remarks are given in Section 5.

## 2 Object detection

For the task of picking a random object from a bin, the bin picking system comprehends all aspects, from identifying the objects to choosing the applicable candidates. The system determines the optimal approach path to pick the object, and controls the robot movements. It requires vision sensors, operating in conjunction with a computer for processing the sensed data. For a different approach [10], we developed a vision-based bin picking system using a single high-resolution camera. The system estimates the pose and distance of the object using geometrically transformed parameters of the local features. In this paper, we introduce a stereo vision based approach to detect a particular object from a bin and identify its position and orientation. Figure 1 shows our vision system used for bin picking, and randomized target objects. The right side of the figure shows our system along with a ring light device.

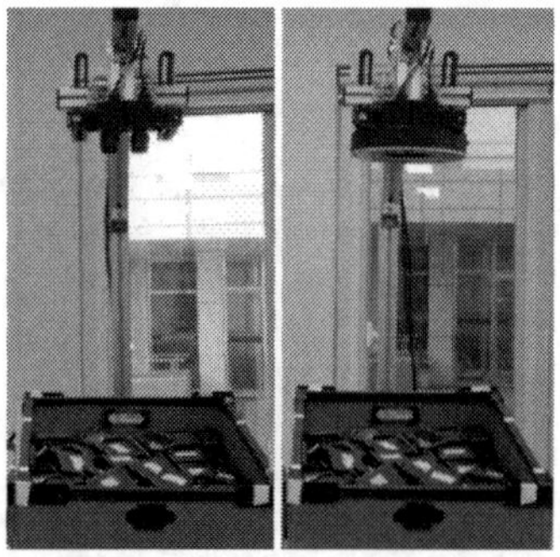

**Figure 1.** Vision system and randomized target objects.

It is important to apply the optimal illumination to acquire high-quality images of the target objects. In the process of applying illumination, it is necessary to consider the influence of the illumination in accordance with the pattern and direction of the light. We estimate the optimal angle formed by the surface of the object and the incident light, and attempt to solve the problem caused by the incidence of shade in a particular lighting direction. We also use a diffuser plate and a polarizing filter along with the lighting itself. Polarization is a general descriptor of light, and contains information on the reflecting objects; in addition, based on its wave property, light can oscillate with more than a single orientation [11]. The application of a polarization filter and a diffuser increases the resolution and contrast for the same light source, and minimizes the occurrence of specular highlights arising from the surface. Sample cases of applying illumination are shown in Figure 2. One is a case of an overexposure of lighting, and the other is an appropriate illumination using a lighting device with a polarizing filter.

(a) Overexposure by lighting

(b) Lighting using polarizing filter

**Figure 2.** Sample case of applying illumination.

Two different images of the target objects are captured using a stereo camera; the amount of noise is then reduced through a pre-processing procedure, and the images are converted into a binary form using different threshold values. The system detects the contours of the objects from the filtered images after applying morphological filters [12], and selects the candidate objects through an analysis of the contour size. A binary mask is generated from the detected contours, and SIFT feature points [13] are extracted. The feature points are extracted from limited local areas using the binary mask described above, and around less than 100 feature points are extracted within the each local area. This has the advantage of improving both the accuracy and speed.

After the feature point detection for each view is conducted, a matching function is applied by comparing the extracted feature points from both the left and right camera images [14], and the feature matching results that are deemed unsuitable based on the predefined conditions are removed. For example, the difference in scale of the feature points must be less than a single step, and the matching points must maintain a one-to-one relationship. If two corresponding points are obtained between the left and right stereo images, the system can calculate the three-dimensional spatial point using the parameter information from the stereo cameras such as optical centre, focal length, and baseline between the cameras. Let $p_l=[x_l,y_l]^T$ be the left matching points, and $p_r=[x_r,y_r]^T$ be the right matching points, the three-dimensional spatial point $P=[x_s,y_s,z_s]^T$ can then be calculated as follows.

$$x_s=(b(x_r-x_c))/d,\ y_s=(b(y_r-y_c))/d,\ z_s=fb/d \quad (1)$$

The disparity is represented as $d=x_l-x_r$, where $[x_c,y_c]^T$ is the primary optical centre, $f$ is the focal length of a camera, $b$ is the baseline or distance between two cameras. $X$ is the X-axis and $Z$ is the optical axis of a camera. Both cameras must have the same focal length. Figure 3 shows the three-dimensional reconstruction geometry based on stereo triangulation.

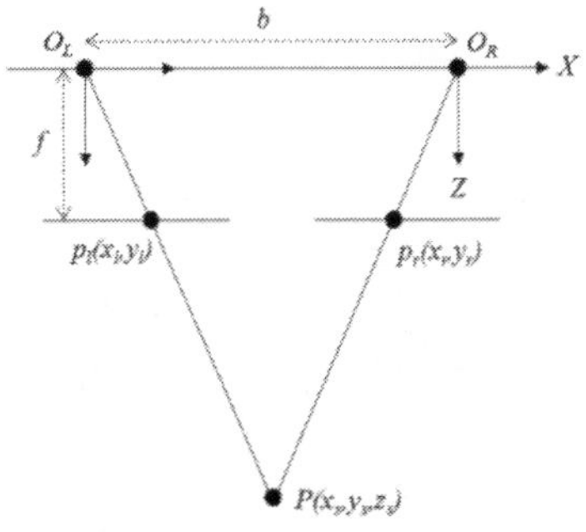


**Figure 3.** Three-dimensional reconstruction geometry.

The equation of a plane is determined using the reconstructed three-dimensional spatial points. Through the calculation of the centroid from the four corner points, the system estimates the position of the target object. It also estimates the orientation of the object based on a calculation of the normal vector of the plane. Let $ax+by+cz+d=0$ be the equation of a three-dimensional plane, and $P=[a,b,c,d]^T$ be the plane vector, the plane equation can be rewritten as $[x\ y\ z\ 1][a\ b\ c\ d]^T=0$. If there are $n$ three-dimensional points, the linear system can be expressed through an extension to an $n$-dimensional matrix. The plane vector can then be calculated using singular value decomposition.

The system estimates the pose of the object. Let the four corners of the detected rectangular area be $p_1$ through $p_4$, vectors $e_1$ through $e_4$ can then be calculated using the intrinsic camera parameter. The centroid of the detected rectangular area is used as the reference picking point for the robot system. For the given plane, the crossing point can be calculated using the plane equation, the normal vector of the plane, and the normal vector of $e_1$ through $e_4$. The centroid can be calculated by dividing the sum of the four crossing points.

The orientation of the object is determined based on the x-, y-, and z-axis vectors of the coordinate system at a corner point of the detected rectangular area.

## 3 Picking

In general, good objects for picking may be considered to be at higher positions. Therefore, the system will choose the best object among the picking candidate objects based on the distance from the camera. To reduce any errors, the candidates are also prioritized based on additional comparisons of the scale, orientation and aspect ratio of the targets. Let $D_n$ be the normalized distance, $O_n$ be the normalized orientation, $S_n$ be the normalized scale factor of the detected object, the probable cost is given by

$$p(\omega_i) = D_n \cdot \delta + O_n \cdot \rho + S_n \cdot \lambda \qquad (2)$$

where $\delta$ is the distance weight factor, $\rho$ is the orientation weight factor, and $\lambda$ is the scale weight factor.

The weight factors are appropriate values determined through a large number of experiments. For all detected objects, the best object to pick up was selected as the object with the lowest cost.

$$\omega_N = \arg\min(p(\omega_i)) \qquad (3)$$

An end effector is a device for picking up objects through robotic manipulation, and serves various purposes. It is attached to the end of a robotic arm and is a part of an industrial robot for interaction with the work environment. The selection of an end effector depends on the application of the industrial robot. Robot grippers are widely used, and support grasping functions between the gripper and the object to be grasped [15]. In general, a gripper can be mechanical, but suction-based grippers are also available. Our dual-arm robot system uses vacuum suction-type end effectors. Therefore, it is important to accurately estimate the position of the centroid of the object. This position is used as the picking point for the robot system. The normal vector of the detected rectangular area can be calculated using the plane equation. The direction of the normal vector is used as the picking orientation for the robot system. The sample results of position and pose estimation for robotic picking are shown in Figure 4.

**Figure 4.** Pose estimation and picking direction.

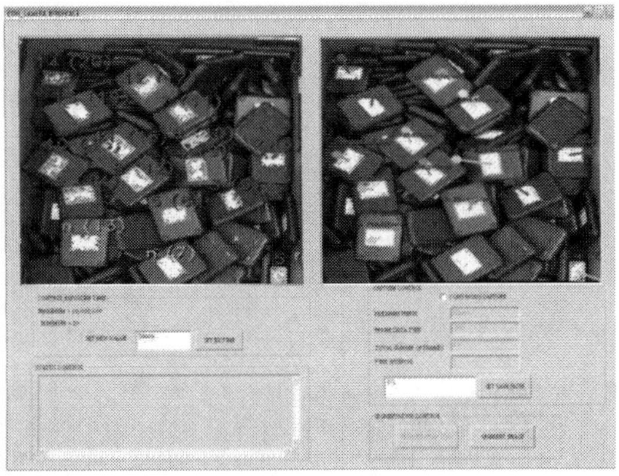

**Figure 5.** Robotic vision client system.

The vision system converts the result of the vision coordinate system into the value of the robot coordinate

DOI: 10.1051/matecconf/20167507003

system, and sends the result information to the dual-arm robot control system. The robot system conducts the picking operation based on the position and orientation information of the target object. The implemented robotic vision client system is shown in Figure 5. The left side of the image viewer area shows a stereo matching and classification result, and the right side shows a measurement result with a camera coordinate system.

## 4 Experiments

We set up a random bin picking test environment using a robotic vision system, and conducted experiments on the proposed method. In the experimental study, charging cradles and travel adapters, which are different cell phone accessories, were used for the objects. The objects were randomly contained in a bin, and were placed in such a way that they partially occluded each other. The objects were located approximately 800 mm away from the camera.

We applied the object detection and position estimation functions, and compared the measured distance and actual distance values. In the distance measurements taken from the input images, the offset value is applied to correct the actual distance. Samples of the experimental results for the charging cradles and travel adapters are shown in Table 1.

**Table 1.** Experimental results.

| Distance | Measure | Error | Distance | Measure | Error |
|---|---|---|---|---|---|
| 807.2 | 807.1 | 0.1 | 815.8 | 814.5 | 1.3 |
| 828.4 | 828.8 | 0.4 | 829.2 | 827.7 | 1.5 |
| 837.2 | 836.6 | 0.6 | 811.6 | 813.4 | 1.8 |
| 829.6 | 828.3 | 1.3 | 826.7 | 827.5 | 0.8 |
| 810.4 | 811.2 | 0.8 | 818.7 | 820.3 | 1.6 |
| 827.1 | 824.9 | 2.2 | 828.5 | 827.6 | 0.9 |
| 812.2 | 812.1 | 0.1 | 830.3 | 830.8 | 0.5 |
| 810.8 | 809.8 | 1.0 | 827.3 | 826.2 | 1.1 |
| Avg. Case 1 | | 0.81 | Avg. Case 2 | | 1.19 |

**Table 2.** Distance accuracy.

| Test Case | Avg.Err.1 | Std.Dev.1 | Avg.Err.2 | Std.Dev.2 |
|---|---|---|---|---|
| 1 | 0.81 | 0.70 | 1.19 | 0.44 |
| 2 | 0.76 | 0.59 | 1.02 | 0.83 |
| 3 | 0.67 | 0.33 | 0.75 | 0.61 |
| 4 | 0.93 | 0.62 | 1.26 | 0.64 |
| 5 | 1.09 | 0.43 | 1.16 | 0.97 |
| Avg. | 0.85 | 0.53 | 1.08 | 0.70 |

In the experiment on the charging cradles, the mean distance error was 0.81, and the standard deviation was 0.7, for eight detected objects with priority. The mean distance error for the travel adapters was 1.19, and the standard deviation was 0.4. The distance accuracy for the travel adapters was relatively low because they are smaller in size and wrapped in plastic. We repeated the experiment on the charging cradles and travel adapters five times. The results of this experiment are shown in Table 2.

As demonstrated through the experiment results above, the average error rate in the distance to the target object was around 1 mm. The reference distance was obtained using a laser distance measurer, which has a measurement accuracy of 1 mm. Therefore, even when assuming that an error of up to 2 mm may occur, there was no problem for the robot to pick up the object in our bin picking environment. The developed bin picking system with a dual-arm robot is shown in Figure 6.

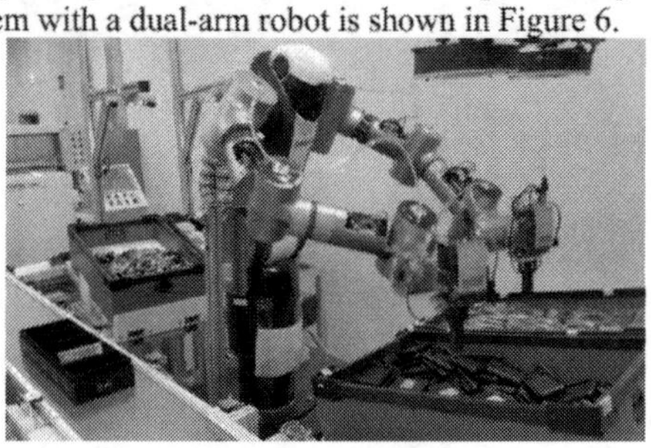

**Figure 6.** Random bin picking.

## 5 Conclusion

One of the most difficult problems in industrial robot applications is bin picking. Although many different solutions to the problems inherent to bin picking have been proposed, a general solution remains unavailable. In this paper, we presented a robotic vision system for randomized bin picking based on stereo vision. The proposed system detects candidate objects from randomized objects, estimates the position and orientation of the target candidates, and determines the best object to pick up for a pick-and-place task of a robot. We developed a pilot system for use in the packaging process of cell phone accessories using dual-arm robots. More research is needed to apply practical solutions to many different real-world situations.

## Acknowledgement

This work was supported by the R&D program of MOTIE & KEIT [10038660, Development of dual-arm robot system based on multi-robot cooperation for cell manufacturing process of IT products]

## References

1. C. Harrison, "A review of automation in manufacturing illustrated by a case study on mixed-mode hot forging," *Manufacturing Review* (2014)
2. K. Rahardja and A. Kosaka, "Vision-Based Bin-Picking: Recognition and Localization of Multiple Complex Objects Using Simple Visual Cues," *Proc. of the Intl' Conf. on Intelligent Robots and Systems*, 1448-1457 (1996)
3. J. Rodrigues, J. Kim, M. Furukawa, J. Xavier, P. Aguiar, T. Kanade, "6D pose estimation of textureless shiny objects using random ferns for bin-

picking," *Proc. of the Intl' Conf. on Intelligent Robots and Systems*, 3334-3341 (2012)
4. M. Liu, O. Tuzel, A. Veeraraghavan, Y. Taguchi, T. Marks, R. Chellappa, "Fast Object Localization and Pose Estimation in Heavy Clutter for Robotic Bin Picking," *The International Journal of Robotics Research*, **31**(8): 951-973 (2012)
5. F. Boughorbel, Y. Zhang, S. Kang, U. Chidambaram, B. Abidi, and A. Koschan, "Laser ranging and video imaging for bin picking," *Assembly Automation*, **23**(1): 53–59 (2003)
6. S. Fuchs, S. Haddadin, M. Keller, S. Parusel, A. Kolb and M. Suppa, "Cooperative Bin-Picking with Time-of-Flight Camera and Impedance Controlled DLR Lightweight Robot III," *Proc. of the Intl' Conf. on Intelligent Robots and Systems*, 4862-4867 (2010)
7. D. Scharstein and R. Szeliski, "High-Accuracy Stereo Depth Maps Using Structured Light," *Proc. of the IEEE Computer Society Conf. on Computer Vision and Pattern Recognition* (2003)
8. C. Papazov, S. Haddadin, S. Parusel, K. Krieger, D. Burschka, "Rigid 3D Geometry Matching for Grasping of Known Objects in Cluttered Scenes," *International Journal of Robotics Research*, **31**(4): 538-553 (2012)
9. S. Park and S. Lim, "Template-Based Reconstruction of Surface Mesh Animation from Point Cloud Animation," *ETRI Journal*, **36**(6): 1008-1015 (2014)
10. K. Kim, S. Kang, J. Lee, J. Kim, "Vision Based Bin Picking for Industrial Robot," *Advanced Science and Technology Letters*, **64**: 14-17 (2014)
11. A. G. Andreou and Z. K. Kalayjian, "Polarization Imaging: Principles and Integrated Polarimeters," IEEE Sensors Journal, **2**(6):566-576 (2002)
12. P. Maragos, "Morpohological Filtering for Image Enhancement and Feature Detection," *Image & Video Processing Handbook (2nd ed.)* (2004)
13. D. Lowe, "Object recognition from local scale-invariant features," *Proc. of the Intl' Conf. on Computer Vision*, 1150-1157 (1999)
14. D. Scharstein and R. Szeliski, "A Taxonomy and Evaluation of Dense Two-Frame Stereo Correspondence Algorithms," *International Journal of Computer Vision*, **47**(1): 7-42 (2002)
15. G. Fantoni, M. Santochi, G. Dini, K. Tracht, B. Scholz-Reiter, J. Fleischer, T.K. Lien, G. Seliger, G. Reinhart, J. Franke, H.N. Hansen, A. Verl, "Grasping devices and methods in automated production processes," CIRP Annals - Manufacturing Technology, **63**(2): 679-701 (2014)

# Research and Implementation of Automatic Fuzzy Garage Parking System Based on FPGA

Kaiyu Wang[1], Zongmin Yu[1], Sanghai Guan[1], Xing Yang[1], Menglin Sheng[1] and Zhenan Tang[1,a]

[1] *Dalian University of Technology, Dalian, Liaoning, 116024, P. R. China*

**Abstract.** Because of many common scenes of reverse parking in real life, this paper presents a fuzzy controller which accommodates front and back adjustment of vehicle's body attitude, and based on chaotic-genetic arithmetic to optimize the membership function of this controller, and get a vertical parking fuzzy controller whose simulation result is good .The paper makes the hardware-software embedded design for system based on Field-Programmable Gate Array (FPGA), and set up a 1:10 verification platform of smart car to verify the fuzzy garage parking system with real car. Verification results show that, the system can complete the parking task very well.

## 1 Introduction

Automatic parking means that after reaching the parking area, according to the information returned by the external sensor and using intelligent algorithm, vehicle-mounted controller guides the car go into the parking space automatically. The development of automatic parking technology can save the driver from the complex and experienced operation process effectively and improve the safety of the parking process greatly.

To improve the stability and intelligence of the automatic parking system, the researchers have proposed a variety of control strategies in recent years. Literature [1-5] proposed the path tracking method to guide parking, it means that the parking route is composed of the defined multistage continuous curvature curve (such as sine curve, five polynomial curve, etc.), and control the cars to drive along the reference path based on precision controller. Literature [6,7,8] use fuzzy controller to solve the problems of control decision of vertical and parallel parking process. Based on fuzzy parking system, literature [9] proposed the attitude adjustment in parking space, making the car move along the center line after drive into the parking space. Literature[10] design three fuzzy controllers which consist of attitude adjustment, steering control and parking decision, and in addition to the on-board controller it design the ground control station, achieving the auxiliary control function in large parking lot. At the same time, some studies combine the fuzzy control and other algorithms to improve the control ability, such as neural networks [11] and genetic algorithms [12].

This paper aims at the common scene of reverse parking in real life to design a fuzzy garage parking system (FGPS) which uses the fuzzy algorithm as the core strategy. The system includes the decision system which is made up of controller, the perceptual system which is made up of sensors and executive system which is made up of steering engine. Based on the coordinate of these three systems, FGPS can complete the parking action. At the same time, FPGA realize the hardware and software embedded design for FGPS, and the paper design a smart car platform to verify the system by real car.

The paper is organized as follows. The section 2 establish the vehicle kinematic equations and design the fuzzy controller of FGPS, and optimize the fuzzy controller based on chaotic-genetic arithmetic, and test the effect of fuzzy controller by MATLAB .The section 3 establish the smart car test platform. The section 4 uses FPGA to make the hardware-software embedded design for FGPS. The section 5 makes the physical verification of smart car. And the section 6 makes a conclusion for the whole passage.

## 2 The design of fuzzy controller

### 2.1 Vehicle kinematic equation

The paper establishes the coordinate system of parking space using the left lower corner of parking space as the original point, and determine the coverage area of the car by the coordinates of four corners of the car. The vehicle adopt the model of front-wheel steering and rear-wheel drive, it means the rear two wheels parallel with the body all the time and the front two wheels can turn left or right. Use the coordinate of the central point of the back shaft (xr,yr) on behalf of the position of the car, and $\phi$ as the

[a] Corresponding author: wkaiyu@dlut.edu.cn

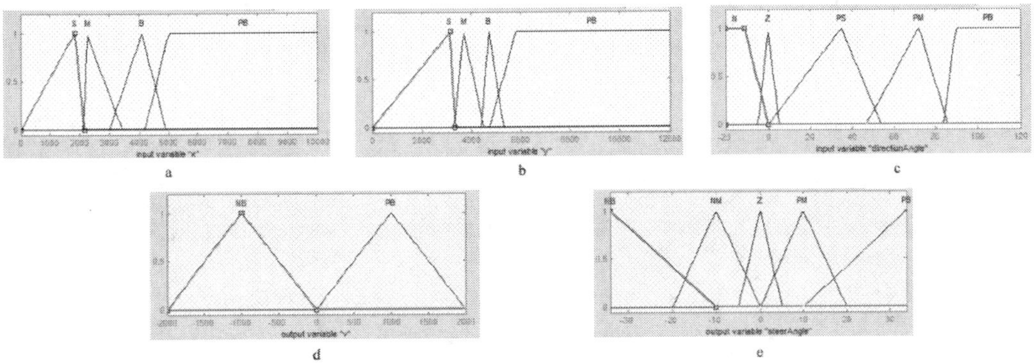


**Figure 2.** a.the membership function of the input variable x, b.the membership function of the input variable y, c.the membership function of the input variable θ , d.the membership function of the output variable v , e.the membership function of the output variable $\phi$

front-wheel steering angle, that means the included angle between the front wheel corner and main axis, and clockwise is positive. Use θ as the body direction angle, that means the included angle between the main axis of vehicle and the X axis of the parking space, and anticlockwise is positive. And use L as the wheelbase. The speed driving backwards is positive. Establish vehicle kinematic model as Fig. 1 and vehicle kinematic equations as Eq. 1.

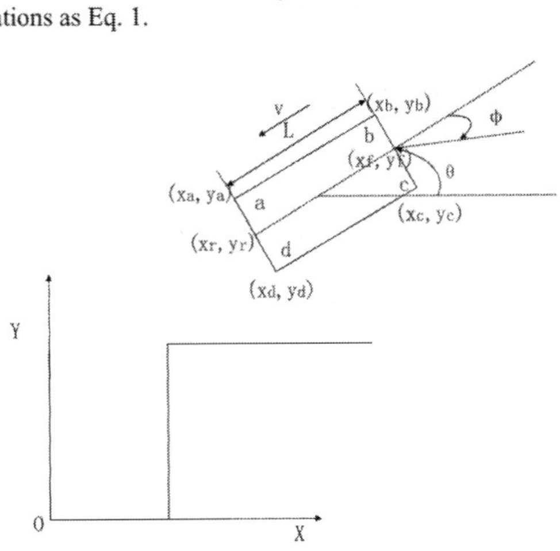


**Figure 1.** Vehicle kinematic model.

$$\begin{cases} \dot{x}_r = -v\cos\theta\cos\phi \\ \dot{y}_r = -v\sin\theta\cos\phi \\ \dot{\theta} = \dfrac{v\sin\phi}{\mathrm{L}} \end{cases} \quad (1)$$

From the equation we know that, we can control the next-time position coordinate and direction angle of the vehicle by controlling the speed v and angle $\phi$.In other words, it controls the vehicle's next-time position and pose.

### 2.2 Design the fuzzy controller

The fuzzy controller has three inputs and two outputs, the inputs are the coordinate of the central point of the back shaft ($x_r$,$y_r$) and the body direction angle θ .The outputs are steering engine corner $\phi$ and the speed of vehicle v .

Using the left lower corner of parking space as the original point and thinking about the area of the parking space and the width of the carriageway, the value range of (x,y) is set in [0,10000mm] and [0,12000mm], the domain of discourse has four levels, they are {S,M,B,PB}; the value range of the body direction angle θ is set in [-20°,120°], the domain of discourse has five levels, they are {N,Z,PS,PM,PB}; the value range of steering engine corner $\phi$ is set in [-34°,34°] ,the domain of discourse has five levels, they are {NB,NM,Z,PM,PB}; the value range of speed v is set in [-2000mm/s, 2000mm/s], the domain of discourse has two levels, they are {NB,PB}. Adopting the combination of trigonometric function and trapezoid function in each variable's value range, the system uses chaotic-genetic arithmetic to optimize the parameter of the membership function and get the shape of the membership function as Fig. 2.

The design of fuzzy inference rules refers to experienced drivers' operating experience,and establishes the form of fuzzy inference rules of fuzzy controller as Table 1 shown.

### 2.3 Simulation and verification

To verify the results of fuzzy control, the paper establishes the simulation model in MATLAB/Simulink. This model consists of four parts; they are the fuzzy controller module which makes the decision, the vehicle kinematics module which control the vehicle move, the constraint module which make the judgment of collision and stop the simulation, and the animation module which output the simulation results. To simplify the simulation, the car is set in the starting parking space. In order to obtain a complete test results, in a possible initial parking area, three points (5619,8022,2.83°), (8012, 8969,8.42°), (3566,6989,-2.5°) which are in the center, on top and on the downside, and have the different initial direction angles are set to test. The testing track figure is shown in Fig. 3.

DOI: 10.1051/matecconf/20167507004

**Table 1.** The fuzzy inference rules after optimized.

| $\phi/v$   $y$ / $x$ | S | M | B | PB |
|---|---|---|---|---|
| S (θ=N) | / | NB/-V | NM/-V | NM/-V |
| M (θ=N) | / | NB/-V | NM/-V | NM/-V |
| B (θ=N) | / | NM/-V | PB/V | PB/V |
| PB (θ=N) | / | NM/-V | PB/V | PM/V |
| S (θ=Z) | / | NB/-V | NM/-V | NM/-V |
| M (θ=Z) | / | NB/-V | NM/-V | NM/-V |
| B (θ=Z) | / | NM/-V | PB/V | PB/V |
| PB (θ=Z) | / | NM/-V | PM/V | PM/V |
| S (θ=PS) | Z/-V | Z/-V | Z/-V | NB/-V |
| M (θ=PS) | PM/-V | Z/V | PB/V | PB/V |
| B (θ=PS) | / | Z/-V | PB/V | PB/V |
| PB (θ=PS) | / | Z/-V | PM/V | Z/V |
| S (θ=PM) | PB/V | NB/-V | PB/V | PB/V |
| M (θ=PM) | PM/V | Z/-V | PM/V | PM/V |
| B (θ=PM) | / | Z/-V | PM/V | Z/V |
| PB (θ=PM) | / | Z/-V | NM/V | NB/V |
| S (θ=PB) | Z/V | PM/V | PM/V | Z/V |
| M (θ=PB) | NB/V | NM/V | NM/V | Z/V |
| B (θ=PB) | / | Z/V | Z/V | NM/V |
| PB (θ=PB) | / | PM/-V | NM/V | NM/V |

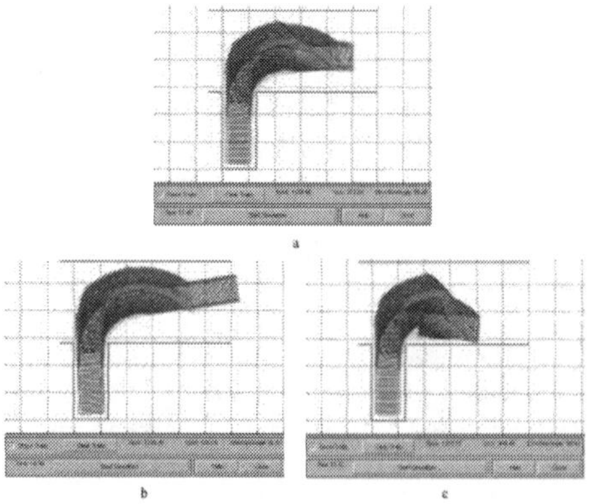

**Figure 3.** Simulation track figure.

Fig. 3 is the track graphs of first, second and third test point. As can be seen from the Fig. 3, the final stop position of the vehicle is almost in an ideal location, the stop pose compared with the condition of 90° verticality has only a least difference of 0.05°, and largest difference is just 1.52°. No significant body tilt and complicated adjustment process express that the design of the fuzzy controller is reasonable and the process of control is effective.

## 3 Design of the smart car verification platform

### 3.1 Global design of hardware platform

In order to verify the design of FGPS by virtual cars, a smart car verification platform whose dimension scale is 1:10 is established, and the global hardware structure of this platform as follows: the platform of smart car ,the Nexys3 board with Xilinx Spartan-6 LX16 FPGA, logic level conversion module, ultrasonic ranging module, speed measurement module with photoelectric encoder, voltage-stabilized power supply module, motor & steering engine drive module and motor and steering engine.

In this hardware system, FPGA is the core processor, ultrasonic ranging module and speed measurement module with photoelectric encoder are sensory perceptual system, motor & steering engine drive module, motor and steering engine are execution system. The sensory perceptual system detects external environment and use logic level conversion module to translate the environment information to 3.3V logic level which can be recognized by Nexys3. This information enter FPGA via Pmods port of Nexys3, and FPGA make decisions based on arithmetic. Then the decision information is transmitted to sensory perceptual system or execution system by Pmods port and logic level conversion module in order to complete the adjustment and detection of model car. The whole parking system is provided a 5V/3.3V voltage by the power supply module.

### 3.2 The distribution of ranging sensor

The paper uses HC-SR04 ultrasonic module as the ranging sensor. In the process of parking, to get car's position and information of body pose real time, each ultrasonic module is needed to collaborate with others. The paper thinks about that the car maybe run on the left

DOI: 10.1051/matecconf/20167507004

or right, so put the ultrasonic modules on car as Fig. 4 shown.18 ultrasonic modules are divided into 6 groups and installed on car body. Because of the detection angle of HC-SR04 is 30°, so each group has 3 sensor and each sensor's detection angle is 30°.In this way, it is not only no mutual interference, but also ensure that the detection range is large enough in parking process.G1,G2,S8,S9 and S11, S12 are used when the car run on the right, and others are used when it runs on the left.

Supposing the parking space is on driving direction's left, when the car begins to find the parking space, it uses G1 and G2 to detect the distance between the obstacle on the right side and itself. The value of each group chooses the least among three sensors'. When start to reverse parking, because of the parking space is confirmed, so the environment map and origin of coordinates is established, and the system only need to pay attention to the position and pose of car. Supposing the value of G1 and G2 is L1 and L2, so the body can be located by Eqs.(2-4). In Eq.3 , $L_{rw}$ is rear track.

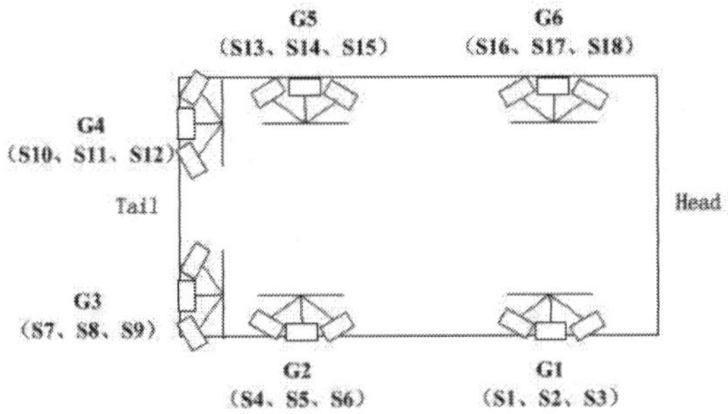


**Figure 4**. The distribution of ultrasonic sensors.

$$\theta = \arcsin\left(\frac{L_1 - L_2}{L}\right) \tag{2}$$

$$y = L_2 + \frac{L_{rw}}{2} \cdot \cos\theta \tag{3}$$

$$\Delta s = v \cdot \Delta t = \Delta x^2 + \Delta y^2 \tag{4}$$

When the coordinate of car goes into the parking space, S8&S9 and S11&S12 are used to detect the distance between car and the bottom of parking space. By this time, the value of S8&S9 and S11&S12 get the larger one, and the positioning method is similar to the Eqs.(2-4).

### 3.3 The other key module

In order to make sure each module get the right voltage, 5V/3.3V logic level conversion module with 74LVC4245A chip is selected. And voltage-stabilized power supply which has a AMS117 chip in it is used to supply 3.3V.The actuator of automatic parking system is composed of 380 motor, MG995 steering engine and other drive modules.

## 4 The embedded design of hardware and software

The system is designed by EDK. Firstly, processor hardware platform which has a 32 bits CPU named MicroBlaze is established by XPS development environment. Then according to the process of parking, the software design is complete by SDK software development environment with C language.

About software design, Fig. 5 and Fig. 6 are the flow charts of the space search process and starting parking process. Based on these flow charts, software system is established by SDK.

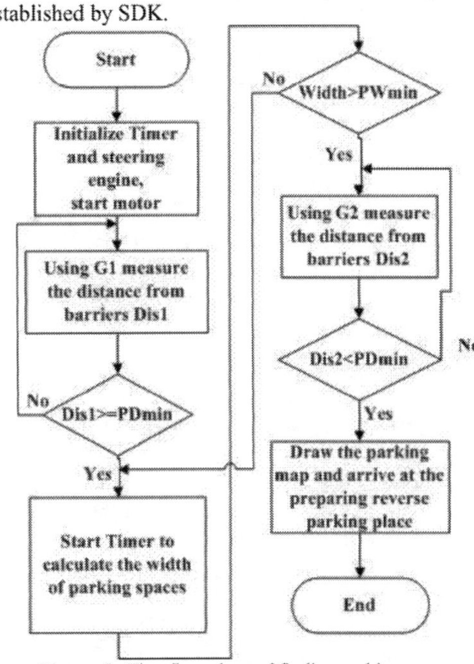


**Figure 5** . The flow chart of finding parking space.

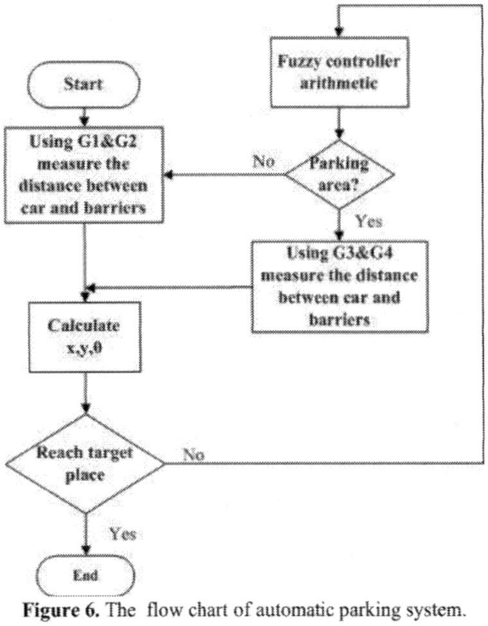


**Figure 6.** The flow chart of automatic parking system.

DOI: 10.1051/matecconf/20167507004

In Fig. 5, PDmin is the minimum depth of the garage and PWmin is the minimum width of the garage. The system uses G1 to find parking space, and get the width of it by the product of the speed and time.Automatic parking system works as Fig.6 shown, using G1&G2 to find parking space, and G3&G4 to determine whether the car goes into the parking space completely.

## 5 Physical verification of smart car

In this section, the parking system has been verified by smart car in a simulative garage. As shown in Fig. 7, 6 pictures in the first row from left to right show the process that the smart finds the garage. At first, the car run slowly along the carriageway and keeps a safe distance from the barrier. When it discovers the garage, keep run until arriving the site which the car must stop and get ready to reverse parking. Then the other picture in the second row from left to right show the whole process of Parking. Finally, the car enter the parking space and the distance between car and both sides of the barriers is suitable, the direction angle is almost 90°. The test shows that the parking system of this paper works in a good state, and the fuzzy control algorithm can complete the parking task perfectly. But in this environment, all the lines of barriers and parking space are totally parallel or perpendicular, so if the parking environment is more complex, the algorithm mentioned in the paper may need to have certain adjustment to adjust it.

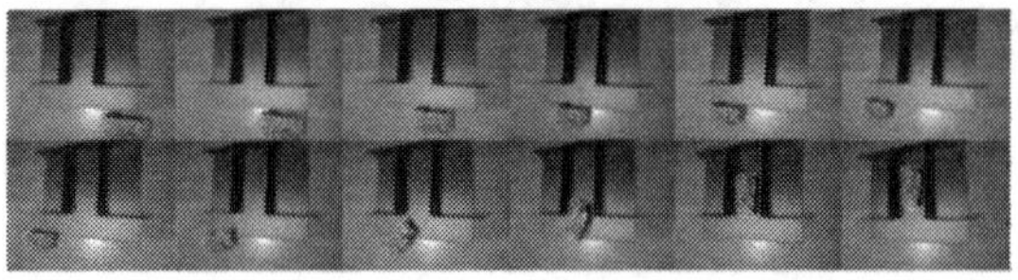

**Figure 7**. The picture of parking test by smart car.

## 6 Conclusion

The paper designs a automatic vertical parking arithmetic, consist of finding parking space automatically and self-adapting control reverse parking, and realize the arithmetic in hardware and software platform by FPGA, and establish the smart car test platform which consist of decision module, sensing module and execution module, and complete the test of this parking system. After debugging, the smart car can complete the parking task safely and efficiently. The system can adjust the body posture automatically when the car meets the barriers; it has good intelligence and security. Because of the design is based on FPGA, so it has the advantage of reconfigurable and system upgrade easily. So the paper has certain practical significance.

## Acknowledgement

The authors are grateful for the support given by the Fundamental Research Funds for the Central Universities (Grant No. DUT15QY28). We also thank reviewers for insightful and helpful suggestions.

## References

1. E. Szadeczky-Kardoss, B. Kiss, Path planning and tracking control for an automatic parking assist system, Springer Tracts in Advanced Robotics, 2008, 44: 175-184.
2. J. Xu, G. Chen, M. Xie, Vision-guided automatic parking for smart car, IEEE Intelligent Vehicles Symposium, Proceedings, Dearbon, 2000: 725-730.
3. T. Hsu, J. Liu, P. Yu, et al., Development of an automatic parking system for vehicle, 2008 IEEE Vehicle Power and Propulsion Conference, Harbin, 2008.
4. T. Lee, C. Lee, Microcontroller based automatic parking system, Proceedings - International Conference on Machine Learning and Cybernetics, Xian, 2012: 875-879.
5. C. H. Chen, C. W. Hsu, C. C. Yao, A novel design for full automatic parking system, 12th International Conference on ITS Telecommunications, Taipei, 2012: 175-179.
6. T. S. Li, S. Chang, Y. Chen, Implementation of autonomous fuzzy garage-parking control by an FPGA-based car-like mobile robot using infrared sensors, Proceedings – IEEE International Conference on Robotics and Automation, Taipei, 2003: 3776-3781.
7. T. S. Li, S. Chang, Y. Chen, Implementation of human-like driving skills by autonomous fuzzy behavior control on an FPGA-based car-like mobile robot, IEEE Transactions on Industrial Electronics, 2003, 50(5): 867-880.
8. T. H. S. Li, S. Chang, Autonomous fuzzy parking control of a car-like mobile robot, IEEE Transactions on Systems, Man and Cybernetics, Part A: Systems and Humans, 2003, 33(4): 451-465.
9. N. Scicluna, E. Gatt, O. Casha, et al., FPGA-based autonomous parking of a car-like robot using Fuzzy Logic Control, 19th IEEE International Conference on Electronics, Circuits, and Systems, Seville, 2012: 229-232.
10. Y. Wang, X. Zhu, Design and implementation of an integrated multi-functional autonomous parking system with Fuzzy logic controller, 2012 American Control Conference, Montreal, 2012: 2928-2933.
11. M. J. Er, C. Deng, Obstacle avoidance of a mobile robot using hybrid learning approach, IEEE Transactions on Industrial Electronics, 2005, 52(3): 898-905.
12. J. De Lope, D. Maravall, Multi-objective dynamic optimization for automatic parallel parking, Lecture Notes in Computer Science, 2005, 3643: 513-518.

MATEC Web of Conferences 75, 07005 (2016)
*ICMIE 2016* DOI: 10.1051/matecconf/20167507005

# Speed and Displacement Control System of Bearingless Brushless DC Motor Based on Improved Bacterial Foraging Algorithm

Xiaoyan Diao, Huangqiu Zhu and Duo Zhang

*School of Electrical and Information Engineering, Jiangsu University, Zhenjiang 212013, Jiangsu Province, China*

**Abstract.** To solve the deficiencies of long optimization time and poor precision existing in conventional bacterial foraging algorithm (BFA) in the process of parameter optimization, an improved bacterial foraging algorithm (IBFA) is proposed and applied to speed and displacement control system of bearingless brushless DC (Bearingless BLDC) motors. To begin with the fundamental principle of BFA, the proposed method is introduced and the individual intelligence is efficiently used in the process of parameter optimization, and then the working principle of bearingless BLDC motors is expounded. Finally, modeling and simulation of the speed and displacement control system of bearingless BLDC motors based on the IBFA are carried out by taking the software of MATLAB/Simulink as a platform. Simulation results show that, speed overshoot, torque ripple and rotor position oscillation are dramatically reduced, thus the proposed method has good application prospects in the field of bearingless motors.

## 1 Introduction

Depending on its outstanding characteristics of good speed regulation, strong ability for overload, high torque of starting/braking, non-friction, high efficiency and reliability, the bearingless BLDC motor has much research values and extensive application prospects in the biomedical field such as blood pumps, high/ultra-high speed centrifugal machines, surgical electric saws and the new energy of flywheel energy storage, etc. [1]. The PID controller is one of the most common controllers with a series of advantages of simple principle, good robustness, strong adaptability and so on [2-3]. But a main problem of the PID controller is that its parameters need manual adjustment and are immutable in the entire control process after regulated [4]. At the same time, the coupling relationships among torque and suspension forces in the bearingless BLDC motor are very complex, and it is easy to cause higher torque ripple, speed overshoot and rotor position oscillation.

To reduce speed overshoot and torque ripple in the speed control system of servo motors, many international and domestic organizations and scholars have proposed many optimization methods for PID controllers, such as genetic algorithm, particle swarm optimization algorithm, and so forth [5]. And these advanced methods have been widely used in many complex engineering problems. Another random search algorithm, BFA, was proposed by American professor Kevin inspired by the foraging method of colibacillus in 2002. The theoretical basis for BFA is that it makes characteristics, concepts and thoughts of bacteria foraging behaviors mathematization and has been shown the unique adaptive advantages in fields of function optimization, nonlinear model identification, electric system and image compression [6]. However, there are some problems in BFA: the total intelligence and random motion of coli-group are considered but the individual intelligence of a coli is ignored in the foraging process; there is still the possibility for dispersing bacteria with the highest fitness value; the swim step is constant in the entire chemotaxis process, it maybe result in too long optimization time and poor accuracy. Thus it is difficult to convergence in some complicated cases for the optimization process and the optimal solution may be placed in the worst position while swimming.

Dynamic decoupling control of bearingless BLDC motors has been widely studied by domestic and foreign scholars and already achieved many major breakthroughs. In [1], an accurate mathematical model of radial suspension force was proposed, the equivalent circuit was established by the permanent magnet equivalent method and a mathematical model with single degree of freedoms for radial suspension force was derived. Decoupling control of bearingless BLDC motors had been realized successfully and the control system has the advantage of simple and easy to realize. Whereas it has the disadvantage of high torque ripples, speed overshoots and rotor position oscillation.

To solve the deficiencies existing in conventional PI controllers of speed and displacement control system for bearingless BLDC motors and conventional bacterial foraging algorithm itself, an IBFA is introduced to the bearingless BLDC motor based on the mathematical model of the control system proposed by [1]. This paper emphatic studies the method of reducing torque ripple, speed overshoot and rotor position oscillation. Firstly, the IBFA is proposed by proper foraging behaviour of population and individual, which has the following characteristics: the optimum bacterial is averted in the dispersal process, swim steps are defined as the fitness function and swim is stopped in chemotaxis cycles. Secondly, the IBFA is used in conventional PI controllers for parameter optimization. The speed and displacement control system is proposed based on the method and this system has the characteristics of online tracking. Finally, modeling and simulation are carried out by taking the software of MATLAB/Simulink as platform. Simulation results show that the proposed method can reduce speed overshoot, torque ripple and rotor position oscillation effectively in the speed and displacement control system of the bearingless BLDC motor.

## 2 Bacterial foraging algorithm

Recently, the search and optimization algorithm based on bacteria foraging behaviour has been used to solve optimization problems. The group can gang-up on larger preys and individuals can obtain protection from predators due to the group characteristics. In the group, each bacterial enjoys the individual intelligence and nutrient source for existence. The bacteria gotten enough food survives in the end. The solution to the problem is corresponding to the bacteria state in the search space, means the fitness value. Bacteria foraging behaviour embodies four elements: chemotaxis, swarming, reproduction and dispersal.

### 2.1 Chemotaxis

Chemotaxis consists of two basic motion, swim and tumbling, by rotation of flagellums. Bacterial direction of forward motion is dependent on the rotation direction of flagellums. Bacteria swims continue along the prior direction as it rotates in a counter-clockwise direction. Similarly, the direction is changed as in a clockwise direction. The position function of bacteria in the process of swim and tumbling can be expressed as

$$\alpha^i(j+1,k,l)=\alpha^i(j,k,l)+C(i)\varphi(i) \quad (1)$$

where $\alpha^i(j, k, l)$ specifies the position of $i$th bacterium of flora at $j$th chemotactic, $k$th reproductive and $l$th dispersal step. $C(i)$ is the size of the step that is taken in the random direction only at the beginning of the algorithm and after that, it is specified automatically. $\varphi(i)$ is the new direction after tumbling and can be represented as

$$\varphi(i)=\beta(i)\frac{\Delta(i)}{\sqrt{\Delta^T(i)\Delta(i)}} \quad (2)$$

where $\Delta(i)$ is the random variable ranging from 0 to 1 in the process of turning direction. Possible values of $\beta(i)$ can be -1 or 1. At the beginning of the algorithm, the initial value of $\beta(i)$ is selected at random and after that, it changes smartly according to the fitness function of the optimization goal. In particular, $\beta(i)$ is always following the pre-defined direction as long as bacteria swims, but cannot get the minimum value of the fitness function in that direction. Provided that we make $J(\alpha)$ as a fitness function for a bacteria at the position $\alpha$, then $C(i)$ can be calculated based on the fitness function $J(\alpha^i)$ for $i$th bacteria

$$C(i)=\max\left(C_{\min},\frac{|J(i)-J_0|}{K_1}\right) \quad (3)$$

where $J_0$ is the global optimum solution and $J(\alpha^i)$ is the fitness value for $i$th bacteria at the position $\alpha$, and there is $K_1>>|J|$. $J(\alpha^i)$ can vary automatically as the optimization process, and its positive constant coefficients should be adjusted properly. From (3), it is observed that $C(i)$ can be adjusted by changing the fitness value and decreases continuously as chemotaxis number increases in the whole process. Thus bacterium has the smaller step with the closer to the global optimum solution.

### 2.2 Swarming

Swarming bacteria can be realized by cell-to-cell signal via an attractant in the foraging process and the fitness function can be expressed as [7]

$$\begin{aligned}&J_{cc}(\alpha_m(j,k,l),\alpha(j,k,l))\\&=k\cdot\left(\sum_{i=1}^{S}\left[-d_a\exp\left(-\omega_a\sum_{m=1}^{p}(\alpha_m-\alpha_m^i)^2\right)\right]\right)\\&+\sum_{i=1}^{S}\left[h_r\exp\left(-\omega_r\sum_{m=1}^{p}(\alpha_m-\alpha_m^i)^2\right)\right]\end{aligned} \quad (4)$$

where $J_{cc}(\alpha_m, \alpha(j, k, l))$ is another fitness function combined with the influence of other bacteria based on the real fitness function $J(j, k, l)$. $d_a$ denotes dosage of the repellent released by bacteria, $\omega_a$ the test value of the repellent signal width, $\omega_r$ the depth of the rejection and $h_r$ the longitudinal rejection value.

### 2.3 Reproduction and dispersal

After entering the stage of reproduction, the superior can be reproduced and the inferior can be eliminated selected by health values. The health value can by represented by the sum of the fitness values in the chemotaxis process,

$$J^i_{\text{health}}=\sum_{j=1}^{N_c+1}J(i,j,k,l) \quad (5)$$

where $J^i_{\text{health}}$ is the health value, that is the sum of the optimal fitness values in the chemotaxis cycles. $N_c$ is the number of chemotaxis. The smaller health value of bacteria means the more nutrition achieved, so it has the greater chance of reproduction. On the other hand, the bigger health value means the bigger possibility of being

 DOI: 10.1051/matecconf/20167507005

eliminated. After reaching the chemotaxis number $N_c$, the health values obtained from (6) are sorted and half of the bacteria with a bigger health value are eliminated. Simultaneously, half of the bacteria with a smaller health value are reproduced to two bacteria. Where one of them is placed in the same location and another is put around the location. It makes the population of bacteria constant in the reproduction process.

When environmental conditions such as nutrient source and temperature variation in the local region, it is possible that the region is no longer suitable for survival of bacteria. Thus bacteria is compelled to migrate to a new location. To simulate this phenomenon, let $P_{ed}$ denote the probability of bacteria dispersed from one location to another location. Because the outside conditions are stable under the laboratory environment that just the unhealthy bacteria could be dispersed, so the faster and more appropriate rate of convergence are favorable to be obtained.

## 3 Improved bacterial foraging algorithm

This section will simulate the behaviors of chemotaxis, swarming, reproduction and dispersal and focus on providing detail realization process of the proposed method.

The proposed algorithm has two main characteristics: According to (2), The operator leads each bacterium to the optimum solution, so that if swimming in a pre-defined direction makes a better solution ($\beta(i)$=1), it continues in the same direction, otherwise the direction is changed in an inverse way by tumbling ($\beta(i)$=-1), thus realizes the intelligent foraging; According to (3), a variable step is introduced by each bacteria's fitness function, thus realizes the adaptive step size. It is beneficial to increase the size of the step for bacterium with higher fitness values. What's more, the step decreases and thereby the accuracy increases simultaneously with the convergence process.

The proposed method is used to search the minimum value of the fitness function $J(a)$. Before presenting the detail the implementation process of the proposed algorithm, the necessary variables are defined as follows: $p$ is the latitude of search range, namely number of parameters to be optimized, $S$ the bacterial number in the entire search region, $N_s$ the maximum step size of swimming, $N_c$ the chemotaxis number, $N_{er}$ the number of reproduction steps, $N_{ed}$ the number of dispersal operation, $S_r$ the number of reproduction in each generation and $P_{ed}$ the probability of dispersal. $P(i, j, k) = \{\alpha^i(j, k, l) | i=1, 2, \ldots, S\}$ is the initial position of each bacterium, respectively.

**Step 1:**

(a) Initialize parameters $j=k=l=0$ and select the position for $i$th bacteria randomly, random variable $\Delta(i)$ in the process of changing direction and the new direction $\beta_0(i)$, where $i=1, 2, \ldots, S$.

(b) The initial variables $j=k=l=0$ is used to calculate the fitness values for each bacteria, and then the sum $J_{sw}$ of all bacteria's fitness values is

$$J_{sw}(i,j,k,l) = J(i,j,k,l) + J_{cc}(\alpha_m(j,k,l), \alpha(j,k,l)) \qquad (6)$$

Let $J^i_{last} = J_{sw}$ and $J_{last} = \min_{i=1,2,\ldots,S}(J^i_{last})$ as the current fitness values. And save these until a better fitness value is found to replace it via a run.

**Step 2:** Dispersal loop count, $l$++.

**Step 3:** Reproduction loop count, $k$++.

**Step 4:** Chemotaxis loop count, $j$++. Bacterium swims and tumbles in the following rules:

(a)Tumble: Generate a vector $\beta(i)$ and a random vector $\Delta(i)$.

(b)Swim: if $m$ is less than $p$, let $m$++;

• Let the initial value of the counter for swim length $N$ equal zero;

•*while*($N$<$N$s), let $\alpha^i_m(j,k,l)_N = \alpha^i_m(i,k,l)_{N-1} + C(i)\varphi(i,m)$ and $N$=$N$+1, where $C(i)$ is the direction of tumbling for each bacteria by (2) and (3).

• Compute $J_{sw}(i, j, k, l)$ by (6). If $J_{sw}(i,j,k,l) < J^i_{last}$, let $\alpha^i_m(j,k,l)_{N+1} = \alpha^i(j,k,l)_N + C(i)\varphi(i,m)$ and $J^i_{last} = J_{sw}(i,j,k,l)$. And use $\alpha^i(j,k,l)_{N+1}$ to compute the new $J_{sw}(i, j, k, l)$; otherwise, if $N$<2, let $\alpha^i_m(j,k,l)_N = \alpha^i(j,k,l)_{N-1} + C(i)\varphi(i,m)$ $\beta(i,m)$=-$\beta(i,m)$ and recompute the new $J_{sw}(i,j,k,l)$. If $J_{sw}(i,j,k,l) < J^i_{last}$, let $\alpha^i_m(j,k,l)_{N+1} = \alpha^i(j,k,l)_N + C(i)\varphi(i,m)$ and $J^i_{last} = J_{sw}(i,j,k,l)$, recompute the new $J_{sw}(i,j,k,l)$; otherwise, let $N$=$Ns$. The *while* statement is over and the next $m$ of *while* loop begins.

If $m$=$p$ and $i$<$S$, then go to the next bacterium $i$=$i$+1 to process the next bacterium.

**Step 5:** if $j$<$N_c$, go to step 4.

**Step 6:** if $l$>$N_{ed}$, the algorithm stops.

**Step 7:** Reproduction.

(a) For the given $k$ and $l$, health values of bacterium calculated by (6) are sorted in ascending order, and the greater the health value means the unhealthier.

(b) Let $S_r$=$S$/2, eliminate $S_r$ bacterium with the bigger health values and reproduce other $S_r$ bacterium. Copies of that are placed in proximity of the same location as their parent.

**Step 8:** If $k$<$N_r$, go to step 3. The next generation will be started in the chemotactic loop in the case of not reaching the number of specified steps.

**Step 9:** After the health values $J_{heath}$ sorted in ascending order, bacterium $i$ ($N_{ed}$, $N_{ed}$+1,…, $S$) is dispersed to a new location, and then go to step 2. After bacterium reaches a new location, it is need to recompute $J_{sw}(i,j,k,l)$ and sort them by health values.

## 4 Mathematical model of bearingless BLDC motor

This paper will model the mathematical model of bearingless brushless DC motors according to [1]. The bearingless BLDC motor has two windings: torque windings and suspension force windings. Where the former provides the torque for the motor rotating and drives external loads; the latter is used to sustain the rotor and realize the purpose of the non-friction and high speed.

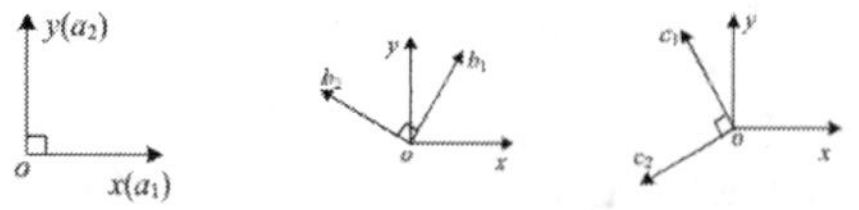


**Figure 1.** Three coordinate systems of suspension forces exist.

Suppose that the rotor position $\theta \in [0, 30°)$, the rotor suspension is controlled by the a1-a2 coordinates defined by Figure 1(a), so the mathematical model of radial suspension force in any position of the bearingless BLDC motor can be expressed as

$$\begin{bmatrix} F_{\mathrm{sa1}} \\ F_{\mathrm{sa2}} \end{bmatrix} = \begin{bmatrix} k_i i_{\mathrm{a1}} + k_x a_1 \\ k_i i_{\mathrm{a2}} + k_x a_2 \end{bmatrix} = k_i \begin{bmatrix} i_{\mathrm{a1}} \\ i_{\mathrm{a2}} \end{bmatrix} + k_x \begin{bmatrix} a_1 \\ a_2 \end{bmatrix} \tag{7}$$

where $i_{\mathrm{a1}}$ and $i_{\mathrm{a2}}$ denote currents in suspension force windings $\mathrm{a}_1$ and $\mathrm{a}_2$, respectively; $a_1$ and $a_2$ denote radial displacement along axis $a_1$ and axis $a_2$, respectively; $F_{\mathrm{sa1}}$ and $F_{\mathrm{sa2}}$ denote radial suspension force along axis $a_1$ and axis $a_2$, respectively; $k_i$ is stiffness coefficient of current and $k_x$ is stiffness coefficient of displacement, there are

$$k_i = \frac{2SB_r l_m N}{(l_m + g)^2} \tag{8}$$

$$k_x = \frac{6B_r^2 l_m^2 S}{\mu_0 (l_m + g)^3} \tag{9}$$

Transform (7) to the coordinates $xoy$, and there is

$$\begin{bmatrix} F_x \\ F_y \end{bmatrix} = \begin{bmatrix} F_{\mathrm{sa1}} \\ F_{\mathrm{sa2}} \end{bmatrix} = k_i \begin{bmatrix} i_{\mathrm{a1}} \\ i_{\mathrm{a2}} \end{bmatrix} + k_x \begin{bmatrix} a_1 \\ a_2 \end{bmatrix} \tag{10}$$

where $F_x$ and $F_y$ denote radial suspension force along axis $x$ and axis $y$, respectively.

In a similar way, mathematical models of radial suspension force under the b1-b2 coordinates and the c1-c2 coordinates can be derived.

Suppose that $m$ denotes rotor quality, without regard to disturbing force from outside (because disturbing force is mainly gravity of rotor), then the motion equations are [8] [9]

$$k_i i_x + k_x x = m \frac{\mathrm{d}^2 x}{\mathrm{d}t^2} \tag{11}$$

$$k_i i_y + k_x y = m \frac{\mathrm{d}^2 y}{\mathrm{d}t^2} \tag{12}$$

where $\Omega$ is rotor velocity, $T_L$ is load torque, $D$ is viscous damping coefficient and $J$ is rotational inertia.

## 5 Modeling and simulation verification

To verify the feasibility and reliability of the proposed method, parameters of $K_{\mathrm{Ps}}$、$K_{\mathrm{Is}}$ and $K_{\mathrm{Pd}}$、$K_{\mathrm{Id}}$ of the speed PI controller and the rotor radial displacement PI controller are carried out with online optimization respectively[10].

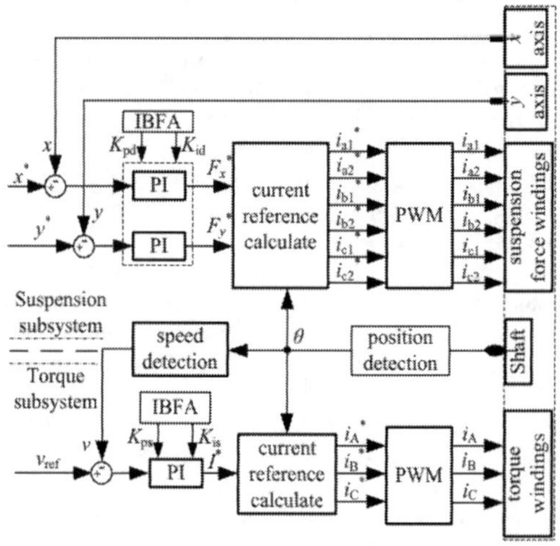


**Figure 2.** Block diagram of the control system for bearingless BLDC motors.

Figure 2 is block diagram of the control system for the bearingless BLDC motor based on the IBFA. The control system consists of position detection, speed detection, current reference calculation, PWM controller, speed PI controller, radial displacement PI controller and the bearingless BLDC motor ontology. Where parameters of $K_{\mathrm{Ps}}$、$K_{\mathrm{Is}}$ and $K_{\mathrm{Pd}}$、$K_{\mathrm{Id}}$ of the speed PI controller and the rotor radial displacement PI controller are carried out with online optimization by the proposed IBFA, respectively. The current reference signal is generated after the speed error via the PI controller, the error obtained by comparison of the current reference and the actual current of the stator is as input of the PWM controller. The suspension force signal is generated after the radial displacement errors via the PI controllers, and then the current reference of radial displacement is obtained by the rotor position[11].

Main parameters are as follows: equivalent impedance of each phase is 0.05 Ω, load torque is constant 5Nm. The initial speed is 850 r/min and the speed jumps to 1350r/min. Simulation time is 0.2 s. The appropriate choice of parameters $N_{\mathrm{er}}$, $N_{\mathrm{c}}$, $N_S$ and $P_{\mathrm{ed}}$ is beneficial to the balance between the algorithm speed and the accuracy, so these values are concerned with the algorithm performance. In this paper, let bacterium number $S = 90$, $N_{\mathrm{ed}} = 3$, $N_{\mathrm{er}} = 2$, $N_{\mathrm{c}} = 1$, $N_S = 5$ and $P_{\mathrm{ed}} = 15\%$。And then, simulation is carried out and the dynamic characteristics of speed, torque and rotor position when the speed changes.

Figure 3 is the response curve of speed. It can be observed that speed rise to the initial speed setpoint 850r/min rapidly during motor starting process and speed overshoot is nearly zero. At time 0.1 s, speed reaches to speed setpoint in such a short time 5ms, and then it is rapidly entering a new steady state.

Figure 4 is the response curve of electromagnetic torque. It can be observed that electromagnetic torque rise to load torque during motor starting process. In order to quickly adapt the speed variation at time 0.1s, torque has a small fluctuation about 5ms, and then it stabilizes quickly. Thus torque has excellent dynamic and static performance, and the torque ripple is small within the entire simulation time except at time of speed mutation.

Figure 5 and Figure 6 are the response curve of radial displacement along the axis $x$ and the axis $y$, respectively. We could see that the rotor position return to zero rapidly when motor starts because the initial status of rotor position is unknown. The rotor position is almost no turbulence throughout the entire process.

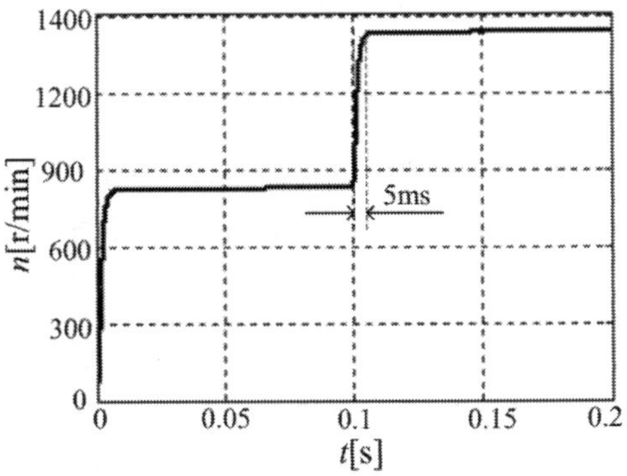


**Figure 3.** The response curve of speed.

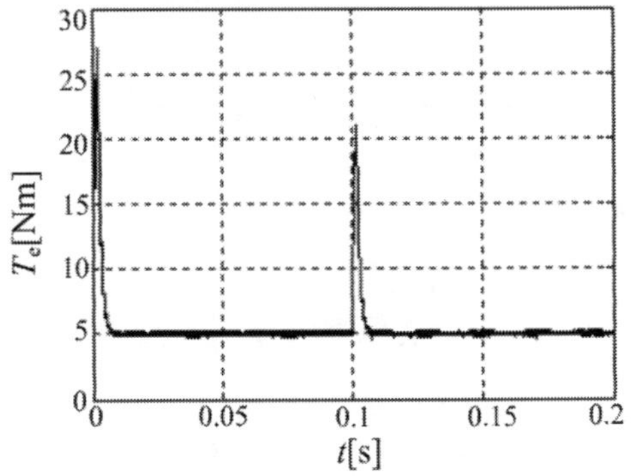


**Figure 4.** The response curve of electromagnetic torque.

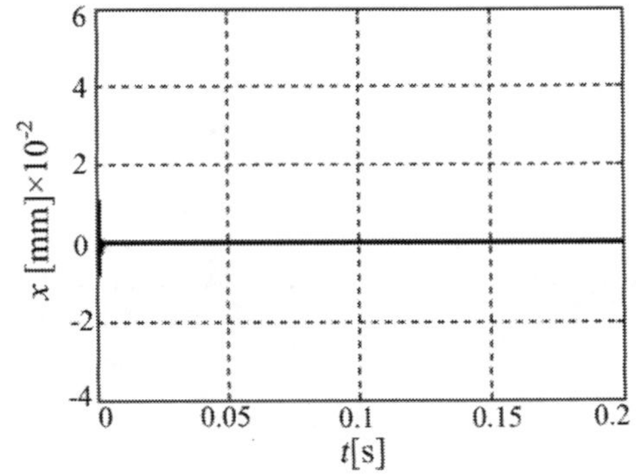


**Figure 5.** The response curve of radial displacement along the axis $x$.

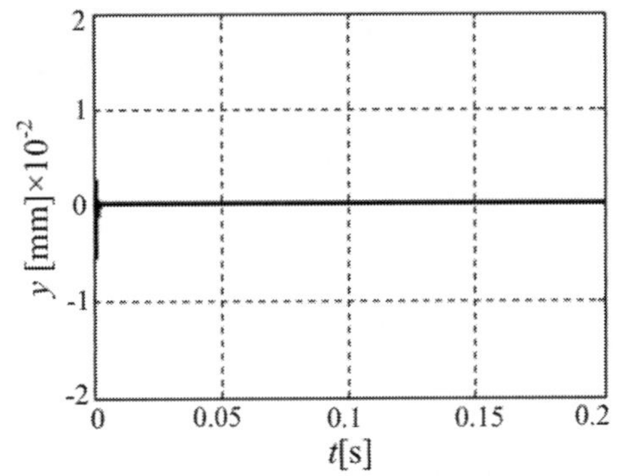


**Figure 6.** The response curve of radial displacement along the axis $y$.

## 6 Conclusion

An improved bacterial foraging algorithm is proposed and applied to the speed PI controller and the radial displacement PI controller for parameter optimization. The principle of bacterial foraging algorithm is described and the proposed bacterial foraging algorithm is concrete introduced. Simulation results show that speed overshoot decreases, torque ripple and radial displacement oscillation are reduced significantly when using the proposed method for parameter optimization of the speed PI controller and the radial displacement PI controller in the bearingless BLDC motor. And the system has good dynamic and static performance. Therefore, the proposed method has a good reference significance and application prospect for improving the speed and displacement control system of bearingless BLDC motors.

## References

1. C.L. Xia, H.W. Fang, Trans. CES, **27**, 25 (2012)
2. L.G. Chen, H.Q. Zhu, Proc. CSEE, **32**, 75 (2012)
3. Z.B. Qin, F.F. Chen, B. Jin, J. ZJ. Univ. **49**, 1503 (2015)
4. G.F. Li, S.F. Pan, X. Wang, Ind. Control. Comput. **27**, 64 (2014)
5. R. Hooshmand, S. Soltani, IEEE Trans. P. S. **27**, 47 (2012)
6. W.F. Huang, W.X. Lin, H.K. Fan, Comput. Eng. Appl. **47**, 82 (2011)
7. E. Daryabeigi, B.M. Dehkordi, Elec. Power Energy Syst. **62**, 364 (2014)
8. H.Q. Zhu, L.G. Chen, X.Y. Diao, *Chinese Control Conference* (Chinese Association of Automation, Hefei, 2012)
9. L.G. Chen, X.D. Sun, H.Q. Zhu, Springer, **97**, 273 (2011)
10. H. Li, F. Wang, Mach. T. Hyd. **36**, 337 (2008)
11. H.Q. Zhu, L. Fang, *IEEE International Power Electronics & Motion Control Conference* (IEEE, Shanghai, 2006)

# Chinese Seal Clasp–Meeting technology and elegance with functional goals

Mario Kong[1] and Vanessa Mendes[2]

[1]*PhD Architect, Professor Faculty of Architecture - University of Lisbon, Researchers and CIAUD-permanent member of the Research Centre for Architecture, Urban Planning and Design FA-UL*
[2]*PhD Student, Faculty of Architecture, University of Lisbon – FA-UL, 1349-055 Polo Universitário – Alto da Ajuda, Lisboa, Portugal*

**Abstract.** This study comes within the framework of a PhD course in Design, which is the continuation of a Masters degree research in Fashion Design, searching for wardrobe technology and innovation, through the use of three-dimensional computer-aided design and printing, with it's use already popular in areas related to product development and design, as well as architecture, translating and adapting, integrating it into the world of Fashion. The intent of this project is to present solutions that promote a new and versatile product, allowing for a different kind of aesthetic. This will be achieved and presented through planning and prototyping, thus creating a new clasp for fashion products. This research will result in a finished clasp, with its purpose and function being tested, in order to achieve concrete results, thus creating an innovative technology to be part of a fashion collection that allows the transformation of the suntan marks left by female swimwear. This will result in the creation of designs and drawings on the body, using shapes inspired by the Oriental Zodiac Signs, as well as colours drawn from the five elements of *Feng Shui*, which are also present in the Western Astrological Signs.

## 1 Introduction

Computer-aided Design for Manufacturing and its potential for creating unique and new types of objects, shapes and designs has been one of the main themes and subjects of this research, as its ability for conceiving printable objects for production allows the achievement of new creative boundaries.

The cultural fusion intended lies in the overall concept and design of the collection, but also finds its way into the little details, that make up part of the whole.

## 2 Design and concept within Chinese seal history

"As an important symbol of traditional Chinese culture, the evolution and development of the Chinese seals reflect the rich connotations of Chinese civilization."[1]

More specifically, the clasps used derive their inspiration from elements present in Chinese culture, and the drawings and the form of ancient seals, merging both form and function.

The look of this chinese seal claps are made after the official seals during the Western Han Dynasty (206 B.C.-9 A.D.)

"...quare, gentle, solemn, elegant, profound and simple..." [1]

**Figure 1.** Seal of Prince Gualing [1].

This clasp seal, based on a full research on the history of these ancient Chinese objects can come in two different designs, just like the two different character cutting techniques that were used in their manufacture.

More specifically, the inspiration for the design of the clasps derives from elements present in Chinese culture, and the drawings and the form of ancient seals, merging both form and function with their inherent conceptual and theoretical values.

According to Rudolf Arnheim: "The artistic imagination can be described appoximately with the discovery of a new form for an old content, or – if you do not want to use the convinient dichotomy between form and content – as a new concept of an old subject".[2]

## 3 Planning before the manufacture

DOI: 10.1051/matecconf/20167508001

Like the industrial production process and design plannig,[3] threre were made several technical drawings of the different aspects and projections of the clasps, rangigng from plans of the superior and inferior faces, lateral sections, as well as transverse and longitudinal ones.

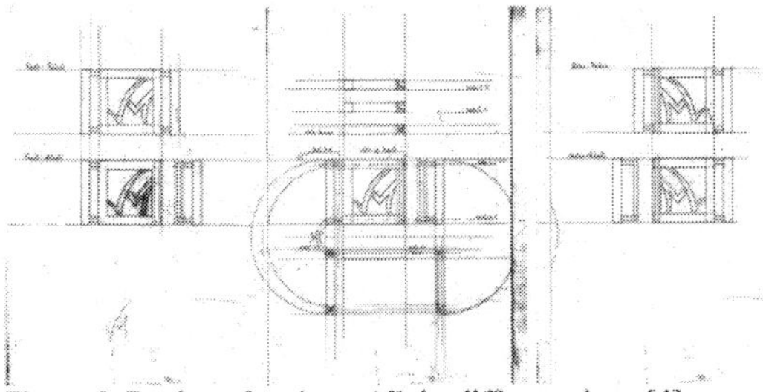
**Figure 2.** Product planning – All the different views [4].

### 3.1 Axonometry

After the multiple different views were drawn, an axonometric drawing was made to clarify its final form based on the previous studies.

**Figure 3.** Hand drawn product planning [4].

## 4 AutoCAD use in 2D prototyping

In order to assure the correct dimension and accuracy of the drawings 2D prototypes were conceived to allow the cutting of the materials using laser technology. These were specifically designed to verify and test the results and effectiveness of the drawing area on the skin.

AutoCAD 2D drawing was made to cut a thin acrylic sheet, 0,5 millimetre thick.

The thickness of the material was chosen due to the laser cutting characteristics, thicker materials are prone to be deformed, due to laser incidence, making the top side thinner than the bottom side of the acrylic sheet.

The choice of the material was also conditioned due to the laser cutting equipment characteristics.

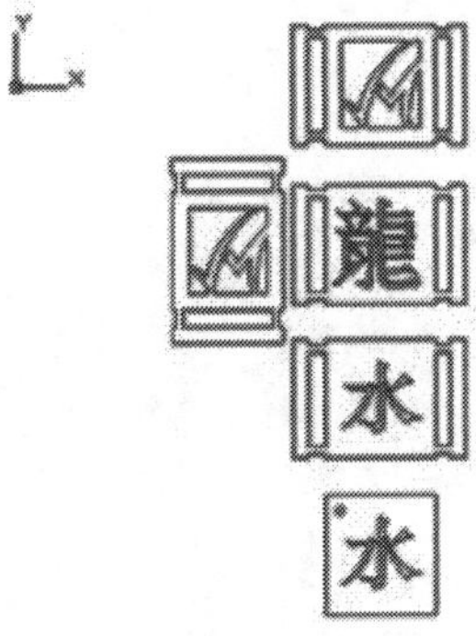

**Figure 4.** 2D AutoCAD printable drawing [4].

## 5 Modelling and Printing in 3D

3D modelling programs have a very important role in achieving these goals, by allowing detailed, yet resistant shapes to be carved.

**Figure 5.** 3D Model ready to print [5].

The possibilities are endless with this printing, depending on the materials used, our own imagination and creativity. – "Creativity is a productive capacity where fantasy and reason are associated…"[2] – There is also the possibility of either gluing or linking several elements together, in order to achieve more complex shapes and objects.

This results in the creation of accessories that correspond to the concept of aesthetics, beauty and seduction which are desired in underwear and beachwear design.

Chinese *Feng Shui* elements are specially relevant when it comes to colour, because "*It is an old concern for mankind, the desire to reproduce the colours of nature in everything that surrounds him. That understands a deep psychological meaning, seeming to be one of mankind's basic needs…*".[6]

Therefore, the clasp's colours are chosen within the range corresponding to each of the *Feng Shui* elements. Water, earth, fire, wood, and metal: black, brown, fire red, light green and white respectively.

After making the 3D print, to test materials and colours, 2D prototypes were intended for printing, to allow and create the perception of how the final product is going to look. It is possible to achieve this in two ways, by creating a mould extracted from a 3D printed prototype or direct 3D printing in the materials intended for use.

**Figure 6.** 3D and 2D models printed side by side [4].

The transparent, or non-pigmented clasp, has the purpose of allowing the sun's rays to go through, leaving no skin mark at all. It is intended to provide the user with the choice of not having a tan mark, having been tested and proven to allow the passage of U.V. rays, leaving only a very light mark on the skin.

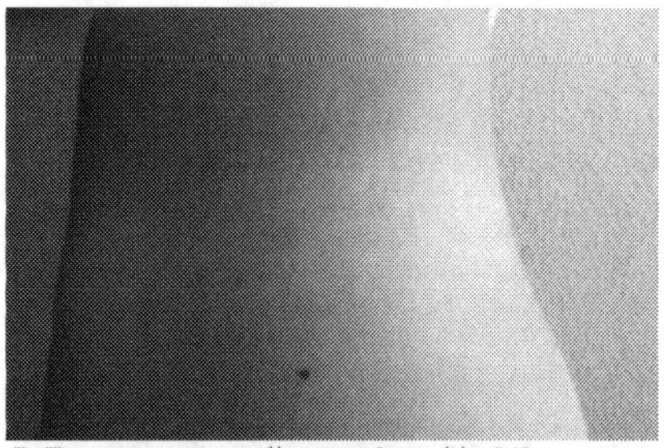

**Figure 7.** Transparent acrylic tested on skin [4].

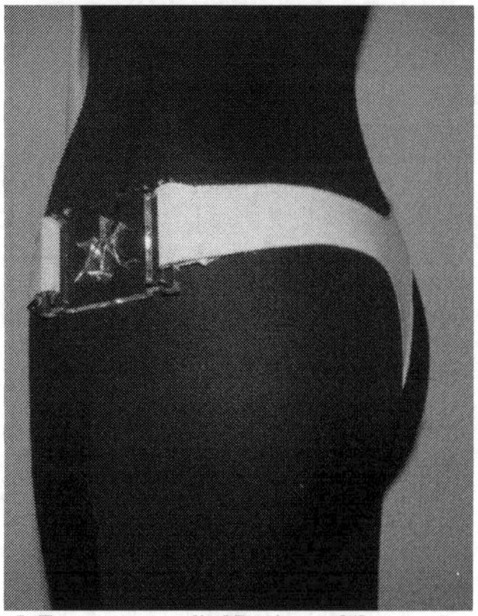

**Figure 8.** Transparent acrylic 2D print test [4].

# 6 Goals and Objectives

This Design PhD research is aimed at Fashion Design of feminine beachwear, making use of *Feng Shui* as a deep source of inspiration, as well as product design technologies.

## 6.1 Fashion Design

The conception of a design for swimwear and beachwear purposes seeks to combine comfort, practicality, function, as well as beauty, seduction and desire associated with them, creating an illusion of nudity, of transparency, of fluidity.

This results in the creation of accessories that correspond to this concept of aesthetics, becoming more than merely functional objects, reinventing them.

For the clasp to become fuctional, the aplication of a layer of silicone, on the side that comes into contact with the users body is going to be considered and studied, in order to provide a greater adherence to the skin, allowing for a more precise tattoing of the skin by holding it in place.

It is a matter of function, and concern with the users confort, thus creating a product that is ergonomic and has the wearers best interests in mind.

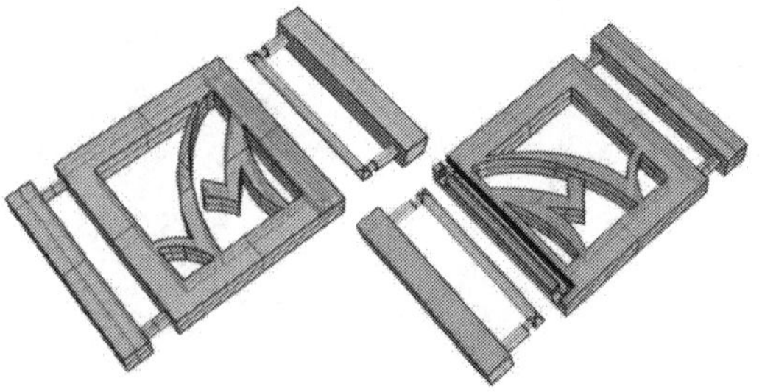

**Figure 9.** Clasp seal front and back view respectively. The blue area is marking the side designed to receive the silicone layer, to be in contact with the user skin for adherence [5].

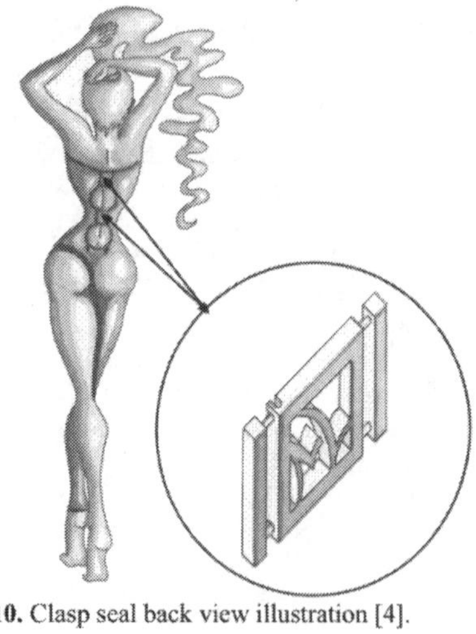

**Figure 10.** Clasp seal back view illustration [4].

DOI: 10.1051/matecconf/20167508001

It will also need an instructional label, so that the user knows how to wear the product correctly, in order to use the clothing to its fullest.

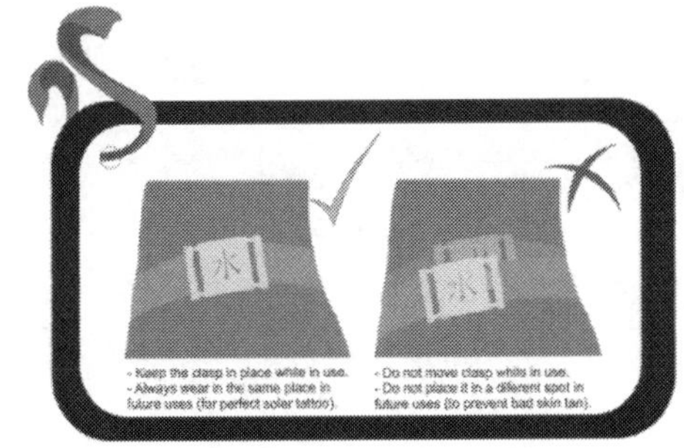


**Figure 11.** *Hanzi* clasp seal instructional label [5].

This instructional label, can be available in both formats: *Hanzi* and Logo, to clarify the correct use of the clasp.

**Figure 12.** Logo clasp seal instructional label [5].

## 6.2 New Tecnology

Designed with materials that allow the blocking and/or passage of ultra-violet lighting, it results in the "printing" of a "solar drawing", on the user's skin through natural sunbathing.[4]

**Figure 13.** Test of the application of the textile technology [4].

Together with the use of new textile technologies and solutions, it's a single part of a whole swimwear product, built to provide a new form of interaction between skin and sun.

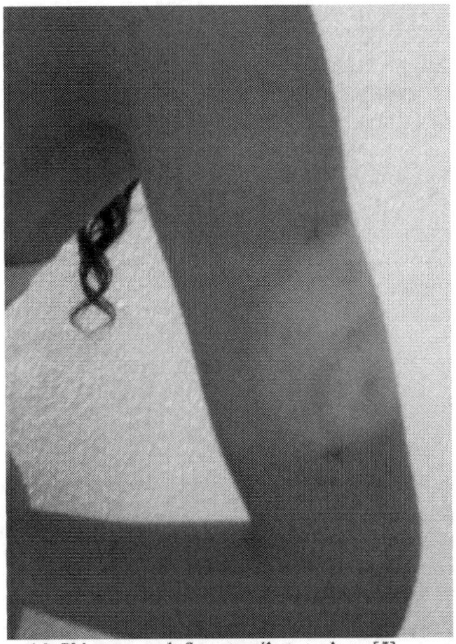

**Figure 14.** Skin tan result from textile tecnology [5].

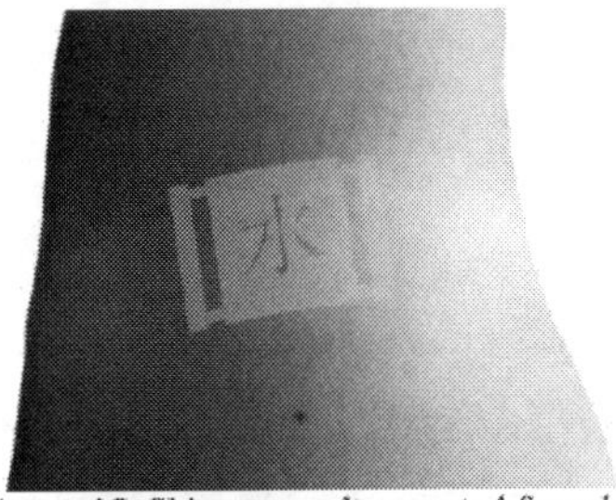


**Figure 15.** Skin tan result expected from *hanzi* clasp seal [5].

**Figure 16.** Skin tan result expected from logo clasp seal [5].

## 7 Final Objectives

The final product is the result of an intensive study of this ancient utensils,[1] created through careful planning and drawing, in order to represent its detail to the fullest.

Elegant, and yet simple in its fluid shape, these clasps create not only a bridge between the Portuguese enjoyment of beach/water activities with the ancient culture of China, but also between the Far East and the West, a fusion of cultures that, hopefully, will result in the enrichment of both.

It is a product intended to invite, entice and intrigue, to attract the Chinese people to enjoy the pleasures of beach going, as well as to introduce to western beachgoers the mysterious and fascinating art of *Feng Shui*, exposing them to the rich history of a captivating and fascinating culture.

As a final goal, this clasp is intended to produce a skin print, to match the textile tattoo print, leaving a Chinese seal imprinted on the user, with a logo, sign or element with a strong reference to Chinese culture.

## References

1. H. Yang, *C. Seals*, pp. II, 25, 60-87 (2012)
2. M. S. M. Kong, *C. F. A. A. Culture*, p. 2 (2013)
3. F. Julián, J. Albarracin, D. P. D. Indústriais, pp. 60-61 (2010)
4. V. Mendes, *S. W. Tattoo*, pp. 48, 50-51, 116 -118, 141, 147-148 (2015)
5. V. Mendes, *S. W. Tattoo* (to be published)
6. M. S. M. Kong, *H. e Proporção*, p. 113 (2012)

# Application of CAD/CFD/RP-technologies for development of LVAD systems

Valentin Morozov, Alexey Zhdanov and Leonid Belyaev

*Department of Mechanical Engineering, Vladimir State University named after Alexander and Nikolay Stoletovs, Vladimir, Russia*

**Abstract.** The important problem of new Left Ventricle Assist Device (LVAD) systems development is to determine their hemodynamic and functional characteristics and reduce the design time. An effective method for achieving this is the use of approaches based on the application of CAD/CFD/PR-technologies. Methods of computer modelling were applied to determine the geometrical parameters of the blood pump, and methods of finite element analysis were used for the optimization of hemodynamic parameters of the LVAD system. The geometrical modelling was carried out in the Pro/Engineer WF5. The CAE-system Ansys was used to analyze the blood flow in the chamber of the pump. The motion of the flexible walls was simulated by normal flow through rigid walls. At the final stage some models were prepared for prototyping. The critical parameters of hemodynamic gave the possibility to optimize the shape of the blood pump. The LVAD system and its components were made using various RP-technologies. The hydraulic tests for the performance investigation and the analysis of implantability for implantable systems were carried out for manufactured prototypes of LVAD systems. The manufactured components of the LVAD system are completely operability and give the possibility to estimate the critical hemodynamic and functional parameters of the LVAD system.

## 1 Introduction

During the design of LVAD systems, complex scientific and engineering problems associated with the construction, simulation and experimental researches of the developed systems are solved. The solution of these problems implies the rise in the cost of the design and increases the time of manufacture of the finished product, based only on the comparative experimental studies of different design versions by a successive approaching of the functional characteristics of the systems to the required parameters that are set in the medical and technical requirements. In this regard, the direction associated with the use of the design approach is promising, giving the possibility of finding the optimal geometric design and the hemodynamic parameters of the LVAD systems already in the designing phase of the product. This approach based on the mathematical modelling of the processes realized during the operation of the LVAD systems allows to optimize the design, to compare the alternatives of the developed systems based on the analysis of the results of computational simulations and bed testing of the samples of LVAD systems fully or partially manufactured using rapid prototyping technologies. In this article it is examined the design of pulsative LVAD systems based on the application of CAD/CFD/RP-technologies which allow to determine the hemodynamic parameters of the system operation at the design stage and reduce the time of production of the finished article.

## 2 Methods

The creation process of the basic components of LVAD systems is a unified interconnected cycle stages. At all stages of the produced intensive exchange of different kinds of information. An important feature of this information is its bidirectional and multi-step iterative refinement based on the nature of the solutions to global and local problems of analysis, synthesis, optimization and compromise. One of the first steps is the creation of three-dimensional geometric model, which is the basis for all the following stages of designing and creating the LVAD systems. Three-dimensional modelling design LVAD systems is implemented in the Pro/ENGINEER WF5 (PTC, Inc, USA). Obtained at this step data has been transferred to the CAE-system for the hemodynamic analysis. The flow simulation inside the blood chamber was led based on actual geometry of the blood chamber (Fig. 1). The computational model represents the internal volume of the blood chamber based on the actual geometry of the main body, valves, radii and transitions. The simulation with one mitral valve position and various operation modes is described in the preliminary work [1].

This work presents the results of the simulation that was performed in order to determine the optimal position of the mechanical tilting disc valve "MIKS" (LLC "Roskardioinvest", Russia) [2]. This valve have a minimal external diameter – 17 mm and a geometric hole area of 1.54 cm$^2$. The starting point of the mitral valve was chosen as follows: the disk valve was oriented in such a way that its major orifice open to the external side of the inner wall, but the strut keeps perpendicular to the diaphragm. The valve rotation was realized clockwise. The modelling was carried out for the valve positions in 0, 30 and 45 degrees. The aortic valve position did not change (Fig. 2).

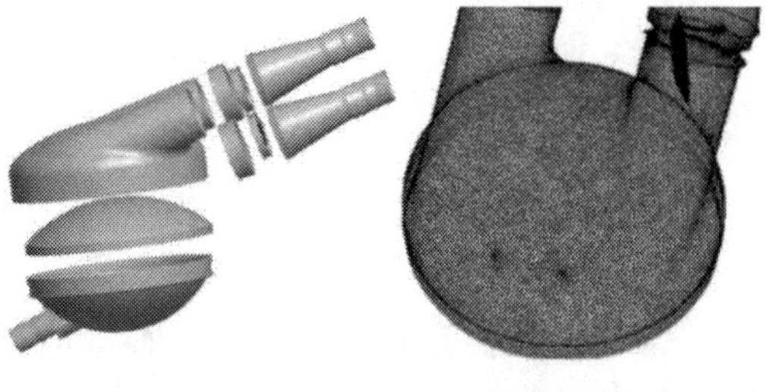

**Figure 1.** Pneumatic LVAD system: 3D model (left) and the mesh of finite-element model for hemodynamic analysis (right). A concentrate mesh was implemented in the valves zones and in the inlet and outlet ports areas.

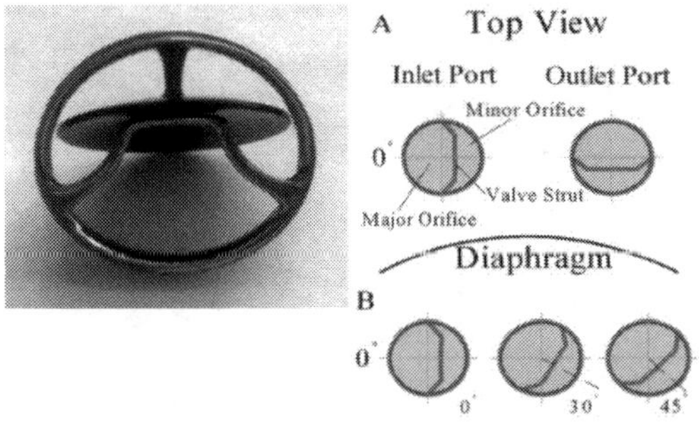


**Figure 2.** Tilting disk valve "MIKS" (left) and examples of the valves orientation during the modelling (A); three options of mitral valve orientation at which the modelling was realized (B) (right).

For the solution of the Navier-Stokes equations in the non-stationary statement, the hybrid (finite-element/finite-volume) system ANSYS CFX (ANSYS, Inc., USA) [3] was used. The step of hemodynamic analysis includes creating a finite element model, the determination of the boundary conditions (Table 1) and the definition of hemodynamic parameters of the flow inside the blood chamber. This study of researches focuses on the hemodynamics inside the chamber of the VAD and the dynamics of the leaflets during opening and closing were not modelled, as they were shown to have only a secondary effect on the flow in the chamber [4].

Besides that, the valve closing and opening times are short compared with the duration of diastole and systole and no significant effect on the flow inside the chamber was observed [5]. The motion of the valves was not modelled. They were held fixed in their fully open or fully closed positions, depending on the phase of the cardiac cycle. During systole, the aortic valve was opened and the mitral valve was closed and during diastole, the mitral valve was opened and the aortic valve was closed. The stroke volume of the system is 30 cc. The pump rate is 75 bpm. To obtain a more accurate solution, a partition was made in accordance with the geometrical complexity of the volume. A concentrate mesh was implemented in the valves zones and in the inlet and outlet ports areas. From the diaphragm side, a change of the drive pressure took place (1), (2) and the deformation of the walls was not included. The modelling assumed a non-slip condition at the boundary.

$$P_{mem.dia} = P_{dia} - P_{dia}(1+t)\sin(\frac{\pi}{T_{dia}}t), \qquad (1)$$

where $t \in [0; T_{dia}]$.

$$P_{mem.sys} = P_{sys} + P_{dia}(1+t)\sin(\frac{\pi}{T_{sys}}t), \qquad (2)$$

where $t \in [0; T_{sys}]$.

$P_{mem.dia}$, $P_{mem.sys}$ – variation of the drive pressure in air chamber in diastole and systole phases; $T_{dia}$, $T_{sys}$ – duration of diastole and systole phases; $P_{dia}$, $P_{sys}$ – diastolic and systolic pressures; $t$ – current time of the phase.

The simulation evaluated the hemodynamic parameters such as velocity of the flow inside the pump chamber, Reynolds and wall shear stresses [6, 7].

**Table 1.** Model Parameters of Hemodynamic Analysis

| Computational model statistic: | |
|---|---|
| Number of nodes | 298 224 |
| Number of elements | 1,011,095 |
| Type of the element | tetrahedral |
| Minimal length of the rib, mm | $10^{-5}$ |
| Maximal length of the rib, mm | $7\cdot10^{-4}$ |
| Flow parameters | |
| Dynamic viscosity, Pa·s | 0.0032 |
| Density, kg/m³ | 1,060 |
| Boundary conditions | |
| Condition at the boundary | no-slip |
| The model of turbulence | k - ε |
| Pump rate, bpm | 75 |
| $\beta = T_{sys}/T_{dia}$ | 1:1 |
| Systolic time, s | 0.4 |
| Diastolic time, s | 0.4 |
| Outlet pressure, mm.Hg | 100 |
| Inlet pressure, mm.Hg | 50 |

## 3 Results and Discussion

The analysis of the modelling results that reflect the change in the flows rate, the Reynolds shear stresses and the wall shear stresses in the blood chamber of the LVAD system, allows to make preliminary conclusions about the presence of stagnation zones, which are areas of potential thrombogenesis and hemolysis. A comparison

 DOI: 10.1051/matecconf/20167508002

of the velocities fields in the diastolic phase was carried out to estimate the influence of the mitral valve orientation on the behaviour of the flow inside the LVAD system (Fig. 3). Using the tilting disc valve "MIKS" in every set position, the flow is directed through the internal walls of the blood chamber and it moves in direction of the aortic valve.

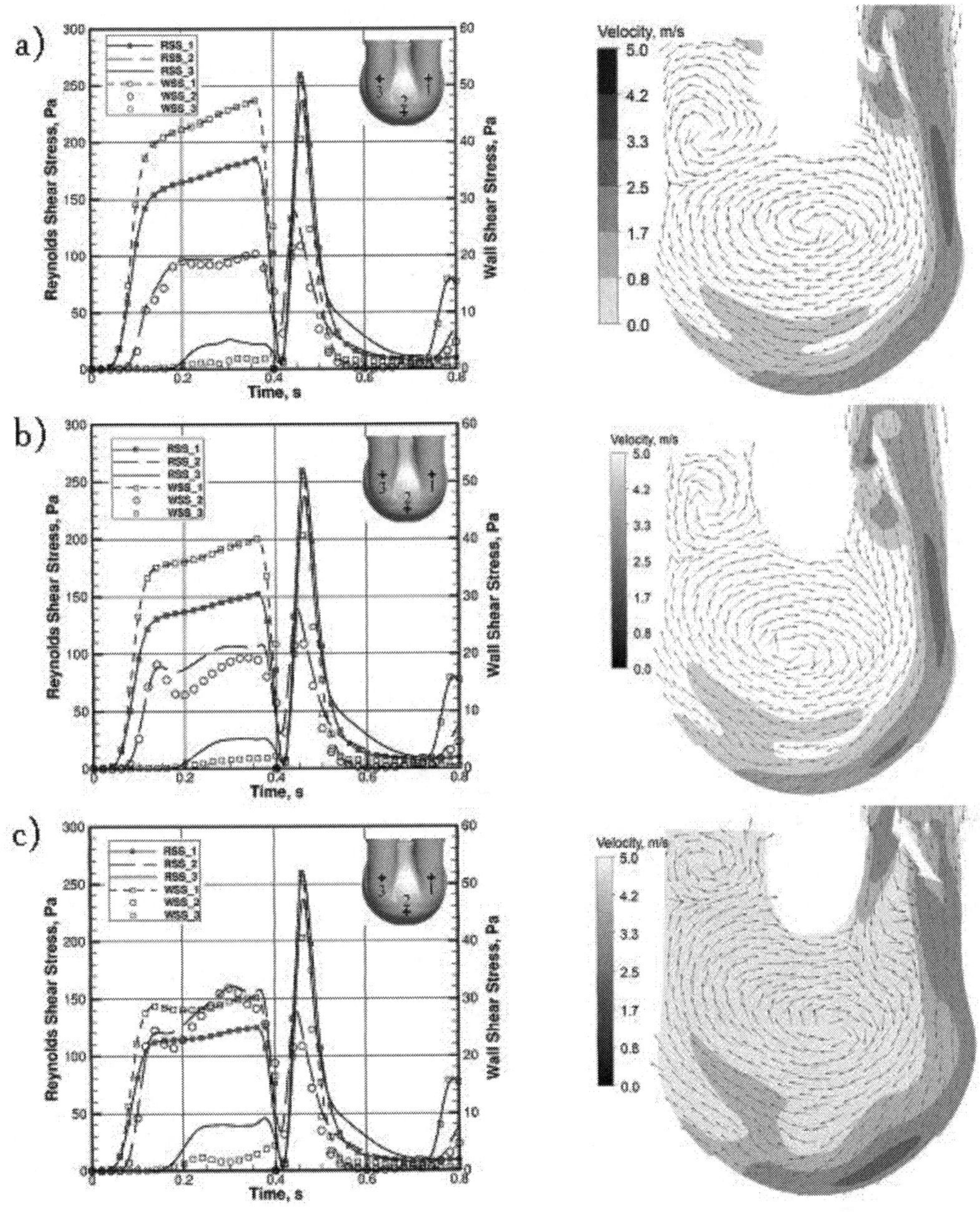


**Figure 3.** Cyclic variation of stresses (left) and velocity (right) magnitudes with inlet orientations of 0° (a), 30° (b), 45° (c) and all of outlet orientations were the same at three indicated points.

The maximal flow rate is observed at its passing through the valve, but exactly through the minor orifice it decreases, falling into the blood chamber. Installing the valves at 0 and 30 degrees it is observed the union of the flows passing through the major and minor orifices, while the flow passing through the minor orifice is directed along the wall located between the inlet and outlet ports, the same for the angle of 45 degrees, and it facilitates its washing. The lack of a good internal wall washing in these areas can contribute to the formation of stagnation zones and thrombus formation. The valve installation at the angle of 45 degrees provides a better internal wall washing of the blood chamber, due to the "direction" of the inlet flow into the chamber. Using the valve, installed at 45 degrees, a maximal swirl flow effect of the liquid is observed inside the blood chamber in the side of the opposite branch. Figure 3 shows the change graph of the Reynolds shear stresses (RSS) and the wall shear stresses (WSS) during the cardiac cycle. Shear stresses are determined in three points: 1 - inlet port, 2 - centre of the blood chamber and 3 - outlet port for every angles of orientation of the valve. The comparison of the shear stresses values arising at different positions of the valve disk showed that the minimal stresses are caused with the valve at the angle of 45 degrees. Placing the valve at an angle of 45 degrees, the shear stresses amplitudes for the first and second points have similar values, while for angles of 0 and 30 degrees the difference between the Reynolds shear stresses and wall shear stresses is more pronounced. For both types of the valves used, the Reynolds shear stresses have bigger values than the wall shear stresses. The lower values of both types of shear stresses arising during the installation of the mitral valve at an angle of 45 degrees, facilitate a less intensive hemolysis of the erythrocytes and improve the hemodynamic indicator of the LVAD system.

Application of the rapid prototyping technologies allowed to obtain as single elements of the device as the complete assembly of the product (Fig. 4) from the materials, of which the properties are similar to the properties of materials [8] planed to apply them during the manufacture of LVAD systems for in-vivo tests. The prototype's development allowed to evaluate the appearance of the designed device, its implantability and also to carry out the hydrodynamic test to evaluate its efficiency without manufacturing items with additional accessories, and hence to obtain a considerable saving in time.

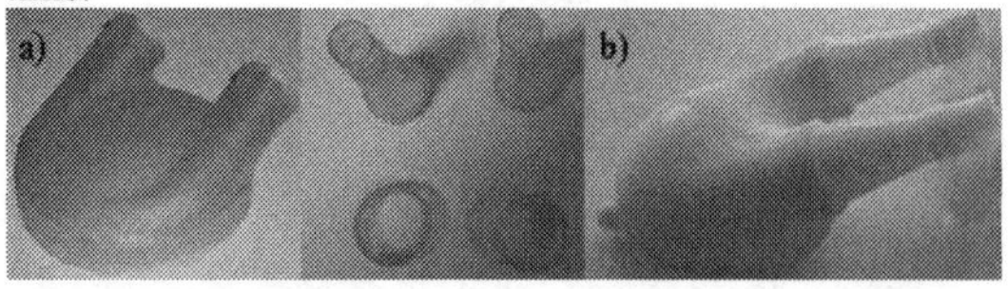


**Figure 4.** The pneumatic LVAD components manufactured using rapid prototyping technologies: housing of the blood pump and connectors (a), LVAD assembly (b).

However, the application of these technologies for manufacturing blood pumps is not appropriate due to the materials used in prototyping, which are not approved by the Ministry of Health of the Russian Federation to apply it in medical equipment.

## 4 Conclusions

In this work it was examined the possibility of application of CAD/CFD/RP technologies during the design of LVAD systems. The main focus is on the evaluation of the influence of the position of the mitral valve in hydrodynamic processes occurring inside the blood chamber during the operation of the LVAD system. The obtained results showed that the use of the tilting disk valve "MIKS" commercially available in Russia oriented at a 45 degrees angle provides a good flow internal wall washing of the blood chamber of the LVAD system. A good wall washing and swirl flow reduce the thrombosis risk during the operation of the LVAD system. The absence of stagnation zones, recirculation flow zones, allow us to talk about an acceptable design of the blood chamber. The application of CAD/CFD/RP-technologies allows accelerating the design process of LVAD systems. The manufactured components of the LVAD system are completely operability and give the possibility to estimate the critical hemodynamic and functional parameters of LVAD system.

## Acknowledgment

Financial support: This research was supported by the Presidential Program of Support for Young Candidates of Science (project no. MK-5860.2015.8).

## References

1. L. V. Belyaev, A. B. Ivanchenko, A. V. Zhdanov, V. V.Morozov, Biomed. Eng., **49**, 1, 24-28 (2015).
2. LLC "Roskardioinvest". 2016. *The mechanical "MIKS" heart valves*. [ONLINE] Available at: http://roscardioinvest.ru/eng/index.php?id_subpart=10.
3. ANSYS, Inc. ANSYS CFX Technical Brief, 2015. Release 14.0. Available at: http://www.ansys.com/staticassets/ansys/staticassets/resourcelibrary/techbrief/cfx-numerics.pdf.
4. M. Rosenfeld, I. Avrahami, S. Einav, J. Biomech. Eng., **124**, 21–29 (2002).
5. I. Avrahami, M. Rosenfeld, S. Raz, S. Einav, Artif. Organs. **30,** 7, 529–538 (2006).
6. K. Okamoto, T. Hashimoto, Y. Mitamura, J. Artif. Organs, **6**, 3, 162 – 167 (2003).
7. E. Okamoto, S. Fukuoka, E. Iwasawa, Y. Mitamura J. Artif. Organs, **4**, 205–213 (2001).
8. V.V. Morozov, A. V. Zhdanov, Mechatronics, automation, control, **11**, 30 – 31, (2008), (in Russian).

# Impact of familiarity on information complexity in human-computer interfaces

Maxim Bakaev

*Economic Informatics dept., Novosibirsk State Technical University, Novosibirsk, Russia*

**Abstract.** A quantitative measure of information complexity remains very much desirable in HCI field, since it may aid in optimization of user interfaces, especially in human-computer systems for controlling complex objects. Our paper is dedicated to exploration of subjective (subject-depended) aspect of the complexity, conceptualized as information familiarity. Although research of familiarity in human cognition and behaviour is done in several fields, the accepted models in HCI, such as Human Processor or Hick-Hyman's law do not generally consider this issue. In our experimental study the subjects performed search and selection of digits and letters, whose familiarity was conceptualized as frequency of occurrence in numbers and texts. The analysis showed significant effect of information familiarity on selection time and throughput in regression models, although the $R^2$ values were somehow low. Still, we hope that our results might aid in quantification of information complexity and its further application for optimizing interaction in human-machine systems.

## 1 Introduction

Human operators so far remain an indispensable part of most control systems, receiving, processing, and outputting information just like any other its component. Although the recognized advantages of humans' in comparison to machines are their integrity, adaptability, and self-development capability, they also remain a somehow limiting factor in semi-automated control progress [1]. Probably the main reason for this is the unresolved problem of organizing optimal interaction between humans and computers (HCI). In particular, the cognitive aspect of HCI remains quite less explored compared to physical (ergonomic) one, as the work of human brain and memory remain somehow of a mystery to the present day.

Naturally, it would be desirable to develop theories describing how humans perform information processing, what are their productivity and constraints in that, so that complexity of user interfaces could be reasonably minimized, which is generally considered to be desirable [2]. In theoretic information science there exist such fields as Computational complexity theory and Analysis of algorithms, but their application in HCI at the moment seems limited. Indeed, complexity in interfaces is more commonly explored on a qualitative level, at best measured as *chunks of information*, per the infamous Miller 7±2 law for working memory capacity. It should be said that the law indeed is an interesting fundamental regularity, which is seemingly applicable in software interfaces as well – e.g. in [3] the recommended number of interactions per component in a component-based system is 5.

As for more human-oriented research, [4] compares effect of complexity (enhancements in software functionality) for elder and regular computer users. An extensive review on complexity is provided in [5], with the emphasis on organization of nuclear plants control rooms. A set of measures for information complexity that can be used to evaluate automation systems is presented in [1], where the authors view it as the combination of three basic factors: numeric size, variety, and relation. The stages of brain information processing are identified as perception, cognition, and action. In [6] they propose operation, transition, and screen complexity, and the quantitative interface complexity measures are based on informational entropy (H) concept, or the information quantity. Quite a straightforward formula:

$$T = 0.03 * H, \qquad (1)$$

where T is time required for receiving and processing information, was mentioned in [7, ref 84], but we feel the constant it employs it is at best averaged value, while each user, task, and concept should be special in HCI.

We also believe that currently limited applicability of Computational complexity theory in HCI is due to significant difference of users, even within a user group. In our current paper we seek to explore the subjective side of the information complexity, by analyzing the impact of *familiarity* (well-known versus unfamiliar information) in human information processing. It should be mentioned that there are clearly different levels for complexity and familiarity in user interfaces, the structure of which so far remains ambiguous. For example, we would identify them as *Task (Functionality) – Thesaurus – Layout – Material – Behaviour*, and so far we focus on the Material level, exploring the impact of familiarity in interface elements (the level corresponding to icons, buttons and other controls, etc.). So, in Section 2 we outline some related methods, while in Section 3 we describe the experiment conducted with real subjects and analyze its implications for the research problem.

## 2 Methods

Familiarity is quite extensively studied in such domains as marketing (product and brand familiarity), sociology, psychology, medicine [8], etc. Although in HCI the studies of familiarity started long before the era of WWW (e.g. [9]), currently the number of corresponding works is rather limited. In a study of effect of user familiarity with a website on its perceived privacy [10], the authors described cognitive and affective functions of familiarity. Among the more interface-related research we'd like to note [11], where the authors employ the concept of cognitive distance, which is important since it expresses the familiarity in a quantitative way. So, next we consider some other approaches and findings that use operable quantitative measures.

### 2.1 Human information input and output

Human, as information processor, can input and output information, and for both processes measures of speed can be introduced ($R_{in}$ and $R_{out}$), as information volume time derivative (or, we may also say, second derivative of knowledge). Naturally, absolute values of the Rs depend upon measurement units, and this adds certain confusion when speed estimations from different research works are measured. All in all, sources seem to agree that the upper limit of human processor to consciously perceive information is $R_{in}<10^2$ bit/s, although human nervous system may process as much as $R_{in}=10^9$ bit/s subconsciously; this fact probably being one of the grounds for popular talks about hidden reserves of human mind.

As for information output, it is said that maximum speed of human's vocal cords operation matches up to $R_{out}=5$ syllables/s in medium-term, which corresponds to $R_{out}=12.5$ sounds/s for English, while for other languages this value may differ. For humans equipped with instruments or tools, the values are generally presented in bits/s or chars/s. Hand-writing would be limited at 3.5 char/s, while average typing speed is about 6.7 char/s, with 10 char/s being the top record, which still seems to be inferior to oral speaking. In any case, maximum speed of human processor as information transmitter is estimated at 40 bit/s [7, pp. 21-22].

The values for conscious information reception seem to be only marginally higher. $R_{in}$ for speaking comprehension is estimated at 3~4 words/s, for reading comprehension – at 2.5 words/s or 18 bit/s, for reading aloud – 30 bits/s, for speedy reading (Russian language) – at up to 11.7 words/s or 80 bit/s. Most sources agree that although general $R_{in}$ for reading is at about 45 bit/s, the capability of human visual perception for long-term work is at about 8 bit/s only [7, p. 23].

The speed can be increased to certain extent by using multi-dimensional signals instead of simple uniform ones – e.g. tone and volume instead of just volume; or size, brightness, and colour, instead of just one varying visual stimulus – but the effect is not additive per each introduced dimension. The ability to excel with signals of increased complexity somehow explains why humans are notably good, in comparison to machines, at image recognition. In one of our previous experiments we found that the higher was the diversity of objects in visual search/selection tasks, the higher throughput (TPS) was demonstrated – from 42 bit/s for objects' vocabulary size = 2 (simplest tasks) to 78 bit/s for vocabulary size = 29 (Russian letters) [12]. It should be noted that recognition speed for familiar letters and digits was previously estimated as 55 bit/s [7, ref 75].

Another factor that is known to increase the speed is training, which in particular allows operators to perceive information on a higher level, processing fewer numbers of chunks/elements. For example, while novice pilots need to individually check every display, expert ones require only a glance on the familiar console to decide if the controlled object, i.e. airplane, is in order. For relatively simple case of incoming textual information flow, it may be represented by introducing the *coefficient of new information density* [7, p. 144-145].

This calls for accounting the degree of information familiarity for a recipient – a qualitative and subjective aspect, and they are generally not convincingly considered in Information Theory-related models.

### 2.2 Human Processor Model

The Human Processor Model (also MHP) that was initially developed in the 80s is essentially a cognitive modeling method predicting cognitive and motor processing times. In this it is based on pre-defined experimental values, and its general structure, which draws similarities between humans and modern computers' architecture, can be outlined as shown on Fig. 1.

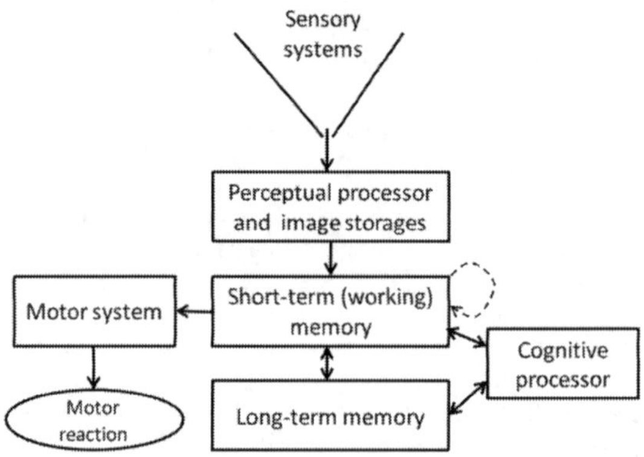


**Figure 1.** Principal components of the Human Processor Model.

It can be seen that the cognitive sub-system (incorporating working memory, long-term memory, and cognitive processor) is the most interlinked and indeed the most complex one. So, although the experimental statistics are quite reliable for senses (e.g. storage time for vision is 0.3-1 s, for hearing is up to 3.5 s), and very sound model exists for the motor system (Fitts' law), the cognitive aspect is quite less developed. The alleged *cognitive processor cycle time* varies in the range of 25-170 ms, but it remains rather unclear how to determine the number of cycles, in which cases information exchange with long-term memory starts and how exactly, what are the particulars for various user groups (e.g. senior people) and contexts of use (e.g. fatigue), etc. Thus we believe the cognitive aspect requires further exploration.

### 2.3 The Hick-Hyman law

Once, W.E. Hick noted that reaction time (RT) when choosing from N equally probable alternatives is proportional to the logarithm of their number:

$$RT \sim k * \log_2(N+1), \quad (2)$$

where k is the rate of gain of information. Afterwards, R. Hyman showed that RT in fact is linearly related to information quantity, i.e. the entropy of the set of stimulus ($H_T$):

$$RT = a_H + b_H * H_T, \quad (3)$$

where $a_H$ and $b_H$ are empirically defined constants. The slope in thus formulated Hick-Hyman law (1), $b_H$, in simplest cases is believed to be equal to 150 ms [13], then the corresponding Hick's rate of gain of information ($b_H^{-1}$) is equal to 6.7 bits/sec. This rate varies significantly due to subjects' age, gender, arousal, fatigue, etc., and also due to training a familiarity with the stimuli. So, it has been shown e.g. by [13] that for unknown alternatives RT increases even linearly with N, but has almost no increase for well-known options - which also calls for introduction of the familiarity factor in the corresponding model. Thus, in the next section we explore the impact of familiarity in the specially designed experiment and outline some notable results.

## 3 The experiment description

### 3.1 Subjects

Our experimental investigation was performed in several sessions during two years, and the subjects were recruited among Master students of Web Design major of a university located in Russia. In total, 33 subjects took part in the experiment, 14 of them being female and 19 being male. Their age ranged from 20 to 42 (M=24.5, SD=5.21). All subjects had normal or corrected to normal vision and high level of experience in using computers and mouse. Before the experiment, data regarding the participants' age and gender were gathered. All subjects participated in the experiment voluntarily, and prior to the experimentation informed consents were obtained. Each subject performed non-recordable test run of trials with random combinations of factors until fully mastering the tasks, to negate the effect of practice. The sessions with three different groups of participants took place with 1-year intervals on same or similar computer equipment (mouse, etc.) and screen sizes and resolutions.

### 3.2 The experiment design and procedure

In the related part of the experiment (more detailed description of the whole experiment can be found in [12]), the subjects were asked to undertake selection tasks. The were presented with a starting square positioned randomly (to negate the effect of movement direction) next to the work area of varying size $S_0$ (256x256, 512x320, 512x512, or 1024x640 pixels), in which N (8, 32, or 128) objects of varying size W (16, 32, or 64 pixels) would appear.

The participants were tasked to click the starting position with a mouse pointer and then, "as fast and as accurately as possible", find the target, move the mouse pointer and click the target – the coordinates of both clicks were recorded. If the second click was outside the target, error was recorded: the participant was taken to a next trial, but if overall error rate exceeded 10%, the message asking to increase accuracy would be displayed. The dependent variables were selection time (ST, between the two clicks) and error rate (E). Errors emerged when clicks were made outside the sought target (i.e. a false alternative was selected or the target just missed), and in case of error rate exceeding rate 10%, the message asking to increase accuracy was displayed between trials.

To measure and record the values of independent and dependent variables, an online application was developed with PHP and MySQL. The subjects used two web browsers, Chrome and Firefox, to negate the possible representation effect, and the performance time was measured with JavaScript to eliminate any server-side delay. Although this measurement method is known to have certain disadvantages, we still considered it appropriate for our rather exploratory study.

### 3.3 The target objects

Since our goals included exploring of the effect of vocabulary and familiarity, we employed different types of objects as the targets: geometric shapes, numbers or Cyrillic letters. The sizes of vocabulary K, corresponding to the number of various possible objects in the type, were 2, 10, and 29 respectively (we excluded several Cyrillic letters that looked too similarly to each other).

Since virtually all our subjects had Russian as their native language, for the Cyrillic letters we used frequencies based on the Russian National Corpus [14] (with the adjustment for the frequencies of the 4 letters excluded from the experiment), see Table 1.

For the digits, strictly speaking, the frequencies would correspond to uniform distribution (10% each), but only for long enough numbers. In our case, however, all the corresponding objects were represented as a single digit, so we chose to use the well-known Benford's law formula for the first digit [15]. We also had the consideration that the subjects likely to pay attention to initial (left-side) digits more often in their life, so their familiarity should follow the first-digit law more closely

 DOI: 10.1051/matecconf/20167508003

than the uniform distribution. The resulting frequencies for digits (with zero excluded from the analysis) are shown Table 2; and we'll be using percentage values rather than decimals, for better readability.

**Table 1.** Observed frequencies for the 29 Russian letters.

| Rank | Letter | Freq. | Rank | Letter | Freq. |
|---|---|---|---|---|---|
| 1 | О | 11.2% | 16 | Я | 2.0% |
| 2 | Е | 8.6% | 17 | Ы | 1.9% |
| 3 | А | 8.1% | 18 | Ь | 1.8% |
| 4 | И | 7.5% | 19 | Г | 1.7% |
| 5 | Н | 6.8% | 20 | З | 1.7% |
| 6 | Т | 6.4% | 21 | Б | 1.6% |
| 7 | С | 5.6% | 22 | Ч | 1.5% |
| 8 | Р | 4.8% | 23 | Х | 1.0% |
| 9 | В | 4.6% | 24 | Ж | 1.0% |
| 10 | Л | 4.4% | 25 | Ш | 0.7% |
| 11 | К | 3.5% | 26 | Ю | 0.6% |
| 12 | М | 3.3% | 27 | Ц | 0.5% |
| 13 | Д | 3.0% | 28 | Э | 0.3% |
| 14 | П | 2.9% | 29 | Ф | 0.3% |
| 15 | У | 2.7% | | | |

**Table 2.** Presumed frequencies for the digits.

| Digit | Frequency |
|---|---|
| 1 | 30.1% |
| 2 | 17.6% |
| 3 | 12.5% |
| 4 | 9.7.% |
| 5 | 7.9% |
| 6 | 6.7% |
| 7 | 5.8% |
| 8 | 5.1% |
| 9 | 4.6% |

## 4 The experiment results

### 4.1 Descriptive statistics

In the experiment, data for 6231 trials were collected, of which 5832 (93.6%) were considered valid (excluded were the ones when subject made erroneous double-click at start, or when the ST exceeded 15000 ms). Table 3 contains mean selection times (ST) and error levels (E) per number of targets (N) and vocabulary sizes (K).

**Table 3.** Mean ST (ms) and E (%) per N and K.

| N \ K | 2 | 10 | 29 | Mean ST, E |
|---|---|---|---|---|
| 8 | 964<br>6.1 | 1477<br>5.4 | 1456<br>3.9 | 1298<br>5.1 |
| 32 | 1043<br>4.2 | 2485<br>3.0 | 2710<br>3.7 | 2080<br>3.6 |
| 128 | 1411<br>4.7 | 5003<br>7.3 | 5570<br>8.2 | 3853<br>6.7 |
| **Mean ST, E** | 1107<br>5.1 | 2688<br>5.0 | 2839<br>4.9 | 2196<br>5.0 |

### 4.2 Regression analysis

In our previous analysis of the experimental data [12], we first built regression for the classic Hick's law, which turned out to be significant ($F_{1,5830}$=1494; p<0.001), but had quite poor $R^2$=0.204 and perplexing negative intercept:

$$ST = -691 + 616 * \log(N) \quad (4)$$

Among several factors potentially reflecting the complexity of interface that we tried in the regression, the "effective interface capacity", defined as number of alternatives multiplied by logarithm of their vocabulary size, showed the best fit:

$$ST = 1123 + 7.92 * N * \log(K) \quad (5)$$

The model showed high significance ($F_{1,5830}$=4160; p<0.001) and had relatively high $R^2$=0.416, compared to the $R^2$=0.666 for Fitts' law model (movement tasks) that we obtained in the control part of the experiment.

In accordance with our current goal, we added frequency as factor to the model. Digits and letters were considered separately, since their frequencies ($F_D$ and $F_L$ respectively, expressed in percentages) are valid only relatively to the same characters:

$$ST_D = 1691 + 8.7 * N * \log(K) - 25.3 * F_D \quad (6)$$

$$ST_L = 1469 + 6.89 * N * \log(K) - 27.2 * F_L \quad (7)$$

The model (6) was highly significant ($F_{2,1734}$=453; p<0.001), with $R^2$=0.343, and the $F_D$ factor was also significant (p<0.001). The model (7) was also highly significant ($F_{2,1900}$=587; p<0.001), with $R^2$=0.382, but the $F_L$ factor was significant only at $\alpha$=0.1 (p=0.094).

### 4.3 Throughput

In our previous work [12] we introduced the concept of throughput for selection tasks (TPS), with Fitts' index of task difficulty substituted for the "effective interface capacity" factor:

$$TPS = \frac{1}{Z}\sum_{i=1}^{Z}\left(\frac{1}{M}\sum_{j=1}^{M}\frac{N * \log(K)_{ij}}{ST_{ij}}\right). \quad (8)$$

We then showed reasonable validity of the TPS concept and discovered that K had significant effect on TPS, which showed highly significant effect (in ANOVA, $F_{2,5829}$=33.1; p<0.001). Mean TPS values for K=2, 10, and 29 were 41.9, 58.6, and 77.6 bit/s respectively. We reasoned that higher values of K led to higher TPS since higher diversity of objects reflect their higher complexity, and humans are prominent at recognizing complex images, rather than quickly and accurately performing simple and typical tasks.

To explore the effect of familiarity on TPS, we built regression models for digits and letters with respective familiarities as the factor:

$$TPS_D = 50.1 + 0.725 * F_D \quad (9)$$

$$TPS_L = 72.5 + 1.487 * F_L \qquad (10)$$

The model (9) was highly significant ($F_{1,1735}$=16.1; p<0.001), and the $F_D$ factor was also significant (p<0.001), although the $R^2$ was very low, at 0.01. The model (10) was also highly significant ($F_{1,1901}$=5.33; p<0.001), the $F_L$ factor was significant only at $\alpha$=0.05 (p=0.021), and the $R^2$ was even lower, at 0.003. As a comparison, in [12] the subjects' age factor had somehow comparable effect on their TPS ($F_{1,5830}$=9.48; p=0.002; $R^2$=0.002).

## 5 Conclusions

Controlling information complexity when designing interfaces in human-machine systems remains a crucial, but unresolved issue, due to lack of agreed quantitative metrics. Effective interface would be one that only presents relevant properties of the controlled object, and the final goal is to allow prompt and minimal-error interaction, and information load on human operators must be not maximized, but optimised according to their processing capabilities. However, there is still ambiguity regarding the ways to measure and optimize information complexity in HCI, in particularly because of its seemingly subject-specific nature.

In our paper we explored the impact of information familiarity on perceived complexity, having compared several existing approaches and finding none satisfactory in this regard. In the analysis of our experimental data, we based on our previous work [12], in which we formulated the concept of selection tasks throughput (8) that utilizes the "interface information complexity" (5) analogously to Fitts' index of difficulty. The results suggest that familiarity of digits and letters sought by the 33 subjects in the experiment did have significant effect on selection time (6), (7). The regression models for TPS (9), (10) also showed that familiarity was significant, although $R^2$ remained quite low.

The values obtained for TPS in the experiment, 42-78 bit/s, seem reasonable, as recognition speed for familiar letters and digits was previously estimated as 55 bit/s [7]. We see limitations of our study in its rather exploratory nature, which in particular lead to somehow low $R^2$ values. Also, in terms in conceptual validity we would agree that it's not so far satisfactory justified that frequency of digits' and letters' occurrence exactly correspond their familiarity for subjects. Still, we hope that our results might aid in defining the quantitative measure of information complexity and its further application for optimizing interaction in human-machine systems.

## Acknowledgements

The reported study was funded by RFBR according to the research project No. 16-37-00184 mol_a.

## References

1. J. Xing. Measures of information complexity and the implications for automation design. No. DOT/FAA/AM-04/17. Fed. Aviation Administration Oklahoma City Ok Civil Aeromedical Inst (2004).
2. A. Holzinger et al. On complexity reduction of user interfaces for safety-critical systems. Multidisciplinary Research and Practice for Information Systems, pp. 108-122 (2012).
3. U. Kumari and S. Upadhyaya. An interface complexity measure for component-based software systems. Int. J. of Com. Applications, 36.1 (2011).
4. Wu, Yu, and C. Slyke. Interface complexity and elderly users: revisited. Proc. SAIS, No. 51 (2005).
5. M.L. Cummings et al. Human-system interface complexity and opacity part i: literature review. Mass. Inst. of Tech., Cambridge, MA (2010).
6. H.G. Kang and P.H. Seong. An information theory-based approach for quantitative evaluation of user interface complexity. Nuclear Science, IEEE Transactions on 45.6, pp. 3165-3174 (1998).
7. V.F. Prisnyakov, L.M. Prisnyakova. Mathematic modelling of information processing by human-machine systems' operator (in Russian). Moscow: Mashinostroenie (1990).
8. L.A. Libby et al. Recollection and familiarity in schizophrenia: a quantitative review. Biological psychiatry 73.10, pp. 944-950 (2013).
9. B.L. Somberg and M.C. Picardi. Locus of the information familiarity effect in the search of computer menus. Proc. of the Human Factors and Ergonomics Society, 27 (9), SAGE (1983).
10. Y. Li. The impact of disposition to privacy, website reputation and website familiarity on information privacy concerns. Decision Support Systems 57, pp. 343-354 (2014).
11. C. Gough, R. Green, and M. Billinghurst. Accounting for user familiarity in user interfaces. Proc. of the 7th ACM SIGCHI New Zealand chapter's international conference on CHI (2006).
12. M. Bakaev, T. Avdeenko. A Quantitative Measure for Information Transfer in Human-Machine Control Systems. Proc. IEEE Int. Siberian Conf. on Control and Communications (SIBCON), IEEE (2015).
13. L.E. Longstreth et al. Exceptions to Hick's law: Explorations with a response duration measure. J. of Exp. Psychology: General, 114, pp. 417-434 (1985)
14. Wikipedia entry on Частотность (Frequency) https://ru.wikipedia.org/wiki/Частотность, 21.03.2016.
15. Wikipedia entry on Benford's law https://en.wikipedia.org/wiki/Benford's_law, 21.03.2016.

MATEC Web of Conferences **75**, 08004 (2016) DOI: 10.1051/matecconf/20167508004
*ICMIE 2016*

# Cluster management of computing resources

V.N. Ruchkin[1], M.N. Makhmudov[1], V.A. Romanchuk[1], V.A. Fulin[1] and B.V. Kostrov[2]

[1]*Ryazan State University, Ryazan, Russian Federation*
[2]*Ryazan State Radio Engineering University, Ryazan, Russian Federation*

**Abstract.** The study suggests management of computing resources through set-theoretic clustering. The research proposes production modelling of knowledge — rules of managing explicit and fuzzy structures, according to their technical specifications: system productivity, number of computing modules, capacity of microprocessor memory, etc. The paper describes an algorithm of "evaluation of management criterion value", according to the additive utility function.

## 1 Introduction

Methods of clustering have found wide use in cases when an a priori definition of data or their use for defining similar objects (samples) is impossible, whether such objects belong to the same group – cluster, or show difference between clusters [1, 2]. The current methods of clustering possess both advantages and disadvantages. That is why a compromise is often necessary between computational complexity, model fitting, and organization of the clusters' structure, depending on the nature of data under analysis and the specific field of their application. Also, the cluster shape depends on the measurement chosen, which influences the visualization and description of a cluster. Due to the uncertain character of the data, the introduction of fuzziness allows modelling smooth change of properties of multi-microprocessor implementations, taking into account qualitative structural relationships, which is especially important for distribution of computing resources [3, 4]. One of the advantages of the fuzzy approach is that the solution of the task of a choice occurs at the same time as the decision-making procedure [8, 9, 13, 14]. Besides, the mathematical tools of fuzzy sets allow for certain optimization, e.g., as max / min operations [14, 15]. As a result, it is possible to state that, to a certain degree, the result is an intellectual optimized system of choice-making [13, 16, 17]. The paper describes multi-criterion fuzzy management of computing resources with the help of linguistic variables, based on cluster analysis proposed by the authors [11-13]. The authors suggest using the set-theoretic approach [8, 9] that allows partitioning the input processing program into non-overlapping sets of equivalence classes, meeting such requirements as reflexivity, symmetry and transitivity, and further to implement cluster management of computing resources.

The structure of the research is as follows. Section 2 discusses set-theoretic clustering for presenting computing resources. Section 3 focuses on production modelling of knowledge - rules of presenting a management system. Section 4 suggests multi-criterion fuzzy choice of a management structure. Section 5 employs the algorithm of "evaluation of management criterion value". Section 6 forms a conclusion.

## 2 Set-theoretic clustering for presentation of computing resources

Cluster systems provide sufficient availability of systems at relatively low costs. Clustering is understood as "implementation of joining together machines, viewed as a unity by the OS, system software, applications, and users" [1]. When machines are clustered, if one of them stops working, the other processors can quickly redistribute the computing inside the cluster, because there is a single management mechanism for the cluster, which allows interactive tasking. However, one disadvantage is that it is not possible to fully manage a task after it is sent to execution by the user. Then, the system should enable the user to check the status of the task under execution, the capacity of memory required, and the processing time of the program's implementation.

The suggested approach is based on algorithms $A^{(j)}, \forall j = \overline{1,N}$ - an organized set of operations, or tuple $A^{(j)} =< O_1, O_q, ..., O_L, O_2, O_1 ..., O_r >$ [8, 9], which, in accordance with the chosen element base chipset and transformation $O_l \to \{MK_m\}; \forall m = \overline{1, K_m}; \forall l = \overline{1, L}$ is provided by data processing program $PR^{(j)} =< MK_1, MK_2, ..., MK_i, ..., MK_M >$, as a tuple of commands.

To implement parallelization and partitioning of the processing program into clusters, it is possible to introduce the notion of equality of data processing clusters $CL_l$ и $CL_k$. $CL_l = CL_k$.

Further, equality of clusters is regarded as equality of their lengths $|CL_l| = |CL_k|$ and their coincidence with an accuracy to command $MK_i(k)=MK_i(l)$, $PR(j)=<MK_i,MK_1,...MK_j>$.

$$CL_l = CL_k = \begin{cases} |CL_l| = |CL_k| \; \forall l,k = 1,N \\ MK_i^{(k)} = MK_i^{(l)}, \forall i = 1,|CL_l| \end{cases} \quad (1)$$

Here, Expression (1) serves as a criterion for the evaluation of the partitioning quality for input program clusters, accurate to a command, which serves as the measurement. If there is a change in transformation $O_l \to \{MK_m\}; \forall m = \overline{1,K_m}; \forall l = \overline{1,L}$, or if the element base changes, the clustering measurement changes, too, leading to different clustering results.

Then, in keeping with the design process, we define a set of all possible structures $S_w \in S$, that allow a $j$ processing program $PR$ to correlate cluster set $CL_l$ $S_w \in S : PR \Rightarrow \{CL_i\}; \forall i = \overline{1,l}; \forall w = \overline{1,W}$ (2).

Expression (2) defines an already distributed structure of the computing system. The research proves that Expression (2) meets the requirements of reflexivity, symmetry and transitivity.

Thus, structure $S_w$ of the data processing computing system under design correlates a set of independent and unequal clusters $CL_l$ to a $j$ processing program $PR$; here, the number of clusters is equal to the number of equivalence classes $L$, $CL_l$, $S_w \in S$, and their repetition factor ($RO_l$) is correlated by the order of their equivalence class ($a_l$):

$$\forall o = \overline{1,N}, PR^{(j)} \Rightarrow \{(CL_l)^q\}, \forall q = 1,a_l; \forall l = 1,L \quad (3)$$

Expression (3) defines an already distributed structure of the computing resource management, is a relationship of the equivalence of partitioning into independent modules, and shows a procedure of explicit parallelism. To preserve the semantics of the input program, we introduce explicit and implicit (fuzzy) parallelism of clusters.

Further, the focus of this research is on possible solutions to task (5): a production model of an expert system.

## 3 Production modelling of knowledge – rules of management system

Solving task (5), we obtain options of structural arrangement for our MPS that define the principal **knowledge — rules of management as productions** of an expert system.

Associations between facts are shown via heuristic rules – expressions of declarative knowledge about relations between objects: where $L$ is the number of equivalence classes, and $|a_l|$ is the order of each class.

According to the model, clusters are used to execute an array of consecutive programs that are not associated with each other; distributed programs (coarse-grain parallelism) of autonomous processes with shared data and parallel programs. Also the distributed programs are the main type of tasks for which clusters are created. Each program of this type consists of a finite number of processes that exchange data via the clusters' communicative environment. For this, the management system acts in accordance with the task and performs operations required to launch and execute a task or a group of tasks (VLIW) during the processing of the computation results.

In some cases, they are package tasks, and their content may be input for other tasks.

As a result, the following options of $\forall L, |a_l|= 1$, knowledge — rules are possible:

**1.** If $\forall L, |a_l|= 1$, and data exchange between fragments in equivalence classes is sequential, it is a pipeline structure.

**2.** If $L = 1, |a_l|= q$ and data exchange management is required simultaneously by all clusters of classes, it is a vector structure (SIMD according to N. Flynn) [8, 9].

**3.** If $L = 1, |a_l|= q$ and management of data exchange between clusters in the classes is sequential, it is a pipeline structure

**4.** If $\forall L, |a_l|= q$ is a pipeline structure and data exchange management is required simultaneously by all clusters of classes, it is a vector structure.

**5.** If $\forall L, |a_l|= q$ and data exchange management in the classes is sequential, the option is a pipeline structure.

**6.** If $\forall L, |a_l|= q$ and data exchange management between clusters in the classes is sequential, with minor exceptions for $l$ class clusters, it is a near-pipeline structure.

**7.** If $\forall L, |a_l|= q$ and data exchange management between clusters in the classes is sequential via a distributed processing program, except for class $l$ clusters, and class $l$ requires data simultaneously, it is a near pipeline-vector structure.

**8.** If $\forall L, |a_l|= q$ and data exchange management between clusters in the classes is sequential via a distributed processing program, but some classes require data simultaneously, it is a pipeline-vector structure.

**9.** If $|a_l|= L$ and does not depend on data transfer, it is an array structure (MIMD).

**10.** If $|a_l|= L$ and data exchange management is required simultaneously by clusters of all classes, but some classes require sequential exchange, it is a near vector-pipeline structure.

**11.** If $|a_l|= L$ and data exchange management is required simultaneously by clusters of all classes, but class $l$ clusters exchange data sequentially, it is a multiple vector-pipeline structure.

**12.** If $|a_l|= L$ and data exchange management is required simultaneously by clusters of all classes, except for some class $l$ clusters, it is a near vector structure.

**13.** If $|a_l|= L$ and data exchange management is

required simultaneously by all clusters of all classes, it is a vector structure.

The management structures analyzed are explicit ones, as they combine elements of explicit parallelism: pipeline, vector, pipeline-vector, or vector-pipeline. In the design procedures, some tasks are parallelized similarly to a vector structure, and other tasks are parallelized like in a pipeline. The result is a set of structures meeting requirements (5), and each requires classification, e.g., through linguistic variables based on fuzzy sets. Such complex and uncertain structures are fuzzy ones, and they are based on implicit parallelism, e.g., with pipeline or vector elements. The correlations obtained are basic for the training of expert systems.

## 4 Multi-criterion fuzzy choice of management structure

In practice, the process of **a management structure** design for clusters in a computing system is a complex multi-criterion choice task, because from the various $S_w \in S$ structures, an expert will make a decision depending on a generalized evaluation of each option, choosing the best one as regards the sum of principal contradictory criteria: 1. performance speed ($1/T_R$), 2. capacity of microprogram memory ($|MP|$), 3. number of microprocessor modules ($N_0$), 4. downtime ($T_{dt}$,), as well as additional criteria: 5. power consumption (P), 6. reliability (Q), 7. adaptability (A), 8. cost (C), etc. As a result, we obtain a set of management structures. To specify the choice of a management system, all the above criteria are ranked in a tuple according to their importance in technical design; this is further referred to as the **choice strategy**.

The authors [8, 14] have suggested the method of multi-criterion choice, based on fuzzy sets, where the characteristic function has a definition range $E=[a,b]$, and the elements are linguistic variables (LVs). An expert evaluates the structures and criteria, using notions of fuzzy logic, or terms $t \in T$ in the $E$ range, each fuzzy variable has a membership function $m_t(e)$, i.e. $t-\{m_t(e),e\}$.

The suggested approach uses two types of **binary** fuzzy variables: 'evaluation of management criterion value' $R_{f|j+1}$ and 'evaluation of management structure effectiveness' $Q_{f|j+1}$, using a **unary** linguistic variable: 'qualitative evaluation of criteria value' $m_{zj}(x)$ and 'evaluation of structure effectiveness value' $m_{zj,si}(x,y)$, accordingly.

The values of binary linguistic variables 'evaluation of management criterion value' и 'evaluation of management structure effectiveness' are computed following the method of a fuzzy algorithm for rules of this type: "if A, then B". Two or more rules of fuzzy expressions are connected with "else": "If A, then B, else C»=$[A][B]+[A]*[C]$, or $V$ ("if $A$, then $B$, else $C$") = $m_A/V(B),1-m_A/V(C)$.

## 5 Algorithm of "evaluation of management criterion value"

Evaluation of the importance of each above criterion is performed in keeping with the strategy design according to the composition rule $Z_j = Z_{j+1} \circ R_{f|j+1}, j=n-1,n-2,...,1$. The membership function for the fuzzy set $Z_j$ will be the result of the max/min operation:

$m_{zj}(x)=\bigcup_{y\in Y} m_{zj+1}(y)Vm_{Rf|j+1}(y,x)$.

For weighing the management criteria importance, fuzzy sets are converted to scalars in a suitable manner, e.g., when the choice of such binary variable values provides maximum use of the membership function. Then scalars of management criteria evaluations are associated with a scale and normalized. As a result, each criterion receives a specific weight $W_l, \forall l=\overline{1,n}; \sum_{l=1}^{n} W_l = 1$.

Similarly, from linguistic estimates of the relation of evaluation of management structure effectiveness, a normalized assessment of effectiveness is found for each option $S_w$, for each $j$ criterion $F_{ij}$.

An integral characteristic of the suggested multi-criterion fuzzy choice is the additive utility function $U_i$, calculated at the final stage of the suggested procedure for evaluation of alternative options of structures of data processing, based on the whole set of criteria. Besides, options are ranked by their value $U_i = \sum_{j=1}^{n} W_j * F_{ij}$.

Basing on the data obtained, the preferred choice of the CS structure $S_w \in S$ is defined, so as to maximize its utility function.

Basing on the described dependencies, an algorithm is suggested, broken into these steps.

1. Choice of carriers and initial distribution for the last criterion..
2. Determination of the strategy of design as a ranked set of given criteria.
3. Calling the process of definition of the linguistic variables, 'evaluation of management criterion value' $R_{f|j+1}$, performing computation according to composition rule $Z_j = Z_{j+1} \circ R_{f|j+1}$, with choice of criteria beginning with the last one $j=n-1,n-2,...,1$.
4. Calling the process of evaluation of quality of criteria value, performing computation according to formula $m_{zj}(x)=\bigcup_{y\in Y} m_{zj+1}(y)\bigcap m_{Rf|j+1}(y,x)$..
5. Obtaining scalar estimates $X_j^k$ of the criteria values, i.e. numeric identifications of those carriers in case of which values of criteria reach the value of one.
6. Obtaining evaluations, according to the formula $H_i = 3(X_j^k + 1) + 1$.
7. Obtaining normalized scales according to the formula: $W_j = \dfrac{H_j}{\sum_{j=1}^{n} H_j}$.

In the general case, a cluster management system should consist of two subsystems: one for the managing node and the other for the computing node – peripheral. So a cluster should have two communications systems: for interaction of components of a distributed program in the cluster, and for organization of interactions between the subsystems of the cluster management system. In this case, the communications system will ensure a maximum throughput and minimize delays in data transfer.

Otherwise, additional load from the management system will reduce the productivity of the cluster. The management subsystem will interrogate peripheral subsystems of the cluster nodes and accept user tasks, sending them to the computing nodes and receiving information about the current state of each cluster. As a result, it will be possible to avoid traffic volume surges when peripheral systems act in miscoordination.

In the general case, a cluster management system performs on a client-server principle, and the management system is the server, while each peripheral system is a client.

## 6 Conclusion

The paper suggests set-theoretic clustering, based on which, production modelling of knowledge – rules of management is performed; it also describes multi-criterion fuzzy choice of a structure for a computing system, according to a required choice strategy. With a view to this, the authors have developed an expert system to analyze fuzzy structure output according to various technical specifications: performance speed $(1/T_R)$, capacity of microprogram memory $(|MP|)$, the number of microprocessor modules $(N_0)$, etc. To implement the management strategies, the authors have developed an algorithm of evaluation of a management criterion value, in order to compute the additive utility function $U_i$ of alternatives in management of computing resources.

## References


1. A.I. Avetisyan, D.A. Grushin, A.G. Ryzhov. Systems of cluster management. // Proceedings of the Institute for System Programming of Russian Academy of Sciences. – 2002. – L. 3.
2. Ruchkin V.N., Romanchuk V.A., Development of algorithms determining the type of structure neuroprocessor system, based on the description of relations of its elements // Informatics and Applied Mathematics: Interuniversity collection of proceedings. Ryazan – 2011. – p. 99.
3. L. A. Zadeh, "Fuzzy logic and its application to approximate reasoning". In: Information Processing 74, Proc. IFIP Congr. 1974 (3), pp. 591–594.
4. L. A. Zadeh, "Computing With Words. Principal Concepts and Ideas". Berlin: Springer, 2012.
5. R. Jordans, L. Jozwiak, H. Corporaal, "Instruction Set Architecture Exploration of VLIW ASIPs Using a Genetic Algorithm", 3rd Mediterranean Conference on Embedded Computing MECO – 2013 Budva, Montenegro, pp.32-35
6. Ruchkin V.N., Romanchuk V.A., Evaluation of the results of modelling of computing systems based on neuroprocessors // Izvestya TulGU. Tekhnicheskiye Nauki. – Tula: Tula State University, 2013. Issue 9, part 2.
7. E. Diken, R. Jordans, R. Corvino, L. Jozwiak, H. Corporaal, and F. A. Chies, "Construction and exploitation of VLIW ASIPs with heterogeneous vector-widths", Microprocessors and Microsystems, vol. 38, no. 8-B, pp. 947–959, November 2014.
8. Romanchuk V.A., Ruchkin V.N. Algorithms of analysis of computing structures based on microprocessors // Vestnik RGRTU. 2012.№2. Issue 40. Pp. 61-67.
9. Ruchkin, V. Romanchuk, R. Sulitsa. Clustering, Restorability and Designing of Embedded Computer Systems Based on Microprocessors", 2nd Mediterranean Conference on Embedded Computing MECO – 2013 Budva, Montenegro, pp. 58-61.
10. Yury Bekhtin, Victor Gurov, Maria Guryeva Algorithmic Supply of IR Sensors with FPN Using Texture Homogeneity Levels 3rd Mediterranean Conference on Embedded Computing MECO – 2014. Budva, Montenegro. pp. 252-255.
11. Grigorenko D.V., Ruchkin V.N. Improving restorability of cluster data processing computing systems. // Tsifrovaya obrabotka informatsii № 1, 2013
12. Ivutin A.N., Larkin E.V. Estimation of Latency in Embedded Real-Time Systems 3rd Mediterranean Conference on Embedded Computing MECO – 2014. Budva, Montenegro. pp. 236-239.
13. V.N. Ruchkin, B.V. Kostrov, A.A. Kolesenkov, E.V. Ruchkina. Anthropogenic Situation Express Monitoring on the Basis of Fuzzy Neural Networks // Proceedings of the 3rd Mediterranean Conference on Embedded Computing MECO, 2014. Budva, Montenegro. pp. 166-169.
14. Ruchkin V.N., Romanchuk V.A., Fulin V.A. Design of computing systems based on fuzzy clustering // Vestnik RGRTU. 2014. №4. Issue 50-1. Pp. 87-93.
15. Romanchuk V.A., Ruchkin V.N. Design of program means of analysis of microprocessor systems // Vestnik RGRTU. 2010. №32. Issue 40. Pp. 61–67.
16. Ruchkin V.N., Fulin B.A. Architecture of computer networks. Moscow: Dialog-MIFI, 2008.
17. Ruchkin V.N., Fulin V.A. Development of a model of a complex neuroprocessor structure // Tsifrovaya obrabotka signalov, 2012, № 4, pp. 70-74.

 DOI: 10.1051/matecconf/20167508005

# Creation of collision data base through the "bottom-up" approach

Alla Brom, Evgeniya Gorlacheva and Irina Omelchenko

*Bauman Moscow State Technical University, 105005, 2-ya Baumanskaya, 5, Moscow, Russian Federation*

**Abstract.** The present epoch of knowledge economy, research and innovation poses unpredictable challenges for high-tech enterprises in aerospace industry. Knowledge and technology have mostly become the key source of creating competitive advantage. However production of successful aerospace products and service is impossible without collisions. The aim of the paper is to elaborate a bottom-up approach that allows reducing collisions by setting up a database of collisions (CDB) that occurs in the life cycle process. The introduced definition of "collision" determines any breach in the implementation process, leading to a deviation in product quality. The classification of collision types is given and the stages of collision management system in aerospace enterprises are described. By means of Risk Priority Numbers (RPN) method the authors have proposed to evaluate the significance of probable production collisions. According to conclusions of the analysis, collision database allows to accumulate, formalize and store invaluable experience of skilled practitioners, especially in unique production process. Other research outcome deals with the understanding that collision database allows detecting business processes where collision occurrence is not always obvious.

## 1. Introduction

The production of rocket and space equipment assumes a large variety of industrial processes and manufacturing operations, which are characterized by their specificity: a high level of workability, a single production, high risks (financial, economic, environmental, etc.), high level of responsibility. Reliability of aerospace technology ensures due to quality engineering and manufacturing processes.

Moreover the Russian aerospace enterprises are continuously updated technological support: there are additional options, progressive design decisions, etc. In aerospace engineering and manufacturing the enforcement requirement is considered to be the continuity and knowledge transfer. The problem of continuity is achieved through the creation of a knowledge management system (KMS) in the aerospace company.

The creation of the KMS from scratch is quite costly because these costs will be recouped over a long period of time. However, space services are now very much in demand. The intensity of the space launches is increasing every year. Table 1 presents data of Russian space launches from 2010 to 2016 [1].

**Table 1.** Total amount of space launches (2010-2016)

| Year | 2010 | 2011 | 2012 | 2013 | 2014 | 2015 | 2016 |
|---|---|---|---|---|---|---|---|
| Total amount of launches | 23 | 24 | 26 | 27 | 26 | 17 | 0 |

The presented data suggest that the production of the satellite can be regarded as mass production. The presence of failed launches demonstrates the need for recording, collecting and processing errors, so called collisions. Reaction velocity to a collision occurred must be sufficiently high so that all admitted ones can be resolved immediately, i.e. in the manufacture of the next product.

## 2. Main part

The paper offers a pragmatic and cost saving approach to the practical realization of the collision data base creation. The proposed approach can be used as small and medium enterprises involved in aerospace projects and large-scale enterprises as well, through existing corporate information systems.

One more accepted perspective in KM area is knowledge influence to innovation growth. This perspective is extremely important in the high R&D intensity sectors which require radically new innovation approaches.

Knowledge acquisition can influence the R&D performance and it's capability to convert resources into innovative products best explains differences in enterprises' innovation performance [2]. Although the above mentioned perspectives are thoroughly elaborated there are many vague areas in KM. One of them is the warehousing and registering collisions in production process based on bottom-up approach.

Even typical and well organized production process could face unpredictable risks and obstacles. Hence the growing interest in high-tech enterprises in recent years associated with the creation of knowledge management system that would have solved the global problem of information formalization. But the process of implementing KMS for high-tech enterprises is rather complicated due to a number of factors:

• Corporate information system (CIS) that brings together developers, designers and production staff on a single platform are rather expensive.

• Creating internal portals that facilitate internal information exchange do not guarantee the quality of products.

• Particular characteristics in processes of unique production creates certain difficulties in the allocation and description of typical algorithms for design, engineering and production.

• There is an acute problem of the experience transfer (young workers lack the intuitive and practical experience, lack of mentoring, etc.), so that once there is a problem of formalization of experience and intuition.

Professional experience and intuition usually appear with the job age. Thus one could pose a question what the professional experience is? Our opinion is that this is the knowledge of previous errors, failures, collisions and consequences to which they lead. In these terms skilled practitioners are more competitive than the novices (inexperienced employees). So any high-tech enterprise faces a problem how to transfer knowledge from experienced professionals to younger employees?
In order to find out a solution of experience and intuition transfer, a concept of collision has been introduced.

Collision is considered as any breach of production process, leading to a deviation in product quality (performance/ non-compliance of required output parameters, etc.).

We discover a collision when aerospace system or its elements:

- doesn't (don't) function at all;
- doesn't (don't) function in a definite period of time;
- functions (function) in not a proper way;
- functions (function) not in full.

The cost of making changes to the project, on average, is hundreds of times less than the cost of making changes to the finished product. Early detection and prevention of collisions at their occurrence place lead to significant cost savings for the enterprise as a whole.

Creation of collision data base (CDB) designed for such conditions will allow formalizing intuitive decisions, both the correct and incorrect, and their consequences, and also warehousing professional experience of employees and preventing failures in production process (fig.1).

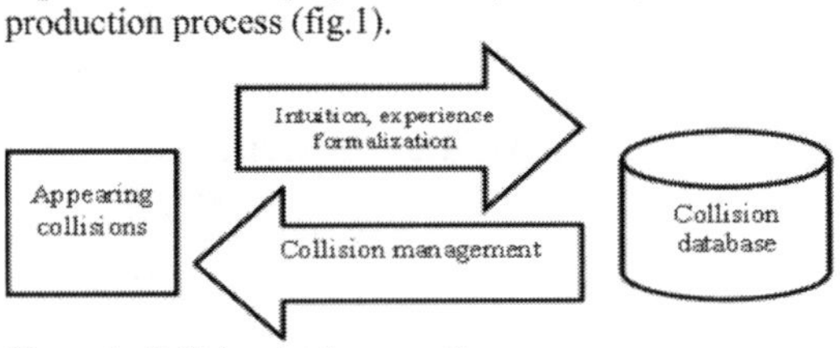


**Figure 1**. Collision database creation

In other words, if the aerospace enterprise hasn't opportunity to create a global knowledge management system that integrates information, experience and intuition (a "top-down" approach), it could accumulate and formalize experience in collisions and their solutions (a "bottom-up" approach).

The "bottom-up" approach allows creating and disseminating knowledge at the operational level, and only after the successful implementation of innovative proposals one can generalize the results to the enterprise level. The proposed approach also takes into account those accumulated production of knowledge held by more experienced employees (fig. 2).

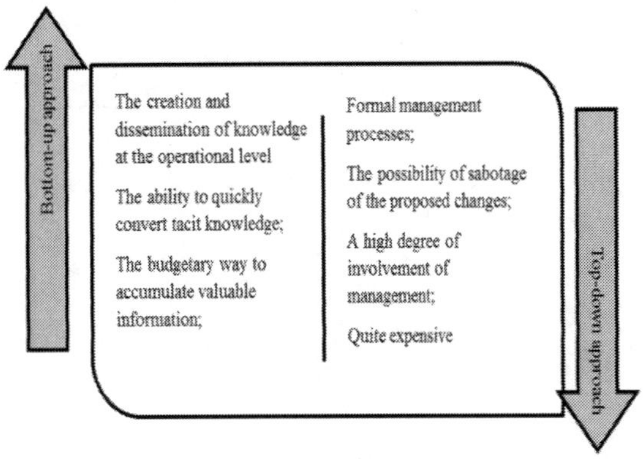


**Figure 2.** Main characteristics of two approaches

One of the key advantages of the "bottom-up" approach is the timely detection and prevention of production errors that occur due to the lack of sufficient professional experience. Throughout the product lifecycle, from design to the finished product, there is a lot of different changes that have a different nature arise, and may adversely affect the quality of the product.

The first step in creating CDB will systematize all possible collision types and their place of occurrence (Table 2).

**Table 2.** Classification of collision types

| Collision types | Probable examples |
|---|---|
| Design collisions | ✓ Incorrect execution of design documentation;<br>✓ Errors in the selection of dimensions and permissible variations;<br>✓ Errors in the selection of materials for the construction, by replacing materials |
| Technological collisions | ✓ Errors to comply with the requirements of the design process and management technical documentation;<br>✓ Technological documentation, allowing incorrect assembly of products;<br>✓ Errors related to the selection of equipment; |

 DOI: 10.1051/matecconf/20167508005

| | |
|---|---|
| | ✓ Errors in the selection of technological base. |
| Production process collisions | ✓ Violation in technology process;<br>✓ Failure of the equipment;<br>✓ Displacement of materials. |
| Human factor collisions | ✓ Errors due to a misunderstanding;<br>✓ Errors due to lack of experience;<br>✓ Errors due to negligence;<br>✓ Errors due to the unexpected factors;<br>✓ Intentional errors. |

Management of CDB is a permanent process that undergoes step by step three stages: warning, monitoring and cessation (fig.3).

**Figure 3.** Management of CDB

Warning is a method by which the product could not be manufactured or assembled incorrectly, a system gives a visual or audible message to the operator to indicate potential danger.

Monitoring is a method of conformity assessment by observation and judgment accompanied by corresponding measurements, testing or calibration.

Cessation is a method that suspends production process intentionally by means of warning horns or alerting services. The main source for CDB is the expertise in form of survey and interviews that allow registering all negative failures.

## 3. Method

Collisions could be considered in two aspects:

- qualitative one, that consists of search and identification of collision causes and their effects, that have negative impact to consumer of aerospace technique;
- quantitative one, that implies the risk evaluation of probable collision. The objective of this evaluation is the identification and priority of potential collisions. This measure allows us to estimate the order of importance of collisions in descending order, to facilitate research and development activities in order of priority, which subsequently eliminate the causes of collisions.

In order to define the importance of appearing collisions in LC manufacturing product one can use RPN method [3].

There are three types of RPN method determined by analysis of an object (fig.4).

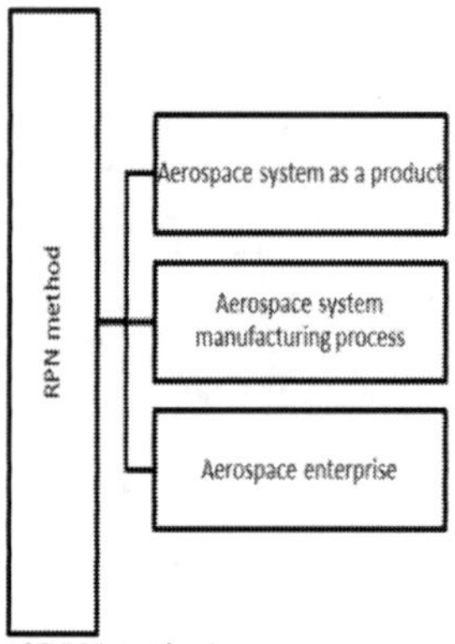


**Figure 4.** Types of RPN method

The first type is the analysis of an aerospace system directly (as a product). The main objective is the determination of collisions in main functions of a system.

The main aims of analysis and calculation of this type:

- Determination of main reliability parameters of an aerospace system;
- Specification of all manufacturing and exploitation restrictions of an aerospace system;
- Determination of critical safety parameters of an aerospace system;
- Taking decisions about changes in design process phase;
- Taking into consideration of manufacturing phases, assembly phases, repairing and transportation.

In order to document all results the authors have elaborated the spreadsheet template (table 3).

**Table 3.** Spreadsheet template

| Collision type | Effect to an aerospace system | Possible causes of collisions | Monitoring |
|---|---|---|---|
| | | | |

The second type of RPN method is the analysis of an aerospace system manufacturing process. It aims to identify collisions in production processes, assembly, installation and maintenance of the product - how process can affect the parameters of the product, their deviations that do not meet certain requirements, possible causes and consequences of the result, as well as warning means to identify and eliminate such collisions.

The main aims of analysis and calculation of this type:

- match critical product characteristics and parameters of the production processes;
- make changes in the structure of production processes and their optimization;
- identify processes shortcomings;
- propose changes in the product concept;

- identify warning means;
- select and follow a plan of action.

The example of a process analysis is given in the table 4.

**Table 4.** Example of process analysis

| Collision type | Effect to a manufacturing process | Possible causes of collisions | Monitoring plan |
|---|---|---|---|
| Collision in assembly phase | Impossibility of launch | Design causes; Equipment | Monitoring at all manufacturing phases |

The third type of RPN method is used for various process control levels of an aerospace enterprise, administration and production: the first level integrates administrative system, information system, human resources management, the last level includes the organization of the basic working units.

While using the Risk Priority Number (RPN) method to assess risk of consequences, the analysis team must:

• Determine the type of collision;

• Rate the severity (S), the likelihood of occurrence (O), and the likelihood of prior detection (D) for each collision

• Calculate the RPN by multiplying three ratings:

RPN = Severity x Occurrence x Detection

The parameters S, O, D range from 1 to 10. Thus, the maximum value of 1,000 RPN. The classification is carried out for the serious consequences of each type of collision shown in table.5.

The RPN method based on expert evaluations shows that human factor collisions may have more negative impact than technological collisions.

The next step in CDB management identifies all the potential collisions in order to assess the more critical ones. As discussed, a collision can be considered as: (i) the real lack of capacity to perform a certain action, or (ii) the presumed/perceived inability to perform it.

## 4. Analysis

**Table 5.** Short Example of RPN method application in collision management

| Severity Ranking Index | | Occurrence Ranking Index | | Detection Ranking Index | |
|---|---|---|---|---|---|
| Rank | Criteria | Rank | Criteria | Rank | Criteria |
| 1 | Undetectable effect on system | 1 | Remote chance for collision | 1 | Almost certain detection of collision |
| 2<br>3 | Low severity impact. Insignificant violation of functioning | 2<br>3 | Low collision rate based on previous designs with low collisions | 2<br>3 | Very high likelihood of detecting collision<br>High likelihood of detecting collision |
| 4<br>5<br>6 | Medium severity. Several tasks can't be done. Production process has to be ceased. The fulfillment of any tasks is possible only after repairing. | 4<br>5<br>6 | Medium collision rate based on similar designs with have some occasional collisions but not in major proportions | 4<br>5<br>6 | Moderately high likelihood of detecting collision<br>Moderate likelihood of detecting collision<br>Low likelihood of detecting collision |
| 7<br>8 | Heavy damage of functioning. Several tasks can't be fulfilled. High level of complexity in functional recovery | 7<br>8 | High collision rates based on similar designs which have been troublesome | 7<br>8 | Very low likelihood of detecting collision<br>Remote likelihood of detecting collision |
| 9<br>10 | Very severe problem involving potential safety problem or major non-conformity | 9<br>10 | Very high collision rates and the collisions will be major occurrence | 9<br>10 | Very remote likelihood of detecting collision<br>Cannot detect collision |

| **Collision type and its example** | **Severity index (S)** | **Occurrence Index (O)** | **Detection Index (D)** | **Risk Priority Number (O)*(S)*(D)** | **Recommended Action (should be elaborated as corporate standards)** |
|---|---|---|---|---|---|
| **Design collisions**<br>✓ Incorrect execution of design documentation; | 4 | 7 | 8 | 224 | Check design documentation |

| | | | | | |
|---|---|---|---|---|---|
| ✓ Errors in the selection of dimensions and permissible variations; | | | | | |
| **Technological collisions**<br>✓ Errors related to the selection of equipment;<br>✓ Errors in the selection of technological base. | 2 | 6 | 5 | 60 | Check technology documentation & process |
| **Production process collisions**<br>✓ Displacement of materials;<br>✓ Failure of the equipment. | 3 | 4 | 5 | 60 | Check production process |
| **Human factor collisions**<br>✓ Errors due to lack of experience;<br>✓ Errors due to negligence; | 9 | 7 | 7 | 441 | Establish motivation factors, implement coaching process |

## 5. Conclusion

The application of RPN method to CDB management looks promising. It is important to open a dialogue between skilled professionals and younger workers. Sharing a common language is extremely powerful in order to establish shared visions of the future. Second, we suggest that RPN analysis is applied systematically in all cases where there are collisions in lifecycle process.

## References

1. Statistics data from Federal space agency // http://www.federalspace.ru/launch/2010
2. B. Urgal et al., Knowledge resources and innovation performance: the mediation of innovation capability moderated by management commitment. Technology Analysis & Strategic Management, 25, (2013).
3. Barringer P. Failure mode and effects analysis, // www.barringer1.com/nov07prb.htm (2009)

# Integration of Multi-Feature Fusion and PLS-DA in Protein Secondary Structure Prediction

Guangzao Huang[1], Meishuang Tang[2], Jinting Guan[1], Sun Zhou[1], Wenbing Zhu[1], and Guoli Ji[1]

[1]*Department of Automation, Xiamen University, Xiamen, Fujian 361005, China*
[2]*Modern Educational Technical and Practical Training Center, Xiamen University, Xiamen 361005, China*

**Abstract.** Protein structure prediction has become one of the central problems in the field of modern computational biology. Protein secondary structure prediction is the basis of the spatial structure prediction of proteins. This paper presents a novel method for protein secondary structure prediction, which integrates multi-feature fusion and partial least square discriminant analysis (PLS-DA). Multi-feature fusion can make full use of the available information of proteins; however, it also leads to high-dimensional and redundant features. Then PLS-DA is utilized to deal with the fused protein data, which can effectively extract features from the protein data and remove the redundant information. Several benchmark datasets are used to verify the performance of the proposed method. The experiment results show that the proposed method gives satisfying prediction results of protein secondary structure compared with existing methods. Therefore the integration of multi-feature fusion and PLS-DA can fully utilize the available protein information, effectively reduce dimension and achieve robust classification in the multi-category analysis of protein secondary structure.

## 1 Introduction

Protein is a kind of important biological macromolecules, which plays many vital roles for almost all kinds of biological phenomena. The function of protein is dependent on its spatial structure, which greatly promotes the study of protein structure [1]. The methods for protein structure prediction can be divided into two categories. One is based on actual experiment method, including X-ray diffraction and Nuclear Magnetic Resonance. The other one is based on computing method, including homology modelling [2], folds recognition [3] and ab initio prediction [4]. Though the actual experiment method is more accurate, it takes a long time and is restricted by the technology and equipment. However, the computing method can overcome these limitations and therefore it has great development potential and space. In the field of biological information engineering, protein secondary structure is the basis of the prediction of protein tertiary structure and it is also the first step of many protein researches. Based on the contents of protein secondary structures, the protein structures are classified into four categories, all-α, all-β, α/β and α+β. At present, the application of machine learning in protein secondary structure prediction has become one of the hottest research areas in bioinformatics.

The feature representation of protein has a huge impact on the secondary structure prediction. The usually adopted feature representation of protein includes Amino Acid Composition (AAC) [5], polypeptide composition, functional domain composition [6], physicochemical features, PSI-BLAST profiles [7] and function annotation information [8]. Obviously, making full use of available protein feature information can effectively improve the prediction of protein structures. To this end, this paper focuses on how to integrate multiple protein information including the PSI-BLAST profiles, PROFEAT and Gene Ontology as the feature representation of protein. However, it makes the protein data contain much more features than that of samples; therefore it is difficult to be processed by traditional discriminant analysis. The partial least squares discrimination analysis (PLS–DA) [9] proposed by Errikson et al. is a robust discriminant analysis method. It is particularly suitable for the data of multicollinearity among the explanatory variables, high dimension and small sample, which is the case of the protein data after being integrated with multiple features. In this paper, we study the performance of the proposed method of integration of multi-feature fusion and PLS-DA in the prediction of protein secondary structures.

## 2 Generating feature vector of protein by multi-feature fusion

In classification, all the information that decides which class a sample falls into is included in the sample feature vector. Therefore, how to create the protein feature vector has a significant influence on the established classification model. In protein secondary structure prediction, in order to make full use of available protein feature information, in this paper the protein feature vector fuses three kinds of commonly used protein information, which are described in the following.

### 2.1 Linear predictive coding of the PSSM

The PSSM (Position Specific Scoring Matrix) [10] of a protein sequence represents homolog information affiliated with its aligned sequences, which is an effective feature for the prediction of protein structure and function. We used the PSI-BLAST program to search the NCBI's non-redundant (NR) database under the parameter setting h=0.001 and j=3 to get the PSSM of each protein sequence. The PSSM matrix elements were standardized between [0, 1] by the following standard sigmoid function

$$f(x)=\frac{1}{1+e^{-x}} \tag{1}$$

Then the linear predictive coding (LPC) scheme was used to turn the PSSM into a fixed-length feature vector.

### 2.2 Gene function annotation features

Gene Ontology (GO) is widely used in the field of bioinformatics, which can describe the function of gene and protein. The GO database used in this paper was downloaded from ftp://ftp.ebi.ac.uk/pub/databases/GO/goa/UNIPROT/gene_association.goa_uniprot.gz (released on Nov 26, 2014). We searched the GO terms of each protein sequence in the GO database and then all the GO terms related to the entire dataset are used to form a feature set:

$$protein_{GO}=\{g_i\}(i=1,\ 2,\ \cdots,\ \mathrm{n}) \tag{2}$$

$$g_i=\begin{cases}1,\ \textit{if its GO terms hit the ith term}\\ \hline 0,\ \textit{if not hit}\end{cases} \tag{3}$$

where $g_i$ is the i-th GO term and $n$ is the number of GO terms of the dataset.

### 2.3 Extracting structural and physicochemical features from PROFEAT

Sequence structural and physicochemical features have been frequently used in the development of statistical learning models for predicting proteins and peptides of different structural, functional and interaction profiles. PROFEAT is a web server for computing commonly used structural and physicochemical features of proteins and peptides from amino acid sequence [11]. In this paper, we acquired 1497-dimension vector of PROFEAT feature for each protein sequence.

## 3 Partial least squares discrimination analysis

Partial Least Squares (PLS) algorithm [12] is a widely modeling method in chemometrics and many other fields in recent years [13]. The PLS algorithm can establish the relationship between the two data block, eliminate the redundant information to achieve the goal of dimension reduction. The PLS regression model has many advantages such as simple, robust, less calculation, no need to remove any explanatory variables and easy to interpret. PLS-DA is based on PLS regression algorithm [14]. When dealing with PLS-DA, the class labels of the samples need to be binary encoded. $X_{n\times m}$ is the feature matrix, $Y_{n\times g}$ is the class matrix, $n$ is the number of samples, $m$ is the number of features, $g$ is the number of classes. The element $y_{ij}$ in $Y_{n\times g}$ represents the relationship between the i-th sample $x_i$ in $X_{n\times m}$ and the j-th class , which is expressed as

$$y_{ij}=\begin{cases}1,\ \text{if}\, x_i\in class\ j\\ 0,\ \text{if}\, x_i\notin class\ j\end{cases} j=1,\,2,\,\cdots,\,\mathrm{g} \tag{4}$$

Then PLS regression model is modeled between $X$ and $Y$ in the usual way. The PLS algorithm decomposes the $X$ and $Y$ as follows:

$$X=TP^{'}+E=\sum_{i=1}^{a}t_ip_i^{'}+E \tag{5}$$

where $T$ is the score matrix of $X$, $t_i$ is the score vector, $P$ is the load matrix, $p_i$ is the load vector, $E$ is the residual matrix and $a$ is the number of latent variables;

$$Y=UQ^{'}+F=\sum_{i=1}^{a}u_iq_i^{'}+F \tag{6}$$

where $U$ is the score matrix of $Y$, $u_i$ is the score vector, $Q$ is the load matrix, $q_i$ is the load vector and $F$ is the residual matrix. PLS regression algorithm extracts the latent variables from X and Y respectively, which satisfies the following conditions: (i) each group of latent variables extracting maximum variation information from the corresponding matrix; (ii) maximize the covariance between the two groups of latent variables. The general regression equation can be written as

$$Y=XB=XW\left(P^{'}W\right)^{-1}Q^{'} \tag{7}$$

where $W$ is the weight matrix of PLS regression algorithm.

For an unknown samples $x_{un}$ , the corresponding predicted value could be calculated by the following formula

$$y_{un}=x_{un}B=x_{un}W\left(P^{'}W\right)^{-1}Q^{'} \tag{8}$$

The predicted value $y_{un}$ is in a real number and it needs to be translated into class attribute. For instance, if the maximum value of $y_{un}$ is in the j-th column, the sample is predicted as the j-th class.

## 4 Materials and Methods

### 4.1 Datasets

Two groups of datasets were adopted to evaluate the proposed method. One group is the high similarity datasets, including Z277 and Z498. The other group is the low similarity datasets, including 1189 and 25PDB. The homology of proteins in low similarity dataset is lower than 40%. The specific information of these datasets is listed in Table 1.

**Table 1.** Datasets used to verify the effect of proposed method

| Dataset | Protein Number | | | | | Feature Number |
|---|---|---|---|---|---|---|
| | All-α | All-β | a/β | α+β | total | |
| Z277 | 70 | 61 | 81 | 65 | 277 | 3387 |
| Z498 | 107 | 126 | 136 | 129 | 498 | 3407 |
| 25PDB | 443 | 443 | 346 | 441 | 1673 | 6621 |
| 1189 | 223 | 294 | 334 | 241 | 1092 | 6137 |

### 4.2 Evaluation method

The k-fold cross validation is used to evaluate the effectiveness of proposed method. The accuracy and overall accuracy are used as the evaluation index, which are formulated as follows:

$$accuracy(i) = \frac{TP_i}{m(i)} \quad (9)$$

$$overallaccuracy = \frac{\sum_{i=1}^{M} TP_i}{N} \quad (10)$$

where $TP_i$ denotes the true positives of proteins in class $i$, $m(i)$ denotes the number of proteins in class $i$, $N$ denotes the total number of proteins.

## 5 Experimental results and discussion

After multi-feature fusion, the protein dataset has some common characteristics such as huge data volume, far less number of samples than the feature dimension and mixed with noise interference. Therefore the discriminant analysis method employed needs to filter out noise, extract the features and create a model for protein secondary structure prediction, which is well suited to PLS-DA. One can see from Table 2 to 5 that the predicted results given by the method of integration of multi-feature fusion and PLS-DA are comparable to the existing methods. Besides, another observation is that protein datasets with low homology are harder to be recognized than those of high homology, especially for the recognition of the protein class of $\alpha+\beta$. This is because PLS-DA is a linear discriminant analysis model; therefore it is not fit for nonlinear classification problem.

PLS-DA can also be used as a dimension reduction method. From Figure 1, it can be seen that the extracted features by PLS-DA can significantly increase the similarity between protein samples within same class and the difference between protein samples from different classes. Therefore, the extracted features by PLS-DA can be combined with other nonlinear classifier, such as SVM to improve the recognition of low homology protein datasets. From Figure 2, it can be seen that PLS-DA is also an effective visualizing technique for protein data. Therefore PLS-DA can amalgamate many actions, such as classification, dimension reduction and data visualization in one for protein secondary structure prediction, and integration of multi-feature fusion further strengthens these advantages.

**Table 2.** Performance comparison of different methods on Z277 dataset.

| Method | Prediction accuracy (%) | | | | |
|---|---|---|---|---|---|
| | All-α | All-β | a/β | α+β | Overall |
| Markov-SVM by Qin et al. [15] | 90.0 | 85.2 | 86.4 | 81.5 | 85.9 |
| PLS-DA (10) | 85.7 | 93.4 | 91.3 | 76.9 | 87.0 |

Note: the number in parentheses represents the optimal number of principal components of PLS-DA.

**Table 3.** Performance comparison of different methods on Z498 dataset.

| Method | Prediction accuracy (%) | | | | |
|---|---|---|---|---|---|
| | All-α | All-β | a/β | α+β | Overall |
| AAC-PSSM-AC by Liu et al. [16] | 94.4 | 96.8 | 97.8 | 93.8 | 95.8 |
| PLS-DA (11) | 93.5 | 96.0 | 97.1 | 88.4 | 93.8 |

**Table 4.** Performance comparison of different methods on 25PDB dataset.

| Method | Prediction accuracy (%) | | | | |
|---|---|---|---|---|---|
| | All-α | All-β | a/β | α+β | Overall |
| MLR model by Xia et al. [17] | 92.6 | 72.5 | 71.7 | 71.0 | 77.2 |
| PLS-DA (9) | 73.8 | 74.3 | 82.7 | 43.1 | 67.7 |

**Table 5.** Performance comparison of different methods on 1189 dataset.

| Method | Prediction accuracy (%) | | | | |
|---|---|---|---|---|---|
| | All-α | All-β | a/β | α+β | Overall |
| Markov-SVM by Qin et al. [15] | 53.8 | 79.3 | 68.3 | 32.0 | 60.3 |
| PLS-DA (10) | 82.1 | 75.2 | 74.3 | 33.2 | 63.8 |

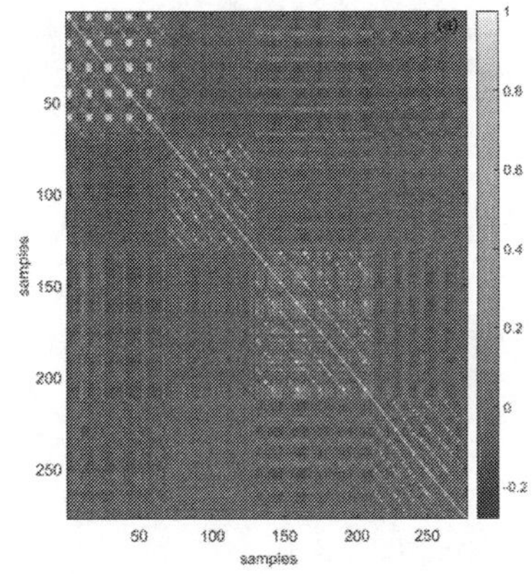

DOI: 10.1051/matecconf/20167508006

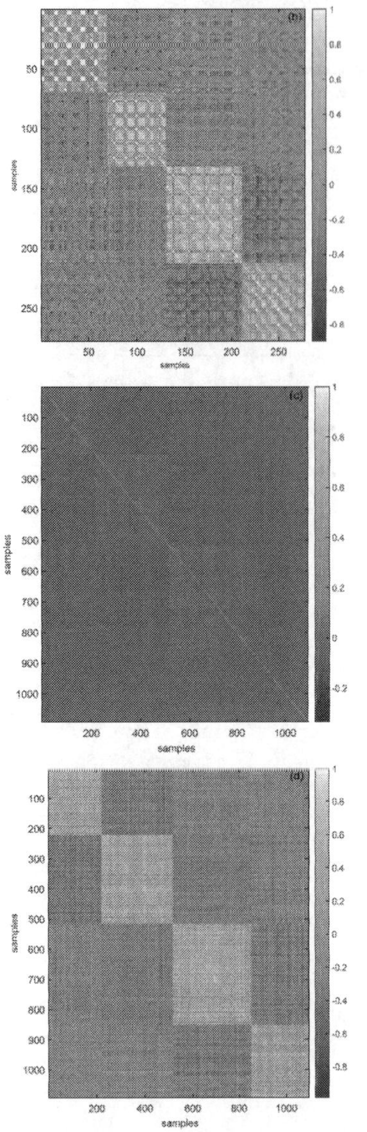


**Figure 1.** Heat maps of correlation coefficient matrixes of the protein datasets. (a) Z277 with raw features; (b) Z277 with 10 features extracted by PLS-DA; (c) 1189 with raw features;(d) 1189 with 10 features extracted by PLS-DA. For (a) (b), samples of 1-70, 71-131, 132-212, 213-277 and for (c) (d), samples of 1-223, 224-517, 518-851, 852-1092 represent the protein structure of All-α, All-β, a/β, α+β respectively.

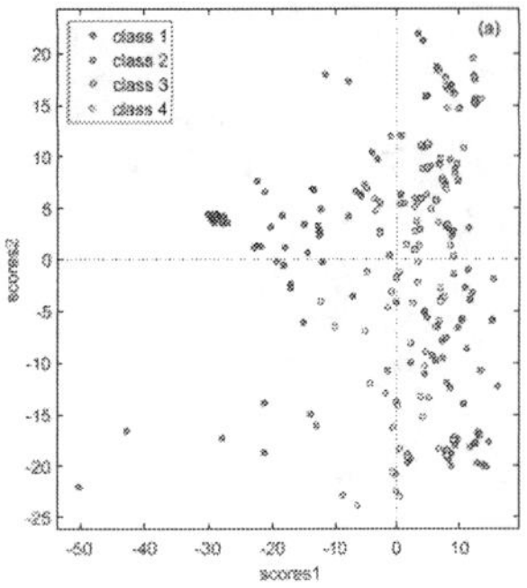


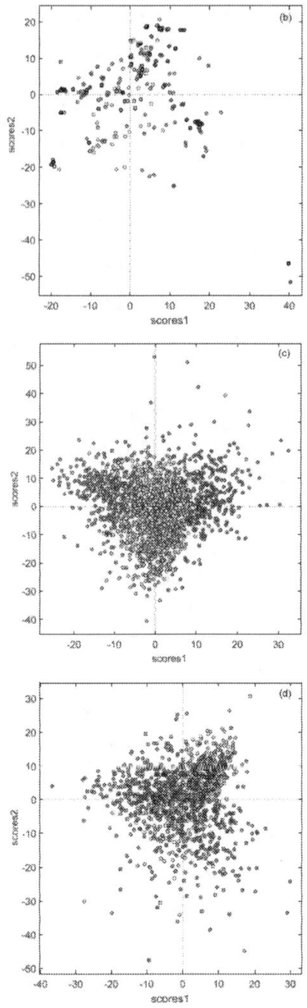


**Figure 2.** The two scores extracted by PLS-DA. (a) Z277; (b) Z498; (c) 25PDB; (d) 1189.

## 6 Conclusion

Multi-feature fusion can achieve the purpose of making full use of available protein information. PLS-DA can effectively extract features and remove redundant information. Integration of multi-feature fusion and PLS-DA can effectively deal with the problem of protein secondary structure prediction. Besides, PLS-DA is a linear discriminant analysis model; therefore it is not fit for nonlinear classification problem, such as recognizing the protein secondary structure of α + β in low similarity dataset. However, PLS-DA can still be used as a dimension reduction method; when it is combined with a regular nonlinear classifier, a more powerful nonlinear classifier can be constructed.

## Acknowledgment

This work was supported by Natural Sciences and Engineering Research Council of Canada, National

Natural Science Foundation of China (No. 61573296, 61304141, 61375077), Fujian Provincial Industry-University-Research Cooperation Major Projects in China (2014H6025), Fujian Province Natural Science Foundation in China (No. 2014J01252), the Specialized Research Fund for the Doctoral Program of Higher Education of China (No. 20130121130004), the Fundamental Research Funds for the Central Universities in China (Xiamen University: No. 2015Y1115, 201412G009, 2014X0217).

## References

1. DE Duve, C., Nature **333**, (1988)
2. Schwede, T., Kopp, J., Guex, N., Peitsch, M.C., Nucleic Acids Res. **31**, 13 (2003)
3. Micsonai, A. et al., Proc. Natl. Acad. Sci. **112**, 24 (2015)
4. Islam, M., Chetty, M., Evol. Comput. IEEE Trans. On **17**, 4 (2013)
5. Chou, K., Proteins Struct. Funct. Bioinforma. **43**, 3 (2001)
6. Rao, H., Zhu, F., Yang, G., Li, Z., Chen, Y., Nucleic Acids Res. **39**, suppl 2 (2011)
7. Song, J., Burrage, K., Yuan, Z., Huber, T., BMC Bioinformatics **7**, 1 (2006)
8. Marchler-Bauer, A. et al., Nucleic Acids Res. **39**, suppl 1 (2011)
9. Eriksson, L., Johansson, E., Kettaneh-Wold, N., Wold, S., Umetrics (2001)
10. Ahmad, S., Sarai, A., BMC Bioinformatics **6**, 1 (2005)
11. Li, Z.R. et al., Nucleic Acids Res. **34**, Web Server (2006)
12. DE Jong, S., Chemom. Intell. Lab. Syst. **18**, 3 (1993)
13. Wold, S., Ruhe, A., Wold, H., Dunn, I., WJ, SIAM J. Sci. Stat. Comput. **5**, 3 (1984)
14. Barker, M., Rayens, W., J. Chemom. **17**, 3 (2003)
15. Qin, Y.-F. et al., Protein Pept. Lett. **19**, 4 (2012)
16. Liu, T., Geng, X., Zheng, X., Li, R., Wang, J., Amino Acids **42**, 6 (2012)
17. Xia, X.-Y., Ge, M., Wang, Z.-X., Pan, X.-M., PloS One **7**, 6 (2012)

# Self-stopping actuating mechanism with high coefficient of efficiency of the forward stroke for linear motion drives

Valentin Morozov and Alexey Zhdanov

*Vladimir State University named after Alexander and Nikolay Stoletovs, Department of Mechanical Engineering, 600000, Vladimir, Russia*

**Abstract.** In this article is examined the an approach to the design of linear motion mechanisms, which consist of self-stopping planetary transmissions of the type $A^3_{H1}$ and $A^1_{H3}$ with transmission values $i = 7 \div 9$ based on the type of self-stopping planetary transmissions and the RSM. That combination allows a reliable self-stopping in backward stroke and a high coefficient of efficiency of the forward stroke (up to 0.95). The dependences of the angles of inclination of the wheel of the self-stopping transmissions from the angle of the original profile were lead. The obtained conditions allow the transmissions work of the type $A^1_{H3}$ and $A^3_{H1}$ in release mode for the back-and-forth stroke.

## 1 Introduction

Recently the roller-screw mechanism (RSM) are widely used in linear motion drives [1-4, 9, 10]. These mechanisms have a high coefficient of efficiency in back-and-forth stroke and the drives that use RSM to hold the load must have different automatic stopping mechanisms, which significantly increases the weight and the dimensions of the drive [6, 11, 12]. The application in such drives of self-stopping sliding screw-nut transmissions and worm gears with a forward stroke efficiency of <0.5 dramatically reduces the efficiency of the drive [5, 6].

Therefore, the search for new design solutions of actuating mechanism to drives with output linear motion is an actual task.

Self-stopping toothed planetary transmissions are promising for use in actuating mechanisms of such drives [8].

The carried out study of cylindrical transmission showed that, in order to achieve the self-stopping in them, it is necessary to use a self-stopping gear with a significant misalignment. [7, 8] As in the spur gears the possible misalignment of the profiles working areas are very limited because of the risk of tapering or undercutting the teeth, if it is necessary, they are replaced by helical or double helical gears, for which is possible a considerably bigger misalignment. In the helical transmissions the misalignment magnitude of the working profiles from the origin is mainly determined by the difference between the angles of inclination of the teeth profiles, which are part of the engagement [7, 8]. Therefore, the task of ensuring the self-stopping of the planetary transmission in a given state determines the necessary angles of inclination of the teeth in recession engagements. At the same time, the normal profiles angles are selected from the same consideration as in conventional transmissions.

## 2 General conditions

The kinematic diagram of the planetary transmission $A^3_{H1}$ with a one-crowned planetary pinion 2 having a external engagement with the central wheel 1 is shown in Fig. 1a, diagram of forces in the engagements of this transmission is shown in Fig. 2. The forces, acting on the cage H and the cage itself are not shown in Fig. 2.

We will write the links equilibrium equations in steady flow in the traction mode of the forward stroke when the wheel 1 is the driving one and the cage H is the driven one:

central wheel 1:

$$M_1 = r_{b1}\left(F_{nt12} + F_{T12}\text{tg}\alpha_{ty12}\right); \quad (1)$$

planetary pinion 2:

$$r_{b2}\left(F_{nt21} + F_{T21}\text{tg}\alpha_{ty21}\right) = r_{b2}\left(F_{nt23} + F_{T23}\text{tg}\alpha_{ty23}\right); \quad (2)$$

cage H:

$$M_H = F_{T21}\left(r_{b1}\text{tg}\alpha_{ty12} + r_{b2}\text{tg}\alpha_{ty21}\right) + F_{nt21}\left(r_{b1} + r_{b2}\right) + \\ + F_{nt23}\left(r_{b3} - r_{b2}\right) + F_{T23}\left(r_{b3}\text{tg}\alpha_{ty32} - r_{b2}\text{tg}\alpha_{ty23}\right) \quad (3)$$

DOI: 10.1051/matecconf/20167509001

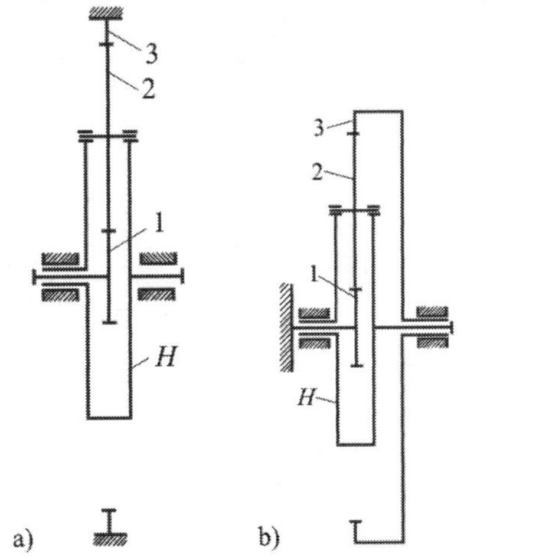


**Figure 1.** Planetary transmission with one-crowned planetary pinions a transmission $A_{H1}^{3}$, b transmission $A_{H3}^{1}$.

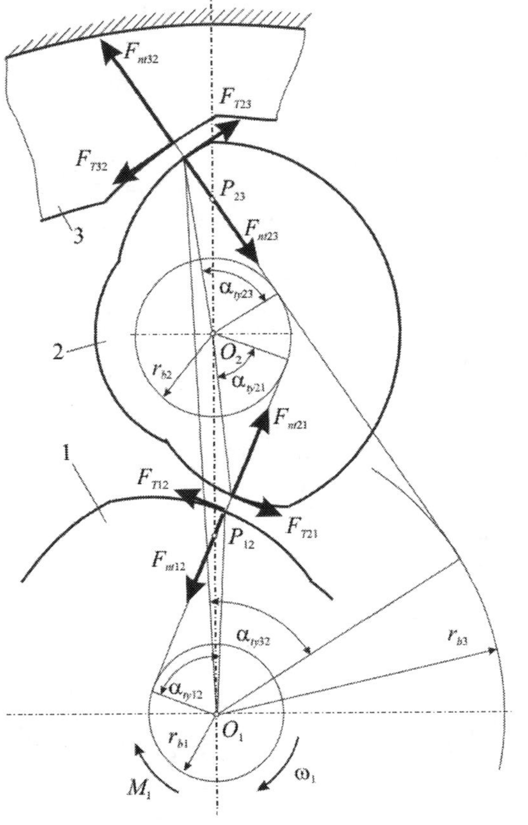


**Figure 2.** Force diagram of the planetary transmission $A_{H1}^{3}$.

We will convert the equation (1) ... (3) to:

$$M_1 = F_{nt12} r_{b1}(1 + k_{12}); \tag{4}$$

$$F_{nt21}(1 + k_{21}) = F_{nt23}(1 + k_{23}); \tag{5}$$

$$M_H = F_{nt21}[r_{b1}(1+k_{12}) + r_{b2}(1+k_{21})] + F_{nt23}[r_{b3}(1+k_{32}) - r_{b2}(1+k_{23})]; \tag{6}$$

where

$$k_{ij} = \frac{F_{Tij}\operatorname{tg}\alpha_{tij}}{F_{ntij}}; \quad i, j = 1, 2, 3. \tag{7}$$

If the friction force is proportional to the normal reactions, then the coefficients $k_{ij}$ are:

$$k_{ij} = \frac{f_{ij}\operatorname{tg}\alpha_{tij}}{\cos\beta_b}; \quad i, j = 1, 2, 3. \tag{8}$$

Solving the system of equations (4) ... (5), we will find the relation of the input and output moments:

$$\frac{M_H}{M_1} = 1 + \frac{r_{b3}(1+k_{21})(1+k_{32})}{r_{b1}(1+k_{12})(1+k_{23})} = 1 + \frac{r_{b3}}{r_{b1}}\eta_{12}\eta_{23}. \tag{9}$$

Multiplying by the transmission ratio, and considering that the main radii are proportional to the number of teeth, we get the known formula [8] for the coefficient of efficiency of forward stroke:

$$\eta_{1H} = \frac{1 + \frac{z_3}{z_1}\eta_{23}\eta_{12}}{1 + z_3/z_1}. \tag{10}$$

From the equations (4) ... (5) and (10) results that the traction mode of the transmission forward stroke $A_{H1}^{3}$ is possible under simultaneous implementation of the conditions:

$$-\frac{z_1}{z_3} < \eta_{23}\eta_{12} < 1; \; k_{12} > -1; \; k_{21} > -1; \; k_{23} > -1. \tag{11}$$

The realization of the conditions (11) substantially does not impose any further constraints during the design of the transmission, since at the real values of the coefficient of friction in the engagements these conditions won't be completed just for the wheels with very large angles of inclination of the teeth (over 80°).

The equations of the links equilibrium in traction mode of the backward stroke differs from the equations (1) ... (3), by the friction forces opposite signs, when H becomes the leading cage:

central wheel 1:

$$M_1 = r_{b1}(F_{nt12} - F_{T12}\operatorname{tg}\alpha_{ty12}); \tag{12}$$

planetary pinion 2:

$$r_{b2}(F_{nt21} - F_{T21}\operatorname{tg}\alpha_{ty21}) = r_{b2}(F_{nt23} - F_{T23}\operatorname{tg}\alpha_{ty23}); \tag{13}$$

cage H:

$$M_H = F_{nt21}(r_{b1}+r_{b2}) - F_{T21}(r_{b1}\operatorname{tg}\alpha_{ty12}+r_{b2}\operatorname{tg}\alpha_{ty21}) + \\ + F_{nt23}(r_{b3}-r_{b2}) - F_{T23}(r_{b3}\operatorname{tg}\alpha_{ty32}-r_{b2}\operatorname{tg}\alpha_{ty23}). \quad (14)$$

Introducing the indications according to the formula (7), we convert the equation (12) ... (14) to:

$$M_1 = F_{nt12} r_{b1}(1-k_{12}); \quad (15)$$

$$F_{nt21}(1-k_{21}) = F_{nt23}(1-k_{23}); \quad (16)$$

$$M_H = F_{nt21}[r_{b1}(1-k_{12})+r_{b2}(1-k_{21})] + \\ + F_{nt23}[r_{b3}(1-k_{32})-r_{b2}(1-k_{23})] \quad (17)$$

Solving the system of equations (15) ... (17), we will find the relation of the input and output moments:

$$\frac{M_H}{M_1} = 1+\frac{r_{b3}(1-\tau_{21})(1-\tau_{32})}{r_{b1}(1-\tau_{12})(1-\tau_{23})} = 1+\frac{r_{b3}}{r_{b1}}\eta_{32}\eta_{21} \quad (18)$$

Multiplying by the transmission ratio, and considering that the main radii are proportional to the number of teeth, we get the formula for the coefficient of efficiency of backward stroke $\eta_{H1}$:

$$\eta_{H1} = \frac{1+\dfrac{z_3}{z_1}}{1+\dfrac{z_3}{z_1\eta_{21}\eta_{32}}} \quad (19)$$

It should be noted that in the formula (19) the coefficients of efficiency $\eta_{21}$ and $\eta_{32}$ of backward stroke of every link are included, but not the coefficients of the forward stroke. Since the coefficient of efficiency of the backward stroke can be significantly different from the efficiency of the forward stroke, especially in the recession engagements, this circumstance is very essential.

From the equations (15) ... (17) and (19) it follows that during the realization of the wheels 1–2 to the recession engagement the conditions of the self-stopping transmission while maintaining the workability in release mode are:

$$-\frac{z_3}{z_1\eta_{32}} < \eta_{21} < 0 \quad (20)$$

From the inequality (20) the parameters $k_{12}$ and $k_{21}$ are expressed, taking into account the friction on the wheels 1 and 2:

$$\eta_{21} = \frac{1-k_{12}}{1-k_{21}} > -\frac{z_3}{z_1\eta_{32}} \quad (21)$$

The parameters $k_{12}$ и $k_{21}$ can be expressed through the generalized coefficient of friction $f_{12}^0$ and $f_{21}^0$ in the engagement of the wheels 1 and 2, and the normal angles $\alpha_{ny1}$ и $\alpha_{ny2}$ of the profiles of the wheels 1 and 2:

$$k_{ij} = \frac{f_{ij}^0 \operatorname{tg}\alpha_{nyi}}{\cos\beta_{yi}\cos\beta_b}, \quad i,j=1,2. \quad (22)$$

Analysis of (21) shows that it limits the maximum permissible value of the parameter $k_{12}$:

$$k_{12\max} < \frac{z_3}{z_1\eta_{32}}(1-k_{21\max}) \quad (23)$$

Thus the value $\beta_{y1}$ must satisfy the condition:

$$\sec\beta_{y1} < \frac{\cos\beta_b}{f_{12\max}^0 \operatorname{tg}\alpha_{ny1}} \frac{z_3}{z_1\eta_{32}}(1-k_{21\max}) \quad (24)$$

The equations of equilibrium of the links in the steady-state mode of release are different from (15) ... (17) only by the sign of the input torque M1. Their solution gives the following relation between the output and the input moments:

$$\frac{M_H}{M_1} = \frac{r_{b3}(1-k_{21})(1-k_{32})}{r_{b1}(1-k_{12})(1-k_{23})} - 1 = -\frac{z_3}{z_1\eta_{32}\eta_{21}} - 1 \quad (25)$$

In this case, when it is necessary to remove the release mode and the mechanism is designed to work only in traction mode of forward stroke, the parameters of the stopping profiles must satisfy the following conditions:

$$\cos\beta_{y1} \le \operatorname{tg}\alpha_{ny1}\sqrt{\sqrt{\frac{1}{4}+\left(\frac{f_{12\min}^0}{\sin\alpha_{ny1}}\right)^2}-\frac{1}{2}}; \quad (26)$$

$$\cos\beta_{y2} \le \operatorname{tg}\alpha_{ny2}\sqrt{\sqrt{\frac{1}{4}+\left(\frac{f_{12\max}^0}{\sin\alpha_{ny2}}\right)^2}-\frac{1}{2}} \quad (27)$$

At the same time the recession condition of the engagements should be saved:

$$\beta_{y1} > \beta_{y2} > \beta_{y3}. \quad (28)$$

Kinematic diagram of the planetary transmission $A_{H3}^1$ with a one-crowned planetary pinion 2 having a external engagement with the central wheel 3 is shown in Fig. 1b. The coefficient of efficiency of backward stroke $\eta_{H3}$ is:

$$\eta_{H3} = \frac{1+\dfrac{z_1}{z_3}}{1+\dfrac{z_1}{z_3\eta_{12}\eta_{23}}}. \quad (29)$$

where $\eta_{12}$ и $\eta_{23}$ are the coefficients of efficiency of backward stroke of the engagements of the wheels 2-1 and 3-2 when the cage is stopped.

During the realization of the wheels 3-2 to the recession engagements, the self-stopping conditions of the transmission that ensure its operation in release mode are:

$$\cos\beta_{y2} > \operatorname{tg}\alpha_{ny2}\sqrt{\sqrt{\frac{1}{4}+\left(\frac{f^{0}_{23\max}}{\sin\alpha_{ny2}}\right)^{2}}-\frac{1}{2}}; \quad (30)$$

$$\cos\beta_{y3} \le \operatorname{tg}\alpha_{ny3}\sqrt{\sqrt{\frac{1}{4}+\left(\frac{f^{0}_{23\min}}{\sin\alpha_{ny3}}\right)^{2}}-\frac{1}{2}}; \quad (31)$$

$$\sec\beta_{y3} < \frac{\cos\beta_{b}}{f^{0}_{23\max} tg\alpha_{ny3}}\frac{z_1}{z_3\eta_{23}}\left(1-\max k_{23}\right). \quad (32)$$

$$\beta_{y3} > \beta_{y2} > \beta_{y1} \quad (33)$$

The conditions that ensure the self-stopping transmission operation $A^{1}_{H3}$ only in traction mode of the forward stroke, similarly to the transmission $A^{3}_{H1}$ are as follows:

$$\cos\beta_{y3} \le \operatorname{tg}\alpha_{ny3}\sqrt{\sqrt{\frac{1}{4}+\left(\frac{f^{0}_{23\min}}{\sin\alpha_{ny3}}\right)^{2}}-\frac{1}{2}}; \quad (34)$$

$$\cos\beta_{y2} \le \operatorname{tg}\alpha_{ny2}\sqrt{\sqrt{\frac{1}{4}+\left(\frac{f^{0}_{23\min}}{\sin\alpha_{ny2}}\right)^{2}}-\frac{1}{2}}; \quad (35)$$

At the same time the recession must be also complied (33).

## 3 Analysis of the research results

The carried out analysis shows that the transmission $A^{3}_{H1}$ can be self-stopping during the backward stroke at the first or second variant maintaining high coefficient of efficiency values of the forward stroke. Fig. 3a shows the graphic of the wheels angles of inclination of the self-stopping transmission from the original profile, constructed according to the formula (24). The existence area of the first variant of self-stopping is the strip between the graphs $\beta_{y1\min}$ and $\beta_{y1\max}$ (two top of the graphic) for the angle of inclination of the tooth of the wheel 1 and the region below the graph $\beta_{y2\max}$ for the angle of inclination of the planetary pinion 2. The existence area of the first variant of self-stopping is the region above the graph $\beta_{y1\min}$ for the angle of inclination of the tooth of the wheel1 and the region above the graph $\beta_{y2\max}$ for the angle of inclination of the planetary pinion 2.

The kind of the existence regions of the first and second variants of self-stopping in the transmission $A^{1}_{H3}$ is similar to that shown in Fig. 3a.

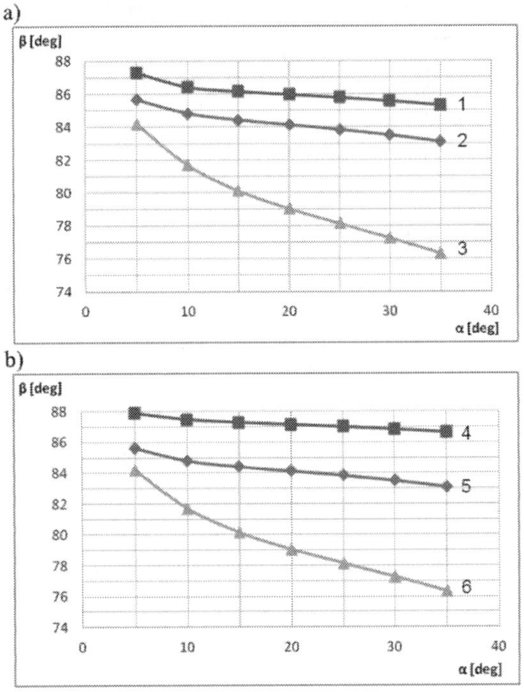


**Figure 3.** Self-stopping region of the planetary transmissions: a – transmission $A^{3}_{H1}$; b – transmission $A^{1}_{H3}$.

$\beta$ – angle of inclination of the wheel tooth; $\alpha$ – normal angle of the profile; 1 – $\beta_{y1\max}$; 2 – $\beta_{y1\min}$; 3 – $\beta_{y2\max}$; 4 – $\beta_{y3\max}$; 5 – $\beta_{y3\min}$; 6 – $\beta_{y4\max}$

The application of self-stopping planetary transmissions $A^{3}_{H1}$ and $A^{1}_{H3}$ with gear ratios $i = 7-9$ and a coefficient of efficiency of forward stroke until 0.98 in conjunction with RSM in the actuating mechanism allows to obtain effective linear motion drives with self-stopping and a coefficient of efficiency of forward stroke until 0.95.

## References

1. SKF Group, *SKF Roller screws: catalogue*, (2014)
2. 2. M.H. Jones, S.A. Velinsky, J. Mech. Design, *Kinematics of roller migration in the planetary roller screw mechanism*, **134**, №6 (2012).
3. 3. S.A Velinsky, B. Chu, T.A. Lasky, J. Mech. Design, *Kinematics and efficiency analysis of the planetary roller screw mechanism*, **131**, №1 (2009).
4. 4. V.V. Morozov, A.V. Zhdanov, ICAMEE, pp. 875-881 (2015)

DOI: 10.1051/matecconf/20167509001

5. V.V. Morozov, V.I. Panyukhin, V.V. Panyukhin, VISU press, *Mechanical transmissions: Coefficient of efficiency and self-stopping: Monograph*, (2002)
6. V.V. Morozov, VISU press, *Roller-screw mechanisms. Kinematic characteristics: Monograph*, (2005)
7. V.V. Morozov, A.V. Zhdanov, Rus. Eng. Res., **35**, №7 pp. 477-480 (2015)
8. V.I. Panyukhin, Nat. Sci. and Tech. Soc. Press, *Self-stopping mechanisms* (1981)
9. V.V. Kozyirev, VISU press, *Design theory and methods of design and study of planetary gear-nut threaded rollers and mechatronic units based on them: Monograph* (2011)
10. P.A Sokolov, O.A Ryakhovskii, D.S. Blinov, A. Laptev, Bulletin Mech. Eng., **1**, pp. 3-14 (2005)
11. S. Ma , G. Liu , J. Zhou, R. Tong, Nanjing Un. Sci. And Tech., *Optimal design and contact analysis for Planetary Roller Screw* (2009)
12. S. Ma, G. Liu, J. Zhou, R. Tong et al., Math. Prob. Eng. *New Study on the Parameter Relationship of Planetary Roller Screws*, (2012)

DOI: 10.1051/matecconf/20167509002

# Numerical modeling of deformation and vibrations in the construction of large-size fiberglass cooling tower fan

Shmakov Arthur Fanisovich, Modorskii Vladimir Yakovlevich

*Federal State-Funded Educational Institution of Higher Professional Education Perm National Research Polytechnic University, Center of high-performance computing systems, Perm, Russia (29 Komsomolsky prospekt, Perm - GSP, Perm krai, Russia, 614990)*

**Abstract.** This paper presents the results of numerical modeling of deformation processes and the analysis of the fundamental frequencies of the construction of large-size fiberglass cooling tower fan. Obtain the components of the stress-strain state structure based on imported gas dynamic and thermal loads and the form of fundamental vibrations. The analysis of fundamental frequencies, the results of which have been proposed constructive solutions to reduce the probability of failure of the action of aeroelastic forces.

## 1 Introduction

In Russia, the cooling towers are about 3500 pieces. Search capabilities significantly (by more than by 2 times) reduction in energy consumption of electric fans cooling towers can be carried out in the direction of optimal aerodynamic profiling of working blades and the use of new materials. This paper presents the results of numerical modeling of the fan blades for cooling towers of large diameter from the point of view of oscillatory processes

Today one of the tools of modern engineering structures is a computer experiment. The global level of numerical studies in this field allows you to simulate the oscillatory processes in complex, large-scale structures with sufficient accuracy. Existing methods of designing cooling towers fans do not provide evaluation of oscillatory processes. However, there are currently no techniques, allowing to simulate the vibration of large cooling tower fans based gas dynamic loads. The work implemented stages of the modeling of stress-strain state of the large structure, taking into account gas-dynamic loads and analyzes the forced and natural frequencies. However, consideration of the above factors is a prerequisite for the correct description of vibrations of structural elements. Accounting gas dynamic and deformation processes even in the quasi-stationary formulation can significantly refine existing techniques and to predict the vibration in a large fiberglass fan, so in a timely manner, at the design stage, to take measures to prevent such occurrences. Furthermore, it will be possible, instead of the traditional use of materials for making the fan blades advanced composite materials based on glass fiber. Applications fiberglass material will allow to create effective aerodynamic blade design to reduce weight and energy usage of the drive motor and thus increase by 10-20% or more efficiency of the cooling tower.

## 2 Discussed problems

The occurrence of vibrations in the cooling tower fans today can't be predicted. For certain values existing cooling tower vibrations increases the likelihood of failure of the fan and thus stop the production line. This can lead to significant losses of the enterprise.

To solve problems of vibration and oscillation of the blades, there are several methods. The first is to determine the natural frequencies of the structure and analysis of the response to harmonic load [1]. The result is a frequency and the natural vibration mode. This information will determine the operating frequency range of designs. Frequency method uses data from a stationary calculation of stress-strain state of the structure. This approach does not take into account the possibility of changes in the gas-dynamic loads and, therefore, does not have enough accuracy.

The second method of estimating parameters of vibrations of the structure is to solve the problem aeroelastic [2]. In this case, problems are solved together gas-dynamic calculation and evaluation of the stress-strain state of the structure. This approach allows you to get a information as close as possible to reality and takes into account the changing load.

Interdisciplinary Problem Solving involves considerable computational complexity and to simplify the method of calculation of large fans cooling towers it is advisable to use the first frequency method, which is presented in this article.

## 3 Physical model

For the physical model was develop for studies describing oscillatory processes in large-sized fiberglass fan [3]:

- material fan blades adopted isotropic;
- field of gas-dynamic loads exported into account the stress-strain state of the construction and taken into account when calculating the natural frequencies;
- oscillatory processes in construction are considered in three-dimensional statement;
- considered the first three modes of natural oscillations, because the occurrence of resonance for them the most dangerous for the structure.

## 4 Mathematical model

Determination of the natural frequencies of vibrations of large-sized fan design, carried out using a mathematical model based on the use of Lagrange II-nd kind of differential equations [3]:

$$\frac{d}{dt}\left(\frac{\partial T}{\partial \dot{x}_i}\right) - \frac{\partial T}{\partial x_i} = -\frac{\partial \Pi}{\partial x_i} - \frac{\partial \Phi}{\partial \dot{x}_i} \quad (1)$$

where: $T$ - kinetic energy of the system, $\Pi$ - potential energy of the system, $i$ – generalized coordinate, $x_i$, - the movement of the i-th element ($i$ = 1, 2, 3), $\Phi$ – dissipative function, $t$ - time.

Kinetic energy of the system is:

$$T = \sum_{i=1}^{3} T_i \quad (2)$$

Potential energy of the system is:

$$\Pi = \sum_{i=1}^{3} \Pi_i \quad (3)$$

Dissipative function is proportional to the velocity of the center of mass:

$$\Phi = \sum_{i=1}^{3} \Phi_i \quad (4)$$
$$1$$

$$\Phi_i = \frac{}{2}\beta_i \dot{x}_i \quad (5)$$

where $\beta_i$ - damping coefficient.

## 5 Algorithm of numerical modeling

For the numerical calculations in the software package ANSYS Workbench engineering studies carried out on the basis of gas-dynamic calculations [4] was created by the geometric design and a model for strength and modal analysis (fig. 1). The computational model consists of four major elements: fan hubs, tubular longeron, fan tracery и blade plating. Due to the existing symmetry in the calculation is sufficient to consider only one blade.

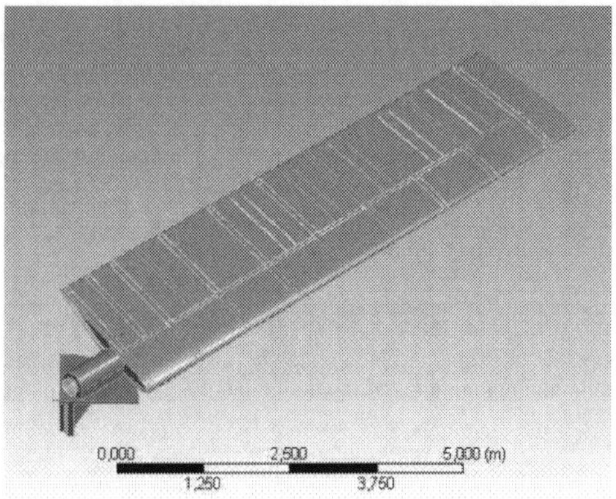


**Figure 1.** Calculated geometry of the large-sized fan blades.

Two variants of settlement construction large size fan blade has been designed For the frequency analysis. The construction of the first embodiment (Construction 1) It consists of a tubular longeron, fan tracery and blade plating. The material all elements of blades is fiberglass. This structure has a low weight and a high specific stiffness. But in this design there are high deflections at the edge of the blade, which may adversely affect the performance of the blade and adhesive joints.

Second Embodiment (Construction 2) is a reinforced structure. The tubular spar mounted steel tube for increased rigidity. To decrease the blade deflection center is supported by a cable.

On the basis of the calculation model is built finite element mesh (fig. 2). Time of construction of models of high performance computing complex PNRPU [5] after the debug computational geometry ranged from 20 to 60 minutes depending on the number of cells. The grid consisted of 216169 computational cells. To build a grid model used the built-in Ansys Workbench Meshing tool.

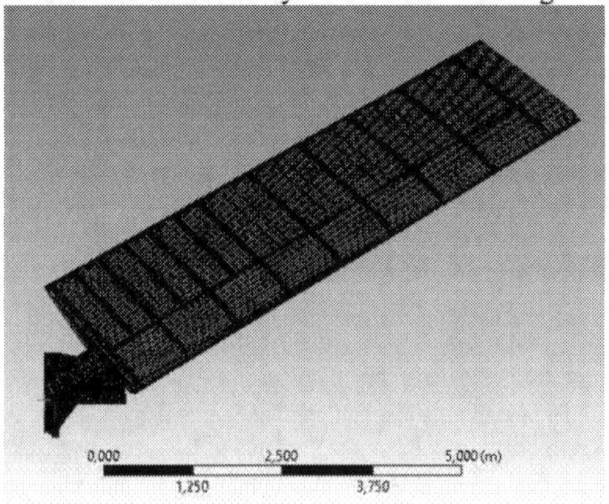


**Figure 2.** Finite element model large size fan blade.

To determine the natural frequencies (modal analysis) fan blade design took into account the action of aerodynamic, gravitational and inertial loads (fig. 3). As the boundary conditions set limits movement of the hub in place on the motor shaft (gear) and the imposition of conditions on the symmetry of the section of the border.

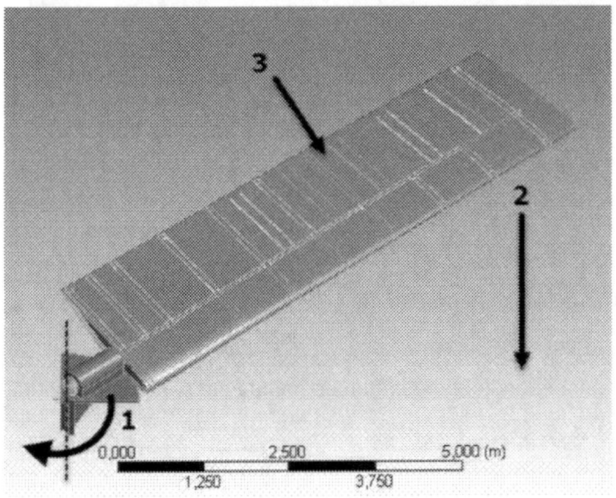


**Figure 3.** Loads for calculation the stress-strain and prestressed modal analysis: 1 – tower fan rotation direction; 2 – direction of action of the Earth's gravity; 3 – aerodynamic load application surface.

## 6 Computational examples and analysis

In Fig. 4 shows the first mode design large size fan blade vibrations. At a frequency of 1.99 Hz natural oscillations observed in the plane of rotation of the fan. This frequency is close to the maximum operating speed of the fan, which can lead to destruction of the blade and resonance. In the second mode at a frequency of 3.23 Hz observed natural oscillations in the vertical plane. This plane may experience vibrations caused by aerodynamic and gravitational forces. You may experience a so-called flutter - self-excited undamped self-oscillations.

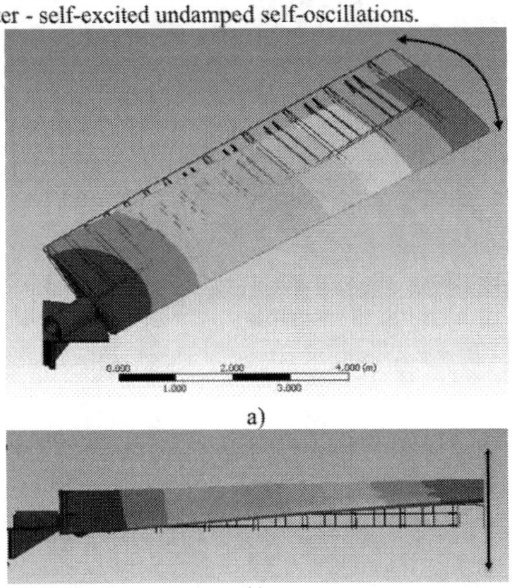

a)

b)

**Figure 4.** Modes of the natural oscillations large size cooling tower fan blades at a frequency of 1.99 Hz (a) and 3.23 Hz (b).

For a more accurate analysis of large size fan cooling tower built resonance diagram for the the first two or three modes of oscillations in the operating speed range (fig. 5). These modes of distribution of energy as much as possible and in the event of resonance can be destroyed blades. As can be seen from the diagram, there are first intersection line with the natural frequency of the second harmonic and higher. The points of intersection are in the operating range, which is critical to the efficiency of the cooling tower fan.

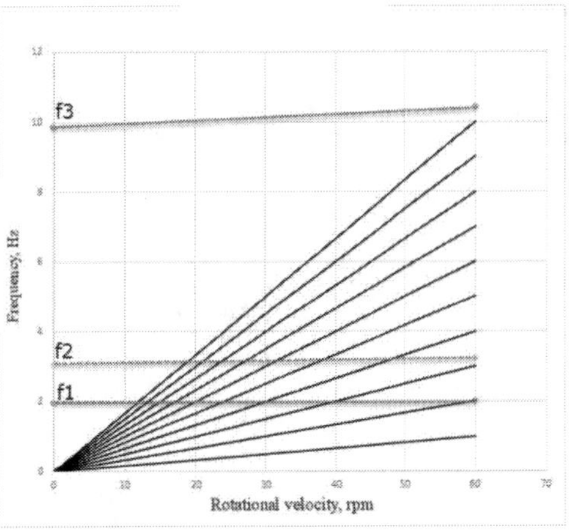


**Figure 5.** Resonance diagram large size fan blades cooling tower for Constructions 1.

Frequency analysis for the blade structure 2 was carried out to increase the natural frequency and resonance diagram built (fig. 6). As can be seen from the diagram, there are the intersection of the first natural frequency of the line with the ninth and higher harmonics. These harmonics are safe for the fan performance.

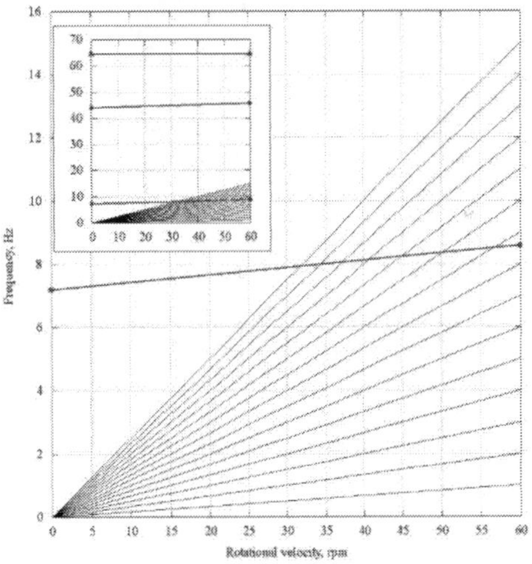


**Figure 6.** Resonance diagram large size fan blades cooling tower for Constructions 2.

According to calculations it was made blade Construction 2. For the manufacture of the blade surface on the engraving and milling machine of «Center for Innovative Engineering Technologies» PNRPU on a reported basis master model of the blade section have been prepared. After you create a master model was fabricated form for displaying the shell portion of the blade. Power blade construction consists of a tubular carrier longeron and ribs mounted thereon. At the final stage the assembly of the blade casing and the power structure. Manufactured the blade is shown in fig 7, a. After manufacture of the blades and their connection to the hub of the fan

DOI: 10.1051/matecconf/20167509002

installation occurs directly on the tower (fig 7, b). During the pilot operation of the sample experienced fan following characteristics achieved: Power consumption was less than 200 kW, instead of project 450 kW; temperature difference cooling water tower rose from 4 °C using a precursor to an analogue to 8 °C, on the prototype.

**Figure 7.** Assembly (a) and installation (b) of large size fan in the cooling tower.

## 7 Conclusions

As a result of this work the following conclusions:

1. To eliminate the resonance effects in the design of the design large size tower fan resonance diagrams necessary to calculate and ensure the absence of oscillations in the operating frequency range.
2. The method of numerical experiments allow to consider gas dynamics and gravitational load when calculating the vibration large size fan design elements has been developed.
3. On the basis of the developed methodology as a result of numerical simulation of the blade design large size tower fan has been designed, which provides the necessary strength, consumables and energy characteristics.

## Acknowledgements

The research work was supported by Russian Science Foundation under Grant No. №14-19-00877.

## References

1. A. Shmakov, V. Modorskii, *Modeling of oscillating processes in elements construction SRM*, Scientific and Technical Volga region Bulletin, **6**, 368-371 (2013)
2. E. Mekhonoshina, V. Modorskii, *On a phase-shift of waves at the medium interface*, Computer Optics, **39(3)**, 385 (2015).
3. A. Klepikovsky, Y. Timofeev, A. Shaiko-Shaikovsky, *The refined mathematical models of estimation natural frequency systems with several degrees of freedom*, Nika (2011) (URL: http://cyberleninka.ru/article/n/utochnyonnaya-matematicheskaya-model-otsenki-sobstvennyh-chastot-kolebaniy-sistem-s-neskolkimi-stepenyami-svobody)
4. A. Shmakov, V. Modorskii. *Energy Conservation in Cooling Systems at Metallurgical Plants*, Metallurgist, **59(9)**, 882-886 (2016)
5. A. Shmakov, V. Modorskii. *Numerical Modeling of Gasdynamic Processes and Processes of Deformation in Compressor of Model Test Bench of the Gas-Distributing Unit*, AMM, **799-800**, 865-869 (2015)
6. M. Dimentberga, S. Kolesnikov, *Vibrations in thechnics. Vibrations machines, structures and their elements* (Moscow, Mashinostroyeniye, 1980)
7. V. Modorskii and other, *The solution of engineering problem on the high-performance computer system of the Perm national research polytechnical university: monograph*. (Perm national research polytechnical university, 2014)

# The Functionally Similar to Adaptive Nonlinear Algorithm for Control of Dynamic Plant with Limited Uncertainty

N. D. Polyakhov, A.D. Stotckaia

*Control Systems Department, Saint-Petersburg electrotechnical university LETI, Saint-Petersburg, Russia*

**Abstract.** A new nonlinear algorithm to control the dynamic plant with limited parametric uncertainty is developed. The algorithm ensures the asymptotic stability and roughness concerning the unaccounted disturbances, and also has almost instantaneous convergence. Simulation studying of operating modes of the synchronous generator with power grid shows the narrow localization (nearly matching) of synchronous generator transient processes in an conditions of significant variations of its parameters with a large multiplicity. Design and implementation of these algorithms is quite simple and does not require precise debugging.

## 1 Introduction

There are many methods of dynamic plants with uncertainty control, including adaptive algorithms that continuously held a leading position. For constructing an adaptive tuning algorithm for controller parameters the method of Lyapunov functions and a variety of gradient methods are frequently used [1].

The core of these algorithms are algorithms in the form of differential equations $\dot{k} = \Phi(e, x)$ , where k - matrix of adjustable parameters, the right side of the equations are defined during the synthesis. Such algorithms include in their structure integrators for updating the configurable parameters. This makes them fundamentally inertia and structurally unstable to unaccounted perturbations because of the presence of stability only in the ideal case of decisions (without disturbances), but in conditions of disturbances action rising drift parameters is appeared.

To eliminate these phenomena, that is, to achieve roughness (robustness) of convergence, algorithms are modified by the introduction of feedback by adjustable parameters (regularization). As a result the algorithms of the form $\dot{k} = \Phi(e, x) - \alpha k$ (α – feedback depth), in which there is an exponential (dissipative) adaptive processes, can be obtained. However, the use of such means for achieving algorithm roughness is associated with the availability of permanent transient processes in the adaptive system: error vector *e* must not be zero.

In order to relax this condition it is necessary to strengthen the parameter $\alpha$ and increase the speed setting that in the multidimensional case is restrictively on stability. In any case it is technically unacceptable.

The paper presents synthesis of a new algorithm in a form $k = F(e, x)$. For its construction Lyapunov functions method is also used. The convergence of processes is almost instantaneous and has an asymptotic nature. Structurally, it looks like a nonlinear algorithm such as parameter feedback, but functionally is efficient with significant (multiple) deviations of the dynamic parameters of control plant.

## 2 Theoretical results

### 2.1 Introductory Notes

Assume that the control plant is specified as:

$$\begin{aligned} \dot{x}(t) &= Ax(t) + Bu(t), \\ x(t_0) &= x_0 \end{aligned} \tag{1}$$

where $x(t)$ – state vector, $u(t) = u^a(t)+g(t)$ – nonlinear control vector, $g(t)$ – input influences vector.

Matrix shown in (1) are described as follows:

$B = \{b_{ij}\}_{n\times m}$ – matrix with known elements;

$A = \{a_{ij}\}_{n\times n}$ – matrix with interval type of uncertainty $\theta$;

$\theta = \left\{\theta \in R^l : \left|\theta_i - \theta_i^*\right| \le \Delta\theta_i,\ i = \overline{1, l}\right\}$, $l = n^2$ ;

$\theta = [a_{11}, a_{12}, a_{13}, \ldots, a_{nn}]$

$\theta_i*$ – rated value.

Construction of the algorithm is represented by the scheme of the system with a reference model [2].

Let us suppose that the equation of reference model is given by:

$$\dot{\hat{x}}(t) = A_0\hat{x}(t) + B_0 g(t), \tag{2}$$

where $\hat{x}(t)$ – $n$-vector of reference model, $A_0$, $B_0$ – reference model matrix corresponding to the desired dynamics.

Let us introduce the control error:

$$e_i(t) = x_i(t) - \hat{x}_i(t), i = \overline{1,n}$$

On the basis of the expressions (1), (2) after several transformations we get the following equation:

$$\begin{gathered}\dot{e}(t) = Ae(t) + Bu^a(t) + \delta, \\ e(t_0) = e_0, \\ \delta(t) = \left[(A - A_0)\hat{x}(t) + (B - B_0)g(t)\right].\end{gathered} \quad (3)$$

For the case when the matrix B is irreversible, modifying the equation (3), so that

$$\dot{e}(t) = Ae(t) + Bu^a(t) + BB^+ + \delta(t).$$

Performing alignment condition

$$\lim_{t\to\infty}(BB^+\delta(t) - \delta(t)) = 0 \quad (4)$$

it leads to the equivalence of the equation (3) and the modified equation.

## 2.2 Lemma

The control law $u^a(t)$ ensures asymptotic stability of the trivial solution of the equation (3) if and only if it is provided by the condition $\lim_{t\to\infty}\delta(t) = 0$.

Indeed, the achievement of an asymptotically stable zero solution e (t) = 0 is associated with the provision of control low $u^a(t)$ the condition $A \xrightarrow{u^a(t)} A_0$, with the result that, as well as in the choice $B = B_0$, we obtain conclusion of the lemma.

Consider the equation (3), when $\delta(t)$=0, for determining the control structure:

$$\dot{e}(t) = Ae(t) + Bu^a(t), e(t_0) = e_0 \quad (5)$$

Let us suppose that the controller description has the form of a feedback:

$$\begin{gathered}u^a(t) = Ke(t), \\ K = \{kij\}_{m\times n}\end{gathered} \quad (6)$$

where $K$ – adjustable parameters matrix.

Let us define elements of adjustable parameters matrix $K(t)$ for the asymptotic stability of the system (1). Suppose that Lyapunov function is in the form [1]:

$$J(t) = \frac{1}{2}e^T(t)e(t)$$

then:

$$\Psi(t) = \dot{J}(t) = e^T(t)\dot{e}(t).$$

To ensure the asymptotic stability of system (5) is sufficient to:

$$\Psi(t) < 0 \quad (7)$$

From the expressions (5) and (6) following equations are obtained:

$$\dot{e}(t) = \Gamma(t)e(t), \ \Gamma(t) = (A + BK). \quad (8)$$

Then, by the last equation:

$$\Psi(t) = e^T(t)\Gamma(t)e(t) = \sum_{i=1}^{n} e_{ii}^2\gamma_{ii} + \sum_{\substack{i,j=1 \\ i\neq j}}^{n}\gamma_{ij}e_ie_j.$$

## 2.3 Lemma

Let us assume that

$$\Psi(t) = \Psi_1(t) + \Psi_2(t) \oplus$$

$$\Psi_1(t) = \sum_{i=1}^{n}\gamma_{ii}e_{ii}^2 \quad \Psi_2(t) < 0, i = \overline{1,n}$$

Then, if

$$\Psi_1(t) < 0, \Psi_2(t) < 0, i = \overline{1,n}$$

the condition of asymptotic stability (6) is satisfied.

Let us take the diagonal elements of $\Gamma(t)$ are constant and negative, that is $\gamma_{ii} < 0$ and then $\Psi_1(t) < 0$ and find the conditions for the inequality $\Psi_2(t) < 0$. Use the substitution [3]:

$$\gamma_{ij}(t) = \alpha_{ij}^{-1}e_i(t)e_j(t), \quad i,j = 1,n \quad i \neq j$$

hence,

$$\Psi_2(t) = \sum_{\substack{i,j=1 \\ i\neq j}}^{n}\gamma_{ij}e_ie_j = \sum_{\substack{i,j=1 \\ i\neq j}}^{n}\alpha_{ij}\gamma_{ij}^2(t) \quad (9)$$

Thus, from (9) it follows that if $\alpha_{ij} < 0$, then $\Psi_2(t) < 0$ and if $\gamma_{ii} < 0$ and constant $\alpha_{ij} < 0$, $i \neq j$, then the asymptotic stability condition (7) of the system (5) is satisfied.

From the expression (8) we obtain the matrix of adjustable parameters $K(t)$), that has the following form:

$$K(t) = B^{-1}(\Gamma - A), \quad (10)$$

if $n = m$ and $B$ has an inverse matrix; when $n \neq m$ matrix $B^{-1}$ is replaced by the pseudoinverse $B^+$.

For the calculation of matrix $K(t)$ "nominal" values of elements $\theta_i$* of the matrix A are used.

## 2.4 Theorem

*The system (1) has:*

1) *asymptotic stability of processes for e (t) with control law* $u^a(t) = Ke(t)$, $K(t) = B^{-1}(\Gamma - A)u$ $K(t) = B^+(\Gamma - A)$, *when the alignment condition (4) performs;*
2) *dissipative processes on the condition of not doing (4), and limitations of all the variables of the system state, but with a regulation radius limit set by selecting parameters of the control law.*

# 3 An example of algorithm using

As a dynamic object with bounded uncertainties, consider a synchronous generator (SG), which is described in certain assumptions equations system deviation [4,5]

$$\Delta\dot{E}_q(t) = -\frac{1}{k_3 T_{d0}}\Delta E_q(t) - \frac{k_4}{T_{d0}}\Delta\delta(t) + \frac{1}{T_{d0}}\Delta E_f(t),$$

$$\Delta\dot{\omega}(t) = -\frac{k_2}{T_j}\Delta E_q(t) - \frac{k_1}{T_j}\Delta\delta(t) + \frac{1}{T_j}\Delta M_M(t),$$

$$\Delta\dot{\delta}(t) = \Delta\omega(t),$$

$$\Delta U(t) = k_5\Delta\delta(t) + k_6\Delta E_q(t),$$

where $\Delta U(t), \Delta\delta(t), \Delta E_q(t), \Delta\omega(t)$ – respectively, voltage deviation, load angle, emf (electric motive force) and the angular frequency; $\Delta M_M(t)$ – of turbine mechanical torque deviation, $\Delta E_f(t)$ – the emf deviation of exciter.

Here, $k_1$ – a coefficient characterizing the change in the electrical power when changing the angle of the rotor, ie, synchronizing torque; $k_2$ – a coefficient characterizing the change in the electrical power when changing the conditions of constant flux in the rotor angle; $k_3$ – a coefficient characterizing the influence of the external resistance, $k_4$ – a coefficient characterizing the demagnetizing effect of changing the angle of the rotor; $k_5$ – a coefficient characterizing the change in voltage on the tires when the generator rotor angle change in conditions of constant emf; $k_6$ – a coefficient characterizing the voltage on the generator, when emf changes in conditions of constant angle.

The study adopted the equivalent system of equations of the object, characterized by the presence of the right sides of the brackets, that can change the sign at parameter variation. This property characterizes the uncertainty in a more complicated and is therefore attractive for the study.

Thus, research equations are

$$\Delta\dot{U}(t) = -\frac{1}{k_3\tau_d}\Delta U(t) + k_5\Delta\omega(t) + \left(\frac{k_5}{k_3\tau_d} - \frac{k_6 k_4}{\tau_d}\right)\Delta\delta(t) + \frac{k_6}{\tau_d}\Delta E_{fd}(t),$$

$$\Delta\dot{\omega}(t) = -\frac{k_2}{k_6 T_j}\Delta U(t) + \left(\frac{k_2 k_5}{k_6\tau_j} - \frac{k_1}{T_j}\right)\Delta\delta(t) + \frac{1}{T_j}\Delta M_M(t),$$

$$\Delta\dot{\delta}(t) = \Delta\omega(t).$$

Simulation of synchronous generator control systems with a nonlinear controller is done in MATLAB/Simulink environment for the parameters of synchronous generator and the reference model listed in Table 1.

**Table 1.** Parameters of synchronous generator and the reference model

| Parameters | Synchronous generator | Reference model |
|---|---|---|
| $k_1$ | 1,773 | 10,9 |
| $k_2$ | 0,854 | 2,7435 |
| $k_3$ | 0,346 | 0,13 |
| $k_4$ | 1,49 | 0,4655 |
| $k_5$ | 0,038 | 0,1 |
| $k_6$ | 0,584 | 1,9512 |
| $\tau_d$, s | 4 | 4 |
| $T_j$, s | 2,5 | 2,5 |

As a result, obtained the following:

$$A = \begin{bmatrix} -0,723 & 0,038 & -0,19 \\ -0,585 & 0 & 0.687 \\ 0 & 1 & 0 \end{bmatrix}$$

$$B = \begin{bmatrix} 0,146 & 0 \\ 0 & 0,4 \\ 0 & 0 \end{bmatrix}$$

$$B^+ = \begin{bmatrix} 6,85 & 0 & 0 \\ 0 & 2,5 & 0 \end{bmatrix}.$$

Nominal values of the elements of the matrix $A$, taking into account peculiarities of the model of an oscillator (change in sign of the difference in brackets) and the presence of uncertainty, are chosen as follows:

$$a_{ij}^* = [-0,723\ \ 0,038\ \ -0,19; -0,585\ \ 0\ \ 5; 0\ \ 1\ \ 0]$$

For the "good" quality of control processes the following nonlinear controller parameters control are chosen:

$\alpha_{12} = -10$, $\alpha_{13} = -3$, $\alpha_{21} = -6$, $\alpha_{23} = -4$.

In addition, the components $k_{11}$, $k_{22}$ are selected arbitrarily in the simulation process (but with negative values): $k_{11} = -18$, $k_{22} = -8$. Then a nonlinear algorithm in accordance to (10) has the following form:

$$k_{12} = 6,85(-10^{-1}e_1e_2 - 0,038),$$

$$k_{13} = 6,85(-3^{-1}e_1e_3 + 0,19),$$

$$k_{21} = 2,5(-6^{-1}e_2e_1 + 0,585),$$

$$k_{23} = 2,5(-4^{-1}e_2e_3 - 5).$$

The equations of nonlinear control (6) is obtained in the form:

$$u_1^a(t) = k_{11}e_1(t) + k_{12}e_2(t) + k_{13}e_3(t)$$

$$u_2^a(t) = k_{21}e_1(t) + k_{22}e_2(t) + k_{23}e_3(t)$$

$$u^a(t) = [u_1^a(t)\ u_2^a(t)]^T = [\Delta E_{fd}\ \Delta M_M(t)]^T,$$

$$e_1(t) = \Delta U(t), e_2(t) = \Delta\omega(t), e_3(t) = \Delta\delta(t).$$

The efficiency of nonlinear algorithm was investigated under the following conditions:

- changing a few values of elements of matrix $A$ (synchronous generator);
- considering of the transient processes of voltage and frequency.

The values of coefficients simultaneously increase or decrease in 5 times. Then:

$$A_1 = \begin{bmatrix} -0,723 & 0,19 & -0,19 \\ -2,925 & 0 & 3,435 \\ 0 & 1 & 0 \end{bmatrix}$$

or

$$A_2 = \begin{bmatrix} -0,723 & 0,0076 & -0,19 \\ -0,117 & 0 & 0,137 \\ 0 & 1 & 0 \end{bmatrix}.$$

The simulation results of synchronous generator transient processes with three sets of matrix elements (reduction/increasing 5 times and the nominal values) in MatLab/Simulink environment shown in Fig. 1 an Fig. 2.

Fig. 1 and Fig. 2 show transient processes of the synchronous generator without adaptive control and under adaptive control. Based on the simulation results, it is clear that the transient response of synchronous generator with controller (the transient time: 3.5 – 4.5 seconds) are located in the vicinity of a sufficiently narrow curves.

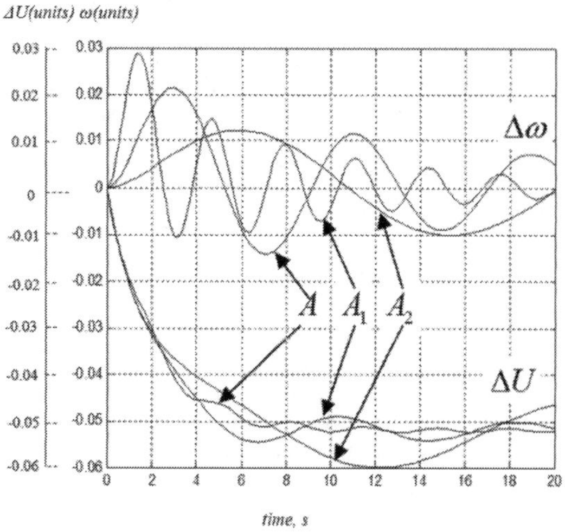


**Figure 1.** Transient processes of the synchronous generator without adaptive control

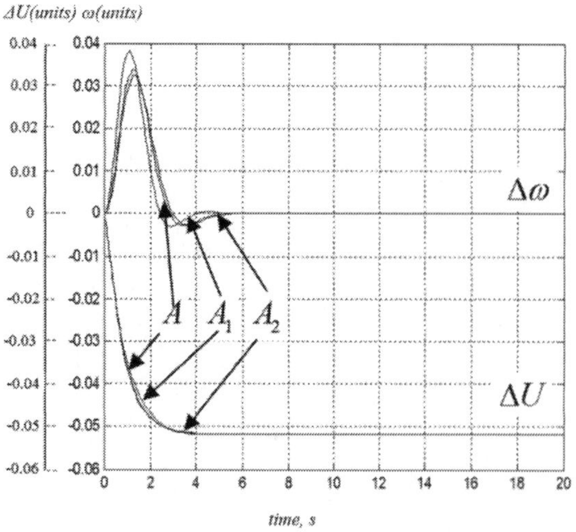


**Figure 2.** Transient processes of the synchronous generator under adaptive control

## 4 Conclusion

The development of automatic excitation controllers are the basic methods of linear control theory in the synthesis of control algorithms, and used only the standard PD - laws. However, in control problems with the uncertainty of a synchronous generator (parametric perturbations), these methods are ineffective.

To stabilize the dynamic behavior of synchronous generator excitation system in different operation conditions most appropriate introduction of adaptive control systems based on models and inertia less algorithms.

The deficiency of adaptive algorithms in schemes with reference / adjustable model and parametric adaptation is that the convergence of the algorithm is exponentially stable in the variables error, but the parametric adaptation processes by Lyapunov indicates unrobust stability convergence parametric adaptation processes.

Synthesized law fast-response parameter adaptation has asymptotic (here exponential) convergence process with zero adaptation time, a robust unaccounted disturbances and large parametric deviations. System simulation study shows effective treatment to parametric deviations even with a change of its sign.

The mode of bifurcation occurrence [6] shows the ability to limit the size of the attractor and even its complete suppression. The algorithm is preferable for practical use, as freely and universally implemented for controlling technical objects.

## Acknowledgment

The research work was partly supported by basic part of Russian Ministry state task (project number 2014/187 – the implementation of government task in the sphere of scientific activity).

## References

1. A. L. Fradkov, "Adaptive control in complex systems: searchless methods" M: Nauka, 1990, 296 p.
2. Yu. A. Bortsov, N. D. Polyakhov, V. V. Putov "Electromechanical systems with adaptive and modal control" L: Energoatomisdat, 1984. 216 p.
3. G. A. Kozhekova "The calculation of the adaptive control systems for synchronous generator" Izvestia KSTU im. I. Razzakova, # 21, 2010.
4. P. Anderson, A. Fuad "Power systems control and sustainability" M: Energia, 1980.
5. N. D. Polyakhov, Ha Anh Tuan Adaptive "Control of synchronous generator based on inertialess parametric adaptation algorithm" // Journal "Electrichestvo", 2014. Vol. 12. - p.47-52
6. N. D. Polyakhov, Ha Anh Tuan "Adaptive control of a synchronous generator in the mode of bifurcation appearance"// Online Journal "Naukovedenie", 2014. Issue 5.

# Modelling of dynamically stable AR-601M robot locomotion in Simulink

Ramil Khusainov, Artur Sagitov, Alexander Klimchik and Evgeni Magid

*ntelligent Robotic Systems Laboratory, Innopolis University, Russian Federation*

**Abstract.** Humanoid robots will gradually play an important role in our daily lives. Currently, research on anthropomorphic robots and biped locomotion is one of the most important problems in the field of mobile robotics, and the development of reliable control algorithms for them is a challenging task. In this research two algorithms for stable walking of Russian anthropomorphic robot AR-601M with 41 Degrees of Freedom (DoF) are investigated. To achieve a human-like dynamically stable locomotion 6 DoF in each robot leg are controlled with Virtual Height Inverted Pendulum and Preview control methods.

## 1 Introduction

Nowadays roboticists actively work on developing, updating, and improving currently existing anthropomorphic robots. One of the critical tasks of robotics is to create a multifunctional robot that would be able to replace people in dangerous work conditions, such as search and rescue operations, working in the field of nuclear safety, chemically or biologically polluted environments, and supporting people during space flights. Such operations may require humanlike skills to overcome obstacles and get through environment, which was designed mainly for a human. That is why it is critically important to develop anthropomorphic robots, which are similar to human body in their size, weight and locomotion characteristics.

Although bipedal robot locomotion has been tackled from different angles for a quite long time, it was only in the last decade of the 20th century that real robots started to demonstrate an acceptably successful walking on two legs. Early solutions involved static walking approach, which targeted to keep the ground projection of the centre of mass (CoM) always inside foot support area, and performed at a very low walking speed and on flat surfaces only [1]. To achieve faster walking speeds, running, and stair climbing, dynamic walking approach permits CoM projection to be outside of the foot support area, while the zero momentum point (ZMP) should always stay within the support area (e.g., [2-4]). To produce dynamically stable gaits the ZMP concept is utilized in different ways and could be roughly classified into two types [5]. One approach considers ZMP as a target reference and applies numerical methods to solve the ZMP equations and to obtain the CoM trajectory, which fulfils some predefined ZMP trajectory [6]. Another approach (called the preview control approach) solves the bipedal walking as a ZMP tracking problem [7]. In this paper we overview two main walking pattern generation principles which are applied for Russian-made AR-601M anthropomorphic robot - a preview control method [8,9] and a virtual height inverted pendulum method [10,11]. Virtual experiments and control algorithms evaluation for AR-601M are performed in Simulink environment.

The rest of this paper is organized as following. Section 2 describes characteristics of AR-601 robot, focusing on pedipulator parameters, and simulation process. Section 3 presents biped robot dynamic balance methods, which are used in our research. Section 4 demonstrates the results of simulating robot locomotion in MATLAB/Simulink environment. Finally, Section 5 presents our conclusions and future work discussion.

## 2 AR-601M robot model

Biped robot AR-601M (Fig. 1) with 41 active degrees of freedom (DoF) has been developed by Russian company "Android Technics". Its virtual model in Matlab/Simulink environment is presented in Fig.1b. During the robot locomotion only 12 pedipulator joints (6 DoF in each pedipulator) are active and are utilized for locomotion control. Each robot leg contains three joint axes in the hip, two joints in the ankle and one in the knee. Mass and size parameters of the robot pedipulators are given in Table 1. The total mass of the robot is 65 kg; further details about AR-601M are available in [10].

**Table 1.** Mass and size parameters of the pedipulators.

| Link | Size parameters (mm) | Mass (kg) |
|---|---|---|
| Thigh | Length : 280 | 7.5 |
| Shank | Length : 280 | 6.9 |
| Foot | L×W×H: 254×160×106 | 3.2 |

Figure 2 demonstrates solid CAD-generated AR-601M robot model with the total number of 59 model parts in SolidWorks environment. The linear dimensions of the model components with a high degree of accuracy correspond to the ones of the real robot. When the robot model is imported from CAD package, its mass distribution is considered as homogeneous, which is not the same as in the real robot because of heterogeneity of the robot parts (for example, the robot hand contains metal parts, printed circuit boards, actuators, wiring harness, etc., and all these components have different weight, unevenly distributed across the hand). Therefore, mass characteristics of the real robot, such as mass, CoM location and moments of inertia for each part, were used in the development of robot model in virtual environment.

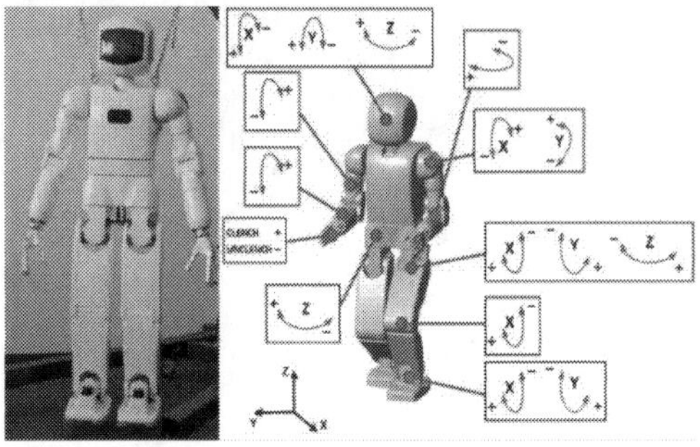

**Figure 1.** Anthropomorphic robot AR-601M

Link Software tool in Simulink SimMechanics library allows to create a mechanical model of the robot in MATLAB SimMechanics environment automatically by importing its solid body CAD model from SolidWorks. Each robot's joint is modelled as a revolute joint with one DoF. Figure 3 demonstrates a block diagram of AR-601M robot leg part simulated in SimMechanics after importing the model from SolidWorks. The figure shows that the leg parts are connected through revolute joints and that joint is controlled by providing a particular input angle. The initial position of the robot is determined so that the legs are fully stretched out, the feet are parallel to the horizontal plane. The choice of the hands initial position is arbitrary and shown in Figure 2.

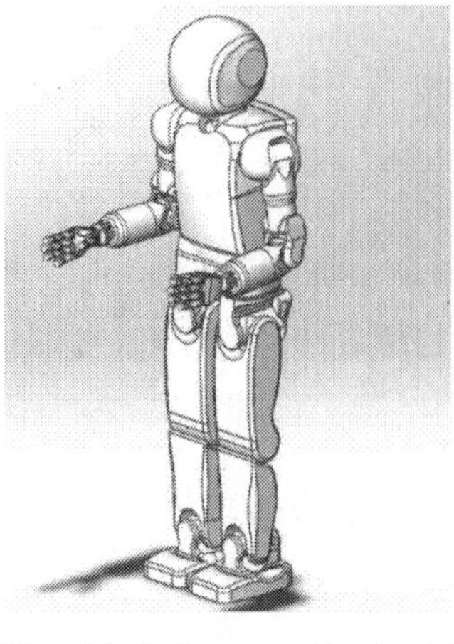

**Figure 2.** CAD model of anthropomorphic robot AR-601M

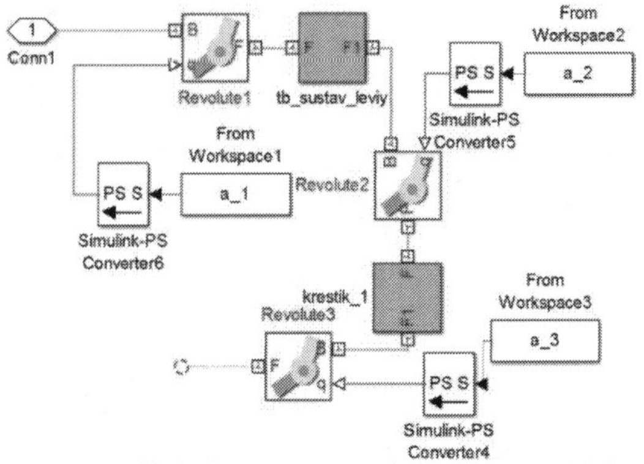


**Figure 3.** A block diagram of the robot leg part, modeled in SimMechanics

## 3 Walking pattern generation principles

### 3.1 Virtual height inverted pendulum method with linear inverted pendulum model

Robot locomotion could be considered as a repetition of a single step motion [3]. We calculated walking trajectory under the following assumptions, which are widely applied in experimental approaches for biped walking [4]:

(i) The swing foot is parallel to the ground
(ii) The upper body is always kept upright
(iii) CoM of the robot model is moving at the constant height
(iv) The swing foot is moving in a cycloid trajectory and its coordinates can be described with the following equations:

$$y(t) = -S\cos(\tfrac{\pi t}{T}) \quad (1)$$

$$z(t) = 0.5H\,(1 - \cos(\tfrac{2\pi t}{T})) \quad (2)$$

$$x(t) = x_0 \quad (3)$$

where S is a single step length, H is a step height, T is a step period, $x_0$ is a distance between the two feet in x-direction. In [4] authors showed that equations for CoM trajectory can be expressed as:

$$\ddot{x}_{CoM} - \frac{g}{\alpha z_{CoM}} x_{CoM} = 0 \quad (4)$$

$$\ddot{y}_{CoM} - \frac{g}{\beta z_{CoM}} y_{CoM} = 0 \quad (5)$$

where coefficients α and β are found experimentally to reduce ZMP error. Therefore, we have the following trajectories for the robot CoM:

$$x(t) = C_1 e^{-w_1 t} + C_2 e^{w_1 t},\ w_1 = \sqrt{\frac{g}{\alpha z_{CoM}}} \quad (6)$$

$$y(t) = D_1 e^{-w_2 t} + D_2 e^{w_2 t},\ w_2 = \sqrt{\frac{g}{\beta z_{CoM}}} \quad (7)$$

where $C_1$, $C_2$, $D_1$ and $D_2$ coefficients are obtained from initial and final values of CoM coordinates. After we define translational and rotational coordinates for the body and the swing leg, the joint angle trajectories are obtained from inverse kinematics problem solution.

Such trajectory assumes that robot has initial velocity. In the virtually simulated environment, to move the robot from a resting state we first calculate trajectory with a very short step length and then gradually increase it up to its maximally possible value where ZMP errors become critical, the robot loses its balance and falls down.

### 3.2 Preview control method

For the virtual robot in our model zero moment point (ZMP) coordinate values were taken as a feedback signal. Using a controller with the feedback allows the robot to adapt to the current situation and to calculate the necessary trajectory. In [14] authors presented Preview Control approach that uses the desired future value of the ZMP coordinates. We obtain the robot trajectory in the frontal and sagittal planes, using this approach. ZMP coordinates for the inverted pendulum model (IPM) are given as:

$$P = X - \frac{Z_c}{g}\ddot{X} \quad (8)$$

where $Z_c$ is the height of CoM, $g$ is the free fall acceleration and X is the x-coordinate of CoM.

Equations for the controller in a discrete time space can be written in the form of:

$$x(k+1) = Ax(k) + Bu(k) \quad (9)$$

$$p(k) = Cx(k) \quad (10)$$

where k is discrete time, x(k) is the value of the state vector (position, velocity, acceleration of CoM), u(k) is the control vector, p(k) is the value of the coordinates of the ZMP. A, B, C, D are the coefficients which are obtained from equation (8), and expansion of the state vector x in Taylor series to find its value in the next time [7]. In the role of a control vector u(k) the third derivative of the coordinates x, $p(k) = \dddot{x}(k)$ is used.

According to [14] the control signal u while using the future value of the desired ZMP position could be calculated as follows:

$$u(k) =$$

$$= -G_I \sum_{i=0}^{k} e(i) - G_x x(k) - \sum_{l=1}^{N_l} G_d(l) p_d(k+l) \quad (11)$$

where e(i) is the error between the calculated and the desired coordinate ZMP, $p_d$ is the desired coordinates of the ZMP, $G_I$, $G_x$, $G_d$ are the proportion coefficients. The first term expresses the cumulative error between the calculated and the desired coordinate of ZMP. The second term is proportional to the current state vector x. The third term takes into account the future values of the desired ZMP position in which the sum runs over the future $N_l$ values.

## 4 Simulation results

MATLAB/Simulink environment was used to model the bipedal robot walking and to test different algorithms in virtual world model. Walking trajectory parameters for experiments are listed in Table 2.

**Table 2.** Simulation parameters

| **Parameter** | **Value** |
|---|---|
| Step height | 0.05 m |
| Step period | 0.5 sec |
| CoM height | 0.75 m |
| Step length | 0.1 m – max. value |

### 4.1. Virtual height inverted pendulum method

Initially, we evaluate the maximal walking speed of the robot virtual model, which could be achieved with the method of linear inverted pendulum model (LIPM). The successful implementation of 100 steps is referred as successful walking trial of the robot. After each stable walking trial, the step length was increased by 0.05 m and algorithm was re-applied again. The step increase process had been repeated until the robot failed to execute a successful stable walking trial. The detailed results of the algorithm testing are available in [15].

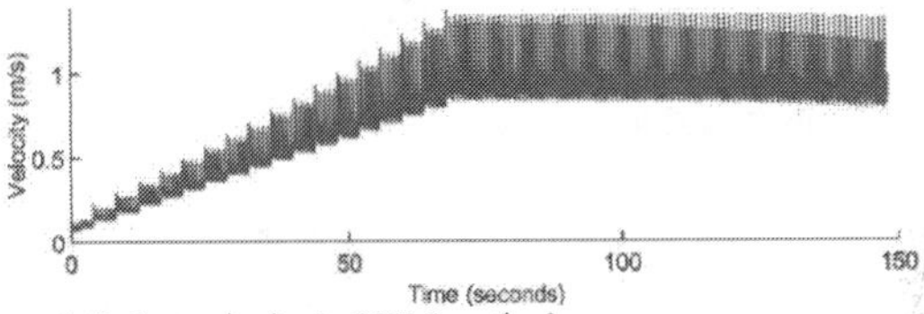


**Figure 4.** Robot velocity in LIPM method

Walking with different CoM height was simulated, and for each height the maximal walking speed was identified. Step period was equal to 0.5 sec. Simulations demonstrated that the maximal step length at which the robot could still perform stable walking is 0.95 m, which corresponds to 0.95 m/s walking speed. CoM forward velocity value is shown in Fig.4. Figure 5 demonstrates ZMP coordinates for a step of 0.95 m length. Support foot center corresponds to (0,0) coordinate. According to the calculated values, the algorithm accumulates approximately a 10 cm error for X coordinate and a 2 cm error for Y coordinate just after a single step.

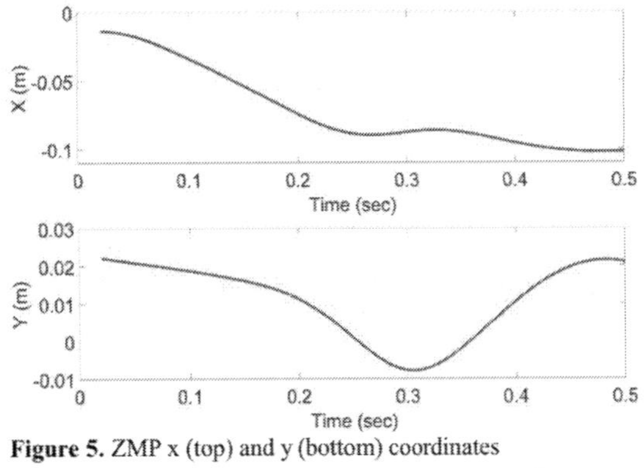


**Figure 5.** ZMP x (top) and y (bottom) coordinates

Next, we vary α and β coefficients in Eq. (4) and (5) around 1 in order to minimize ZMP errors. Figures 6-7 show the dependence of maximal error of ZMP coordinate on α for *x*-coordinate and on β for *y*-coordinate. It was empirically detected that for *x*-coordinate the optimal value of α coefficient should lie within [0.6, 0.8] interval, and for *y*-coordinate the optimal

DOI: 10.1051/matecconf/20167509004

value of β is equal to 0.9. Walking simulation with these coefficients showed that such optimization of trajectories allows to increase a maximal speed of 1.1 m/s accordingly while keeping stable walking for at least 100 sequential steps.

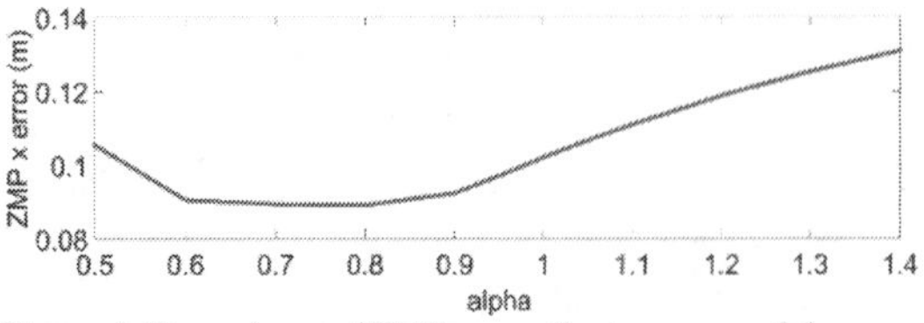


**Figure 6.** Dependence of ZMP x-coordinate error on alpha

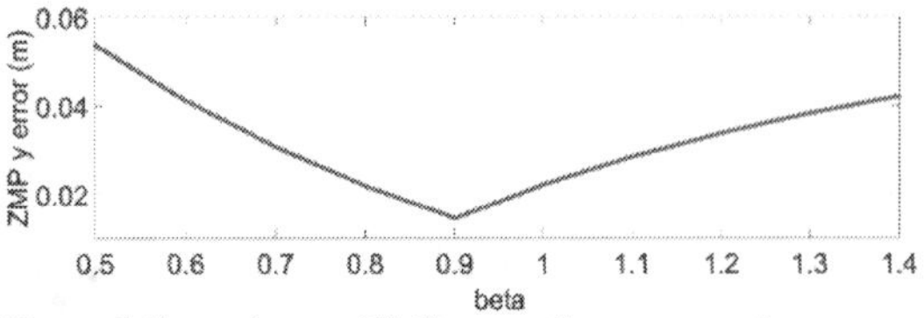


**Figure 7.** Dependence of ZMP y-coordinate error on beta

Finally, we estimated the peak torque value in ankle and knee joints that the robot motors should generate in order to afford the robot locomotion according to the calculated trajectory. Figure 8 shows the calculated torque values for the ankle and the knee joints.

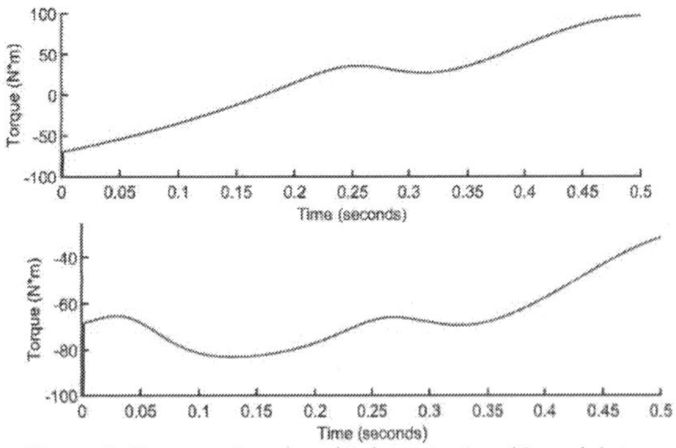


**Figure 8.** Torque values in ankle joint (top) and knee joint (bottom).

The maximal value is approximately 100 N*m for the ankle and 80 N*m for the knee joint. With regard to the technical specifications of real AR601-M robot, its motors can generate maximal values of 50 N*m for the knee and 20 N*m for the ankle joint. Unfortunately, this means that such high maximal speeds could not be achieved for the real robot. Therefore, we are limited to use trajectories with lower maximum speed, which are dynamically stable according to the virtual model simulations.

## 4.2 Preview control method

Simulations using preview control method were also performed in MATLAB/Simulink environment. In contrast to the virtual height approach, preview control method considers feedback of ZMP error. Therefore, there is no need to make 100 steps, as in VHIPM method, in order to conclude if the walking is stable. If certain stable movement is achieved, robot will not fall down later under condition that there is no external disturbance.

Desired ZMP coordinate, calculated ZMP coordinate value and CoM position for *x* and *y* coordinates are presented in Fig. 9-10. The results correspond to maximal speed of 0.5 m/s. Fig. 9 demonstrates that the calculated ZMP *x*-coordinate values are close to the desired ZMP coordinate within an acceptable error. ZMP coordinate leaves supporting polygon only for a very limited time and, as a result, robot's walking is kept stable. The *x*-coordinate of CoM smoothly increases, which corresponds to the forward locomotion of the robot. The same is for ZMP *y*-coordinate, which calculated value is shown in Fig. 10. The *y*-coordinate of CoM oscillates around zero, which corresponds to side movement from one foot to another. Further increase of step length or decrease of step time result in critical ZMP errors. High ZMP errors for larger speeds occur because of a simple dynamical model of robot, according to which the controller is constructed. Therefore, full robot dynamics should be taken into account during controller development in further applications in order to achieve higher walking speeds.

Torque value in supporting foot knee joint is shown in Fig. 11. Maximum value is approximately 100 N*m, which is higher than motor nominal value. That means that joint trajectories of the real robot may differ from the virtually calculated ones, and we should consider related effects, such as errors in swing foot position.

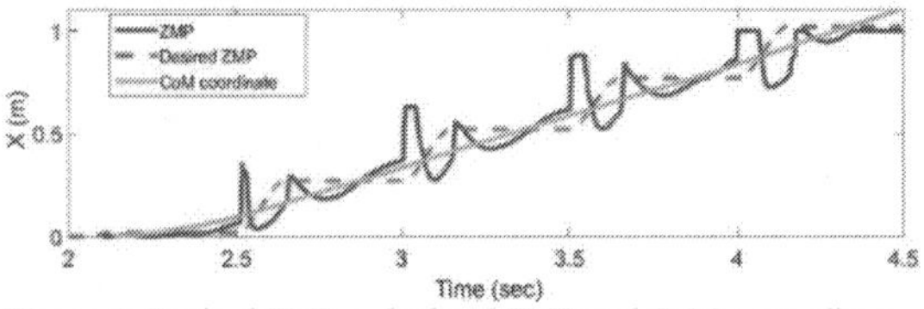


**Figure 9.** Desired ZMP, calculated ZMP and CoM x coordinate

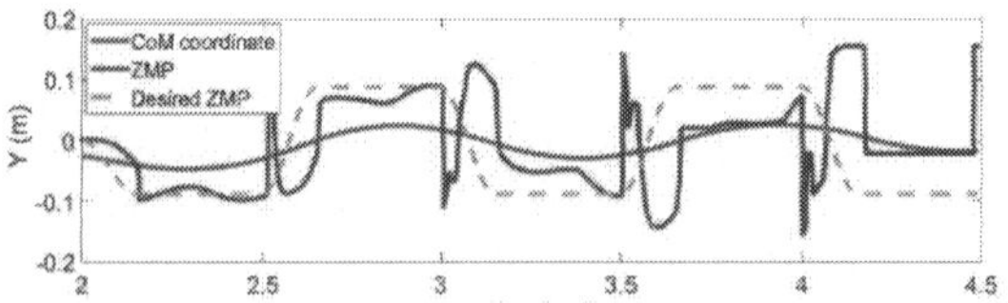


**Figure 10.** Desired ZMP, calculated ZMP and CoM y coordinate

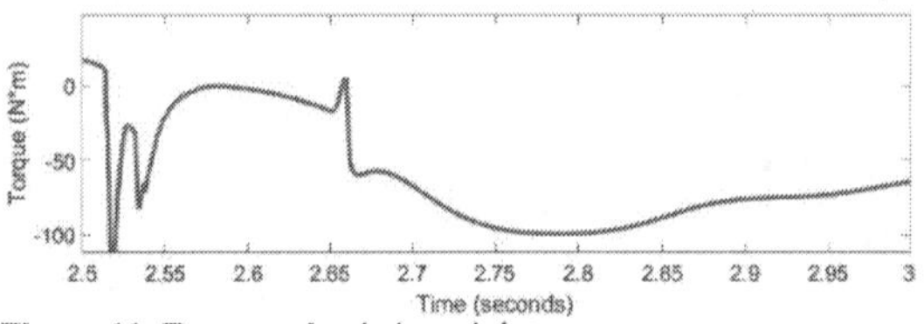


**Figure 11.** Torque value in knee joint

DOI: 10.1051/matecconf/20167509004

## 5 Conclusions

At the current stage of our research we had modelled dynamically stable AR-601M robot locomotion with VHIPM and preview control methods. Experiments with a simulated robot model were performed in MATLAB/Simulink environment. VHIPM method allows to calculate the stable locomotion path quickly and effectively. The main weakness of this algorithm is the lack of a feedback, which results in significant accumulation of errors with time and limits its application only for an ideal mechanism on a perfect flat floor.

The preview control method successfully avoids this disadvantage, but does not allow to reach maximal walking speed because it does not consider full dynamics of the robot motion. Thus, maximal walking speed with preview control method is lower than in virtual height method. Therefore, as a part of our future work we schedule to improve preview control method accordingly to take into account full dynamics of the robot.

## Acknowledgments

This research has been supported by Russian Ministry of Education and Science as a part of Scientific and Technological Research and Development Program of Russian Federation for 2014-2020 years (research grant ID RFMEFI60914X0004) and by Android Technics company, the industrial partner of the research.

## References

1. C. Shih Ascending and descending stairs for a biped robot. *Systems, Man and Cybernetics, Part A: Systems and Humans, IEEE Transactions* **29(3)**, 255-268. (1999).
2. K. Nishiwaki and S. Kagami Simultaneous planning of com and zmp based on the preview control method for online walking control. *IEEE-RAS International Conference on Humanoid Robots* 745-751, (2011).
3. K. Nishiwaki, S. Kagami, Y. Kuniyoshi, M. Inaba and H. Inoue Online generation of humanoid walking motion based on a fast generation method of motion pattern that follows desired zmp. *Intelligent Robots and Systems IEEE/RSJ International Conference* **3** 2684-2689 (2002).
4. B. Gabbasov, I. Danilov, I. Afanasyev and E. Magid ZMP trajectory from human body locomotion dynamics evaluated by Kinect-based motion capture system. *The 11-th International Joint Conference on Computer Vision, Imaging and Computer Graphics Theory and Applications*, Italy.
5. Z. Li, N. Tsagarakis and D. Caldwell Walking pattern generation for a humanoid robot with compliant joints. *Autonomous Robots* **35(1)**, 1-14 (2013).
6. S. Kagami, T. Kitagawa, K. Nishiwaki, T. Sugihara, H. Inoue A fast dynamically equilibrated walking trajectory generation method of humanoid robot. *Autonomous Robots*, **12(1)**, 71–82 (2002)
7. S. Kajita, F. Kanehiro, K. Kaneko and K. Fujiwara Biped walking pattern generation by using preview control of zero-moment point. *IEEE International Conference on Robotics and Automation*. **2** 1620–1626 (2003)
8. H. Audren, J. Vaillant, A. Kheddar, A. Escande, K. Kaneko and E. Yoshida Model preview control in multi-contact motion-application to a humanoid robot. *IEEE/RSJ International Conference on Intelligent Robots and Systems*, 4030-4035 (2014).
9. S. Shimmyo, T. Sato and K. Ohnishi Biped walking pattern generation by using preview control based on three-mass model. *IEEE Transactions on Industrial Electronics*, **60(11)**, 5137-5147, (2013).
10. R. Khusainov, I. Shimchik, I. Afanasyev and E. Magid Toward a human-like locomotion: Modelling dynamically stable locomotion of an anthropomorphic robot in Simulink environment. *Proceedings of the 12th International Conference on Informatics in Control, Automation and Robotics*, Colmar, Alsace, France, **2**, 141-148 (2015)
11. T. Ha and H. Choi An effective trajectory generation method for bipedal walking. *Robotics and Autonomous Systems*, **10** 795-810 (2007).
12. J. Denk, G. Schmidt, Synthesis of walking primitive databases for biped robots in 3-D environments. *IEEE International Conference on Robotics and Automation*. **1** 1343-1349 (2003).
13. A. Yonemura, Y. Nakajima, A. Hirakawa and A. Kawamura Experimental approach for the biped walking robot MARI-1 *6th International Workshop on Advanced Motion Control*, 548–553 (2000).
14. T. Katayama, T. Ohki, T. Inoue and T. Kato, Design of an optimal controller for a discrete-time system subject to previewable demand. *Int. J. Control.* **41** 677–699 (1985)
15. R. Khusainov, I. Shimchik, I. Afanasyev and E. Magid Anthropomorphic robot modelling with virtual height inverted pendulum approach in Simulink: step length and period influence on walking stability. *The 2016 International Conference on Artificial Life and Robotics*, Japan.

DOI: 10.1051/matecconf/20167509005

# Golf cart prototype development and navigation simulation using ROS and Gazebo

Ilya Shimchik[1], Artur Sagitov[1], Ilya Afanasyev[1], Fumitoshi Matsuno[2] and Evgeni Magid[1]

[1]Intelligent Robotic Systems Laboratory (LIRS), Innopolis University, 420500, Universitetskaya 1, Innopolis, Russia
[2]Department of Mechanical Engineering and Science, Graduate School of Engineering, Kyoto University, Kyoto, Japan

**Abstract.** This paper presents our approach to development of an autonomous golf cart, which will navigate in inaccessible by regular vehicles private areas. For this purpose, we have built a virtual golf course terrain and golf cart model in Gazebo, selected and modernized ROS-based packages in order to use them with Ackermann steering vehicle simulation. To verify our simulation and algorithms, we navigated the golf cart model from one golf hole to another within a virtual 3D golf course. For the real world algorithms' verification, we developed a small-size vehicle prototype based on Traxxas radio-controlled car model, which is equipped with an on-board controller and sensors. The autonomous navigation of Traxxas-based vehicle prototype has been tested in indoor environment, where it utilized sensory data about environment and vehicle states, and performed localization, optimal trajectory computation and dynamic obstacles' recognition with adjusting the route in real time.

## 1 Introduction

The growth in the amount of research on unmanned vehicle technology for the past decade shows significant interest to autonomous vehicle control and navigation. Almost all major car manufacturers conduct their own research in this field, and even non-automotive companies like Google and Uber perform investigations toward rapid commercialization of autonomous vehicle technology, where an intelligent autopilot will replace drivers to transport people on public roadways [1]. However, due to slow evolution of autonomous cars related regulations, and inability of existing technologies to handle complex aspects of driving (e.g., poor road conditions or unsafe navigation of some human drivers), fully autonomous vehicles are still far from being commercially available. Nevertheless, modern conventional cars are more commonly being equipped with features that make them semi-autonomous. Most high-end cars from famous brands bear radars, cameras and other sensors that allow having safe and convenient features like automatic cruise control, automatic parking, lane keeping and automatic braking. The next step in this domain is to transfer all driver's functions to a vehicle, beginning from a semi-autonomous mode when a driver should be ready to take control over a car at every moment and aiming to a fully autonomous mode. Already available on the market Tesla cars are equipped with a semi-autonomous system which can drive the car at highway speeds and in stop-and-go traffic mode, set the speed correspondingly to traffic conditions, steer around curves within a lane and change a lane on driver request. While large car manufacturers expand testing of autonomous cars, different universities research teams concentrate on specialized vehicles automation, for example, such as self-driving golf carts [2-5].

Our long-term goal deals with development of autonomous golf cart (AGC) for passenger transportation. Although, while within our simulation we consider AGC navigation through golf course oriented on golf players' conveyance, operation areas for the AGC prototype are not limited to golf purposes and may also include parks, tourist tours, airports, and recreation zones as well as potential application in search and rescue outdoor scenarios within urban areas (e.g., under conditions of chemical or nuclear pollution which do not significantly affect urban infrastructure). The AGC implementation contains three main stages: (1) golf cart simulation and testing of navigation algorithms in a simulated virtual environment; (2) building a prototype on a base of a radio-controlled car model, and (3) transferring the developed systems to a full sized golf cart and testing it in a real world environment. In this paper we present the first two stages of our research.

The rest of the paper is organized as follows. Section 2 describes the system setup: the basic system components for the AGC configuration, and our results on development of vehicle prototype based on the Traxxas radio-controlled car model. Section 3 presents the AGC model simulation in Gazebo. Section 4 introduces a set of ROS-based algorithms for 3D cart model locomotion and first results on golf cart navigation within a 3D golf course in Gazebo. Finally, we conclude and discuss our future research work in Section 5.

DOI: 10.1051/matecconf/20167509005

## 2 System setup

Our long-term goal is to build an autonomously navigated golf cart, which will be controlled by Robot Operating System (ROS) [1] based set of algorithms. This section describes the golf cart configuration and our efforts in development of an autonomous vehicle prototype based on Traxxas radio-controlled car model as a first approximation of AGC implementation.

### 2.1 An autonomous golf cart configuration: sensors, control and systems

This subsection describes the key AGC hardware components, which are necessary for environmental data acquisition and control of the vehicle. These key components include sensors, a driving system, a human-vehicle interface system and an on-board computer, which solves computational, safety and decision-making tasks. Based on the presented in [5, 6] concepts we specify the main functional autonomous vehicle system components, which will be used in our AGC, as follows:

- **Driving system** provides vehicle locomotion control and consists of motors, a steering wheel and brake pedals (the two later components give a human driver a possibility to take control of the autonomous vehicle at any moment).
- **Power system** supplies with electrical power all AGC systems, e.g. motors, on-board computer, sensors, etc.
- **Computational platform** is an on-board computer, which performs functions of sensor fusion, environment model building, case-based reasoning, path planning, and computing appropriate control signals for AGC goals.
- **Safety system** provides a functionality of emergency stop and alarm warnings, followed by immediate passing of AGC control back to a human driver at any moment; it includes such special equipment as light and sound warning systems, and emergency stop "red buttons" to switch off the autopilot mode.
- **Sensory system** provides AGC with necessary information about surrounding environment state and internal vehicle state. Table 1 presents in details the sensory kit, which is responsible for real-time data acquisition, vehicle localization and navigation, and a dynamic world model creation.
- **Human-vehicle interface** system provides a graphical user interface to interact with a passenger (e.g., a golf player) by selecting a next waypoint and to present a current golf cart status and AGC system states.
- **Decision-making system** is a sophisticated intelligent system, which uses as its input a current world model and its predicted states, and performs next AGC actions planning.

[1] Robot Operating System (ROS) is a set of software tools and libraries that helps to build various robotic applications, ros.org

**Table 1.** Sensory kit for the autonomous golf cart.

| Sensors | Description |
|---|---|
| Front Bumper LIDAR | Detects wheel level obstacles in front of the golf cart and inclination type of road surfaces (flat/ascending/descending). |
| Front Roof LIDAR | LIDAR is placed with a negative pitch angle to a travelling plane and detects front obstacles before the golf cart, determining also road's local relief details. |
| GPS/GLONASS navigation system | Identifies the golf cart location in absolute coordinate system. |
| Camera | Recognizes dynamical obstacles and moving objects. |
| Inertial measurement unit (IMU) | Measures car's dynamical characteristics (angles, velocities and accelerations), using a set of 3D accelerometers, gyroscopes and magnetometers. |
| Encoders | Receives odometry data from wheels' angular position. |
| Side sonars | Detects side obstacles from short-distance proximity sonars. |

### 2.2 A vehicle prototype development

As the first step toward AGC, we have been developing a small-size vehicle prototype based on Traxxas 7407 radio-controlled (RC) car model[2] (Fig.1), which will be used to verify and test our ROS-based navigation algorithms. The Traxxas 7407 model, which is a 1:10 scaled model of a typical car, becomes widespread among RC car enthusiasts, as far as it has powerful electric drive platform, allowing to achieve the maximum speed up to 100 km/h. The model is supplied with Velineon 3500 brushless sensorless DC motor, NiMH Battery 3000mah 8.4V, Velineon VXL-3s Electronic Speed Control (ESC), and the Traxxas TQi 2.4GHz radio controlling system.

We have modernized the basic vehicle model with on-board computer and sensors. As a computational platform we have used the low power NVidia Jetson TX1 embedded computer, supporting CUDA technology with on-board parallel computation. We have also utilized the 9 Degrees of Freedom (DOF) Razor IMU[3], incorporated MEMS triple-axis gyro, triple-axis accelerometer and triple-axis magnetometer for nine degrees of inertial measurements. We mounted an optical incremental Lynxmotion Quadrature Motor Encoder to the electric motor, and installed ZED stereo camera, Logitech web-camera (1280x800 pixels, 25 fps) and Hokuyo UTM-30LX-EW LIDAR. We have used both LIDAR and stereo camera to get a point cloud data, constructing a map and detecting obstacles. Figure 2 presents our current version of the model for prototyping. While the final decision of visual sensory information fusion and sensor utilization is still on the way and requires

[2] Traxxas radio-controlled vehicle models are electric models of racing car, which are spread for unprofessional racing sport competitions. They are produced by Traxxas company (USA): https://traxxas.com
[3] The 9DOF Razor IMU (Inertial Measurement Unit) from the company Sparkfun, www.sparkfun.com

additional testing, we have obtained the first results of the autonomous RC car navigation with a web-camera. The vehicle utilized web-camera data about environment and sensory information on vehicle states, and performed localization, optimal trajectory computation and dynamic obstacles' recognition with adjusting its route in real time while performing a white line following of an ellipsoid-like figure in indoor laboratory environment.

**Figure 1.** The Traxxas radio-controlled model (courtesy of Traxxax Company).

**Figure 2.** The vehicle model for algorithms prototyping.

## 3 Golf cart's 3D model simulation in Gazebo

We began developing the AGC navigation algorithm with creation of a proper simulation model, using Gazebo 2.2 for building both the AGC model and simulation environment. Gazebo [4] is an integrated with ROS graphical engine, which simulates physics, builds a 3D robot model using its URDF description[5], and then allows running vehicle control algorithms in a developed environment. To control AGC model we used ROS Indigo [7] software, where we have modified the basic algorithms of ROS Navigation Stack framework[6]. The ROS Navigation Stack permits implementing navigation algorithms for differential drive robots. It receives data from odometry, sensors streams, and computes commands of safe velocity toward a determined goal position, which then will be executed by a car driving system. The algorithms, which prove their effective work within the simulation, will be further applied for the real AGC, and ROS structure and functionality helps to simplify such process significantly.

Figure 3 presents a full-scale AGC simulation model in Gazebo. As far as Gazebo allows simulating a car-like steering system, we have created the Ackerman steering vehicle model that corresponds to standard golf cart steering. Gazebo also allows obtaining sensors' simulation such as IMU, LIDARs, encoders, cameras, and stereo cameras. After finishing the steering system we have integrated these sensors into our AGC model and created a golf cart course simulation. Figure 4 demonstrates the golf cart moving within a developed golf field environment in Gazebo simulator.

**Figure 3.** The full-scale golf cart simulation model in Gazebo. Blue area indicates the range of LIDAR scanning.

**Figure 4.** The model of simulated golf cart with sensors and ground relief simulation.

[4] Gazebo is 3D robot simulator: www.gazebosim.org

[5] URDF is a Unified Robot Description Format, which is an XML format for robot model representing within ROS.

[6] ROS-based Navigation Stack package: wiki.ros.org/navigation

## 4 ROS-based golf cart navigation in Gazebo environment

### 4.1 ROS-based algorithms for navigation

For our AGC simulation we have modified the existing ROS packages to use them with Ackermann steering robot. The following algorithms were utilized:

- **Global planner** is used to set a global path for the car within a point-to-point navigation task. Default ROS package implements Dijkstra's global path planning algorithm [8].
- **Local planner**, which is realized in a search-based planning library (SBPL) of domain-independent graph searches and is based on a lattice planner [9].
- Direction Cosine Matrix (DCM) algorithm [10] for **filtration of IMU data**.
- **Robot localization** is based on AMCL ROS package[7] adaptation to the AGC model. This package represents 2D probabilistic robot localization during its locomotion in a simulated environment and implements adaptive Monte Carlo localization approach [11] with a particle filter to track robot location on a given map [12].

This set of algorithms has been implemented within our AGC simulation for successful vehicle navigation to a predefined goal pose within a 2D map. Further, we started to verify and configure this set of algorithms on the car prototype, which was presented in Section 2.2.

### 4.2 Results of golf cart navigation simulation

As we have mentioned in Section 3, we used the standard package of ROS Navigation Stack, which includes a set of algorithms for localization, local and global planning, and controlling the mobile robot. This package is designed for a differential drive and holonomic drive robots, but it does not perform well for a car-like steering robot. We have modified the Navigation Stack package in order to work with Ackermann Steering vehicle. The main changes were as follows:

- Adding local planner (teb_local_planner).
- Supplementing with specific motion primitives for a car-like robot.
- Adapting specific for car-like commands to control the vehicle.

We have used the package teb_local_planner[8], as a local planner, which relies on the method called Timed Elastic Band [13-15]. This approach is extended to capture the specific constraints and objectives, which are appropriate for emergency manoeuvres in critical situations [16]. It can locally optimize the robot's trajectory with respect to trajectory execution time, separation from obstacles (i.e., path safety) and compliance with kinodynamic constraints at a runtime.

[7] Adaptive Monte Carlo Localization (AMCL) simulation package: wiki.ros.org/amcl
[8] ROS-based Timed Elastic Band (TEB) local planner package: wiki.ros.org/teb_local_planner

With this set of packages and algorithms, we can navigate golf cart through a golf course from one golf hole (marked with a flag) to another, as shown in Fig. 5.

**Figure 5.** Simulation of a golf cart navigation from one golf hole to another in a virtual golf course environment in Gazebo.

## 5 Conclusion and Future plans

We are focusing on a long-term goal of building a full-size autonomous golf cart, which will be controlled and navigated with ROS-based algorithms. We have divided the AGC implementation into three main stages: (1) golf cart simulation and testing of navigation algorithms in the simulated environment; (2) small-size car model prototyping, and (3) a full-size golf cart prototype supplying with sensors and systems, and its navigation in a real environment. This paper presents the current state of research, which dealt with the first two stages.

We have completed the virtual golf course relief as a simulated environment and built AGC model in Gazebo. Then we selected and further developed a set of ROS-based algorithms for navigation of a car-like robot within the simulated environment. The original ROS package algorithms, which include global and local planners, algorithms of data filtration and robot localization, were further modernized in order to be applied with Ackermann steering vehicle simulation. As the result of our simulation, we have successfully navigated the golf cart model from one golf hole to another in the simulated golf course environment in Gazebo. Finally, we have developed a small-size vehicle prototype based on Traxxas radio-controlled car model, which has been used to verify and configure the set of ROS-based algorithms in the real world conditions. Although we have not approved the final configuration of vision sensors yet, we have obtained the first results of the autonomous vehicle prototype navigation along a white line of an ellipsoid shape within indoor laboratory environment using web-camera data processing.

Our future work will concentrate on transferring navigation algorithms from Gazebo simulation to the real model, and adapting algorithms for new vision sensors: LIDAR and stereo camera, with selection of one, which optimally suits our tasks. After the verification of all algorithms on the scaled model, we will implement them into a full-size autonomous golf cart prototype.

## Acknowledgements

The research was supported by a Russian Foundation for Basic Research grant (Project No. 15-57-06010).

## References

1. S. L. Poczter, M. J. Luka, "The Google Car: driving toward a better future?", J. Business Case Studies, **10**, 1, 7-14 (2014)
2. C. Tan, and K. Tham, "Autonomous Vehicles, Next Stop: Singapore", Journeys, 5-11 (2014)
3. D. Allender, J. Kuehn, A. Miller, B. Wellman, J. Young, "Autonomous Golf Cart Navigation Using ROS", CPE 450 (2011)
4. S. Pendleton, T. Uthaicharoenpong, Z.J. Chong, et al., "Autonomous Golf Cars for Public Trial of Mobility-on-Demand Service", IEEE IROS, 245-251 (2015)
5. D. Gaynor, T. Latham, I. Anderson, and C. Johnson, "Autonomous Golf Cart", Proc. ASEE North-Central Section Conf. (2013)
6. M. Silva, L. Garrote; F. Moita, M. Martins, U. Nunes, "Autonomous electric vehicle: Steering and path-following control systems", Proc. 16th IEEE Mediterranean Electrotechnical Conf., 442-445 (2012)
7. S. Ivaldi, J. Peters, V. Padois, F. Nori, "Tools for simulating humanoid robot dynamics: A survey based on user feedback", 14th IEEE-RAS Int. Conf. on Humanoid Robots, 842-849 (2014)
8. D. Fan, P. Shi, "Improvement of Dijkstra's algorithm and its application in route planning", 7th Int. Conf. on Fuzzy Systems and Knowledge Discovery (FSKD), 1901-1904 (2010)
9. M. Likhachev, D. Ferguson, "Planning long dynamically-feasible maneuvers for autonomous vehicles", Int.J. on Robotics Res. (IJRR), **28**, 8, 933-945 (2009)
10. D. Choukroun, H. Weiss, I. Bar-Itzhack, Y. Oshman, "Direction Cosine Matrix Estimation from Vector Observations using a Matrix Kalman Filter", IEEE Trans. on Aerospace and Electronic Systems, **46**, 1, 61-79 (2010)
11. W. Li, Y. Zhao, Y. Song, Z. Yang, "A Monte-Carlo based stochastic approach of soccer robot self-localization", Conf. on Human System Interactions, 915-920 (2008)
12. H. Wu, A.C. Sankaranarayanan, R. Chellappa, "Online Empirical Evaluation of Tracking Algorithms", IEEE Trans. on PAMI, **32**, 8, 1443-1458 (2010)
13. C. Rösmann, W. Feiten, T. Wösch, F. Hoffmann, and T. Bertram, "Trajectory modification considering dynamic constraints of autonomous robots", Proc. 7th German Conf. on Robotics, Germany, Munich, 74–79 (2012)
14. C. Rösmann, W. Feiten, T. Wösch, F. Hoffmann, and T. Bertram, "Efficient trajectory optimization using a sparse model", Proc. IEEE European Conf. On Mobile Robots, Spain, Barcelona, 138–143 (2013)
15. C. Rösmann, F. Hoffmann, and T. Bertram, "Planning of Multiple Robot Trajectories in Distinctive Topologies", Proc. IEEE European Conf. on Mobile Robots, Lincoln, UK (2015)
16. M. Keller, F. Hoffmann, T. Bertram, C. Hass, and A. Seewald, "Planning of Optimal Collision Avoidance Trajectories with Timed Elastic Bands", Proc. 19th IFAC World Congress, Cape Town, South Africa, 9822-9827 (2014)

MATEC Web of Conferences **75**, 09006 (2016) DOI: 10.1051/matecconf/20167509006
*ICMIE 2016*

# Review Results on Wing-Body Interference

Vladimir Frolov[1,a]

[1]*Samara National Research University, Design and Aircraft Construction Department, 443086 Samara, Russia*

**Abstract.** The paper presents an overview of results for wing-body interference, obtained by the author for varied wing-body combinations. The lift-curve slopes of the wing-body combinations are considered. In this paper a discrete vortices method (DVM) and 2D potential model for cross-flow around fuselage are used. The circular and elliptical cross-sections of the fuselage and flat wings of various forms are considered. Calculations showed that the value of the lift-curve slopes of the wing-body combinations may exceed the same value for an isolated wing. This result confirms an experimental data obtained by other authors earlier. Within a framework of the used mathematical models the investigations to optimize the wing-body combination were carried. The present results of the optimization problem for the wing-body combination allowed to select the optimal geometric characteristics for configuration to maximize the values of the lift-curve slopes of the wing-body combination. It was revealed that maximums of the lift-curve slopes for the optimal mid-wing configuration with elliptical cross-section body had a sufficiently large relative width of the body (more than 30% of the span wing).

## 1 Introduction

An analysis of a lift for the wing-body combinations nowadays still plays an important role in a theoretical research of an aerodynamic and a preliminary design of modern aircraft.

Aircraft designers have been interested in problems of the aerodynamic wing-fuselage combination in aviation and missile technology since the aircraft occurrence. Initially, investigations were focused on the experimental study of the specific wing-body combinations [1-6]. The next years, there were theoretical papers, in which first mathematical models of the wing-fuselage interference were offered. Exact solution even for the linearized problem of the ideal incompressible flow around arbitrary shape wings in the presence of the fuselage is a difficult task, since it is necessary to solve the three-dimensional Laplace equation for the velocity potential which satisfies the boundary conditions on the surface of the combination and at infinity. The first exact solution of this problem was obtained in Golubinsky's paper [7]. The first mathematical models were based on using disposition of vortices inside the fuselage [8], on an approximate representation of integral equations [9], on the thin body theory [10-21], on the strips method [14, 15, 22] and on the calculation of the lift by the found in Trefftz's plane using velocity potential [23-26] or stream function [27]. The distribution of lift along the wing span can be obtained using methods based on the consideration of the velocity potential in the Trefftz's plane [22-26, 28-30]. It is possible to note an important result that was first obtained theoretically by Multhopp [22] and was confirmed experimentally by Jacobs and Ward [1]. This result is that the lift-curve slope of the wing-fuselage combination at a certain relative value diameter of the fuselage $\bar{D} = d_f / b$ (where $d_f$ is the diameter of the fuselage; $b$ is the wing span) exceeds the same value for the isolated wing of the same geometry which is used in the wing-body configuration. Here are some examples that confirm this fact. It was obtained that for the scheme mid-wing monoplane with cylindrical fuselage $\bar{D} = 0.14$ and trapezoidal wing with aspect ratio $\boldsymbol{AR} = 4.83$ and taper ratio $\lambda = 2{,}38$ the value lift-curve slope of the wing-fuselage combination $C_{L_\alpha \mathrm{W,B}}$ exceeds the same value $\bar{C}_{L_\alpha W}$ for the isolated wing of the same geometry which is used in the wing-body configuration by approximately 5% (4.75% – an experiment; 4.92% – theory) [22]. It was obtained also that for the wing-body combination №13 in the experimental Jacobs's and Ward's paper [1] relative increase in the lift-curve slope of the wing-fuselage combination was a slightly greater (≈6.49%) compared with the isolated wing of the same geometry. In Korner's book [6] you can read that the lift-curve slope of the wing-fuselage combination is some 5% higher than that of the wing alone. Excess value of the lift-curve slope of the wing-fuselage combination above the same magnitude for the isolated wing of the same geometry is also noted in the theoretical papers [31, 32]. In paper [32] on the basis of approximate method for the scheme mid-wing monoplane $\left(AR = 16; \lambda = 1; \bar{D} = D_f / b \approx 0{,}25\right)$ was received an

[a] Corresponding author: frolov_va_ssau@ mail.ru

increase of the lift-curve slope of the wing-fuselage combination compared with the isolated wing with the same geometry approximately 19%. Theoretical calculations the lift-curve slope of the wing-fuselage combination are devoted also book [33] and papers [34-36]. This review is by no means covers all papers on the interference of the wing and fuselage. Author's book [37] contains more a detailed bibliography on the problems of the lift of the wing-body combination.

The main purpose of this paper is to demonstrate conformity of computational results obtained by the author with the known experimental and theoretical results of other authors, as well as to give results of solving optimization problems for lift wing-body configurations.

## 2 Mathematical model of the wing-body interference

Mathematical model of the wing-body interference included two methods: 1) a discrete vortices method (DVM) and 2) 2D potential model for cross-flow around fuselage [37].

The original three-dimensional problem (Fig. 1) is divided into two: two-dimensional problem of the flow around cross section of the fuselage (Fig. 2) and three-dimensional problem for the isolated wing. The method consists in the fact that in the 2D problem of flow cross-section of the fuselage is added a pair of discrete point vortices. The vortices are the consequence arising lift on the wing. According to Zhukovsky's theory about lift of the wing, any lifting surface can be replaced by an equivalent П-shaped vortex, its free vortices at low angles of attack lie in the plane of the wing and extends to infinity. In this model is offered that each console part of the wing replaced by a single П-shaped vortex lying in the plane of the wing. In Fig. 1, the П-shaped vortex shown in the left wing console. The intensity and the coordinate of the free vortex can find from the constraint equations after will be calculated lift isolated wing by DVM.

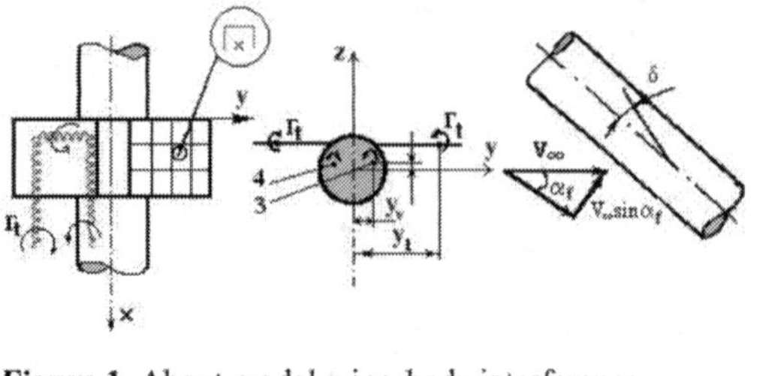

**Figure 1.** About model wing-body interference

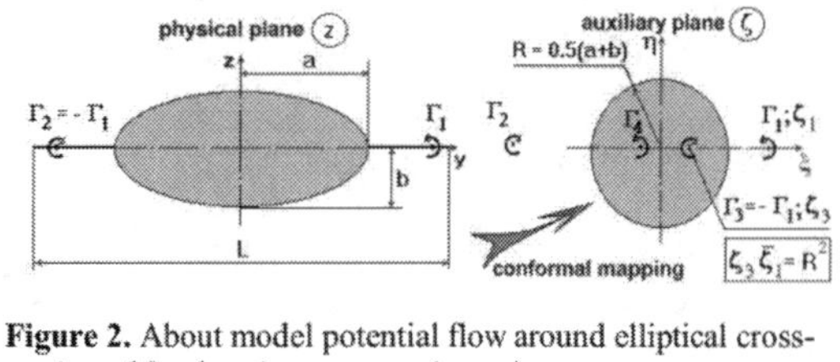


**Figure 2.** About model potential flow around elliptical cross-section of fuselage in present pair vortices

In order to satisfy the boundary conditions of impermeability on the surface of the body cross section for canonical body vortices can use inversion method (Fig. 2) and for the arbitrary two-dimensional cross section can use the panel method. Fig. 3 shows an example of the potential flow solutions for flow around elliptical cross-section of fuselage in present pair vortices.

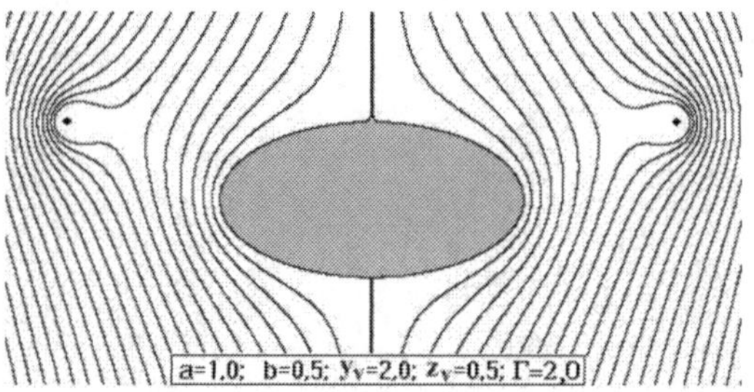


**Figure 3.** Example potential flow around elliptical cross-section of fuselage in present pair vortices

In this formulation, the problem is reduced to solving the following system of algebraic linear equations

$$\sum_{i=1}^{L}\mathbf{\Gamma}_i\left(\mathbf{A}_{ij}\cdot\mathbf{n}_j\right)=-\left(\mathbf{F}_j\cdot\mathbf{n}_j\right),\quad j=1,...L,\qquad(1)$$

where $\mathbf{\Gamma}_i$ is a column vector composed of unknown intensities of attached vortices; $\mathbf{n}_j$ is the unit normal vector to the $j$-th control point on the wing surface; $\mathbf{A}_{ij}$ is the matrix of the aerodynamic influence or the matrix of the induced velocities at the control points of the wing surface from all system of horseshoe vortices (left and right consoles for the isolated wing); $\mathbf{F}_j$ is column vector of the velocity induced in the $j$-th the control point on the wing surface by incoming flow and the flow from the cross section of the fuselage that includes either inversion of the vortices or sources and **sinks** providing satisfying conditions impermeability on the surface body from the free vortices left and right wing; $L$ is the number of control points (collocation points), equal to the number of attached vortices on the right wing console.

In the linear formulation, for small angle of attack of the wing-body combination $(\alpha\ll1)$ and small wing tilt angle (angle of inclination wing) $(\delta\ll1)$ solution can be writing as a linear function of the angle of attack and wing tilt angle

$$\mathbf{\Gamma}_i=\mathbf{\Gamma}_i^{\alpha}\cdot\alpha+\mathbf{\Gamma}_i^{\delta}\cdot\delta\qquad(2)$$

wherein right parts system of algebraic linear equations (1) can be represented as

$$\mathbf{F}_j=\mathbf{F}_j^{\alpha}\cdot\alpha+\mathbf{F}_j^{\delta}\cdot\delta\qquad(3)$$

where $\mathbf{\Gamma}_i^{\alpha},\mathbf{\Gamma}_i^{\delta},\mathbf{F}_j^{\alpha},\mathbf{F}_j^{\delta}$ are derivatives of the $\mathbf{\Gamma}_i,\mathbf{\Gamma}_i,\mathbf{F}_j,\mathbf{F}_j$ on the angle of attack $\alpha$ and wing tilt angle $\delta$.

To solve the problem the wing-body interference with the fuselage of an arbitrary cross-section for calculating the right parts of the system (1) is proposed to use the panel method that leads to the solution of algebraic linear equations (4)

$$[\mathbf{A}][\boldsymbol{\sigma}]=[\mathbf{R}] \qquad (4)$$

where $\boldsymbol{\sigma}$ is a column vector

$$\boldsymbol{\sigma}=\boldsymbol{\sigma}^{\alpha}\alpha+\boldsymbol{\sigma}^{\delta}\delta \qquad (5)$$

Let's give the final formula for the components of the induced velocity for example for case of circular cross-section of the fuselage in $j$-th control point of the wing panel

$$V_{nj}=V_{zj}\cos\delta=\cos\delta\left\{V_{\infty}\sin\alpha_B\left[1+\frac{R^2\left(y_j^2-z_j^2\right)}{\left(y_j^2+z_j^2\right)^2}\right]-\frac{\Gamma_t}{2\pi}\left[\frac{y_j-\tilde{y}_v}{\left(y_j-\tilde{y}_v\right)^2+\left(z_j-\tilde{z}_v\right)^2}-\frac{y_j+\tilde{y}_v}{\left(y_j+\tilde{y}_v\right)^2+\left(z_j-\tilde{z}_v\right)^2}\right]\right\}, \qquad (6)$$

where $V_{nj}$ is a normal velocity component to the surface at the $j$-th control point wing panel; induced velocity component along the $OZ$-axis of the cylinder in crossflow; $\left(y_j, z_j\right), \left(\tilde{y}_v, \tilde{z}_v\right)$ are coordinates of the control point and the inversion vortex point (see Fig. 1), respectively. Coordinates inversion vortices are by Milne-Thomson's theorem about the circle [38]. The intensity $\Gamma_t$ and the coordinate $y_t$ of the free vortex can be found on the connection equations [37, 39].

Thus, the problem of the wing-fuselage interference is reduced to the solution (1) to the right side (6) or right-hand sides, obtained by solving the linear algebraic equation system (4), which provides a solution of the problem for the potential flow around an arbitrary contour on the basis of the panel method. An agreement of the velocity field on the surface wing and on the surface fuselage is provided by successive iterations, each of which is reduced to the solution of systems linear algebraic equations (1) with corrected right sides (6). The solution of the problem can be got if as the zero-iteration was selected the solution for isolated wing only. The proposed model it allows get solution for the linear formulation for small angles of attack and the angle of the wing installation two problems at once, which will be called: αα-problem (fuselage and wing have the same angle of attack, angle of the wing tilt angle equal zero); δ0-problem (the fuselage has a zero angle of attack and the wing has δ tilt angle). In the linear case, the formula for the coefficients of the normal forces of the wing and the body are of the form (7)

$$\begin{aligned} C_{L\,\mathrm{W(B)}} &= C_{L_\alpha\,\mathrm{W(B)}}\alpha + C_{L_\delta\,\mathrm{W(B)}}\delta, \\ C_{L\,\mathrm{B(W)}} &= C_{L_\alpha\,\mathrm{B(W)}}\alpha + C_{L_\delta\,\mathrm{B(W)}}\delta, \\ C_{L_\alpha\,\mathrm{W,B}} &= C_{L_\alpha\,\mathrm{W(B)}} + C_{L_\alpha\,\mathrm{B(W)}},\ C_{L_\delta\,\mathrm{W,B}} = C_{L_\delta\,\mathrm{W(B)}} + C_{L_\delta\,\mathrm{B(W)}}. \end{aligned} \qquad (7)$$

The values $C_{L_\alpha\,W(B)}, C_{L_\alpha\,B(W)}$ are obtained from the solution αα-problem, and the values $C_{L_\delta\,W(B)}, C_{L_\delta\,B(W)}$ – from the solution δ0-problem.

## 3 Review of calculation results

Figs. 4-9 are presented comparing the calculation results obtained by the above theoretical model with calculations by the DVM for case αα-problem [33, 35, 36]. The rectangular, triangular, and swept wings were considered. It may be noted is enough good agreement of calculated data.

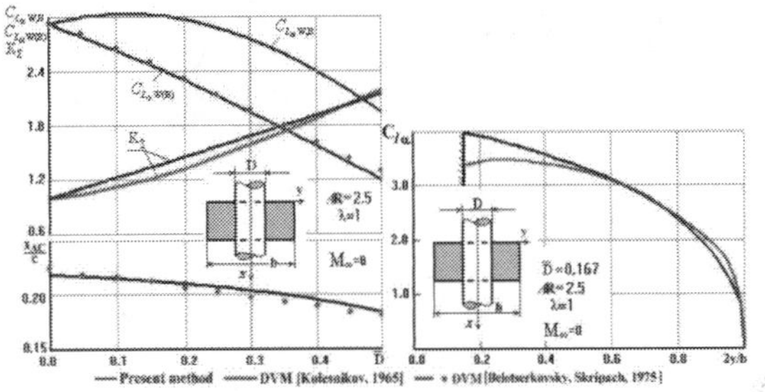

**Figure 4.** Theoretical results lift-curve slopes vs dimentiones diameter of the fuselage (a); Distribution lift-curve slopes along span of the rectangular wing for case midwing-body combination (b)

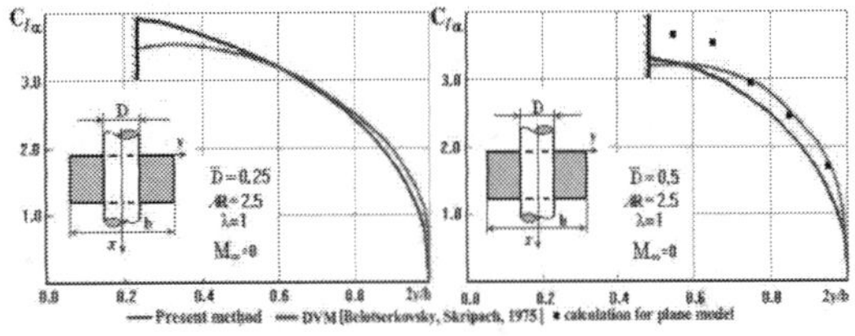

**Figure 5.** Distribution lift-curve slopes along span of the rectangular wing for case midwing-body combination

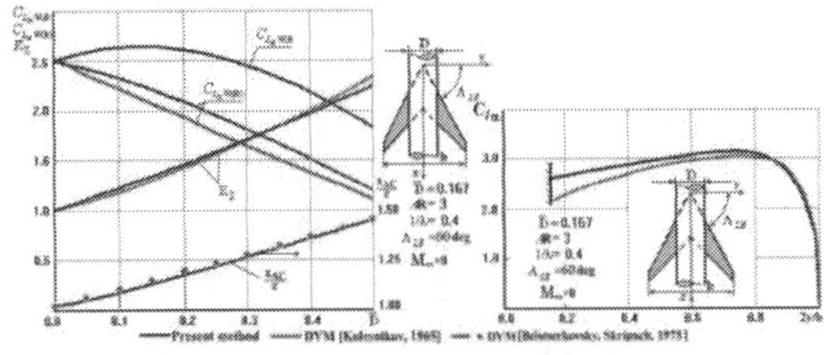

**Figure 6.** Theoretical results lift-curve slopes vs dimensionless diameter of the fuselage (a); Distribution lift-curve slopes along span of the swept wing for case midwing-body combination (b)

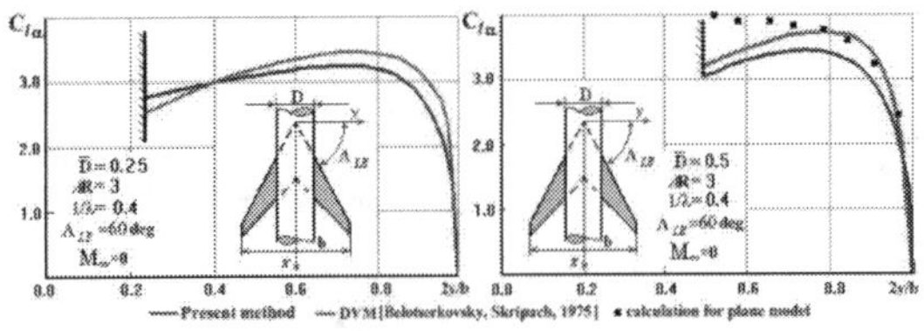

**Figure 7.** Distribution of the lift-curve slopes along span of the sweep wing for case midwing-body combination

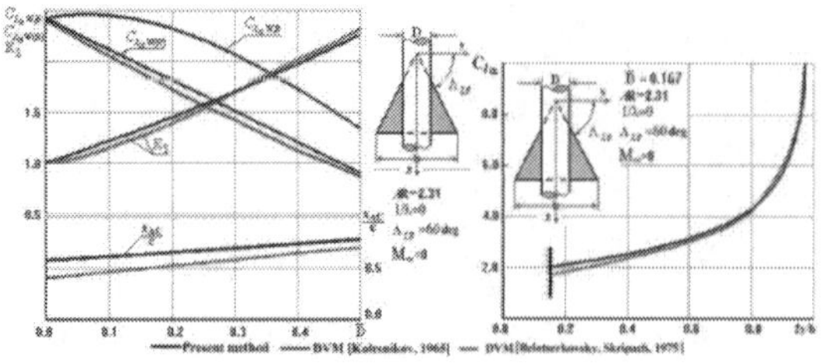


**Figure 8.** Theoretical results lift-curve slopes vs dimensionless diameter of the fuselage (a); Distribution lift-curve slopes along span of the triangular wing for case midwing-body combination (b)

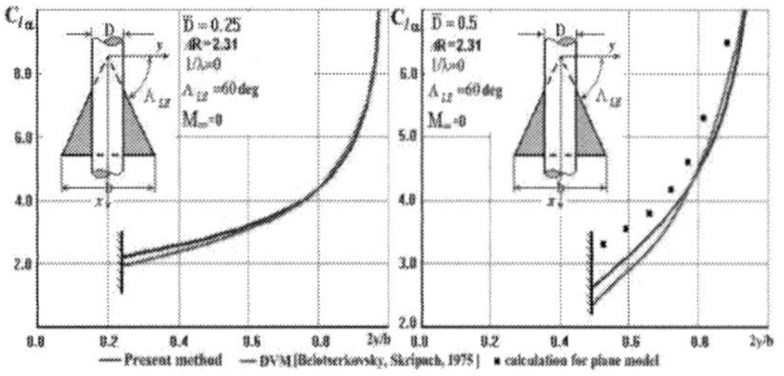


**Figure 9.** Distribution of the lift-curve slopes along span of the triangular wing for case midwing-body combination

Figs. 4, 6 and 8 also shows the changing in the aerodynamic centre coordinates measured from the beginning of the mean aerodynamic chord. The value $K_\Sigma$ in these figures is defined by the formula

$$K_\Sigma = \frac{C_{L_\alpha \mathrm{W.B}}}{C_{L_\alpha \mathrm{W}}}, \tag{8}$$

where $C_{L_\alpha \mathrm{W}}$ is a lift-curve slope for isolated wing for isolated wing, composed of two consoles.

Fig. 10 is presented comparing the calculation results obtained by the above theoretical model with calculations by the numerical method of singularities for case δ0-problem [31].

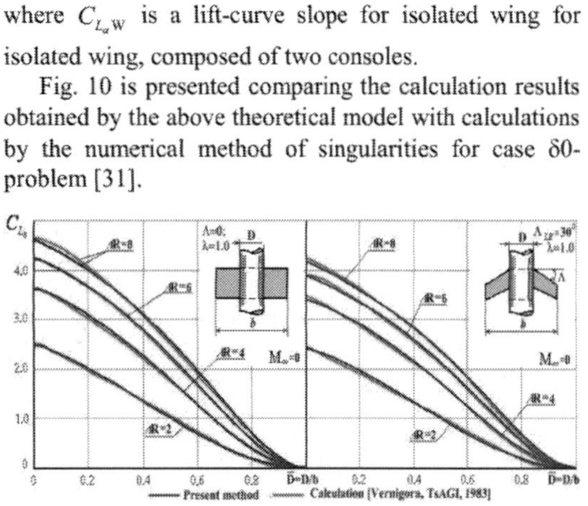


**Figure 10.** Theoretical results lift-curve slopes vs dimensionless diameter of the fuselage for case δ0-problem for case midwing-body combination

Figs. 11-13 shows a comparison of calculated data for the mathematical model described above and the calculated and experimental data of other researchers [40-45].

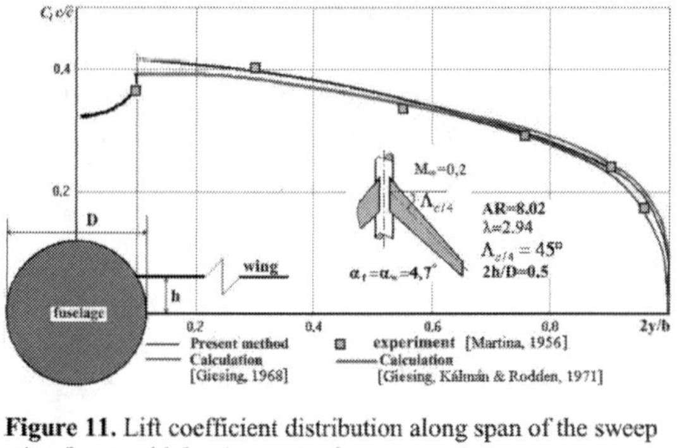


**Figure 11.** Lift coefficient distribution along span of the sweep wing for case high-wing monoplane

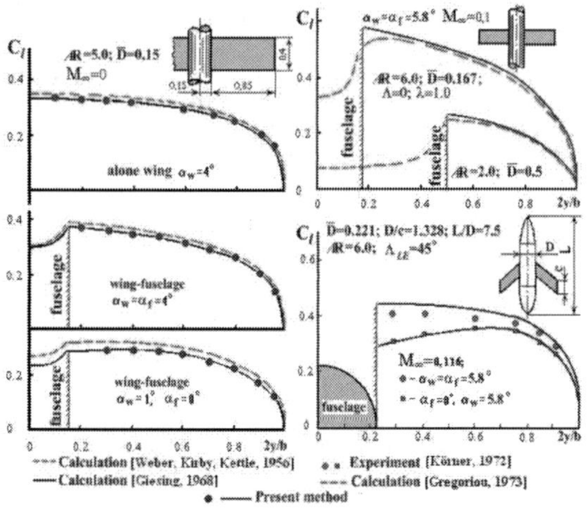


**Figure 12.** Distribution lift coefficient along span of the midwing-body combination

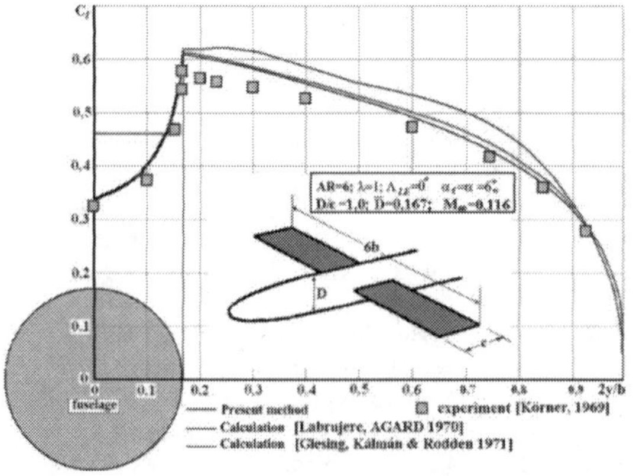


**Figure 13.** Distribution lift coefficient along span of the rectangular wing for case midwing-body monoplane

An influence of compressibility on values of theoretical lift-curve slopes can see in Figs. 14 and 15. Fig. 14 show results for midwing-body monoplane configuration for rectangular and triangular wings and Fig. 14 show results for high wing-body monoplane configuration for rectangular and triangular wings also [34].

DOI: 10.1051/matecconf/20167509006

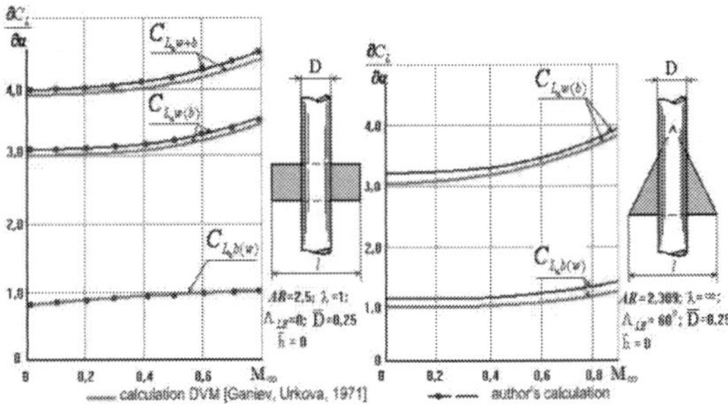


**Figure 14.** Theoretical results lift-curve slopes vs Mach number

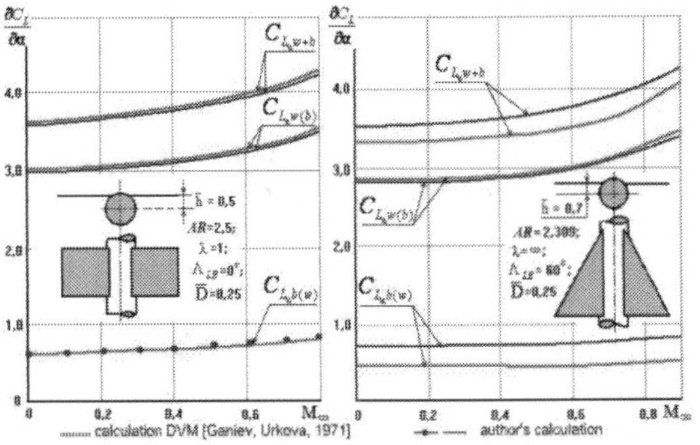


**Figure 15.** Theoretical results lift-curve slopes vs Mach number

## 4 Optimization problem for lift-curve slope of for midwing-body monoplane configuration

The review given above it was noted that in a number of theoretical and experimental papers devoted to the wing-body interference, revealed the presence of a maximum in the dependence $\partial C_L/\partial\alpha = f\left(D/b\right)$ . The carried out calculations on the offered mathematical model also confirms this fact. Generally, the maximum lift-curve slopes for wing-body combination may depend on the cross-sectional shape of the fuselage and shape of the wing. In this paper solutions of the optimization problem for the wing-body combinations with circular and elliptical cross-sections and trapezoidal unswept wings are considered.

The optimization problem is formulated as a nonlinear programming problem as follows:

$$max\, C_{L_\alpha W,B}(\mathbf{X}),\quad \mathbf{X}\in E^n \tag{9}$$

where $\mathbf{X}=\left[x_1,x_2,0\right]^{\mathrm{T}}$ is a vector of the project parameters connected with geometrical characteristics of the wing-body configuration by formulates

$$\bar{D}=\frac{D}{b}=\frac{1}{x_1^2+1},\quad \frac{1}{\lambda}=\frac{1}{x_2^2+1}, \tag{10}$$

where $x_1$ and $x_2$ are auxiliary variables. The problem (8), (9) is a problem of unconditional optimization, for which there are $\bar{D}\in[0;1]$, $(1/\lambda)\in[0;1]$ and $x_1\in[-\infty;+\infty]$, $x_2\in[-\infty;+\infty]$.

## 5 Results of the optimization problem for lift-curve slope for midwing-body monoplane configuration

Figs. 16, 17 show results of the optimization problem for lift-curve slope for midwing-body monoplane configuration with circular cross-section fuselage vs the aspect ratio of the rectangle wing. In Fig. 16 is used the notation

$$\bar{K}_\Sigma=\frac{C_{L_\alpha W,B}}{\bar{C}_{L_\alpha W}}, \tag{11}$$

where $C_{L_\alpha W,B}$ is the lift-curve slope of the wing-body combination same as (8), $\bar{C}_{L_\alpha W}$ is a lift-curve slope of the isolated wing in which it is included part of the occupied fuselage.

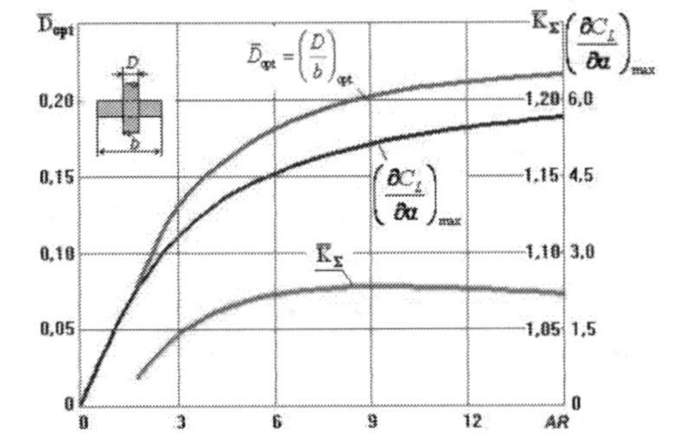

**Figure 16.** Optimal dimensionless diameter of the mid-wing configuration with circular cross-section body vs the aspect ratio of the rectangular wing

Fig. 17 show results of the optimization problem for lift-curve slopes for midwing-body monoplane configuration with elliptical cross-section fuselage vs the aspect ratio of the rectangular wing. It was revealed that maximums of the lift-curve slopes for the optimal mid-wing configuration with elliptical cross-section body had a sufficiently large relative width of the body (more than 30% of the span wing).

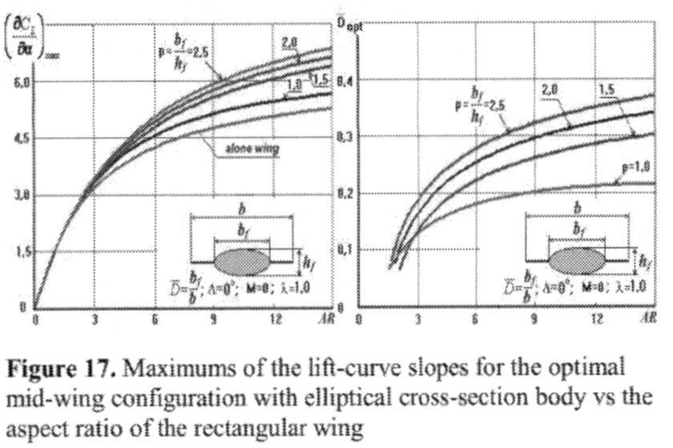


**Figure 17.** Maximums of the lift-curve slopes for the optimal mid-wing configuration with elliptical cross-section body vs the aspect ratio of the rectangular wing

DOI: 10.1051/matecconf/20167509006

Fig. 18(a) shows the influence of compressibility on optimal dimensionless diameter vs the aspect ratio of the rectangular wing. Fig. 18(b) also shows statistics for modern aircraft. It is possible to be noted one point (red colour) for the project of fifth-generation aircraft (project M-60, Russia). The project M-60 differs from all other aircraft by the presence of wide fuselage. One can predict that with increasing wing aspect ratio to a value equal 15, the optimal ratio of the width of the fuselage to the wing span can reach approximately 20%.

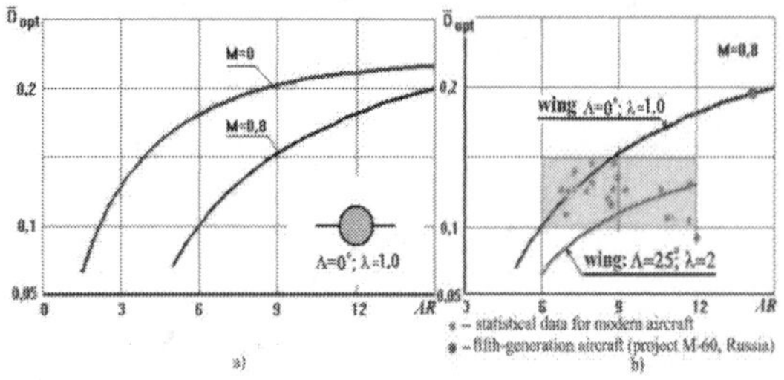


**Figure 18.** Effect compressibility on dimensionless optimal diameter of the fuselage for case rectangular midwing-body combination

## 6. Conclusions

The paper gave a brief overview of the works on the wing-body interference. The theoretical approximate model for calculating of the lift-curve slope for wing-body configuration with elliptical cross-section fuselage was considered. The optimization solution of the problem was presented. The present results of the optimization problem for the wing-body combination allowed to select the optimal geometric characteristics for configuration to maximize the values of the lift-curve slopes of the wing-body combination. It was revealed that maximums of the lift-curve slopes for the optimal mid-wing configuration with elliptical cross-section body had a sufficiently large relative width of the body (more than 30% of the span wing).

## References

1. E.N. Jacobs, K.E. Ward, NACA Report, **540**, (1935)
2. W.C. Pitts, J.N. Nielsen, G.E. Kaattari, NACA Report, **1307**, (1957)
3. W. Schneider, AGARD CP, **35**, (1968)
4. D.A. Kirby, A.G. Hepworth, R&M, **3747**, (1971)
5. H. Körner, ZFF, **20**, (1972)
6. S.F. Korner, *Fluid-Dynamic Lift* (1985)
7. A.I. Golubinsky, Proceedings TSAGI, **810**, (1961)
8. B.F. Lebedev, Proceedings TsAGI, **719**, (1958)
9. A.A. Dorodnitsyn, Proceedings TsAGI, **546**, (1944)
10. V.V. Keldysh, Proceedings TsAGI, **759**, (1959)
11. V.V. Keldysh, Scientists Notes TsAGI, **VI**, 5, (1975)
12. V.V. Keldysh, Scientists Notes TsAGI, **VIII**, 3, (1977)
13. A.H. Flax, J. Aircraft, **11**, 5, (1974)
14. A.H. Flax, J. Aeron. Sc., **20**, 7, (1953)
15. A.H. Flax, H.R. Lawrence, *Third Anglo-American Aeron. Conf.* (Brighton, 1951)
16. H.R. Lawrence, A.H. Flax, J. Aeron. Sc., **21**, 5, (1954)
17. J.N. Nielsen, *Missile aerodynamics* (McGraw-Hill Book Company, Inc., New York-Toronto-London, (1960)
18. J.R. Spreiter, NACA TN, **1662**, (1948)
19. J.R. Spreiter, NACA TN, **1897**, (1949)
20. J.R. Spreiter, NACA Report, **962**, (1950)
21. J.R. Spreiter, A.H. Sacks, NACA Report, **1296**, (1956)
22. C. Ferrari, *Aerodynamic components of aircraft at high speeds, Vol. VII. High Speed Aerodynamics and Jet Propulsion* (Princeton University Press, Princeton, New Jersey, 1959)
23. E. Karafoli, *Aerodynamics of Aircraft Wing*, AN USSR Publishing, (1956)
24. U.A. Ruhgov, Proceedings TsAGI, **759**, (1959)
25. J. Lennertz, NACA TM, **400**, (1927)
26. J. Lennertz, *In: Durand W. F. (editor) Aerodynamics. Vol. IV*, Springer, Berlin, (1934)
27. A.A. Nikolsky, Proceedings TsAGI, **2122**, (1981)
28. J. Weber, ARC CP, **1331**, (1975)
29. J. Weber, ARC CP, **1332**, (1975)
30. J. Weber, M.G. Joyce, ARC CP, **1334**, (1975)
31. V.N. Vernigora, Proceedings TsAGI, **2176**, (1983)
32. V.I. Kholyavka, *Interference of aircraft parts* Kharkov. Aviation. Inst., Kharkov, (1967)
33. S.M. Belotserkovsky, B. K. Skripach, *Aerodynamic derivatives wing and the aircraft at subsonic speeds*, Nauka, Moscow, (1975)
34. F.I. Ganiev, M. E. Urkova, Proceedings Air Force Academy named after Zhukovsky, **1302**, (1971)
35. G.A. Kolesnikov, Proceedings TsAGI, **810**, (1961)
36. G.A. Kolesnikov, Proceedings TsAGI, **954**, (1965)
37. V.A. Frolov, *Methods for Calculating Lift for Wing-Body Combination*, LAP Lambert Academic Publishing, Saarbrucken, (2011)
38. L.M. Milne-Thomson, *Theoretical Aerodynamics* Constable and Company, London, (1958)
39. M.R. Mendenhall, J. N. Nielsen, NASA CR, **2473**, (1975)
40. J.P. Giesing, McDonnell-Douglas Rep. Douglas Aircraft Co., Report DAC, Vol. I, **67212**, (1968)
41. J.P. Giesing, W.P. Kalman, W.P. Rodden, USAF FDL-TR-**71-5**, Part I and Part II, (1971)
42. A.P. Martina, NACA TN, **3730**, (1956)
43. H. Körner, Zeitschrift für Flugwissenschaften, **20**, (1972)
44. G. Gregoriou, BMVg-FBWT-**73-33**, (1973)
45. T.E. Labrujere, W. Loeve, J.W. Slooff, AGARD-CP-**71**, (1971)

MATEC Web of Conferences **75**, 09007 (2016) DOI: 10.1051/matecconf/20167509007
*ICMIE 2016*

# Study of EBSD Experiment Parameters Influence on Computation of Polycrystalline Pole Figures and Orientation Distribution Function

Anastasia O. Antonova[1], Pavel S. Dzhumaev[2] and Tatyana I. Savyolova[1]

[1] *Moscow Engineering Physics Institute (State University), Department of Applied Mathematics, Kashirskoe sh. 31, Moscow, 115409 Russia*
[2] *Moscow Engineering Physics Institute (State University), Department of Material Science, Kashirskoe sh. 31, Moscow, 115409 Russia*

**Abstract.** Mathematical model for a polycrystalline specimen and EBSD experiment is proposed. As the measurement parameters, the scanning step and the threshold disorientation angle are considered. To study the impact of the measurement parameters Pole Figures and Orientation Distribution Function of model specimen and corresponding ones, calculated from model EBSD measurements, are compared. The real EBSD experiment was also performed. The results of the model experiment are correlated with such detected in the real EBSD data. The most significant results are formulated in the given work.

## 1 Introduction

Recent advances in electron back scattering diffraction (EBSD) [1] have opened up new opportunities in the study of material structure and texture. The problem of choosing measurement parameters remains insufficiently investigated, while these parameters have a large effect on measurement results. It was indicated than an unsuitable choice of measurement parameters leads to incorrect results. For example, an insufficient number of orientation measurements results in low accuracy of computed Pole Figures (PFs) [2]; disorientation angle measurements may lead to inconsistent average grain size values [3]; a small step of measurements leads to a considerable increase in time required for conducting an experiment (see [4]). For solving the problem of EBSD parameters choosing we need the opportunity to compare the specimen data and obtained EBSD data. Accordingly, it is of interest to study this issue with the help of the mathematical simulation of an experiment. If the real specimen characteristics are close to ones of the model specimen the EBSD experiment model helps to make a decision about experiment parameters.

Below, we develop a two-dimensional mathematical model of a specimen and a subsequent EBSD experiment. Specimen measurements are modeled for various experiment parameters. PF and orientation distribution functions (ODF) for model specimen and corresponding ones, calculated from model EBSD data, are compared for each set of experiment parameters. The substantial result of analysis are presented. The real EBSD experiment was also performed. The results of the model experiment (see [5,6]) are correlated with such detected in the real EBSD data.

## 2 Carrying out a model experiment

A specimen is a set of grains

$$\{(x_i, g_i, \omega_i)\ i = 1,2, \dots N-1, (x_N, g_N)\}, \qquad (1)$$

where $x_i$ is the size of the $i$th grain, which is equal to the diameter of a circle whose area is equal to that of the grain's cross section; $g_i = (\varphi_i, \theta_i, \psi_i)$ is the orientation of the $i$th grain corresponding to the Euler angles $-\pi \leq \varphi_i, \psi_i < \pi, 0 \leq \theta_i \leq \pi$ ; $\omega_i = (g_i, g_{i+1})$ is the disorientation angle between neighboring grains; and $N$ is the number of grains in the specimen. Grain sizes are chosen as random variables distributed satisfying a gamma distribution [5].

A set of grain orientations with a distribution function in the form of central distribution function (CND) [7] with a given parameter $\varepsilon > 0$ is generated by the special Monte-Carlo method with the number of convolutions equal to $n = 6$. Grains (Eq. 1) are separated by boundaries, which are variables uniformly distributed in $[0, 2\gamma/10)$, where $\gamma > 0$ is such that the area of the boundaries is 10% of the grains' area.

To simulate the EBSD experiment the following measurement parameters are specified: $h$ is the scanning step, and $\omega_0$ is a threshold disorientation angle for determining two different neighboring grains $g_i$ and $g_{i+1}$ from the disorientation angle between them.

Measurements are made at the lattice points $\Delta_k = kh, k = 1,2, \dots, N_1$ on the surface of the simulated specimen. If the point hits the boundary, it is assigned an orientation corresponding to the Euler angles equal to the arithmetic mean of the angles of two nearest determined orientations [7].

The resulting orientations are used to calculate the disorientation angles $\widehat{\omega}_k = (\hat{g}_k, \hat{g}_{k+1})$, $k = 1, \ldots, N_1 - 1$. If $\widehat{\omega}_k \leq \omega_0$, then the *k*th and $(k + 1)$th steps belong to a single grain. If *n* measurements are inside the *j*th grain, then the grain size is set equal to $\tilde{x}_j = nh$ and the Euler angles are calculated as the arithmetic mean of the corresponding *n* measurements. If $\widehat{\omega}_k > \omega_0$, then a new grain begins at the point $(k + 1)h$. As a result of the measurements, we obtain a set of grains

$$\{(\tilde{x}_i, \tilde{g}_i, \widetilde{\omega}_i), i = 1,2, \ldots \widetilde{N} - 1, (\tilde{x}_{\widetilde{N}}, \tilde{g}_{\widetilde{N}})\}. \quad (2)$$

## 3. Results of the model experiment

Following the scheme described in Section 2, we modeled specimen, in which the boundaries amount to about 10% of the total grain area, with the average grain size equal to 2.0μm. The number *N* of grains in a specimen was set equal to 1000. The grain orientations in a specimen were modeled by the special Monte-Carlo method for CND on *SO*(3) with texture sharpness $\varepsilon = 1/8$ (see [7]). The following EBSD measurement parameters were chosen for a given specimen: $h = 0.5\ \mu\text{m}, 1.0\ \mu\text{m}, 2.0\ \mu\text{m}$ ; $\omega_0 = 5°, 10°$ and $20°$. Below, the results of observations concerning the ODF and PF changing according to EBSD measurement parameters are given.

Firstly, let us emphasize that for ODFs at $\omega_0 = 20°$ and, in some cases, for $\omega_0 = 10°$, the maximum is observed to increase [5]. This behavior of the ODF is similar to that in the presence of dependent orientations [7], which arise at $\omega_0 = 20°$ because of merging a large number of grains. What about the coordinates of the maximum we can not reasonably confirm whether they remain unchanged or not while the variable ψ is fixed. In the cased of strongly textured specimens, the ODF should be recovered with a small threshold disorientation angle. For the specimens under consideration, as such an angle, we can use $\omega_0 = 5°$.

A decrease in the scanning step insignificantly improves the ODF recovered from individual orientations. At the same time, the statistical sample is reduced substantially. For more detail on the effect of a sample on ODFs recovered from individual orientations, see [8].

Some more remarks have to be made about the scanning step. To obtain more accurate shapes and sizes of grains, the step is usually chosen to be 1/10–1/5 times as large as the average grain size in the specimen (see [8]). This step corresponds to $\delta_{less\_h} < 5\%$, where $\delta_{less\_h}$ is the fraction of grains whose sizes are less than scanning step. acceptable results are obtained even with a step corresponding to $\delta_{less\_h} \approx 25\%$ [5]. In this case, the sample turns out to be several times as large.

According to results of specimen PF and PF of modeling measurements comparing it was obtained that for strongly textured specimens threshold disorientation angle is appeared to be more significant than scanning step; for weakly textured specimens, conversely, measurements are strongly affected by the scanning step [6]. The calculated deviation of measured PF from model specimen Pf in $L_1$ metrics and RP-factor for measured PF verify statement mentioned. We should also notice that the PF maximum and its position are unstable to EBSD measurements parameters. Instability grows with weakening of specimen texture sharpness. it shows up in the shift of PF maximum position or second maximum appearing.

The more grain size distribution approaches the uniform one, the stronger instability and the less accurate PF calculation [6].

## 4. Correlation of the model experiment results with real EBSD data

Material for the investigation was a ferritic-martensitic steel EP823 (12Cr-Mo-W-Si-V-Nb-B) treated by the nitrogen gas high-temperature pulsed plasma flows (HTPPF). HTPPF surface modification were carried out by means of the Z-pinch type experimental plasma installation "Desna-M" [9]. This treatment results in the formation of the microcrystalline surface layer with a mean grain size of 2 mm. This layer has a single phase bcc type (a-Fe) equiaxed grain structure with a homogeneous distribution of alloying elements. An EBSD data were acquired using the HKL CHANNEL5 software (Oxford Instruments, UK) coupled to an EVO 50 XVP scanning electron microscope (ZEISS, Germany)..

EBSD measurement were performed with scanning step sizes $h = 0.5\ \mu\text{m}, 1.0\ \mu\text{m}, 2.0\ \mu\text{m}$ and then for each step size meaning the different threshold disorientation angle meanings were considered: $\omega_0 = 5°, 10°$ and $20°$.

Below, the results of analysis of the obtained ODF and PF from EBSD orientation data according to EBSD measurement parameters are given.

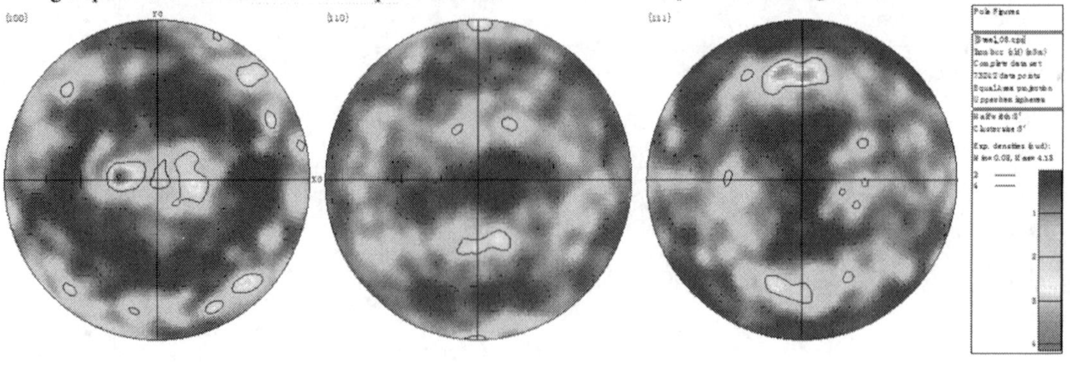

 DOI: 10.1051/matecconf/20167509007

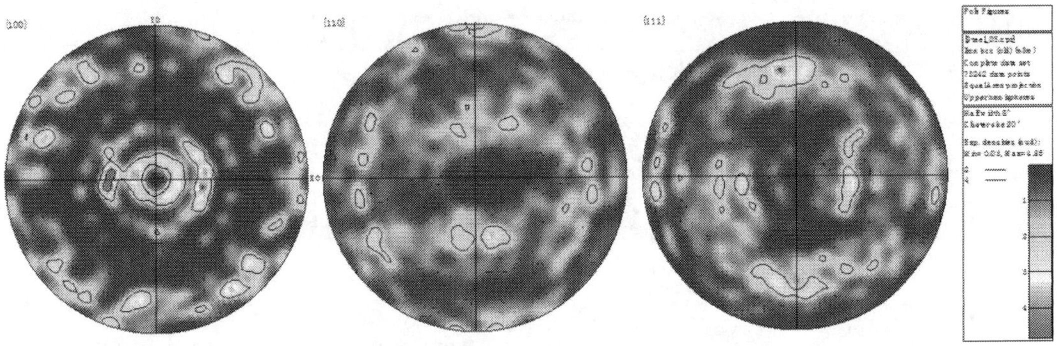

**Figure 1.** PF for fixed $h = 0.5$ µm, (a) for $\omega_0 = 5°$ and (b) for $\omega_0 = 20°$

First of all, it is worth noting that according to ODF and PF images measurements are strongly affected by the threshold disorientation angle (Fig. 1), the influence of step size on ODF and PF evaluation is not so significant (Fig. 2). The given statement is in good agreement with model experiment results.

Fig.1 shows that $\omega_0$ can change PF plot essentially. It should be noted that considerably different scanning steps $h = 2.0$ µm and 0.5 µm give close PF plots (Fig.2). Herewith the time expended in EBSD measurement increases by a factor of several tens under step size growth (Table 1).

**Table. 1.** Measurement time

| $h$ [µm] | Number of points measured | $t_{meas}$ [h] |
|---|---|---|
| 2.0 | 4524 | 0.16 |
| 1.0 | 18251 | 0.63 |
| 0.5 | 73242 | 2.56 |

The behavior of PF and ODF maximums was also considered. Table 2 presents PF maximums according to different EBSD parameters.

**Table 2.** Meanings of PF maximums

| $h$ [µm] | $\omega_0$ [°] | | |
|---|---|---|---|
| | 20° | 10° | 5° |
| 2.0 | 5.56 | 4.85 | 4.82 |
| 1.0 | 4.93 | 4.11 | 4.23 |
| 0.5 | 4.85 | 4.02 | 4.15 |

Let us notice that PF maximum increases under step size growth for each fixed $\omega_0$, but dependence PF maximum on threshold disorientation angle is appeared not so clear: maximum can increase or decrease with $\omega_0$ growth. Such behavior points out instability of PF calculation to EBSD parameters that is obtained in model experiment (Section 1).

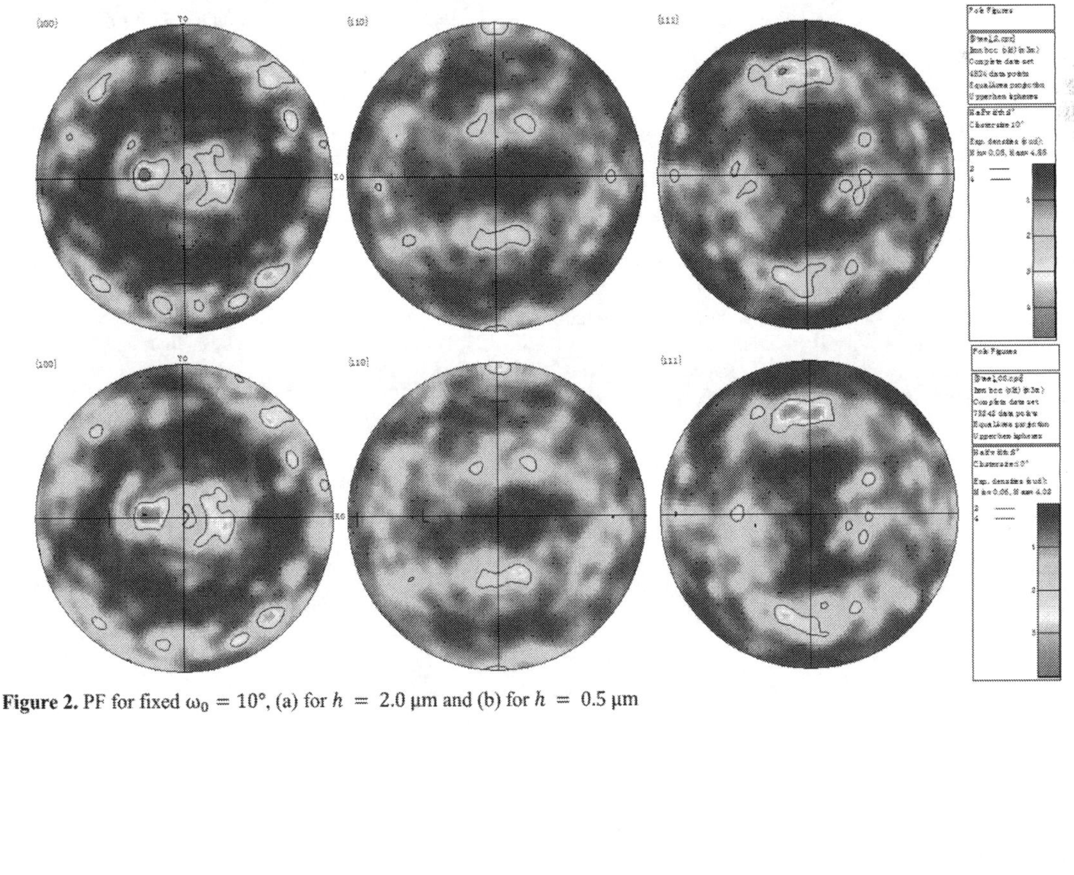

**Figure 2.** PF for fixed $\omega_0 = 10°$, (a) for $h = 2.0$ µm and (b) for $h = 0.5$ µm

 DOI: 10.1051/matecconf/20167509007

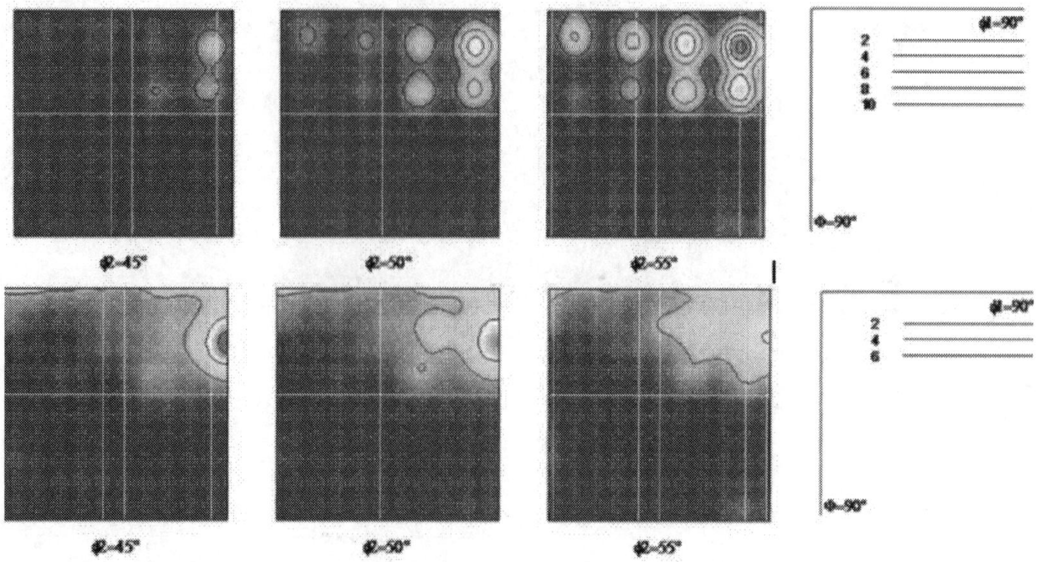

**Figure 3.** Contour lines of ODF cross section for fixed $h = 0.5$ μm, (a) for $\omega_0 = 20°$ and (b) for $\omega_0 = 10°$

Relying on the contour lines of ODF cross section (Fig. 3) we receive the sufficient maximums growing (almost twice) for $\omega_0 = 20°$ versus ones for $\omega_0 = 5°$ and $10°$. Angles $\phi1, \Phi, \phi2$ (Fig. 3) correspond to Euler angles $\varphi, \theta, \psi$ respectively. Besides unsuitable meaning of $\omega_0$ may changed ODF character. This observation was also found out during the processing of model experiment results [5].

## 5 Summary

The comparison of PF and ODF calculation results for model and real EBSD experiment found out strong similarity and gave us the ability to use our model for study of choosing measurement parameters. For the single-phase strong textured specimen with small amount of boundaries the threshold disorientation angle should not be more than 10° and the step size equal to half an average grain size is suitable. The step size being less than a quarter of average grain size can result in unjustifiably high time of measuring. Moreover, step size should not less a quarter of average grain size as it can result in unjustifiably high time of measuring in view of inessential result improvement.In general researchers need to carry out the corresponding modeling for each specimen with new characteristics and further model experiment with parameters varying to choose appropriate ones.In the future the improvement of experiment and specimen model as well as investigation of wider class of materials using the model developed are planned.

## References

1. A. J. Schwartz, M. Kumar, B. L. Adams, and D. P. Field, *Electron Backscatter Diffraction in Materials Science*, 2nd. ed. Springer, New York, (2009).
2. E. Guilmeau, C. Henrist, T. S. Suzuki, Y. Sakka, D. Chateigner, D. Grossin, and B. Ouladiaf, ICOTOM **14**, 179, (2005).
3. S. Piazolo, V. G. Sursaeva, and D. J. Prior, ICOTOM 14, 213 (2005).
4. M. V. Sypchenko, T. I. Savyolova, Cristallogr. Rep. **55** (4), 546 (2010).
5. A.O. Antonova, T. I. Savyolova, Comput. Math. Math. Phys. **55** (2), 317 (2015).
6. A.O. Antonova, T. I. Savyolova, Cristallogr. Rep. **61** (3), 1 (2016).
7. T. I. Savyolova, T. M. Ivanova, and M. V. Sypchenko, *Methods for Solving Ill-Posed Problems in Texture Analysis and Their Applications,* Mosk. Inzh._Fiz. Inst., Moscow, (2012) [in Russian].
8. N. Bozzolo, F. Gerspach, G. Sawina, and F. Wagner, J. Microscopy. **227**, 275 (2007).
9. V.A. Gribkov, F.I. Grigoriev, B.A. Kalin, V.L. Yakushin, *Advanced Radiation-Beam Technologies of Materials Treatment*, Krugly God, Moscow, (2001). [in Russian].

MATEC Web of Conferences **75**, 09008 (2016) DOI: 10.1051/matecconf/20167509008
*ICMIE 2016*

# Numerical analysis of nonlinear oscillatory processes in columns at drilling oil and gas well

Lelya Khajiyeva[1], Batyrgan Mardonov[2], Sirozhiddin Amanov[2] and Almatbek Kydyrbekuly[1]

[1]*al-Farabi National University, 71, al-Farabi ave., 050040 Almaty, Kazakhstan*
[2]*Tashkent Institute of Textile and Light Industry, 5, Shahjahan str, 100100 Tashkent, Uzbekistan*

**Abstract.** The paper describes the behavior of the system of geometrically nonlinear drill columns under the action of dynamic loads. The authors used the finite element method to solve the boundary value problems. The forms of the beam in the transverse direction were expressed by cubic Hermite polynomials. It has been found out that nonlinearity of the column may substantially change the character of oscillatory processes in its transverse direction. The loss of stability may occur during slip kicking at high values of nonlinearity.

## 1 Introduction

Nonstationary oscillations of geometrically nonlinearly distributed systems make complicated problems in mechanics of deformed solid bodies and the theory of vibrations. Studying of such problems enabled scientists to reveal a number of poorly studied aspects including physically and geometrically non-linear processes accompanied by various types of drilling hazards (loss of column stability, pipe breaks, etc.) as well as wave and oscillatory processes in the elements of the dynamic drill system [1], [2].

The dynamic impact on the column can be produced by the engine torque and repeatedly varying loads during drilling with cutting bits, axial tensile loads from the drilling rig and compressive loads accompanied by vibrations during bore bit operations in the bottomhole, etc. The behavior of the column under the dynamic load depends on the dynamic characteristics of the column design, its frequencies and modes in longitudinal, transverse and torsional oscillations. It was found out that the impact of the torque on the behavior of the column is determined not by its magnitude but by a possible deviation of the column from the state of static equilibrium. In this case the twisted column loses its stability not as a result of static bending but as result of a flutter, when the energy supplied to the column is converted into the energy of transverse oscillations with the amplitude increasing in time.

Theoretically, the column must be considered as a non-linear mechanical system with an infinite number of degrees of freedom. However, there are no analytical methods that would enable specialists to study the dynamics of such a system and to evaluate the influence of vibrations on its strength and stability during drilling. In practice, the drill column usually loses its rectilinear form due to the loss of stability. Stability of drill columns was considered in [3], [4], where the influence of a concentrated force, the column weight, the internal and external pressure, centrifugal forces, deformation properties of the pipe material and other factors on column stability was studied. Construction and specificity of operation of drilling columns make it possible to represent them as long heavy stretched-compressed tubular rods experiencing constant action of longitudinal and transverse loads. In the study of dynamic stability of drill columns under the action of dynamic loads, new effects caused by the nature of loads arise [3]. Such effects, in particular, include fast loading of the column by the axial compressing load. If the rate of increase in the pressing force is rather high, the cross section of the drill column does not have time to move in the direction normal to the axis of the column. Because of this, the compressive force can reach the first critical value and even exceed it considerably earlier than bending reaches significant values. As in this dynamic process the compressive force can exceed not only the first but also higher critical values, higher forms of stability loss can be expected. This feature of the dynamic loading process is important because it causes a significant increase in the carrying capacity of the column. If the rate of increase of the pressing force is much smaller than the speed of propagation of longitudinal waves in the drill column, the time interval during which the wave front reaches the lower end of the column is small as compared to this time. In such a case we can assume that the process is quasi-stationary and the compressive force at any given moment of time in all sections of the column changes immediately. If the compressive load is applied during a short period of time, which corresponds to an impact plunge of the column, it is necessary to follow the process of force transfer along the length of the column, and thus to take into account the inertia forces of the column elements corresponding to the longitudinal movement.

To obtain quantitative solutions of nonlinear problems in the elasticity theory it is often necessary to use numerical methods. One of such approaches based on the idea of piecewise approximation of continuous fields is the finite element method. The simplicity and similarity of this method make it a convenient tool for solving a wide class of nonlinear problems [5]-[7].

## 2 Mathematical model

Let us consider vibrations of drilling rods (columns) under the action of time-varying compression loads. It is assumed that the column is a heavy rod. The deformed state of the drill column is assumed to be geometrically nonlinear, i.e. finite deformations are considered to be acceptable. Let us determine the potential energy of bending and longitudinal displacement of the rod in the case of non-linear deformation U [3], [8], [9]:

$$U=\frac{EI}{2}\int_0^l\left[\frac{\partial^2(v_l-v_0)}{\partial x^2}\right]^2dx+\int_0^l\left[\left(\frac{\partial u}{\partial x}\right)^2+\left(\frac{\partial u}{\partial x}\right)\left(\frac{\partial v_l}{\partial x}\right)^2+\frac{1}{2}\left(\frac{\partial v_l}{\partial x}\right)^4\right]dx, \qquad (1)$$

the work $A_l$ of the external force $P$ in the vertical displacement, the work $A_2$ of the weight of the rod when the rod deviates from its equilibrium position are:

$$A_l=\frac{P}{2}\int_0^l\left[\left(\frac{\partial v_l}{\partial x}\right)^2-\left(\frac{\partial v_0}{\partial x}\right)^2\right]dx+\frac{Pl}{EF}, \qquad (2)$$

$$A_2=-\frac{q}{2}\int_0^l[x+l(i-1)]\left(\frac{\partial v_i}{\partial x}\right)^2dx. \qquad (3)$$

The kinetic energy $T$ is

$$T=\frac{qF}{2g}\int_0^l\left\{\left(\frac{\partial v_l}{\partial t}\right)^2+\left(\frac{\partial u}{\partial t}\right)^2\right\}dx, \qquad (4)$$

where $v_0(x)$ is the initial form of bending, $v_l=v_l(x,t)$ is the total bending of the rod, $u=u(x,t)$ is the longitudinal displacement of cross-sections of the rod, $E$ is the Young module for the rod material, $l$, $F, I$ are the length, the cross-sectional area and the moment of inertia, $q$ is the linear weight of the rod.

The total potential energy is: $Э=U-A_l-A_2$ [3]. Let us define the Lagrangian as $L=T-Э$. Taking $q=q(t)$ as a generalized coordinate, we can write the Lagrange equation of the second type:

$$\frac{d}{dt}\left(\frac{\partial L}{\partial \dot q}\right)-\frac{\partial L}{\partial q}=0. \qquad (5)$$

It means that in the energy method it is necessary to determine the expressions for all potentials and the kinetic energy in terms of generalized coordinates. The deflections of the beam $v_0=v_0(x)$ and $v_l=v_l(x,t)$ and the longitudinal displacement $u=u(x,t)$ can be represented by arbitrary functions satisfying corresponding initial and boundary conditions.

## 3 Discretization of model

Let us consider the following boundary value problem :

$$v_l(x,t)=0,\ \frac{\partial v_l(x,t)}{\partial x}=0,\ \frac{\partial u}{\partial x}=\frac{P(t)}{EF}\ \text{for}\ x=0 \qquad (6)$$

$$v_l(x,t)=0,\ \frac{\partial v_l(x,t)}{\partial x}=0,\ u=u_0(t)\ \text{for}\ x=l \qquad (7)$$

Finite element method.is used to solve of boundary value problem. The length of the drilling rod is divided into $n$ finite elements ($n+1$ nodes) of the same length $a$. The form of the beam in the transverse direction is expressed by the cubic Hermite polynomials:

$$Э_1=1-2\xi^2+2\xi^3,\ Э_2=a(\xi-2\xi^2+\xi^3), \qquad (8)$$

$$Э_3=3\xi^2-2\xi^3,\ Э_4=a(-\xi^2+\xi^3),\ \xi=x/a. \qquad (9)$$

The longitudinal displacement of each element can be expressed in terms of quadratic functions: $N_1=1-3\xi+2\xi^2$, $N_2=2\xi^2-\xi$.

Therefore, for an arbitrary element the bending of the beam can be written as $v=Э_1q_i+Э_2q_{i+1}+Э_3q_j+Э_4q_{j+1}$, where $q_i$, $q_j$ are deflections, $q_{i+1}$, $q_{j+1}$ are angles of rotation in the nodes of elements. The total number of unknowns for the chosen $n$ elements is $2n+2$.

The longitudinal displacement of cross-sections of the element is expressed as $u=N_1r_i+N_2r_j$, where $r_i$, $r_j$ are longitudinal displacement in the nodes of the element. The number of unknown displacements of nodes is equal to $n+1$. Thus, in the general case the chosen system of finite elements has $3n+3$ degrees of freedom, corresponding to bendings $v=q_i$, $v=q_j$, angles of rotation $v'=q_{i+1}$ $v'=q_{j+1}$ and longitudinal displacements $u=r_i$, $u=r_j$. Further, these coordinates are called generalized. Taking into account the boundary conditions (6) - (7), the number of unknowns becomes equal to $k=3n+3-6=3(n-1)$ ($n\geq 2$). Let us introduce the following notations: $v_{i,i+1}$ is the beam bending in the element located between $i$ and $i+1$ nodes and $u_{i,i+1}$ are longitudinal displacements. According to the boundary conditions (6) - (7) $v_{i,i+1}$ can be expressed as:

$$v_{1,2}=Э_3q_3+Э_4q_4, \qquad (10)$$

$$v_{2,3}=Э_1q_3+Э_2q_4+Э_3q_5+Э_4q_6, \qquad (11)$$

$$v_{3,4} = \vartheta_1 q_5 + \vartheta_2 q_6 + \vartheta_3 q_7 + \vartheta_4 q_8, \quad (12)$$

. . . . . ,

$$v_{i,i+1} = \vartheta_1 q_{2i-1} + \vartheta_2 q_{2i} + \vartheta_3 q_{2i+1} + \vartheta_4 q_{2i+2} \; (i = 2, n-1), \quad (13)$$

. . . . . . . . . ,

$$v_{n-1,n} = (1 - 3\xi^2 + 2\xi^3) q_{2n-3} + \vartheta_2 q_{2n-2}. \quad (14)$$

For longitudinal displacements and $u_{i,i+1}$ is obtained:

$$u_{1,2} = (1 - 4\xi^2) r_1 + N_2 \frac{Pa}{EF}, \quad (15)$$

$$u_{2,3} = (-3r_1 + \frac{Pa}{EF}) N_1 + N_2 r_3, \quad (16)$$

$$u_{3,4} = N_1 r_3 + N_2 r_4, \quad (17)$$

. . . . . . . . . ,

$$u_{i,i+1} = N_1 r_i + N_2 r_{i+1}, \; (i = 3..n-1) \quad (18)$$

. . . . . . . . . . . . . . ,

$$u_{n,n+1} = N_1 r_n + N_2 u_0. \quad (19)$$

Substituting expressions $v = v(\xi, t)$ and $u(\xi, t)$ into formulae (1)-(3), determine the potential energy of deformation, the potential load P, the potential weight force and kinetic energy through the Hermite polynomials.

In case of the dynamic drill system with concentrated masses (clutches), the energy of the beam and concentrated masses $m_i$ can be expressed as:

$$T_i = \frac{qF}{2g} \int_0^l \left\{ \left( \frac{\partial v_i}{\partial t} \right)^2 + \left( \frac{\partial u_i}{\partial t} \right)^2 \right\} dx + \frac{1}{2} m_i \left( \frac{\partial u_i(0,t)}{\partial t} \right)^2, \quad (20)$$

where $u_i = u_i(t)$ is the displacement of the $i$ – th mass, and $u_n(l,t) = u_0(t)$ ($u_0(t)$ is the displacement of the lower end of the $n$ -th beam).

The sum of potentials of the beam weight, compression $P$ and concentrated masses is calculated by the formula

$$V_i = -\frac{q}{2} \int_0^l [x + l(i-1)] \left( \frac{\partial v_i}{\partial x} \right)^2 dx - m_i g u_i(0,t). \quad (21)$$

To determine the longitudinal displacement of the rod we will use Kirchhoff's hypothesis [12], according to which the average elongation $\varepsilon_{ix}$ of each drilling rod does not dependent on coordinates and depends only on time. Now we obtain the following expression for the deformation [10]:

$$\varepsilon_{ix} = \frac{\partial u_i}{\partial x} + \frac{1}{2} \left( \frac{\partial v_i}{\partial x} \right)^2$$
$$\varepsilon_{ix} = \frac{1}{l} \left[ u_i(l,t) - u_i(0,t) + \frac{1}{2} \int_0^l \left( \frac{\partial v_i}{\partial x} \right)^2 dx \right] \quad (22)$$

The longitudinal displacement of the $i$ - th beam is expressed as

$$u_i = r_{i-1} \frac{l - x}{l} + r_i \frac{x}{l} \; (i = 2...n), \quad (23)$$

where $r_{i-1} = r_{i-1}(t)$ and $r_i = r_i(t)$ are displacements of the initial and final cross-sections of the $i$ - th pipe (stalk), with $r_n = u_0(t)$. On the boundaries of pipe connections the conditions of continuity of efforts must be fulfilled:

$$EF \frac{du_i}{dx} = EF \frac{du_{i+1}}{dx} = k[u_{i+1}(0) - u_i(l)], \; EF \frac{du_1}{dx} = -P_0.$$

Taking into account Kirchhoff's hypothesis the expressions for potential and kinetic energy of the $i$-th beam with regard to (21) and (23) are written as:

$$U_1 = \frac{EI}{2} \int_0^l \left( \frac{\partial^2 v_i}{\partial x^2} \right)^2 dx + \frac{EF}{2l} \left[ -\frac{P_0 l}{EF} + \frac{1}{2} \int_0^l \left( \frac{\partial v_i}{\partial x} \right)^2 dx \right]^2, \quad (24)$$

$$U_i = \frac{EI}{2} \int_0^l \left( \frac{\partial^2 v_i}{\partial x^2} \right)^2 dx + \frac{EF}{2l} \left[ r_i - r_{i-1} + \frac{1}{2} \int_0^l \left( \frac{\partial v_i}{\partial x} \right)^2 dx \right]^2, \quad (25)$$

$$T_i = \frac{qF}{2g} \int_0^l \left( \frac{\partial v_i}{\partial t} \right)^2 dx + \frac{1}{2} m_i \dot{r}_{i-1}^2, \; (i = 1..n, \; \dot{r}_0 = 0). \quad (26)$$

Substituting expressions (24) - (26) into (5), we obtain a system of $3n-3$ nonlinear equations, which can be used to determine generalized coordinates $q_i = q_i(t)$, $r_i = r_i(t)$. For example, for $n = 3$ we obtain the following system:

$$\begin{aligned}
&\rho l^2 (a_{11} \ddot{\bar{q}}_1 + a_{13} \ddot{\bar{q}}_3 + a_{14} \ddot{q}_4) = \\
&-(b_{11} \bar{q}_1 + b_{12} q_2 + b_{13} \bar{q}_3 + b_{14} q_4) + F_1(\bar{q}_1, q_2, \bar{q}_3, q_4, \bar{r}_1, \bar{r}_2, t), \\
&\rho l^2 (a_{22} \ddot{q}_2 + a_{23} \ddot{\bar{q}}_3 + a_{24} \ddot{q}_4) = \\
&-(b_{12} \bar{q}_1 + b_{22} q_2 + b_{23} \bar{q}_3 + b_{24} q_4) + F_2(\bar{q}_1, q_2, \bar{q}_3, q_4, \bar{r}_1, \bar{r}_2, t), \\
&\rho l^2 (a_{13} \ddot{q}_1 + a_{23} \ddot{\bar{q}}_2 + a_{33} \ddot{q}_3) = \quad (27) \\
&-(b_{13} \bar{q}_1 + b_{23} q_2 + b_{33} \bar{q}_3 + b_{34} q_4) + F_3(\bar{q}_1, q_2, \bar{q}_3, q_4, \bar{r}_1, \bar{r}_2, t), \\
&\rho l^2 (a_{14} \ddot{\bar{q}}_1 + a_{24} \ddot{q}_2 + a_{44} \ddot{q}_4) = \\
&-(b_{14} \bar{q}_1 + b_{24} q_2 + b_{34} \bar{q}_3 + b_{44} q_4) + F_4(\bar{q}_1, q_2, \bar{q}_3, q_4, \bar{r}_1, \bar{r}_2, t), \\
&m_1 l \ddot{\bar{r}}_1 = -EF(\bar{r}_1 - \bar{r}_2) + F_5(\bar{q}_1, q_2, \bar{q}_3, q_4, \bar{r}_1, \bar{r}_2, t), \\
&m_2 l \ddot{\bar{r}}_2 = -EF(2\bar{r}_2 - \bar{r}_1 - \bar{u}_0) + F_6(\bar{q}_1, q_2, \bar{q}_3, q_4, \bar{r}_1, \bar{r}_2, t),
\end{aligned}$$

where $\rho = q / g$, $\bar{q}_1 = q_1 / l$, $\bar{q}_3 = q_3 / l$,
$\bar{r}_1 = r_1 / l$, $\bar{r}_2 = r_2 / l$, $\bar{u}_0 = u_0 / l$,
$a_{11} = 52/35$, $a_{13} = 9/35$, $a_{14} = -13/210$,
$a_{22} = 4/105$, $a_{23} = 13/210$, $a_{24} = -1/70$,
$a_{33} = 52/35$, $a_{44} = 4/105$,
$b_{11} = 48EJ/l^2 + 24ql/5 + P_0/5$, $b_{12} = 2ql/5$,
$b_{13} = -24EJ/l^2 - 18ql/5 - 12P_0/5$,
$b_{14} = 12EJ/l^2 + ql/5 + P_0/5$,
$b_{22} = 16EJ/l^2 + 8ql/15 + 8P_0/15$,
$b_{23} = -12EJ/l^2 - 2ql/5 - P_0/5$,

DOI: 10.1051/matecconf/20167509008

$b_{24} = 4EJ / l^2 - ql / 10 - P_0 / 5,$

$b_{33} = 48EJ / l^2 + 48ql / 5 + 24P_0 / 5,$

$b_{34} = 2ql / 5,\ b_{44} = 16EJ / l^2 + 16ql / 15 + 8P_0 / 15,$

$F_i$ are nonlinear components of derivatives $\frac{\partial U_i}{\partial q_i}$, $\frac{\partial U_i}{\partial r_1}$, $\frac{\partial U_i}{\partial r_2}$ $(i = 1...4)$.

## 4 Numerical results

In calculations the following values $n = 3$, $k = 1/м$, $v_0 = 2м/c$, $l = 1000м$, $D_1 = 140мм$, $D_2 = 128мм$, $E = 2 \cdot 10^5$ MPa, $\rho = \rho_0 F$, $\rho_0 = 7000кг/м^3$, $m_1 = 40кг$, $m_2 = 40кг$, $P_0 = 485H$ were used.

Figure 1 shows the curves of displacement of the bottom part of the column obtained using the formula $u_0 = \frac{ln(kv_0 t + 1)}{k}$, and longitudinal displacement of the concentrated mass $m_2$ in time.

It has been found out that the concentrated mass and the lower part of the column make not only the translational movement but also the oscillatory motion. The calculations show that concentrated masses produce a negligible impact on the longitudinal displacements of ring junctions.

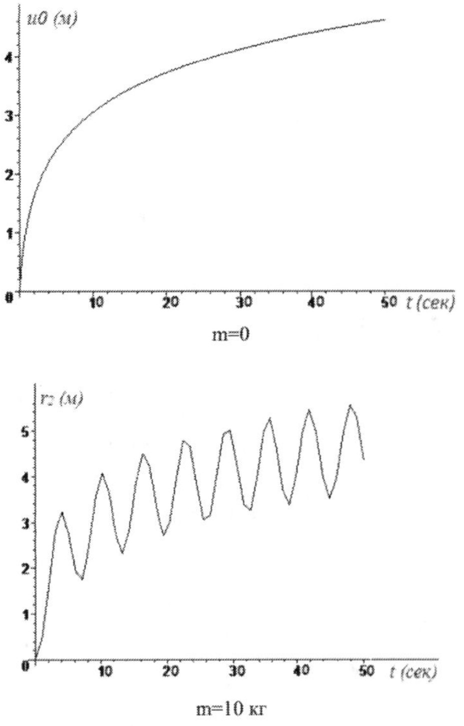


**Figure 1.** Displacements of the bottom of the column $u_0(t)$ *(m)* and the concentrated mass $r_2(t)$ *(m)* as a function of time *t (s)*.

Fig. 2 and 3 show the dependence of the transverse displacements of junctions $q_1$ and $q_3$ on time for various values of masses $m_1 = m_2 = m$.

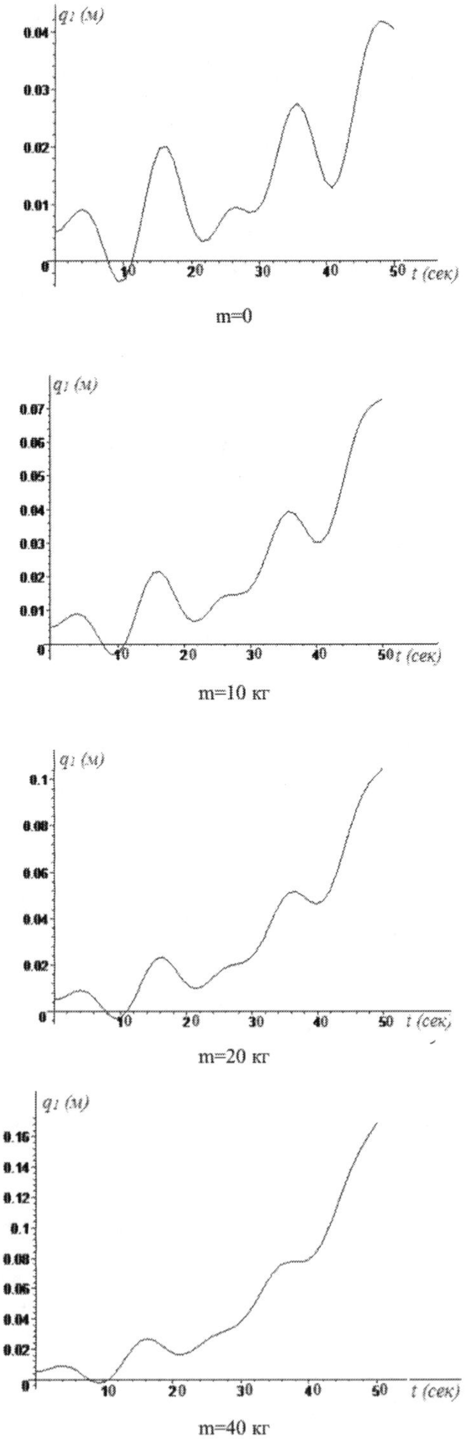


**Figure 2.** Transverse displacement of the first junction $q_1$ (m) as a function of time *t* (s) for various mass values *m* (kg).

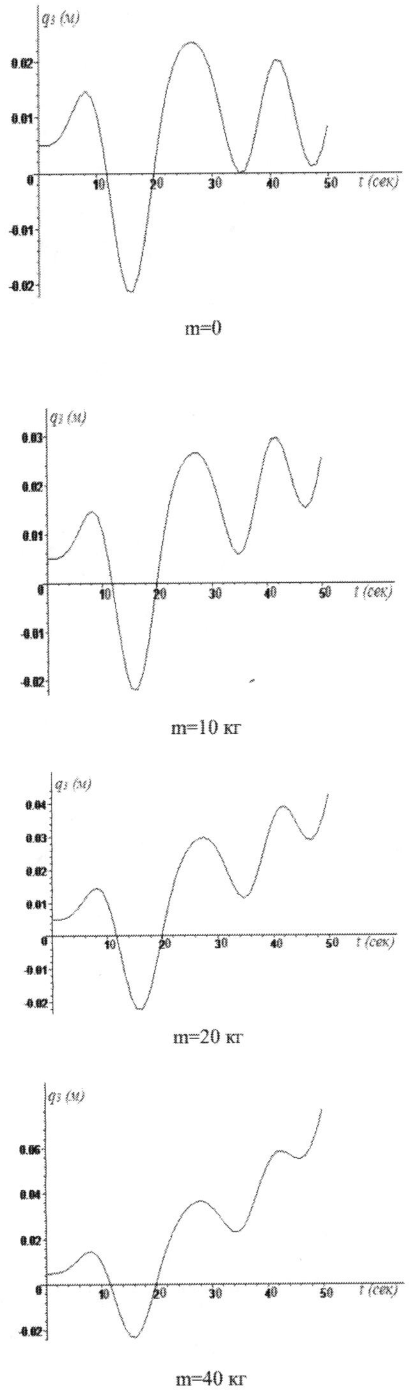


**Figure 3.** Transverse displacement of the second junction $q_j$ (m) as a function of time $t$ (s) for various masses $m$ (kg).

The analysis of the curves shows that the presence of concentrated masses in junctions of column parts causes changes in frequencies and an increase in the amplitude of their vibrations. This indicates a change in the type of transverse vibrations in sleeve joints between stalks of the column when masses of sleeve joints are taken into account.

## 5 Conclusion

An analysis of the dynamics of the drill column under the action of variable axial forces has shown that the presence of concentrated masses practically does not affect the amplitude of longitudinal vibrations. However, they may have a significant impact on the transverse vibrations of the drilling rod. They cause an asymmetric distortion in the bending of the drilling rod, which means that it may lose its stability.

The duality of the deformed state of the drilling rod is explained by its geometric nonlinearity. The loss of stability may occur during slip kicking at high values of nonlinearity. However, the probability of transition to an unstable state in the initial moments of time is reduced.

## Acknowledgment

The research work was partially supported by the scientific grant from the Ministry of Education and Science of the Republic of Kazakhstan (Project № ГР112РК01496).

## References

1. B.M. Mardonov, I. Mardanova, Poisk, Series of nat. and techn. sci. **1**, 217-220 (2001)
2. B.M. Mardonov, A.B. Barayev, N.M. Akhmetov, *Applied problems of mechanics of drilling oil and gas wells* ( Alem, Shymkent, 2013)
3. A.S. Volmir, *Stability of elastic systems* (Fizmatgiz, Moscow, 1963)
4. I.L. Barsky, T.V. Bredikhina, I.N. Genkina, M.A. Tchaikovskaya, N.E. Shurova, Papers of VNIIBT, **60**, 38-43 (1985)
5. O. Zenkevich, *Finite Element Method in technology* (Mir, Moscow, 1975)
6. I.F. Obraztsov, L.M. Savelyev, H.S. Khazanov H.S. *FEM in problems of structural mechanics of aircrafts* (High School, Moscow, 1985)
7. J. Oden, *Finite elements in nonlinear continuum mechanics* (Mir, Moscow, 1976)
8. M.J. Turner, E.H. Dill, H.C. Martin, R.J. Melosh, J.Aerospace Sce. **27,** 97-102 (1960)
9. R.W. Glough, *2d Conf. Electronic Compilation,* J.Struct. Div. ASCE, 345-378 (1960)
10. V.V. Novozhilov, *Fundamentals of nonlinear elasticity theory* ( OGIZ, M-L., 1948)
11. L.A. Khajiyeva, *Models of nonlinear deformable systems. Practical application* (Kazak universiteti, Almaty, 2004)
12. G. Kauderer, *Nonlinear mechanics* ( M. Publ. house. Foreign, IL, Moscow, 1960)

# Assessment of impact of borehole heat exchanger design on heat extraction/rejection efficiency

V.F. Gornov [1], N.V.Peskov [1, 2], G.P. Vasilyev [1] and M.V. Kolesova [1]

[1]*JSC «INSOLAR-INVEST», Bol'shaya Filevskaya str. 32-3, Moscow, 121433 Russia*
[2]*Lomonosov Moscow State University, Leninskie Gory 1, str. 52, Moscow, 119991 Russia*

**Abstract.** The article considers the impact of design of borehole heat exchanger (BHE) as one of the main elements of a geothermal heat pump system on its efficiency in the ground heat extraction/rejection. Four BHE modifications are considered: coaxial with metal and polyethylene outside tube as well as single and double U-shaped structures of polyethylene tubes. Numerical modeling resulted to data on the efficiency of these BHE modifications for rejection heat into ground (heat pump system in cooling mode), and ground heat extraction (heat pump system in heating mode). Numerical values were obtained and BHEs were ranked according to their efficiency in both operation modes. Besides, additional calculations were made for the most common modification - double U-shaped design - in the ground heat extraction mode for various tube sizes with various wall thicknesses.

## 1 Introduction

The problem of increasing the specific heat extraction/rejection of ground heat exchangers is currently one of the most important for the future development of geothermal heat pump systems. Many scientists around the world are looking for ways to improve the efficiency of ground heat exchangers. So, there are a number of articles focused on optimizing the existing modifications of ground heat exchangers. For example, article [1] deals with the effect of various grouting materials on BHE thermal efficiency. The article considers grouting solutions based on graphite and traditional bentonite. Article [2] considers possible applications of phase-change fluid, namely $CO_2$, in U-shaped BHEs. Article [3] is focused on the impact of ground heat exchanger tube arrangement across BHE section on its thermal resistance, and the authors of the article [4] proposed improvement of the double U-shaped model wherein the coolant is supplied through three tubes, and is raised through one tube.

In addition, there is a considerable amount of research dedicated to developing new designs of BHE. Article [5] proposes the new BHE design that enhances heat transfer from the ground to the BHE coolant due to natural convection of intermediate coolant filling the BHE cavity instead of "conventional" heat-conducting material. Article [6] considers the new design of a metal ground heat exchanger in the form of an aluminum tube with external finning. Article [7] is also focused on developing new designs of ground heat exchangers, but in contrast to the previous study, the authors suggest to use the U-shaped heat exchanger with internally finned tubes.

The authors of this article have set themselves the task to compare efficiency of coaxial and U-shaped BHEs both for rejection of heat to the ground and for its extraction.

## 2 Numerical Analysis

### 2.1 Considered physical model

The two most commonly used modifications of BHEs are coaxial and U-shaped.

Coaxial BHEs represent a large diameter pipe, inside which a smaller diameter pipe is located so that the coolant supplied through the inside pipe is then introduced into the annular channel, raising there and exchanging heat with the surrounding ground through the wall of the larger (outside) tube. Coaxial BHEs can be made of both metal and polyethylene, where the inner tube is usually made of polyethylene as a material with a lower thermal conductivity in order to minimize the thermal interference of descending and ascending flows of coolant (thermal short circuit effect).

U-shaped models are mainly made of plastic tubes and used in two modifications - with one or two U-shaped loops within a single borehole.

In assessing the impact of the design features of BHEs on the rate of heat extraction/rejection efficiency, four BHE models presented in Figure 1 were analyzed:

- Coaxial model, polyethylene external tube with an outside diameter of 160 mm and wall thickness of 9.5 mm;
- Coaxial model, steel external tube with an outside diameter of 160 mm and wall thickness of 7 mm;
- Single U-shaped model, polyethylene tube with an outside diameter of 32 mm and wall thickness of 3 mm;
- Double U-shaped model, polyethylene tube with an outside diameter of 32 mm and wall thickness of 3 mm.

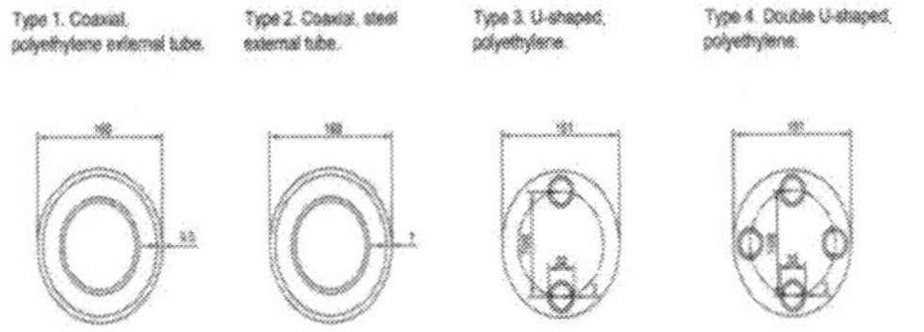


**Figure 1**. Analyzed BHE models

In modelling, all BHEs are placed in the same conditions, for this purpose data of geological surveys conducted in Martemianovo, Moscow region, were used as the ground design parameters. Data used in the calculations are shown in Table 1.

**Table 1.**

| Ground density, kg/m$^3$ | Ground thermal conductivity, W/(m*°C) | Ground volumetric heat capacity, kJ/(m$^3$*°C) |
|---|---|---|
| 1766 | 1,71 | 2778 |

The length of ground heat exchangers in the calculation is assumed to be 45 m at a drilling depth of 50 m. It is their effective depth, minus the top part, where a distributor well and piping will be located.

## 2.2. Modeling strategies

Various models of BHE were placed in the same conditions to assess the specific heat extraction/rejection value. Thus it was assumed that outside surface of coaxial BHEs has a good contact with the surrounding ground, and in U-shaped models grouting material has the same thermal properties as the main ground. At the same time the effects of thermal short-circuit were not considered in order to simplify the calculations. Insignificance of these effects, for example, in coaxial models is shown in [8].

Numerical study to assess the impact of structural features of BHEs on the rate of specific heat extraction/rejection were conducted on following model. A horizontal circle at the depth $h$ with center point in the center of a borehole is considered. Radius of the circle $R$ is chosen to ensure that the ground temperature change on its border due to BHE operation is small enough to neglect it. In that flat region, formed by this big circle and small circles, representing cross-sections of BHE tubes, cut off from it, equation for thermal conductivity is considered:

$$c\rho \frac{\partial T}{\partial t} = \lambda \Delta T \qquad (1)$$

As a boundary conditions zero heat flux at outer border and convective heat transfer at BHE tubes are taken. At initial moment ground is considered as undisturbed and having its natural temperature. Initial boundary value problem for equation (1) is solved by finite elements method. Internal temperature of coolant in BHE tubes is constant and depending on depth $h$.

## 2.3. Heat rejection

Calculations were performed for ground temperature equal to 8°C and coolant temperature equal to 30°C. The calculation was performed for the forecast period of 180 days. By this time the specific heat rejection value reaches a certain value, different from the initial one, that subsequently changes only slightly. The calculation results are shown in Fig. 2. "Minus" sign denotes heat is rejected into the ground.

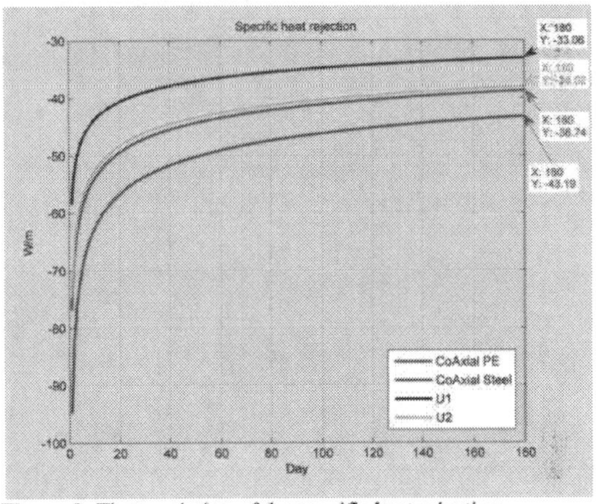


**Figure 2.** Time variation of the specific heat rejection.

The calculation results presented in the charts show that the specific heat rejection rate by 180th day is distributed as follows:

1. Coaxial steel BHE 43.19 W/m;
2. Double U-shaped 38.74 W/m;
3. Coaxial polyethylene BHE 38.02 W/m;
4. Single U-shaped BHE 33.06 W/m.

These results show that in the mode of heat release into the ground, the considered BHE models have similar enough characteristics and the difference between the maximum and minimum values of the specific heat rejection is only 23.5%. The use of the outside tube of heat conductive material (steel) in coaxial models as compared to the modification with polyethylene outside tube allows growth of the specific heat rejection of 13.6% and use of a double U-shaped model as compared with a

 DOI: 10.1051/matecconf/20167509009

single U-shaped model allows growth of 17.2% with almost doubled heat exchange surface.

### 2.4 Heat extraction

In order to compare the efficiency of heat extraction for the considered BHE models, the specific heat flow from the ground to the ground heat exchanger was estimated, and the input parameters were the same BHE geometric characteristics and ground parameters as for the previous calculation. BHE coolant temperature was assumed to be minus 8°C.

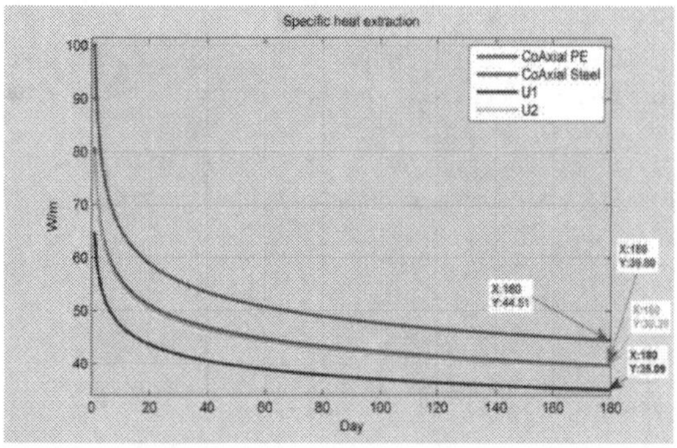


**Figure 3.** Time variation of the specific heat extraction.

The calculation results presented in the charts show that the specific heat extraction rate by 180th day is distributed as follows:

1. Coaxial steel BHE
   44.51 W/m;
2. Double U-shaped
   39.80 W/m;
3. Coaxial polyethylene BHE
   39,39 W/m;
4. Single U-shaped BHE
   35,09 W/m.

These results also show that BHE design features play an important role in the heat extraction mode.

Comparing the results of this calculation with the calculation performed for the heat rejection mode it can be seen that BHE models are arranged in the effectiveness ranking in the same manner, and in this case variation is comparable and amounted to 21.2% (23.5% in heat rejection mode). Thus the use of outside tube made of heat-conducting material (steel) in coaxial models in comparison with the use of polyethylene outside tube allows to increase the specific heat extraction by 13% that is close to the same figure obtained for the heat release mode (13.6%). The difference in the efficiency of use of double U-shaped structure as compared with a single U-shaped structure in this mode was almost 13.4% that is slightly less than in heat rejection mode (17.2%).

### 2.5 Effect of changes in geometric parameters of U-shaped BHEs on the heat extraction efficiency

Additionally, the calculations of efficiency of dual U-shaped ground heat exchangers as the most common were conducted, with smaller wall thickness and larger tube diameter than in the already discussed modification. Thus, the following models of polyethylene tubes for a ground heat exchanger were chosen for calculations: 32x2, 40x2 and 40x3. The results of calculating the time variation of the specific heat flow for these models of ground heat exchangers are given in Figure 4 and in Table 2.

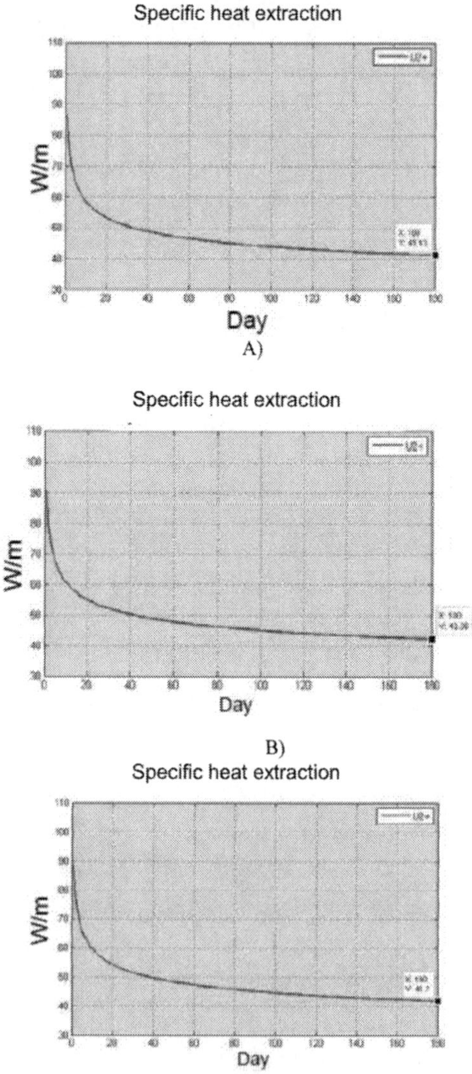


A). Double U-shaped BHE 32x2;

B). Double U-shaped BHE 40x2;

C). Double U-shaped BHE 40x3.

**Figure 4.** Time variation of the specific heat extraction.

The calculation results show that change in the geometry of tubes constituting the ground heat exchanger affects the specific heat extraction, but this effect is relatively small. With increasing tube diameter with the

 DOI: 10.1051/matecconf/20167509009

same wall thickness, the specific heat recovery is predictably growing. In transition from diameter of 32 mm to 40 mm, variation was 4.6% for tubes with a wall thickness of 3 mm and 2.7% for tubes with a wall thickness of 2 mm. These data differ somewhat from the results obtained by authors of the study [9], where it was observed some reduction in efficiency of BHEs with 40 mm tube diameter compared to BHEs with 32 mm tube diameter. The explanation for this fact is that in the article [9] authors analyzed heat exchangers with tubes having a "standard wall thickness", and the thickness has not been specified. It is possible to assume that the "standard wall thickness" of 32 mm diameter tube was smaller than of 40 mm diameter tube that affected the final result. In our case, both tubes had the same wall thickness, so the results showed an increase in efficiency of BHEs with tube diameter increase.

**Table 2.**

| Ground heat exchanger tube dimensions, mm | Specific heat extraction, W/m |
|---|---|
| 32x3 | 39,9 |
| 32x2 | 41,13 |
| 40x3 | 41,7 |
| 40x2 | 42,26 |

In addition, the specific heat extraction growth is observed while reducing the wall thickness and keeping the outside diameter. For 40 mm diameter tubes, wall thickness reduction from 3 to 2 mm lead to the heat flow increase by 1.3%, and for 32 mm diameter tubes - by 3.3%.

## 3 Discussion

As a result of analysis, it was found that the impact of the BHE design on its efficiency varies from 20 % to 25 % depending on operation mode.

It is expected, that in heat rejection mode BHE efficiency would be significantly lower than in heat extraction mode, because these processes have different associated physical effects. Thus, in case of heat rejection into the ground, drying of the ground mass adjacent to the ground heat exchanger occurs with a corresponding reduction in its heat conductivity, which causes a sharp drop in thermal characteristics of ground heat exchangers. In this work we did not account for any effects of phase transitions of ground moisture, so instead of smaller influence of BHE design on its efficiency in heat rejection mode if compared to heat extraction mode we got it even bigger (23.5 % vs. 21.2 %). It shows that models, not dealing with ground humidity and ground moisture phase transitions could lead to incorrect results in calculations and in final design of geothermal heat pump system. This conclusion is in good agreement with [10] where influence of groundwater on heat transfer efficiency.

In studying the effect of changes in geometric parameters of U-shaped ground heat exchangers on heat recovery efficiency, a relatively small impact of increase of outside diameter and reduction of wall thickness of tubes constituting the ground heat exchanger, on the parameters of the specific heat recovery has been shown. The reason is that by the end of heating season, when the temperature of the ground immediately adjacent to the ground heat exchanger has markedly decreased relative to its initial level, the efficiency of heat transfer is significantly affected not so much by heat conductivity of the material from which the heat exchanger is made, as heat conductivity of the ground itself.

## 4 Conclusions

The research presented in this article led to the following conclusions regarding the impact of BHE model on efficiency of the ground heat extraction and rejection:

a) BHE efficiency depends significantly on its construction, both in heat extraction and heat rejection modes.
b) Considered BHE models in both modes of operation in terms of efficiency can be ranked as follows (in decreasing order):
   I. Coaxial model, steel external tube with 160 mm outside diameter;
   II. Double U-shaped model, polyethylene tube with 32 mm outside diameter;
   III. Coaxial model, polyethylene tube with 160 mm outside diameter;
   IV. Single U-shaped model, polyethylene tube with 32 mm outside diameter.
c) The use of outside tube made of heat-conducting material (steel) in coaxial models in comparison with the use of polyethylene outside tube allows to increase the specific heat extraction by 13% that is relatively close to the value, obtained for heat rejection mode (13.6%).
d) The difference in the efficiency of use of double U-shaped structure as compared with a single U-shaped structure in heat extraction mode was 13,4 % that is somewhat less than in heat rejection mode (17.2%).
e) Efficiency of coaxial BHE with polyethylene external tube appeared to be very close to efficiency of double U-shaped BHE in both modes for chosen geometry and dimensions.
f) It is necessary to take into account the ground humidity and its change during the operation of ground heat exchangers, in process of calculations and characterization of ground heat exchangers, especially in the heat rejection mode.
g) The ability to control the humidity of ground adjacent to the ground heat exchanger can be a good tool to improve BHE efficiency, reduce their size and/or number, as well as to optimize the geothermal heat pump system as a whole.

## Acknowledgements

The research was conducted by JSC "INSOLAR-INVEST" with the financial support of the Ministry of

Education and Science of Russia. Unique identifier of the project RFMEFI57915X0115.

## References

1. Selçuk Erol 1, Bertrand François. Efficiency of various grouting materials for borehole heat exchangers. Applied Thermal Engineering 70 (2014), 788-799.
2. Rita Mastrullo, Alfonso W Mauro, Laura Menna, Giuseppe P Vanoli. A model for a borehole heat exchanger working with CO2. 68th Conference of the Italian Thermal Machines Engineering Association, ATI2013. Energy Procedia 45 (2014), 635 – 644.
3. S. Emad Dehkordi, Robert A. Schincariol, Stanley Reitsma. Thermal performance of a tight borehole heat exchanger. Renewable Energy 83 (2015), 698-704.
4. Yong Liu, Yuqiao Zhang, Shengqiang Gong, Zhenfei Wang, Huiting Zhang. Analysis on the Performance of Ground Heat Exchangers in Ground Source Heat Pump Systems based on Heat Transfer Enhancements. Procedia Engineering 121 (2015) 19 – 26.
5. S. Focaccia, F.Tinti.An innovative Borehole Heat Exchanger configuration with improved heat transfer. Geothermics 48 (2013), 93–100.
6. Ahmet Koyun, Hakan Demir, Zakir Torun. Experimental study of heat transfer of buried finned pipe for ground source heat pump applications International Communications in Heat and Mass Transfer 36 (2009), 739–743.
7. Benamar Bouhacina, Rachid Saim, Hakan F., Oztop, Numerical investigation of a novel tube design for the geothermal borehole heat exchanger. Applied Thermal Engineering 79 (2015), 153-162.
8. E. Zanchini, S. Lazzari, A. Priarone. Effects of flow direction and thermal short-circuiting on the performance of small coaxial ground heat exchangers. Renewable Energy 35 1255–1265 (2010).
9. Yang Li, Qingsong An, Liangxu Liu, Jun Zhao. Thermal performance investigation of borehole heat exchanger with different U-tube diameter and borehole parameters. Energy Procedia 61 (2014) 2690 – 2694.
10. Jin Luo, Joachim Rohn, Wei Xiang, Manfred Bayer, Anna Priess. Experimental investigation of a borehole field by enhanced geothermal response test and numerical analysis of performance of the borehole heat exchangers, Energy 84 (2015), 473-484

# Fourth-order Perturbed Eigenvalue Equation for Stepwise Damage Detection of Aeroplane Wing

Chun Nam Wong[a], Kongzheng Tang, Wuyin Jin and Gaoqiao Tu

*Lanzhou University of Technology, School of Mechanical and Electronical Engineering, Lanzhou, Gansu 730050, China*

**Abstract.** Perturbed eigenvalue equations up to fourth-order are established to detect structural damage in aeroplane wing. Complete set of perturbation terms including orthogonal and non-orthogonal coefficients are computed using perturbed eigenvalue and orthonormal equations. Then the perturbed eigenparameters are optimized using BFGS approach. Finite element model with small to large stepwise damage is used to represent actual aeroplane wing. In small damaged level, termination number is the same for both approaches, while rms errors and termination d-norms are very close. For medium damaged level, termination number is larger for third-order perturbation with lower d-norm and smaller rms error. In large damaged level, termination number is much larger for third-order perturbation with same d-norm and larger rms error. These trends are more significant as the damaged level increases. As the stepwise damage effect increases with damage level, the increase in stepwise effect leads to the increase in model order. Hence, fourth-order perturbation is more accurate to estimate the model solution.

## 1 Introduction

To solve physical problem without exact solution, perturbation theory is established to obtain their approximate solution. A general approach is carried out using various perturbation order expansions on its equation. Adopting the advanced modeling method of element stiffness matrix modification, the order of the structure stiffness matrix was kept invariable in establishing the model of intact and damaged structures by Yu et al. [1]. First-order eigenvalue perturbation theory is introduced to obtain the eigenvalues and eigenvectors of the damaged structure. In the first-order perturbed eigenvalue equation of Wang et al. [2], modal strain energy was used as damage indicator in the damage detection of simply supported beam. Second-order perturbation technique was developed by Chen et al. [3] for damage detection of simply supported beam. When the extent of damage was small, first-order perturbation equation was adopted. Meanwhile the extent was large, second-order perturbation equation was employed to improve the identification precision.

A general order perturbation method involving multiple perturbation parameters was developed by Wong et al. [4] for eigenvalue problems with changes in the stiffness parameters. Symmetric perturbation solutions and eigenparameter sensitivities of different orders were derived. The perturbed eigenvalue problem was established from the perturbations of stiffness matrix, eigenvector, and eigenvalue. Then stiffness parameters of fixed–fixed modular beam was estimated from this equation using the DFP approach [5]. The perturbed orthonormal equation was generated from the perturbation of the eigenvectors and eigenvalues to obtain the *k*-th skew-symmetric coefficients[6].

In this effort, third and fourth-order perturbation equations are established. Complete set of perturbation terms including orthogonal and non-orthogonal coefficients are computed using perturbed eigenvalue and orthonormal equations. Then the perturbed eigenparameters are optimized using BFGS approach. Limited number of eigenparameters are used to detect different level stepwise damages within aeroplane wing. Finite element model with variational elastic moduli is used to represent the aeroplane wing.

## 2 Derivation of fourth-order perturbation equation

The stiffness parameters of the healthy structure prior to any damage are denoted by $G_{hi}\,(i=1,2,...,m)$, where $m$ is the number of the stiffness parameters. Structural damage is characterized by reductions in the stiffness parameters. The estimated stiffness parameters of the damaged structure before each iteration are denoted by $G=[G_1,G_2,\cdots,G_m]^T$ , and its stiffness matrix, which depends linearly on $G_i$, is denoted by $K=K(G)$. The eigenvalue equation of the structure with stiffness parameters $G_i$ is

$$K\phi^k=\lambda^k M\phi^k \tag{1}$$

where $M$ is the constant mass matrix, and $\lambda^k,\phi^k$ ,are the k-th eigenvalue and mass-normalized eigenvector

[a] Corresponding author: znhuang@yahoo.com

DOI: 10.1051/matecconf/20167509010

respectively. It is noted that $\lambda_k = \omega_k^2$, where $\omega_k$ is the k-th natural frequency of the structure. The normalized eigenvectors of Eq.(1) satisfy the orthonormality relations

$$\left(\phi^k\right)^T M\phi^u = \delta_{ku}, \quad \left(\phi^k\right)^T K\phi^u = \lambda^k \delta_{ku} \tag{2}$$

where $1 \le u \le N$ and $\delta_{ku}$ is the Kronecker delta. Before the first iteration, the initial stiffness parameters of the damaged structure are assumed to be $G_i^{(0)} = \sigma_i G_{hi}$ ( $i = 1,2,...,m$ ), where $0 < \sigma_i \le 1$ , and the eigenvalue equation in Eq.(1) corresponds to that of the structure with stiffness parameters $G_i^{(0)}$ . If there is no prior knowledge of the integrity of the structure, one can start the iteration from the stiffness parameters of the "healthy" structure and set $\sigma_i = 1$ . Let $G_{di}$ ( $i = 1,2,...,m$ ) denote the stiffness parameters of the damaged structure. The eigenvalue equation of the damaged structure is

$$K_d \phi_d^k = \lambda_d^k M \phi_d^k \tag{3}$$

where $K_d = K\left(G_d\right)$ is the damaged stiffness matrix, and $\lambda_d^k$ , $\phi_d^k$ are the $k$-th eigenvalue and mass-normalized eigenvector respectively. As $K$ is a linear function of $G_d$ , first-order Taylor expansion of $K_d$ is

$$K_d = K + \sum_{i=1}^{m} \frac{\partial K}{\partial G_{di}} \delta G_{di} \tag{4}$$

where $\delta G_{di} = G_{di} - G_i$ ( $i = 1,2,...,m$ ) are the changes in the stiffness parameters. Note that based on its finite element model, the system stiffness matrix of a distributed structure satisfies Eq.(4) because the higher-order derivatives of the stiffness matrix with respect to the elemental stiffness parameters vanish.

Let the $k$-th eigenvalue and mass-normalized eigenvector of the damaged structure be related to $\lambda^k$ and $\phi^k$ through

$$\begin{aligned} \lambda_d^k = \lambda^k + \sum_{i=1}^{m} \lambda_{(1)i}^k \delta G_i + \sum_{i=1}^{m}\sum_{j=1}^{m} \lambda_{(2)ij}^k \delta G_i \delta G_j + \ldots \\ + \sum_{i=1}^{m}\sum_{j=1}^{m} \ldots \sum_{t=1}^{m} \lambda_{(p)ij...t}^k \delta G_i \delta G_j \ldots \delta G_t + \varepsilon_\lambda^k \end{aligned} \tag{5}$$

$$\begin{aligned} \phi_d^k = \phi^k + \sum_{i=1}^{m} z_{(1)i}^k \delta G_i + \sum_{i=1}^{m}\sum_{j=1}^{m} z_{(2)ij}^k \delta G_i \delta G_j + \ldots \\ + \sum_{i=1}^{m}\sum_{j=1}^{m} \ldots \sum_{t=1}^{m} z_{(p)ij...t}^k \delta G_i \delta G_j \ldots \delta G_t + \varepsilon_\phi^k \end{aligned} \tag{6}$$

where $\lambda_{(1)i}^k, \lambda_{(2)ij}^k \cdots \lambda_{(p)ij...t}^k$ are the coefficients of the first, second, …, and $p$-th order perturbations for the eigenvalue. $z_{(1)i}^k, z_{(2)ij}^k \cdots z_{(p)ij...t}^k$ are the coefficient vectors of the first, second, …, and $p$-th order perturbations for the eigenvector. $\varepsilon_\lambda^k$ and $\varepsilon_\phi^k$ are the residuals of order $p+1$. Note that the numbers in the parentheses in the subscripts of the coefficients and coefficient vectors indicate the orders of the terms. By the Taylor expansion, one has for any $p \ge 1$ ,

$$p!\lambda_{(p)ij...t}^k = \frac{\partial^p \lambda^k}{\partial G_i \partial G_j \cdots \partial G_t}$$

$$p! z_{(p)ij...t}^k = \frac{\partial^p \phi^k}{\partial G_i \partial G_j \cdots \partial G_t} \tag{7}$$

By Eq.(7), $\lambda_{(p)ij\cdots t}^k$ and $z_{(p)ij...t}^k$ are symmetric in the $p$ indices, $i, j$, …, and $t$. The right-hand sides of Eq.(7) are the $p$-th order sensitivities of the eigenvalues and eigenvectors with respect to the stiffness parameters. Using the normalization relations of the eigenvectors, $\phi^k$ and $\phi_d^k$ , and symmetry of the coefficient vectors in Eq.(6), as indicated earlier, one obtains the perturbed orthonormal equation:

$$\begin{aligned} 1 &= (\phi_d^k)^T M \phi_d^k \\ &= (\phi^k + \sum_{i=1}^{m} z_{(1)i}^k \delta G_i + \sum_{i=1}^{m}\sum_{j=1}^{m} z_{(2)ij}^k \delta G_i \delta G_j + \cdots + \\ &\quad \underbrace{\sum_{i=1}^{m}\sum_{j=1}^{m} \cdots \sum_{t=1}^{m}}_{p\ \text{summations}} z_{(p)ij\cdots st}^k \delta G_i \delta G_j \cdots \delta G_s \delta G_t + \cdots)^T \\ &\quad M(\phi^k + \sum_{i=1}^{m} z_{(1)i}^k \delta G_i + \sum_{i=1}^{m}\sum_{j=1}^{m} z_{(2)ij}^k \delta G_i \delta G_j + \cdots + \\ &\quad \underbrace{\sum_{i=1}^{m}\sum_{j=1}^{m} \cdots \sum_{t=1}^{m}}_{p\ \text{summations}} z_{(p)ij\cdots st}^k \delta G_i \delta G_j \cdots \delta G_s \delta G_t + \cdots) \end{aligned}$$

$$\begin{aligned} &= 1 + \sum_{i=1}^{m} [(\phi^k)^T M z_{(1)i}^k + (z_{(1)i}^k)^T M \phi^k] \delta G_i + \\ &\sum_{i=1}^{m}\sum_{j=i}^{m} \frac{1}{R_{ij}} \{2!(\phi^k)^T M z_{(2)ij}^k + [(z_{(1)i}^k)^T M z_{(1)j}^k \\ &+ (z_{(1)j}^k)^T M z_{(1)i}^k] + 2!(z_{(2)ij}^k)^T M \phi^k\} \delta G_i \delta G_j + \\ &\sum_{i=1}^{m}\sum_{j=i}^{m}\sum_{l=j}^{m} \frac{1}{R_{ijl}} \{3!(\phi^k)^T M z_{(3)ijl}^k + 2![(z_{(1)i}^k)^T M z_{(2)jl}^k \\ &+ (z_{(1)j}^k)^T M z_{(2)il}^k + (z_{(1)l}^k)^T M z_{(2)ij}^k] + \\ &2![(z_{(2)jl}^k)^T M z_{(1)i}^k + (z_{(2)il}^k)^T M z_{(1)j}^k + (z_{(2)ij}^k)^T M z_{(1)l}^k] \\ &+ 3!(z_{(3)ijl}^k)^T M \phi^k\} \delta G_i \delta G_j \delta G_l + \cdots + \underbrace{\sum_{i=1}^{m}\sum_{j=i}^{m} \cdots \sum_{t=s}^{m}}_{p\ \text{summations}} \end{aligned}$$

$$\begin{aligned} &\frac{1}{R_{ij\cdots st}} \{p!(\phi^k)^T M z_{(p)ij\cdots st}^k + (p-1)![(z_{(1)i}^k)^T M z_{(p-1)jl\cdots st}^k \\ &+ (z_{(1)j}^k)^T M z_{(p-1)il\cdots st}^k + \cdots + (z_{(1)t}^k)^T M z_{(p-1)ij\cdots s}^k] + \\ &2!(p-2)![(z_{(2)ij}^k)^T M z_{(p-2)l\cdots st}^k + (z_{(2)il}^k)^T M z_{(p-2)j\cdots st}^k \\ &+ \cdots + (z_{(2)st}^k)^T M z_{(p-2)ij\cdots r}^k] + \cdots + (p-2)!2! \end{aligned}$$

DOI: 10.1051/matecconf/20167509010

$$\begin{aligned}&[(z^k_{(p-2)l\cdots st})^T Mz^k_{(2)ij}+(z^k_{(p-2)j\cdots st})^T Mz^k_{(2)il}+\cdots\\&+(z^k_{(p-2)ij\cdots r})^T Mz^k_{(2)st}]+(p-1)![(z^k_{(p-1)jl\cdots st})^T Mz^k_{(1)i}+\\&(z^k_{(p-1)il\cdots st})^T Mz^k_{(1)j}+\cdots+(z^k_{(p-1)ij\cdots s})^T Mz^k_{(1)t}]+\\&p!(z^k_{(p)ij\cdots st})^T M\phi^k\}\delta G_i\delta G_j\cdots\delta G_s\delta G_t+\cdots\end{aligned}\tag{8}$$

where the superscript $T$ denotes transpose. Expanding $z^k_{(3)ijl}$ using normalized eigenvectors of Eq.(1) as basis vectors, one has

$$z^k_{(3)ijl}=\sum_{u=1}^{n}P^k_{(3)ijlu}\phi^u\tag{9}$$

where $P^k_{(3)ijlu}$ is the coefficient of the $u$-th term and the number in the parentheses in its subscript corresponds to that of $z^k_{(3)ijl}$. Equating the coefficients of the $\delta G_i\delta G_j\delta G_l$ terms in Eq.(8) and using symmetry of $M$ and $z^k_{(3)ijl}$ yields

$$P^k_{(3)ijlk}=-\frac{2!}{3!}\left[(z^k_{(1)i})^T Mz^k_{(2)jl}+(z^k_{(1)j})^T Mz^k_{(2)li}+(z^k_{(1)l})^T Mz^k_{(2)ij}\right]\tag{10}$$

Expanding $z^k_{(4)ijlr}$ using normalized eigenvectors of Eq.(1) as basis vectors, one has

$$z^k_{(4)ijlr}=\sum_{u=1}^{n}P^k_{(4)ijlru}\varphi^u\tag{11}$$

where $P^k_{(4)ijlru}$ is the coefficient of the $u$-th term and the number in the parentheses in its subscript corresponds to that of $z^k_{(4)ijlr}$. Equating the coefficients of the $\delta G_i\delta G_j\delta G_l\delta G_r$ terms in Eq.(8) and using symmetry of $M$ and $z^k_{(4)ijlr}$ yields

$$\begin{aligned}&P^k_{(4)\ ijlrk}=-\frac{1}{48}\{6[(z^k_{(1)i})^T Mz^k_{(3)jlr}+(z^k_{(1)j})^T Mz^k_{(2)ilr}+\\&(z^k_{(1)l})^T Mz^k_{(3)ijr}+(z^k_{(1)r})^T Mz^k_{(3)ijl}]+8\left[(z^k_{(2)ij})^T Mz^k_{(2)lr}+\right.\\&(z^k_{(2)il})^T Mz^k_{(2)jr}+(z^k_{(2)ir})^T Mz^k_{(2)jl}+(z^k_{(2)jl})^T Mz^k_{(2)ir}+\\&\left.(z^k_{(2)jr})^T Mz^k_{(2)il}+(z^k_{(2)lr})^T Mz^k_{(2)ij}\right]+6\left[(z^k_{(3)jlr})^T Mz^k_{(1)i}+\right.\\&\left.(z^k_{(3)ilr})^T Mz^k_{(1)j}+(z^k_{(3)ijr})^T Mz^k_{(1)l}+(z^k_{(3)ijl})^T Mz^k_{(1)r}\right]\}\end{aligned}\tag{12}$$

Substituting Eqs.(4-6) into Eq.(3) yields the perturbed eigenvalue equation:

$$\begin{aligned}&\left\{K+\sum_{i=1}^{m}\frac{\partial K}{\partial G_i}\delta G_i\right\}\left\{\phi^k+\sum_{i=1}^{m}z^k_{(1)i}\delta G_i+\right.\\&\left.\sum_{i=1}^{m}\sum_{j=1}^{m}z^k_{(2)ij}\delta G_i\delta G_j+\sum_{i=1}^{m}\sum_{j=1}^{m}\sum_{l=1}^{m}z^k_{(3)ijl}\delta G_i\delta G_j\delta G_l+\cdots\right\}\\&=\left\{\lambda^k+\sum_{i=1}^{m}\lambda^k_{(1)i}\delta G_i+\sum_{i=1}^{m}\sum_{j=1}^{m}\lambda^k_{(2)ij}\delta G_i\delta G_j+\right.\\&\left.\sum_{i=1}^{m}\sum_{j=1}^{m}\sum_{l=1}^{m}\lambda^k_{(3)ijl}\delta G_i\delta G_j\delta G_l+\cdots\right\}M\\&\left\{\phi^k+\sum_{i=1}^{m}z^k_{(1)i}\delta G_i+\sum_{i=1}^{m}\sum_{j=1}^{m}z^k_{(2)ij}\delta G_i\delta G_j+\right.\\&\left.\sum_{i=1}^{m}\sum_{j=1}^{m}\sum_{l=1}^{m}z^k_{(3)ijl}\delta G_i\delta G_j\delta G_l+\cdots\right\}\end{aligned}\tag{13}$$

Equating the coefficients of the $\delta G_i\delta G_j\delta G_l$ terms in Eq.(13) to obtain the third-order coefficient equation. Pre-multiplying by $(\phi^k)^T$ and using Eqs.(1,2) yields

$$\begin{aligned}\lambda^k_{(3)ijl}&=\frac{2!}{3!}(\phi^k)^T[\frac{\partial K}{\partial G_i}z^k_{(2)jl}+\frac{\partial K}{\partial G_j}z^k_{(2)li}+\frac{\partial K}{\partial G_l}z^k_{(2)ij}\\&-\lambda^k_{(1)i}Mz^k_{(2)jl}-\lambda^k_{(1)j}Mz^k_{(2)li}-\lambda^k_{(1)l}Mz^k_{(2)ij}]\end{aligned}\tag{14}$$

Pre-multiplying the coefficient equation by $(\phi^v)^T$, where $1\le v\le N$ and $v\ne k$, and using Eqs.(1,2,9) yields

$$\begin{aligned}&P^k_{(3)ijlv}=\frac{2!}{3!(\lambda^k-\lambda^v)}(\phi^v)^T[\frac{\partial K}{\partial G_i}z^k_{(2)jl}+\\&\frac{\partial K}{\partial G_j}z^k_{(2)li}+\frac{\partial K}{\partial G_l}z^k_{(2)ij}-\lambda^k_{(1)i}Mz^k_{(2)jl}\\&-\lambda^k_{(1)j}Mz^k_{(2)li}-\lambda^k_{(1)l}Mz^k_{(2)ij}-\lambda^k_{(2)jl}Mz^k_{(1)i}\\&-\lambda^k_{(2)li}Mz^k_{(1)j}-\lambda^k_{(2)ij}Mz^k_{(1)l}]\end{aligned}\tag{15}$$

By Eqs.(10,15) we have determined $z^k_{(3)ijl}$. Equating the coefficients of the $\delta G_i\delta G_j\delta G_l\delta G_r$ terms in Eq.(13) to obtain the fourth-order coefficient equation. Pre-multiplying by $(\phi^k)^T$ and using Eqs.(1,2) yields

$$\begin{aligned}&\lambda^k_{(4)\ ijlr}=\frac{1}{4!}(\varphi^k)^T\left[3!\left(\frac{\partial K}{\partial G_i}z^k_{(3)jlr}+\frac{\partial K}{\partial G_j}z^k_{(3)ilr}+\right.\right.\\&\frac{\partial K}{\partial G_l}z^k_{(3)ijr}+\frac{\partial K}{\partial G_r}z^k_{(3)ijl}-3!\left(\lambda^k_{(1)i}Mz^k_{(3)jlr}+\lambda^k_{(1)j}Mz^k_{(3)ilr}+\right.\\&\left.\lambda^k_{(1)l}Mz^k_{(3)ijr}+\lambda^k_{(1)r}Mz^k_{(3)ijl}\right)-2!2!\left(\lambda^k_{(2)ij}Mz^k_{(2)lr}+\right.\\&\lambda^k_{(2)il}Mz^k_{(2)jr}+\lambda^k_{(2)ir}Mz^k_{(2)jl}+\lambda^k_{(2)jl}Mz^k_{(2)ir}+\lambda^k_{(2)jr}Mz^k_{(2)il}\\&\left.+\lambda^k_{(2)lr}Mz^k_{(2)ij}\right)-2!2!\left(\lambda^k_{(2)lr}Mz^k_{(2)ij}+\lambda^k_{(2)jr}Mz^k_{(2)il}+\right.\\&\left.\lambda^k_{(2)jl}Mz^k_{(2)ir}+\lambda^k_{(2)ir}Mz^k_{(2)jl}+\lambda^k_{(2)il}Mz^k_{(2)jr}+\lambda^k_{(2)ij}Mz^k_{(2)lr}\right)\end{aligned}\tag{16}$$

Pre-multiplying the coefficient equation by $(\phi^v)^T$, where $1\le v\le N$ and $v\ne k$, and using Eqs.(1,2,11) yields

$$P^k_{(4)ijlrv}=\frac{1}{4!(\lambda^k-\lambda^v)}(\varphi^v)^T\Bigg[3!\Bigg(\frac{\partial K}{\partial G_i}z^k_{(3)jlr}+\frac{\partial K}{\partial G_j}z^k_{(3)ilr}+\frac{\partial K}{\partial G_l}z^k_{(3)ijr}+\frac{\partial K}{\partial G_r}z^k_{(3)ijl}\Bigg)-3!\big(\lambda^k_{(1)i}Mz^k_{(3)jlr}+\lambda^k_{(1)j}Mz^k_{(3)ilr}+\lambda^k_{(1)l}Mz^k_{(3)ijr}+\lambda^k_{(1)r}Mz^k_{(3)ijl}\big)-2!2!$$
$$\big(\lambda^k_{(2)ij}Mz^k_{(2)lr}+\lambda^k_{(2)il}Mz^k_{(2)jr}+\lambda^k_{(2)ir}Mz^k_{(2)jl}+\lambda^k_{(2)jl}Mz^k_{(2)ir}+\lambda^k_{(2)jr}Mz^k_{(2)il}+\lambda^k_{(2)lr}Mz^k_{(2)ij}\big)-3!\big(\lambda^k_{(3)jlr}Mz^k_{(1)i}+\lambda^k_{(3)ilr}Mz^k_{(1)j}+\lambda^k_{(3)ijr}Mz^k_{(1)l}+\lambda^k_{(3)ijl}Mz^k_{(1)r}\big)\Bigg] \quad (17)$$

Using Eqs. 12,17 we have determined $z^k_{(4)ijlr}$. To investigate the applicability on different orders perturbed eigenvalue equations, various stepwise damaged cases are introduced to wing model. Then these equations are optimized by BFGS algorithm[7].

## 3 Geometric dimension of aeroplane wing

XX-XXX series aeroplane belongs to air-jet type passenger aeroplane. It is a short distance twin turbo-engine aeroplane. It meets the requirements of middle and short distance flights, reliable, simple and fast, with economical operational and maintenance characteristics. Passenger capacity is around 120, while maximum flight distance is 5648 miles. Main geometric dimensions of aerodynamic wing are discussed here. Wing span $W_s$ is the length between left and right aeroplane wings. Wing chord is the chord length along the body. Besides the rectangular wing, the wing chord is different at different position. At wing tip, wing chord is $b_0$, while at wing root, wing chord is $b_1$. Fig.1 shows the ordinarily dimension used in wing chord is the mean geometric chord length $b_m$ calculated by $b_m=(b_0-b_1)/2$.

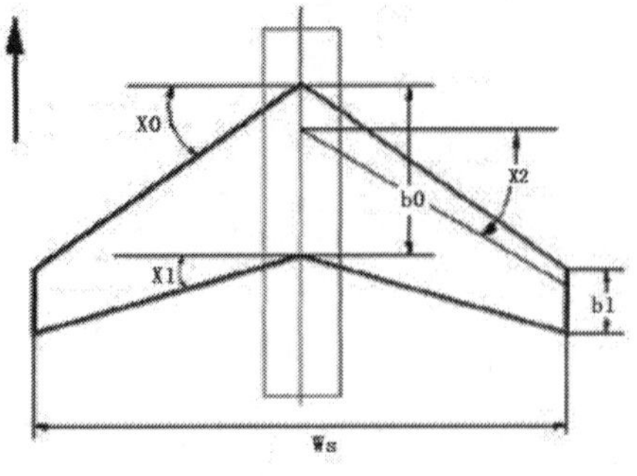


**Figure 1.** Span-chord dimensions of aeroplane wing.

Another geometric dimension is the span-chord aspect ratio. Indicated as $A_{sc}$, its calculation formula is $A_{sc}=W_s/b_m$. Simultaneously, aspect ratio can be represented as the ratio between square of wing span and area of wing surface. When $A_{sc}$ increases, wing lift force coefficient increases, and its drag force increases simultaneously. Therefore, small $A_{sc}$ is utilized in high speed aeroplane wing.

## 4 Stepwise damage detection of variational wing model

Finite element model of the wing with variational chord sections is used to model its planar transverse vibrations. Length of the model is $L_t=15.97m$. Linear variations are imposed on the dimensions of different wing chords by this length. Thus $4.65\le b_i\le 2.91m$, $0.937\le h_i\le 0.234m$ where $i$ is the chord number. For example, at the middle chord, width $b_m=3.78m$, thickness $h_m=0.63m$, area moment of inertia $I_m=b_m h_m^3/12=7.86\times10^{-2}\,m^4$, mass density $\rho=2715\,kg/m^3$ and healthy Young's modulus $E_m=1.0\times10^{10}\,N/m^2$. Then its healthy elastic modulus is $Wc_m=E_m I_m$. The wing is divided into $N_e=6$ elements with length of each element being $l_e=L_t/N_e$, and there are $N_e+1$ nodes. First five eigenparameter pairs are used with $N=5$. The force moment diagrams of the wing subjected to shear force and bending moment are illustrated in Fig.2. From the diagram, one can the region around wing root and aeroengine takes up the peak loads. This accounts for the peak damage level at first three chords. Thus the extra random damage range of 0.2-0.3 is situated at first three chords, while at second three chords extra random damage range of 0-0.2 is assumed. They are namely small, medium and large damaged levels for 0.1, 0.4 and 0.7 respectively.

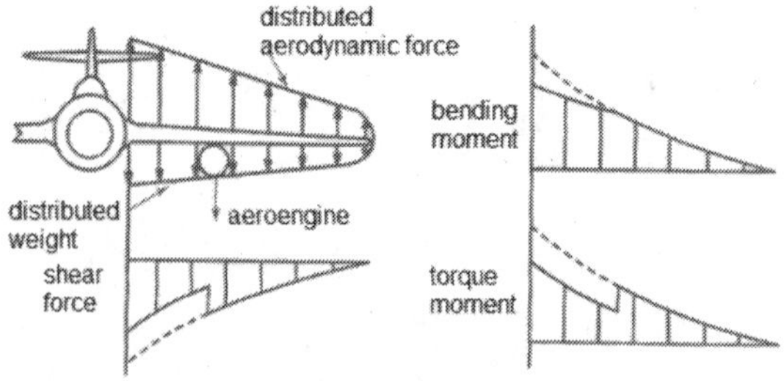


**Figure 2.** Load curves of variational wing model.

Three damaged levels are implemented to cater for the whole stepwise damaged range. The first case illustrates the small stepwise damaged level of $0.1<\sigma_i\le 0.4$. Using fourth-order perturbation, the d-norm drops rapidly from $2.92\times10^{-1}$ to $2.66\times10^{-2}$ in second iteration. Afterwards, it converges gradually to the $2.13\times10^{-5}$ termination level. For the elastic moduli of wing chords, $Wc_1$ and $Wc_2$ change rapidly in two to three iterations as shown in Fig.3. The other chords $Wc_3$ to $Wc_6$ follow the similar patterns. Then they converge smoothly to the correct values in 16th iteration with rms error of

0.174%. Similar iteration patterns with same termination number are observed for the third-order perturbation. However, it's rms error is slightly larger at 0.178%, while it's termination d-norm is slightly higher at $2.26\times10^{-5}$.

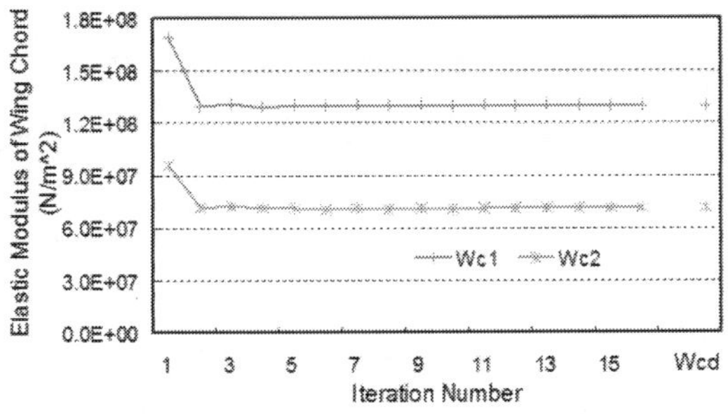


**Figure 3.** Elastic moduli of wing model in small stepwise damage detection ( $0.1<\sigma_i\le0.4$ ).

Medium stepwise damaged level of $0.4<\sigma_i\le0.7$ is shown in second case. Convergence rate is faster for the fourth-order perturbation case, in which termination number is 18. From Fig.4, the d-norm drops rapidly from 1.26 to $4.12\times10^{-2}$ in second iteration. Then it converges gradually to the $2.03\times10^{-5}$ termination level. In the initial region, the Wcs change rapidly. Gradual convergence patterns are observed until 18[th] iteration. It's rms error is 0.192%. Larger termination number of 21 is recorded for third order perturbation case, while its iteration patterns are similar with lower d-norm of $1.26\times10^{-5}$. Slightly smaller rms error of 0.180 % is resulted.

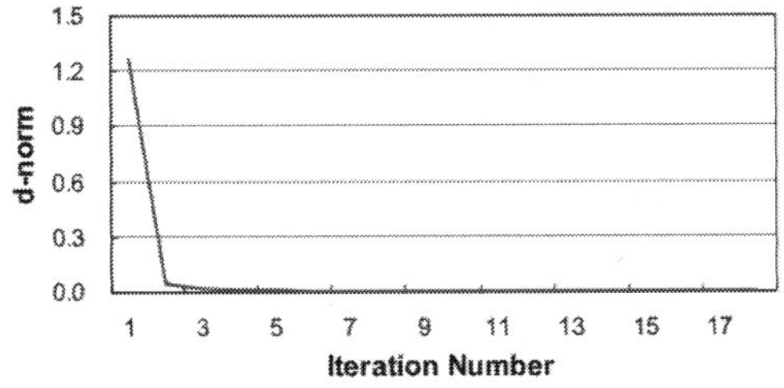


**Figure 4.** d-norm of wing model in medium stepwise damage detection ( $0.4<\sigma_i\le0.7$ ).

Third case illustrates the large stepwise damaged level of $0.7<\sigma_i\le1.0$. For fourth-order perturbation, the d-norm drops rapidly from 19.2 to 3.44 in second iteration. Afterwards, it converges to the $1.87\times10^{-5}$ % termination level gradually. Considering the elastic moduli of wing chords, $Wc_3$ to $Wc_6$ change rapidly in two to three iterations (Fig.5). Similar patterns are obtained for $Wc_1$ and $Wc_2$. Then they converge smoothly to the correct values in 52[th] iteration with rms error of 0.265%. In the third order perturbation, termination number is larger at 76. It's rms error is larger at 0.465 %, while it's termination d-norm is exactly the same.

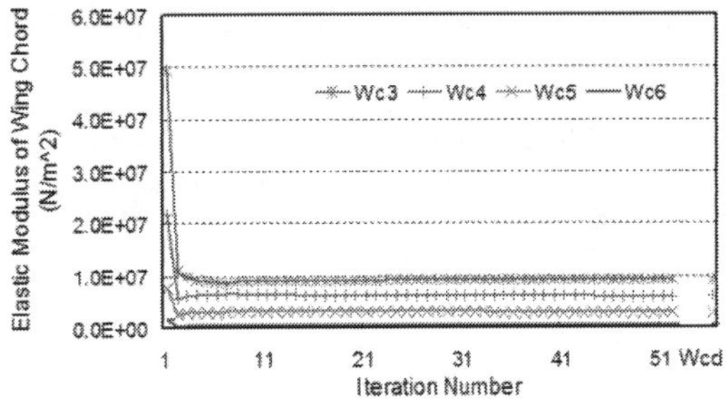


**Figure 5.** Elastic moduli of wing model in large stepwise damage detection ( $0.7<\sigma_i\le1.0$ ).

## 5 Summary

Finite element model of aeroplane wing is constructed. Third and fourth-order perturbation equations with orthogonal and non-orthogonal coefficients are computed using perturbed eigenvalue and orthonormal equations. In small stepwise damaged level, termination number is the same for both approaches, while rms error and termination d-norm are very close. For medium stepwise damaged level, termination number is larger for third-order perturbation with lower d-norm and smaller rms error. In large damaged level, termination number is much larger for third-order perturbation with same d-norm and larger rms error. This trend is more significant as the damaged level increases. This indicates that the increase in stepwise effect leads to the increase in model order. Hence, fourth-order perturbation is more accurate to approximate the model solution.

## Acknowledgements

This research is supported by the National Natural Science Foundation of China under grant numbers 50905028 and 51275554.

## References

1. L. Yu, J. Jiang, Y. Yan, Q. Liu, Eigenvalue perturbation theory with applications to small structural damage detection. Chinese Journal of Applied Mechanics, **23**, 2 255-258 (2006)
2. X. Wang, G. He, Effective structural damage detection method based on modal strain energy, Journal of the China Railway Society, **27**, 5 92-95 (2005)
3. H. Chen, W. He, B. Wang, J. Li, Study on structure damage detection based on perturbations of frequency and mode shapes, Engineering Mechanics, **27**, 12 244-249 (2010)
4. C.N. Wong, W.D. Zhu, G.Y. Xu, On an iterative general-order perturbation method for multiple structural damage detection, Journal of Sound and Vibration, **273**, 363-386 (2004)
5. C.N. Wong, H. Huang, J. Xiong, T. Hu, H.L. Lan, Perturbed eigenvalue problem with the DFP quasi-

DOI: 10.1051/matecconf/20167509010

Newton approach for damage detection of fixed-fixed beam, Mathematics and Mechanics of Solids, **16**, 2 228-247 (2011)
6. C.N. Wong, A.A. Barhorst, General-order perturbation with skew-symmetric approach for structural health monitoring of modular beam, Journal of Vibration and Control, **15**, 12 1783–1807 (2009)
7. C.N. Wong, J. Xiong, Y.J. Zhao, Second order perturbation with BFGS optimization for damage detection of fixed-fixed beam, *ICQR2MSE 2011*, Xian, China, 908-914 (2011)

MATEC Web of Conferences **75**, 09011 (2016) DOI: 10.1051/matecconf/20167509011
*ICMIE 2016*

# Research on the Comparison of Solar Power Generation with Different Technology Paths

Yuan Ren[1], Dan Nie[2] and Wei Yan[2]

[1]*State Grid Shanxi Electric Power Company, Tanyuan Shanxi 033001, China*
[2]*School of Economics and Management, North China Electric Power University, Changping District, Beijing 102206, China*

**Abstract.** To effectively cope with global climate change and the environment pollution problem, all countries are promoting low-carbon development actively. China is rich in solar energy resources, which has important influence on the sustainable development. Photovoltaic and thermal power generation were compared by constructing the comprehensive comparison model which is based on the interval weight and set pair analysis method. Then, their superiorities can be judged. The results of the study can provide strategic advice for investors in the solar-thermal and photovoltaic power generation to make decisions, which has far-reaching significance for them.

## 1 Introduction

Chinese Academy of Social Sciences, China Meterological Administration jointly issued the "Green book of climate change: climate change report (2013)" that pointed out in recent 50 years, haze weather days in China generally showed an increasing trend. The number of fog days is significantly reduced, whereas the number of haze days increased significantly. Persistent haze process has been greatly increased [1], [2]. China has officially proposed that carbon dioxide emissions will reach the peak in 2030. To reduce carbon emissions, Chinese government plans to make non fossil energy consumption account for about 20% in energy consumption in 2030. According to estimates, in order to achieve this goal, China needs to increase 800-1000GW nuclear energy, wind, solar and other zero emission power generation capacity. The potential of China's solar power generation installed capacity is about 16000GW, and power generation potential is about 42000TWh/year [3]. This means that even in the future, after all the fossil energy depletion, China still has a wealth of stable solar power resources far greater than self-sufficiency [4]-[6].

Solar power generation mainly has solar photovoltaic power generation and solar thermal power generation two types. According to the research of domestic and foreign scholars, though solar thermal power generation is now behind the photovoltaic power generation, it will catch up from behind and become the direction of the future development of solar energy [7]-[11]. Based on the comparison and evaluation index system of the solar power generation technology, a comprehensive comparison model based on the interval weight and set pair analysis method is established. After fitting the estimates to index into the model, we got some results that can provide both strategic advice for enterprises and theoretical guidance for the development of China's solar energy industry.

## 2 Composition of evaluation index system

Through detailed analysis of solar power technology, this paper selects the three aspects of technical feasibility, economic rationality and social friendly index to establish a comparative solar power evaluation index system of different technical paths, as shown in Table 1.

### 2.1 Evaluation of solar power generation with different technology paths

The research framework of solar photovoltaic power generation and thermal power generation is shown in Figure 1.

#### *2.1.1 Determine the weight set*

The importance of photovoltaic power generation and photovoltaic power generation is not consistent with the two main body of the enterprise investors and the state, therefore, from the point of view of the two main body of the enterprise investor and the state to determine the weight.

(1) Establish interval number complementary judgment matrix. According to the index, the experts use the 0.1-0.9 nine scale method to compare the index, and establish the interval number complementary judgment matrix. The enterprise comprehensive comparison matrix

is BC, and the national comprehensive comparison matrix is BN, which are expressed as follows.

$$B_C = \begin{bmatrix} 0.500,0.500 & 0.385,0.455 & 0.522,0.582 \\ 0.545,0.615 & 0.500,0.500 & 0.625,0.665 \\ 0.418,0.478 & 0.335,0.375 & 0.500,0.500 \end{bmatrix}$$

$$B_N = \begin{bmatrix} 0.500,0.500 & 0.515,0.558 & 0.415,0.458 \\ 0.442,0.485 & 0.500,0.500 & 0.325,0.355 \\ 0.542,0.585 & 0.645,0.675 & 0.500,0.500 \end{bmatrix}$$

Calculate the weight vector in the form of interval numbers. After the interval number complementary judgment matrix is obtained, the weight vector expressed in the form of interval numbers is calculated. The weight vector of the index of comprehensive comparison is respectively calculated as follows:

$$\omega_C = ([0.301,0.355],[0.358,0.411],[0.268,0.312])^T$$

$$\omega_N = ([0.310,0.346],[0.274,0.306],[0.365,0.401])^T$$

Establish the possibility degree matrix. Using the weight vector represented in the form of interval numbers, the possibility degree matrix can be calculated. The possibility degree matrix of the comprehensive comparison PC, PN are as follows:

$$P_C = \begin{bmatrix} 0.500 & 0.000 & 0.886 \\ 1.000 & 0.500 & 1.000 \\ 0.114 & 0.000 & 0.500 \end{bmatrix}$$

$$P_N = \begin{bmatrix} 0.500 & 1.000 & 0.000 \\ 0.000 & 0.500 & 0.000 \\ 1.000 & 1.000 & 0.500 \end{bmatrix}$$

**Table 1.** Comparative evaluation index system of solar energy power generation.

| Target layer | Criterion layer | Sub criterion layer |
|---|---|---|
| Comparative evaluation index of solar energy power generation | technical feasibility | Light intensity coincidence rate |
| | | Land use efficiency |
| | | Energy conversion rate |
| | | Grid connected stability |
| | | Industrialization level |
| | Economic rationality | Original total investment |
| | | Average cost of power generation |
| | | Internet Pricing |
| | | Project construction period |
| | | Support policy |
| | Social friendliness | environmental effect |
| | | Energy saving effect |
| | | Employment opportunities |
| | | Desertification control |
| | | Popularization of electric |

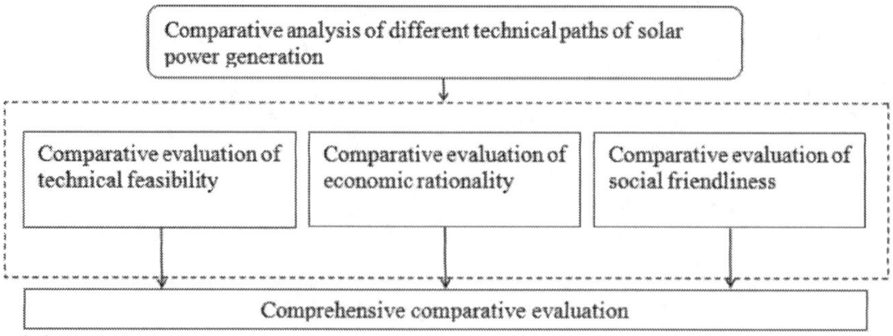


**Figure 1.** Research framework.

(2) Determine the weight vector. After obtaining the possibility degree matrix, the weight vector is obtained. The weight vector of the index of comprehensive comparison $\omega_C$, $\omega_N$ are as follows:

$$\omega_C = (0.308,0.555,137)^T$$

$$\omega_N = (0.333,0.111,556)^T$$

Calculate the connection degree and result. According to the formula (1) (2), the connection of the comprehensive evaluation index is calculated, and the results are shown in Table 2.

**Table 2.** The connection degree of comprehensive comparative evaluation index

| | a | b1 | b2 | b3 | c |
|---|---|---|---|---|---|
| μ1 | 0.373 | 0.343 | 0.067 | 0.184 | 0.032 |
| μ2 | 0.077 | 0.166 | 0 | 0.087 | 0.670 |
| μ3 | 0.308 | 0.0385 | 0.071 | 0.394 | 0.189 |
| ωC1μ1 | 0.115 | 0.106 | 0.021 | 0.057 | 0.010 |
| ωC2μ2 | 0.043 | 0.092 | 0 | 0.048 | 0.372 |
| ωC3μ3 | 0.042 | 0.005 | 0.010 | 0.054 | 0.026 |
| ωN1μ1 | 0.124 | 0.114 | 0.022 | 0.061 | 0.011 |
| ωN2μ2 | 0.009 | 0.018 | 0 | 0.010 | 0.074 |
| ωN3μ3 | 0.171 | 0.021 | 0.040 | 0.219 | 0.105 |

$$\mu_{ij}=\begin{cases} 1+0i_1+0i_2+0i_3+0j & x>s_1 \\ \frac{x-s_2}{s_1-s_2}+\frac{s_1-x}{s_1-s_2}i_1+0i_2+0i_3+0j & s_2\le x<s_1 \\ 0+\frac{x-s_3}{s_2-s_3}i_1+\frac{s_2-x}{s_2-s_3}i_2+0i_3+0j & s_3\le x<s_2 \\ 0+0i_1+\frac{x-s_4}{s_3-s_4}i_2+\frac{s_3-x}{s_3-s_4}i_3+0j & s_4\le x<s_3 \\ 0+0i_1+0i_2+\frac{x-s_5}{s_4-s_5}i_3+\frac{s_4-x}{s_4-s_5}j & s_5\le x<s_4 \\ 0+0i_1+0i_2+0i_3+1j & x<s_5 \end{cases} \quad (1)$$

$$\mu_{ij}=\begin{cases} 1+0i_1+0i_2+0i_3+0j & x<s_1 \\ \frac{s_2-x}{s_2-s_1}+\frac{x-s_1}{s_2-s_1}i_1+0i_2+0i_3+0j & s_1\le x<s_2 \\ 0+\frac{s_3-x}{s_3-s_2}i_1+\frac{x-s_2}{s_3-s_2}i_2+0i_3+0j & s_2\le x<s_3 \\ 0+0i_1+\frac{s_4-x}{s_4-s_3}i_2+\frac{x-s_3}{s_4-s_3}i_3+0j & s_3\le x<s_4 \\ 0+0i_1+0i_2+\frac{s_5-x}{s_5-s_4}i_3+\frac{x-s_4}{s_5-s_4}j & s_4\le x<s_5 \\ 0+0i_1+0i_2+0i_3+1j & x>s_5 \end{cases} \quad (2)$$

The result is obtained according to the formula (3).

$$\mu=\sum_{i=1}^{3}\omega_i\mu_i=\sum_{i=1}^{3}\omega_i a_i+\sum_{i=1}^{3}\omega_i b_{i1} i_1+\sum_{i=1}^{3}\omega_i b_{i2} i_2+\sum_{i=1}^{3}\omega_i b_{i3} i_3+\sum_{i=1}^{3}\omega_i c_i j \quad (3)$$

$\mu_C=0.199+0.203\,i1+0.030\,i2+0.159\,i3+0.408j$

Comprehensive comparison result in the perspective of corporate investors is $\triangle_C=-0.119<0$

$\mu_N=0.304+0.154\,i1+0.062\,i2+0.290i3+0.190j$

Comprehensive comparison result in the perspective of nation is $\triangle_N=0.027>0$

Through the above results, we can see that for investors in the enterprise, the comprehensive benefit of investment in photovoltaic power generation is better than that of investment in thermal power generation; however, from the national point of view, investment in solar thermal power generation comprehensive benefit is better than that of the photovoltaic power generation, though the advantage is not obvious.

From the current comparison results, compared to photovoltaic power generation, thermal power generation does not have the advantage for investors in the enterprise, but has weak advantage for the nation. Although the solar thermal power generation started late, there is a larger space for further development. With the national attention to the development of new energy and investment volume increased, the technical level of solar thermal power generation and application promotion will achieve a great leap forward. So, we can predict the future of the solar thermal power generation will be bright.

## 3 Conclusion

By comparing the results of the comparative evaluation model, the paper obtains the results of the comprehensive comparison of the feasibility, economic rationality and social friendly indexes of the photovoltaic power generation and thermal power generation technology. Results show that under the current situation, solar thermal power generation is greatly superior to solar thermal power generation in the technical feasibility, but slightly weaker in terms of social friendliness and greatly weaker in the economic rationality. The results of the comprehensive comparison show that for corporate investors, photovoltaic power generation is better than the thermal power generation, whereas for the nation, thermal power generation has a slight advantage. This paper make comparisions from the aspects of technical feasibility, economic rationality and social friendly, and calculated from different angles. The results both for the enterprises or to the nation can play a guiding role.

## Acknowledgment

This study is supported by National Natural Science Foundation of China (Granted No. 71471058) and the Beijing Education Committee of co-construction project.

## References

1. Chinese Academy of Social Sciences, China Meterological Administration. Green book of climate change: Climate change report (2013). Beijing, (2013)
2. S. Yang, F. Deng. Solar energy and wind power generation technolog. Beijing: Publishing House of Electronics Industry, (2013)
3. J. Shi. The mathematical model of photovoltaic power generation cost analysis. Solar Energy, (2012), **2**:53–58
4. S. Y. Cao, Q. H. Li, B. B. Huang, X. L. Wang, Q. K. Wang. Economic analysis and development forecast

DOI: 10.1051/matecconf/20167509011

of photovoltaic power generation technology. Electric Power, (2012), **45**(8): 64–68
5. Royal Institution of Great Britain. Solar thermal power generation. London: International Solar Energy Society, UK Section, (1978)
6. M. K. Slam, M. Hosenuzzaman, M. M. Rahman, et al. Thermal performance improvement of solar thermal power generation, Clean Energy and Technology (CEAT), 2013 IEEE Conference on. IEEE, **2013**:158-162
7. T. Trainer, Limits to solar thermal energy set by intermittency and low DNI: Implications from meteorological data. Energy Policy, (2013), **63**:910-917
8. A. Poullikkas, Economic analysis of power generation from parabolic trough solar thermal plants for the Mediterranean region—A case study for the island of Cyprus. Renewable and Sustainable Energy Reviews, (2009), **13**:2474–2484
9. I. Purohit, P. Purohit. Techno-economic evaluation of concentrating solar power generation in India. Energy Policy, (2010), **38**:3015–3029
10. V. S. Reddy, S. Kaushik. State-of-the-art of solar thermal power plants—A review. Energy Reviews, (2013), **27**:258–273
11. Z. K. Wang. Application and development of solar thermal power generation technology. Glass, **2012**(6): 30–35

# Genetic Algorithm Applying to Calculate the Great Wall Soil Water Electrical Conductivity

Zhang Wanxu, Yan Yang, Jiang Bo, Wang Lin and Liu Cheng

*School of Information Science and Technology, Northwest University, 710127, Xi'an, Shaanxi Province , China*

**Abstract.** In this paper, we report a simple but effective algorithm to estimate the undetermined coefficients in the formula of soil water electrical conductivity value by the soil apparent electrical conductivity method. By doing so, the coefficients can be accurately obtained to further calculate the soil water electrical conductivity. According to the process of calculating the soil solution electrical conductivity based on the apparent conductivity of soil, the error of the calculation equation is effectively calculated by the two order curve fitting method. Meanwhile the genetic algorithm is adopted to reduce the influence of the data error generated by the proposed algorithm, and improve the accuracy of the data as well. The experimental results demonstrate that the solution by the proposed algorithm can be effectively obtained and the error change is in one direction simultaneously. With these advantages, we can use soil water electrical conductivity to calculate the content of soil salt.

## 1 Introduction

There are many factors that affect the apparent conductivity of soil, such as soil salt content, water content, temperature, organic matter content and geological structure [1]. But the main factors are soil salt content, water content and temperature. By the composition of the soil, the soil apparent conductivity is mainly composed of the soil surface electrical conductivity and the soil water electrical conductivity [2]. The soil salinity is directly evaluated by the soil water electrical conductivity, and the relationship in $EC_a$-$EC_w$-$\theta$ is the theoretical basis for the estimation of the soil salinity. According to Ref [3], the analytical relationship between the soil apparent conductivity $EC_a$ and the soil water electrical conductivity $EC_w$ is as follows:

$$EC_a = EC_s + T\theta \cdot EC_w \quad (1)$$

where $EC_S$ is the soil surface electrical conductivity ($mS/m$). $\theta$ is volumetric soil water content ($m^3/m^3$). $EC_w$ is the soil water electrical conductivity ($mS/m$). $T$ is transmission factor which can generally be expressed as:

$$T = a\theta + b \quad (2)$$

where $a$ and $b$ are undetermined coefficients.

This paper mainly rsearches the estimation of the undetermined coefficients in the formula of soil solution conductivity, so that the estimated coefficients can be more accurate to calculate the soil water electrical conductivity.

## 2 Common method to estimate the soil water electrical conductivity

The mathematical definition of curve fitting is use continuous curve to approximately describe the function relation between the coordinates represented by a group of discrete points on the plane, and it is a method to approximately fit the discrete data by analytic expression. Curve fitting is usually called 'pulling curve' [4], a method that the existing data is introduced into a mathematical equation through mathematical method

Combine the formulas (1) and (2), we can have:

$$\frac{EC_a}{EC_w} = a\theta^2 + b\theta + c \quad (3)$$

where $c = \frac{EC_s}{EC_w}$.

According to this formula, $\frac{EC_a}{EC_w}$ can be treated as quadratic polynomial $\theta$. After measuring a large number of data about $EC_a$, $\theta$ and standard $EC_w$, we can use the polynomial curve fitting method to estimate the soil water electrical conductivity by using the principle of least square method.

Although this method has strong applicability and simple calculation process, but the calculation accuracy is poor, and it may trap into local optimum, which can not accurately calculate the calculation formula of the soil water electrical conductivity. To overcome the drawbacks of the above algorithm, this paper presents the application of genetic algorithm to calculate the formula of the soil water electrical conductivity.

# 3 Calculation formula of soil water electrical conductivity based on genetic algorithm

## 3.1 Genetic algorithm

Genetic algorithm is a global optimization algorithm for random search [5]. Basis on the individual gene expression, it simulates the phenomenon of reproduction through genetic operator, crossover and mutation in genetic operator, and then optimal individual [6]. Genetic algorithm can deal with multiple individuals in the population at the same time, that is, it can evaluate the multiple solutions in the search space, reduce the risk of falling into the local optimal solution, and be implemented easily [7]. Genetic algorithm uses the probability of transition rules not the deterministic rule to guide its search direction, with the property of self-organization, self-adaption and self-learning. The basic flow chart of genetic algorithm [9] is shown in Figure 1.

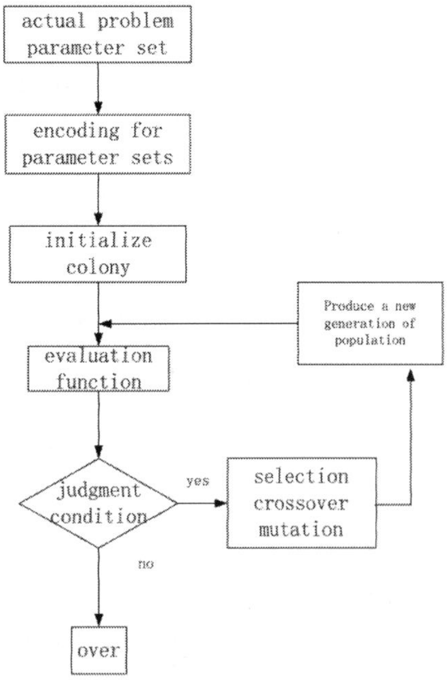


**Figure 1.** Basic flow chart of genetic algorithm.

## 3.2 The genetic algorithm steps

The purposes of introducing the genetic algorithm into apparent soil electrical conductivity method are (1) to optimize the coefficients of the equation, and (2) to more accurately calculate the soil water conductivity. The specific steps of the algorithm are as follows:

**Step 1:** Determining the population size $M$. Random generation of initial population including $M$ individuals $W=(S_1,S_2,\ldots,S_M)$ determines a range of data, and three coefficients were used as three genes $L$, $P$ and $N$, encoded with 8 bit binary number to form a chromosome. The length of each chromosome is:

$$S = L + P + N \tag{4}$$

Gene composition of each chromosome is determined randomly, where the initial value of each individual in the population satisfies $S_i=L_m+P_n+N_l$. ($i$=1,2,…,$M$ $m=n=l=\sqrt[3]{M}$).

**Step 2:** Determining the appropriate evaluation function of the individual. Split chromosomes into three obtained corresponding coefficients equation, inputs reference samples for calculation and gets a training output value $\hat{y}_i$. The fitness function of the population is:

$$f(S_i) = \frac{|y_i - \hat{y}_i|}{y_i} \quad i = (1,2,3,\cdots,M) \tag{5}$$

where $\hat{y}_i$ denotes training output value, $y_i$ denotes desired output.

**Step 3:** Using Roulette method to select operator. The selection of chromosomes in each generation population is based on the proportion of fitness. And the selection of the probability can be expressed as:

$$p_i = \frac{f_i}{\sum_{i=1}^{M} f_i} \quad \text{where } f_i = \frac{1}{f(S_i)} \tag{6}$$

**Step 4:** Using single point crossover method to cross the chromosome. Single individuals $S_1$ and $S_2$ are crossed according to crossover probability in the gene of $k$ ($k$=1,2,3).

Before crossed

$$S_1 = L_{m1} + P_{n1} + N_{l1}$$

$$S_2 = L_{m2} + P_{n2} + N_{l2}$$

After crossed

$$S_1 = L_{m1} + P_{n2} + N_{l1}$$

$$S_2 = L_{m2} + P_{n1} + N_{l2}$$

The value of crossover mainly is 0.25-0.75.

Select the $k$ Gene ($k$=1,2,3) of the $i$ individual to operate with the $P_m$ probability variation.eg:

Before $\quad S_i = L_m + P_n + N_l$

After $\quad S_i = L_m + P_{n1} + N_l$

Range of values $P_m$ is 0.01-0.2.

**Step 5:** Decomposing the optimal solution of genetic algorithm into three undetermined coefficients of the equation. Put these into the equation 3 and calculate the soil water electrical conductivity. And the procedure of genetic algorithm is shown in Figure 2.

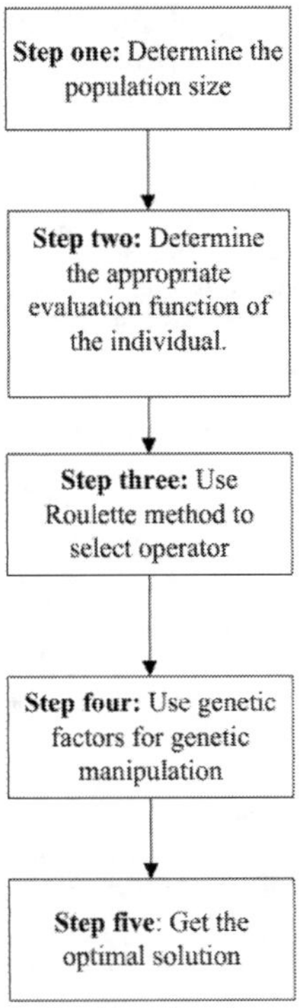


**Figure 2.** The genetic algorithm procedure

# 4 Experimental results and analysis

## 4.1 Experimental results

Several groups of different regions were selected to measure the field, then used the host computer to deal with data. Figure 3 is the common method to estimate the soil water electrical conductivity comparing with standard value.

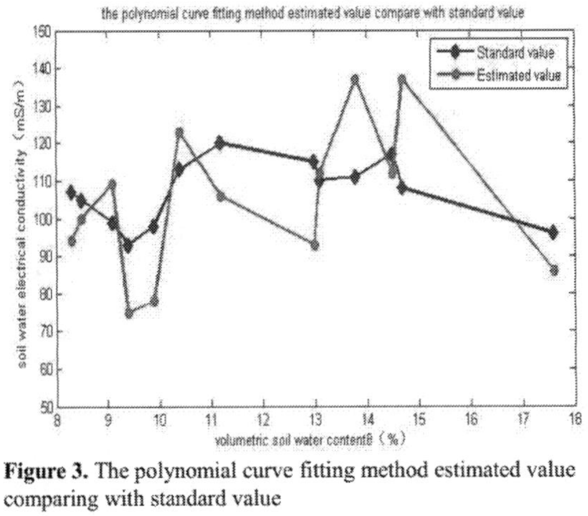


**Figure 3.** The polynomial curve fitting method estimated value comparing with standard value

Figure 4 is based on genetic algorithm to calculate the soil water electrical conductivity comparing with standard value.

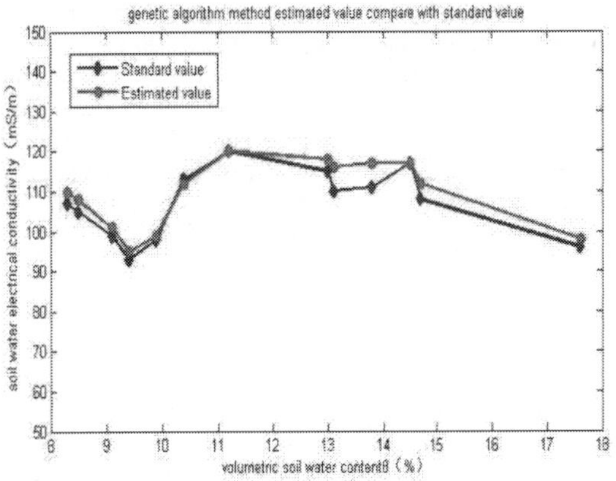


**Figure 4.** Genetic algorithm method estimated value comparing with standard value

## 4.2 Experimental analysis

From Figure 3, the accuracy of the two curve fitting is largely dependent on the initial sample size. The gap between the soil water conductivity calculated value and the standard value is obvious, and no change tend can be observed. From Figure 4, compared with the standard value curve, the calculated value is suitable, and the change trend is more obvious. This indicates that by using the optimization of genetic algorithm the calculation value of the equation is more accurate.

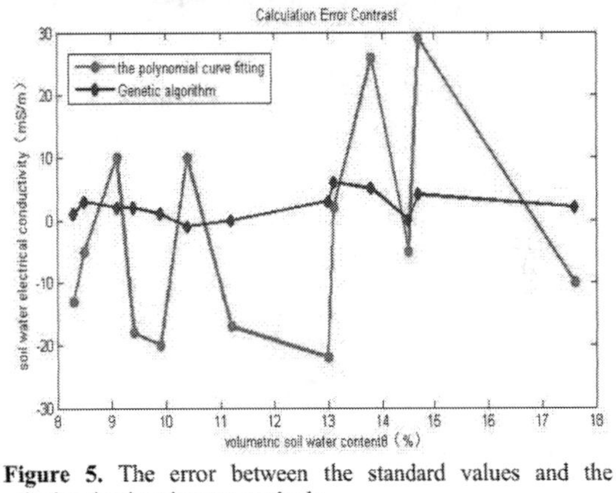


**Figure 5.** The error between the standard values and the calculated values by two methods

Figure 5 shows the error between the standard values and the calculated values by two methods. From the picture we can clearly see, soil water electrical conductivity error by using common method to calculate is larger, and there is no law of error variation. But the error fluctuation of the value obtained by genetic algorithm optimization is relatively small, and error change tend is in one direction.

# 5 Conclusion

This paper discussed the equations for calculating the soil water electrical conductivity by using soil apparent electrical conductivity. In order to eliminate the defects of common method, the genetic algorithm method was adopted to calculate the equation. According to experimental results, the equation derived from the genetic algorithm method had a significant improvement in terms of accuracy of the calculation, and avoided the calculation falling into the local optimum. So we could get accurate salinity data from the soil solution electrical conductivity.

## Acknowledgment

This work was supported by the National Natural Science Foundation of China under Grant 61503300, Natural Science Foundation of Shaanxi Province of China under Grant 2014JQ8327, China Postdoctoral Science Foundation under Grant 2014M560801, Scientific Research Program Funded by Shaanxi Provincial Education Department under Grant 14JK1750, Science Foundation of Northwest University under Grant 13NW40, and Foundation of Key Laboratory of Space Active Opto-Electronics Technology of Chinese Academy of Sciences under Grant AOE-2016-A02.

## References

1. Rhoades JD. Electrical conductivity methods for measuring and mapping soil salinity. Adv. Agron., **46**: 201-251
2. Rhoades JD, Manteghi NA, Shouse PJ, Alves WJ. Estimating soil salinity from saturated soil–paste electrical conductivity. Soil Sci. Soc. Am. J., **53**: 428-433, (1989)
3. Rhoades JD, Manteghi NA, Shouse PJ, Alves WJ. Soil electrical conductivity and soil salinity: New formulations and calibrations. Soil Sci. Soc. Am. J., **53**: 433-439 (1989)
4. Wang He-xin, Cheng Shi-zhou. The least square method of curve fitting. Journal of Inner Mongolia University for Nationalities
5. Ma Yongjie, Yun Wenxia, Research progress of genetic algorithm. **29**(4):1201-1206, (2012)
6. Li Min-qiang, Kou Ji-song, Lin Dan, et al. Basic theory and application of genetic algorithm. BEI Jing: Science Press, (2002)
7. Yang Xiao-hua, Yang Zhi-feng, Y1n Xi-nan, et al. Chaos gray-coded genetic algorithm and it application for pollution source identifications in convection-diffusion equation. Communications in Nonlinear Science and Numerical Simulation, **13**(8): 1676-1688, (2008)
8. Murat A, Novruz A. Development a new mutation operator to solve the traveling salesman problem by aid of genetic algorithms. Expert Systems with Applications, **38**(3): 1313-1320, (2011)
9. Kusum T. Manoj T. A new crossover operator for real coded genetic algorithms. Applied Mathematics and Computation, **188**(1): 895-911, (2007)
10. Ming Lang, Wang Yu-ping. Convergent Rate of enetic Algorithms with Arbitrary Encoding. Systems Engineering Theory& Practice, **26**(3):88-93. (2006)
11. Yu Zhi-gang, Song Shen-min, Duan Guang-ren. Research on the mechanism and convergence of genetic algorithm. Control and Decision, **20**(9):971-980, (2005)
12. Ferreir C. Gene expression programming: A new adaptive algorithm for solving problems. Complex Systems, **13**(2):87-129, (2001)
13. Ingvalson R.D., Sensitivity of time domain reflctometry measurements to laternal varition in soil watercontent. 5011Sei.SocAmer.Proe., **34**:57 (1970)
14. Zhang Ling, Zhang Bo. Research on genetic algorithm mechanism. Journal of software, **11**(7):945-952 (2000).
15. Yin Zuo-yun. Research on Logistic-curve fitting method. Application of Statistics and Management, **21**(1):41-46 (2002)
16. Bao Jian, Zhao Jian-yong, Zhou Hua-ying. Research on the method of curve fitting based on BP network. Computer Engineering and Design, **26**(7):1840-1841 (2005)

# Study of Key Non-dimensional Parameters for Wave Drag Reduction with High-Frequency Repetitive Laser Pulse Energy Depositions

Zexu Qing[1,2], Yanji Hong[1], Diankai Wang[1] and Bin Zhang[2]

[1]State Key Laboratory of Laser Propulsion & Application, Equipment Academy, 101416 Beijing, China
[2]School of Aeronautics and Astronautics, Shanghai Jiao Tong University, 200240 Shanghai, China

**Abstract.** The problem of wave drag reduction with high-frequency repetitive laser pulse energy depositions is multivariable. Three key non-dimensional parameters, *non-dimensional energy*, *non-dimensional depositing position* and *Mach number*, were constructed from a number of original variables by using Buckingham pi theorem. Influences of these non-dimensional parameters on energy deposition performance, namely drag reduction and energy deposition efficiency, were investigated numerically by solving three-dimensional Navier-Stokes equations with an upwind scheme. Optimizing method of non-dimensional energy and non-dimensional depositing position is proposed. Drag reduction and energy deposition efficiency have exponential relationships with non-dimensional energy; Drag reduction and energy deposition efficiency have quadratic relationships with non-dimensional depositing position. Drag reduction has exponential relationship with freestream Mach number and energy deposition efficiency has quadratic relationship with Mach number. Non-dimensional laser energy and non-dimensional depositing position should be optimized synthetically for a given freestream.

## 1 Introduction

At the present state of art, realizing practicable hypersonic flight is hampered by great drag wave, which even accounts for more than half of total drag in some cases. High-frequency repetitive laser pulse energy deposition is a promising technique to reduce supersonic wave drag with the merit of good controllability, high energy density and energy efficiency. The disturbed region (core) with lower density and higher temperature, generated by the laser-induced optical breakdown, interacts with the bow shock. A mitiagation of the bow shock accompanied by drag and heat flux reduction can be observed.

Experimental study of a single laser pulse energy deposition was performed over a sphere in a supersonic flow with Mach number of 3.45 by Adelgren et al. [1]. Erdem et al. [2] mitigated the bow shock by single-pulse energy and high voltage discharge. Study conducted by Kim et al. [3] shown that a drag reduction about 21% corresponding to energy deposition efficiency of ~10 could be obtained at Mach number of 1.92 utilizing the laser whose maximum frequency and power was up to 50 kHz and 400 W respectively. Numerical results indicated the existence of optimal depositing position [4].

P. K. Tret'yakov realized steady high-frequency repetitive laser pulse energy depositions for wave drag reduction for the first time [5]. It was shown that drag reduction increased with increasing of laser frequency in the premise that laser average power kept nearly unchanged. A series of experiments were conducted in a Mach-1.92, in-draft wind tunnel by A. Sasoh's group. It was shown that energy deposition efficiency monotonically decreased with increasing laser average power and drag reduction was proportional to laser average power [6]. Riggins et al. [7] investigated the drag reduction utilizing a full Navier-Stokes CFD code. According to their study, drag force was reduced to value as low as 30% associated with energy deposition efficiency of up to 33 at Mach numbers of 6.5 and 10 without optimization. Wave drag reduction by a single pulse energy deposition was calculated in sphere model by S. Taguchi et al. [8] and Y. Ogino [9]. The drag reduction was proportional to the freestream Mach number under the constant energy deposition condition in Taguchi's study. Existence of optimal depositing position was also proved in Taguchi's work. Ogino determined that the reduced energy was proportional to the square of the freestream Mach number. It is seen that the energy deposition efficiency increases with increasing of Mach number from V. A. Levin's work [10]. Nearly linear (but very shallow) decreasing trend in the percentage of drag reduced with increasing Mach number (lower than 6) was shown in T. Taylor's study [11].

Parameters for wave drag reduction with high-frequency repetitive laser pulse energy depositions are so numerous, including laser energy magnitude, energy depositing position, Mach number, pressure and temperature of freestream, size and shape of the body etc., that it is necessary to construct key non-dimensional parameters to reduce the difficulty and cost of experiments. Based on these key non-dimensional parameters, a series of numerical experiments were carried out and more general laws about the effects of these parameters on energy deposition performance were obtained.

## 2 Dimensional analysis

There are two important criterions in evaluating energy deposition performance, namely the drag reduction $\eta$ and energy deposition efficiency $S$, defined as:

$$\eta = \frac{\Delta D}{D_{\text{ref}}} = \frac{D_{\text{ref}} - D_{\text{mod}}}{D_{\text{ref}}}$$
$$S = \frac{(D_{\text{ref}} - D_{\text{mod}}) \cdot V_\infty}{P} = \frac{D_{\text{ref}} \cdot V_\infty}{P} \cdot \eta \quad (1)$$

where $\Delta D$, $D_{\text{ref}}$ and $D_{\text{mod}}$ are the drag modification, baseline drag without laser energy deposition and modified drag, respectively; $P$ is the laser power and $V_\infty$ is the freestream velocity.

In the premise that optical breakdown took place, we expect $\eta$ and $S$ to depend on blunt body diameter $D$, freestream velocity $V_\infty$, freestream density $\rho_\infty$, freestream temperature $T_\infty$, freestream specific heat at constant pressure $c_p$, laser power $P$, distance from body nose to breakdown $L$, kinetic viscosity $\mu$. Without any a priori knowledge about the variation of $\eta$ and $S$, we can use common sense to write

$$\eta = f(D, V_\infty, \rho_\infty, T_\infty, P, c_p, L, \mu)$$
$$S = g(D, V_\infty, \rho_\infty, T_\infty, P, c_p, L, \mu) \quad (2)$$

Due to four fundamental dimensions and eight physical variables, there are four dimensionless $\pi$ products. Choose $D$, $V_\infty$, $\rho_\infty$ and $T_\infty$ as the arbitrarily selected sets of four physical variables. These $\pi$ products are

$$\pi_1 = \frac{P}{D^{\alpha_1} V_\infty^{\beta_1} \rho_\infty^{\gamma_1} T_\infty^{\kappa_1}} \qquad \pi_2 = \frac{L}{D^{\alpha_2} V_\infty^{\beta_2} \rho_\infty^{\gamma_2} T_\infty^{\kappa_2}}$$
$$\pi_3 = \frac{c_p}{D^{\alpha_3} V_\infty^{\beta_3} \rho_\infty^{\gamma_3} T_\infty^{\kappa_3}} \qquad \pi_4 = \frac{\mu}{D^{\alpha_4} V_\infty^{\beta_4} \rho_\infty^{\gamma_4} T_\infty^{\kappa_4}} \quad (3)$$

The results of dimensional analysis is

$$\eta = f\left(\frac{P}{D^2 V_\infty^3 \rho_\infty}, \ \frac{L}{D}, \ c_p \frac{T_\infty}{V_\infty^2}, \ \frac{\mu}{D V_\infty \rho_\infty}\right) \quad (4)$$

Non-dimensional energy $\varepsilon$ is defined as the ratio of laser power $P$ and freestream enthalpy flux $\dot{H}$

$$\varepsilon = \frac{P}{\dot{H}} = \frac{P}{\rho_\infty c_p T_\infty V_\infty D^2} \quad (5)$$

Introducing Mach number and Reynolds number, we have

$$\eta = f\left(\varepsilon, \ \frac{L}{D}, \ Ma, \ Re\right)$$
$$S = f\left(\varepsilon, \ \frac{L}{D}, \ Ma, \ Re\right) \quad (6)$$

In Equation (6), $\eta$ and $S$ are expressed as general functions of non-dimensional energy $\varepsilon$, non-dimensional depositing position $L/D$, freestream Mach number and Reynolds number $Re$. Due to the little influence of viscosity on wave drag, $Re$ is ignored and three key non-dimensional parameters are obtained.

## 3 Study of key non-dimensional parameters

### 3.1 Computational methodology

Three-dimensional compressible Navier-Stokes equations were solved with an upwind scheme to simulate the flow. The calculation adopted second-order accurate scheme based on the finite volume method and the domain decomposition of the structural grid. An implicit difference method was employed. An upwind scheme based on a rotating Riemann solver was used to calculate the inviscid flux. A perfect gas model was used. Laser energy was deposited instantaneously with frequency of 150 kHz because frequency has little influence on energy deposition performance when frequency is high enough. Evolution of the plasma was ignored. Laser energy was deposited in the area with a diameter of 1 mm. two-dimensional blunt body model with width of 20 mm was adopted. Mach number of freestream was 2.5 except for Section 3.4. Static pressure and temperature of the freestream were 58 kPa and 130 K respectively. The structured grids were generated with commercial software GridgenV15, and the total number was $9\times10^4$. No-slip wall conditions were applied on the surface of the blunt body. The grid and definition of the boundary were shown in Figure 1. The CFD code was developed on our own.

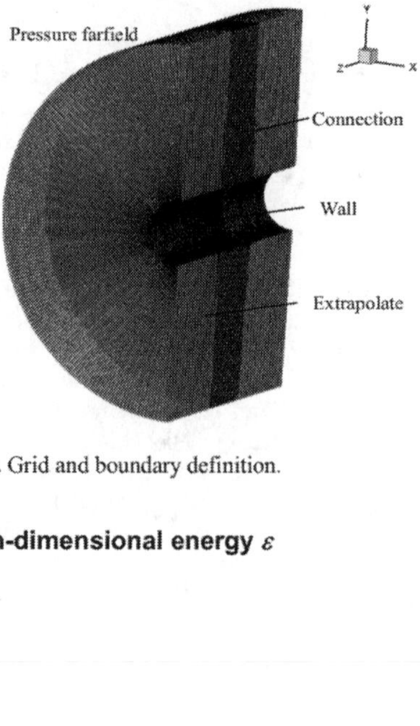


**Figure 1.** Grid and boundary definition.

### 3.2 Non-dimensional energy $\varepsilon$

To study the influence of non-dimensional energy on performance of laser energy deposition, non-dimensional depositing position and Mach number are fixed to 1 and 2.5 respectively. Flow fields with different $\varepsilon$ were shown in Figure 2. The hot core becomes bigger and density becomes lower when more laser energy is deposited, bringing more drag reduction. However, bigger area of the hot core means that the influence is moving away from stagnation point, which results in the decrease of $S$.

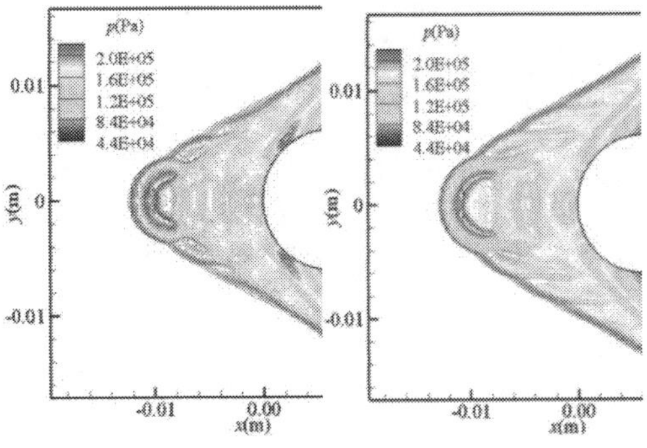


$\varepsilon$=18% $\varepsilon$=54%

**Figure 2.** Flow fields with ε =18% and ε =54%.

As shown in Figure 3, three blunt bodies with different diameters go through similar performance trend with the increase of non-dimensional energy and the energy deposition performance is finely discriminating which demonstrates the reliability of key non-dimensional parameters. With increasing of $\varepsilon$, $\eta$ increases and $S$ decreases.

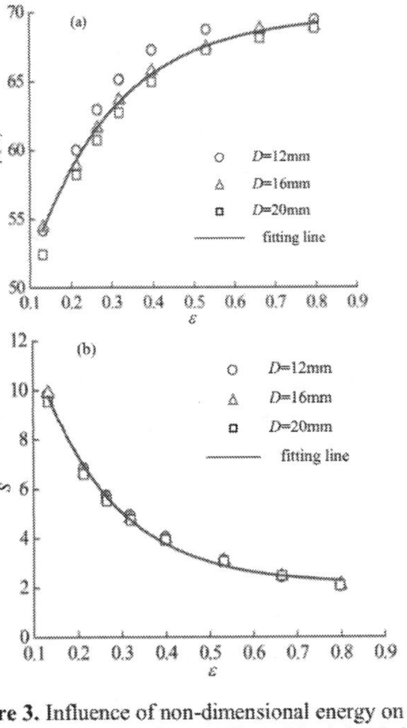


**Figure 3.** Influence of non-dimensional energy on performance of laser energy deposition at $L/D$=1, $Ma$=2.5: (a) drag reduction; (b) energy deposition efficiency.

Fitting data points of Figure 3 with exponential curves, we have

$$\eta = -29.49e^{\frac{-\varepsilon}{0.21}} + 69.80 \quad S = 16.52e^{\frac{-\varepsilon}{0.17}} + 2.13 \qquad (7)$$

From Equation (7), it is found that $\eta$ seems to converge to 69.8% and $S$ converges to 2.13 as $\varepsilon$ goes to infinity for the given calculation condition. $\eta$ and $S$ can be determined approximately by Equation (7) if $\varepsilon$ locates in the range of 0.1~0.8. Keeping Equation (6) in mind, it's reasonable to infer that constants in Equation (7) have relations to $L/D$ and $Ma$. While more numerical experiments are needed to reveal those relationships.

### 3.3 Non-dimensional depositing position *L/D*

Influence of $L/D$ on performance of laser energy deposition is shown in Figure 4. The diameter of blunt body is chosen to be 10mm. Freestream condition is the same as what mentioned above. Large or small $L/D$ makes the compression wave front caused by interaction of blast wave and bow shock too close to the body surface, so energy deposition performance is not good enough, as shown in Figure 5. Cases with different $\varepsilon$ is also included, symbolized by different colours.

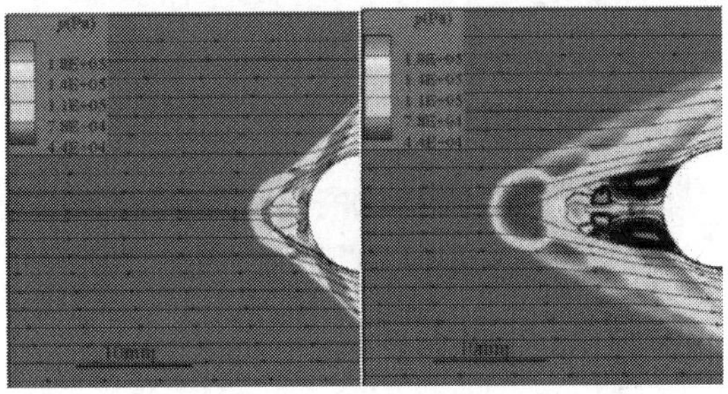

$L/D$=0.5 $L/D$=1.5

**Figure 4.** Flow fields with different $L/D$, $\varepsilon$=0.066.

Data points in Figure 5 were fitted with quadratic curves. The expressions have the form of

$$\eta, S = A(L/D)^2 + B(L/D) + C \qquad (8)$$

So ( − $B/2A$) represents the optimal $L/D$ with the highest $\eta$ and $S$. A positive correlation of ( − $B/2A$) with $\varepsilon$ can be found in Figure 6.

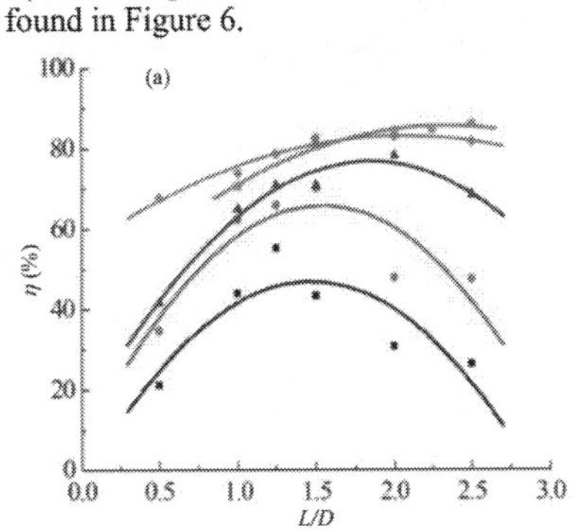

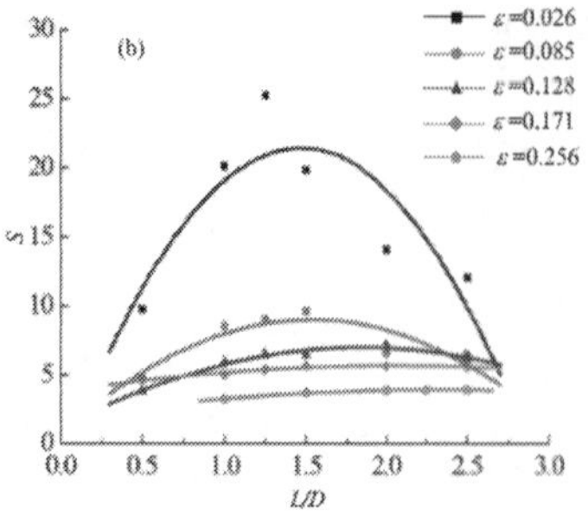


**Figure 5.** Influence of non-dimensional depositing position on performance of laser energy deposition at $P$=1000W, $D$=10mm, $Ma$=2.5: (a) drag reduction; (b) energy deposition efficiency.

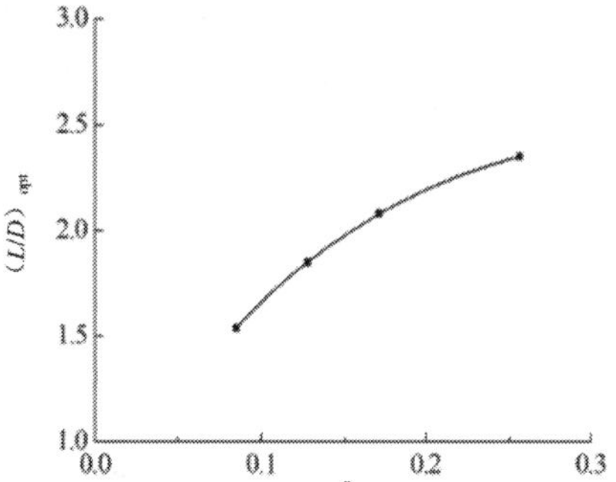


**Figure 6.** Relationship of non-dimensional energy with optimal depositing position.

### 3.4 Mach number *Ma*

Fixing $P$=1000W, $D$=12mm, $L/D$=1, numerical experiments with diffenent $Ma$ are conducted. Figure 7 shows two typical flow fields. As shown in Figure 8, for the constant laser power $P$, it is commonsensible that $\eta$ decreases with the increase of $Ma$, which can be described roughly by exponential equation, meaning that more laser energy is needed to obtain enough drag reduction in high Mach number situation.

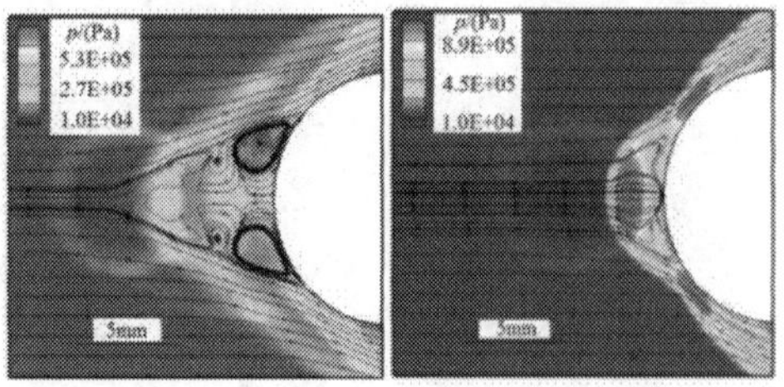


**Figure 7.** Pressure contour and streamlines of different $Ma$: left, $Ma$=3; right, $Ma$=5.

On the other hand, with increasing of $Ma$, $S$ increases firstly and then decreases, like a quadratic relation. $S$ reaches the maximum when $Ma$ is ~5.5. If $Ma$ is too large, the generation of baroclinic vortex will be limited due to the decrease of shock standoff distance, resulting the decrease of $\eta$ and $S$. If $Ma$ is too small, meaning the low flow speed, the hot core will expand too large before interacting with bow shock, resulting in the decrease of $S$.

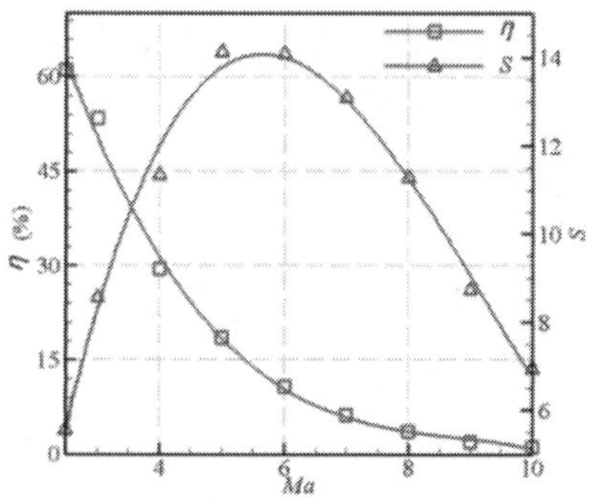


Figure 8. Influence of freestream $Ma$ on performance of laser energy deposition

## 4 Summary

Buckingham pi theorem is a powerful tool in analyzing numerical experiments of wave drag reduction with high-frequency repetitive laser pulse energy depositions. Dimensional analysis shows that non-dimensional energy $\varepsilon$, non-dimensional depositing position $L/D$ and Mach number $Ma$ are the key non-dimensional parameters. $\varepsilon$ has an exponential relation with $\eta$ and $S$; $L/D$ has a quadratic relation with $\eta$ and $S$; $Ma$ has exponential and quadratic relation with $\eta$ and $S$ respectively. Influence of these three key non-dimensional parameters on laser deposition performance intertwines each other and integrated optimization method is proposed.

## References

1. R. G. Adelgren, H. Yan, G. S. Elliott, D. D. Knight, T. J. Beutner, A. A. Zheltovodov, AIAA J., **43**, 2 256-269 (2005)
2. E. Erdem, L. Yang, K. Kontis, AIAA 2011-1027 (2011)
3. J. H. Kim, A. Matsuda, T. Sakai, A. Sasoh, AIAA 2010-5104 (2010)
4. L. N. Myrabo, Yu. P. Raizer, M. N. Shneider, Beamed energy propulsion: Third international symposium on beamed energy propulsion, *AIP Publishing*, 485-498 (2005)
5. P. K. Tret'yakov, G. N. Grachev, A. I. Ivanchenko, et al., Dokl. Phys., **39**, 6 415-416 (1994)
6. A. Sasoh, Y. Sekiya, T. Sakai, J. H. Kim, A. Matsuda. AIAA J., **48**, 12 2811-2817 (2010)
7. D. Riggins, H. F. Nelson, E. Johnson., AIAA 98-1647 (1998)
8. S. Taguchi, N. Ohnishi, M. Furudate, et al, AIAA 2007-1235 (2007)
9. Y. Ogino, N. Ohnishi, S. Taguchi, K. Sawada, Phys. Fluids., **21**, 6 (2009)
10. V. A. Levin, N. A. Afonia, V. G. Gromov, AIAA-99-4967 (1999)
11. T. Taylor, AIAA 2004-131 (2004)

# Research on and design of key circuits in RFID tag chip for container management

Wenjie Wang

*Tianjin Transportation Vocational College, China*

Abstract: This paper introduces the design of semi-passive RFID tag chip capable of monitoring container safety. A system framework complying with requirements by ISO/IEC 18000-6C is firstly presented, and then differences from the key units of common passive chip, such as switch-state monitoring circuit, power management unit and anti-shake design in baseband processor, are elaborated. The main function of such a chip is to record the container opening frequency during transportation. Finally, the realizations of each unit's function are simulated.

## 1. Introduction

With the development of RFID technology, mere identification of target objects can not satisfy the various application needs; thus, embedding a sensor unit into the RFID tag chip to monitor the environment of objects has great significance in the application fields, such as logistics and supply chain management. In recent years, several RFID tag chips inserted with temperature sensors have been proposed [1]-[4]. To meet the specific requirement in the application fields, tag chips with various special features have become a research focus and developing trend of the RFID technology. However, no such RFID tags that can monitor the transportation safety and the logistics industry where RFID is widely used urgently needs have been developed.

## 2. System Framework

Fig. 1 is the block-diagram of the chip system [5], [6], including Power Management Unit (PMU), Switch-state Monitoring Unit (SSM), ASK demodulator, modulator, POR, oscillator, baseband processor and memory. The PMU consists of recertification circuit, voltage regulator circuit, and wake-up circuit. The rectification circuit converts the received RF energy into DC voltage which will then be stabled by the voltage regulator circuit at 1V and 1.8V to support for the operation of other units; while the wake-up circuit controls the power supply based on the strength of chip's received RF energy, which increases the effective working distance and concomitantly minimizes the battery consumption. SSM monitors the state of the container switch through initiating battery activation tag and sending a signal "Son" when the switch state changes. Then the baseband processor starts to count when receiving the signal "$S_{on}$", and stores the counted values in the memory; after that, the processor will send signal "$P_{ctl}$" to switch off the battery, following which the chip will turn to sleep. Functions and designs of the recertification circuit, voltage regulator circuit, ASK demodulator, modulator, oscillator and POR are the same as those of passive chip. What different from the common passive chip are the designs of PMU, baseband processor and SSM, which will be elaborated in the following.

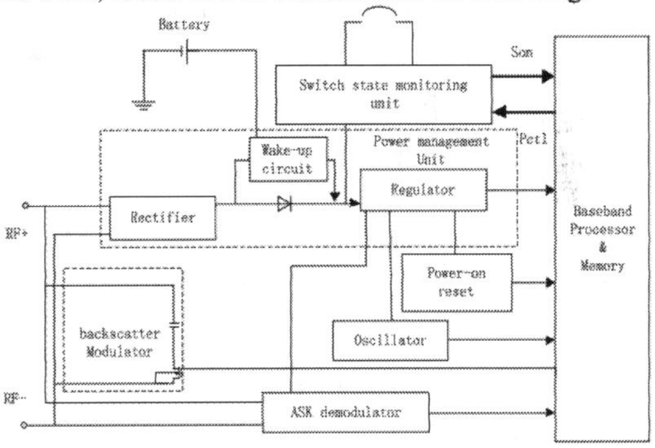


**Figure 1**. Block diagram of a special label system for container management

## 3. Ultra-low-consumption Switch-state Monitoring Circuit

Fig. 2 (a) is a schematic diagram of the switch-state monitoring circuit with cores of switching tube T1 and control logic. Switch S, representing the state of the container switch, jointly controls the state of the battery with T1. When S and T1 are simultaneously turned on, the battery will provide power to chip to make it work. Inverters Inv2 and Inv3 can transform the levels of signals, changing the signal $S_{on}$ level to 1V. Other control logic units are directly powered by the battery. Therefore, the high level of signals "$P_{ctl}$" and "$S_{on}$" is 1V, and 3V for others (CR2032 battery voltage).

Fig. 2 (b) shows the sequence diagram of the switch-state monitoring circuit. When the switch S disconnects, the chip cannot get battery power with $P_{ctl}$ of low level, $V_{en}$ of high level, $V_g$ and $V_a$ of same low level, so T1 is in conducting state. When the switch S closes, $V_{en}$ is low and the latch D stops working and T1 remains conductive, then the battery can supply power to the chip to make it work through the switches S and T1, and the level of $S_{on}$ becomes higher, baseband processor starts counting and send $P_{ctl}$ to turn off T1 to switch off the battery, and the tag turns to sleep [7]. Thus, whenever the switch state is changed, the chip only works several milliseconds for counting in the whole process, i.e. $V_b$ is at a high level in Figure. 2 (b), and only control circuit of SSM keeps working during the remaining time, which greatly extends the service life of the battery.

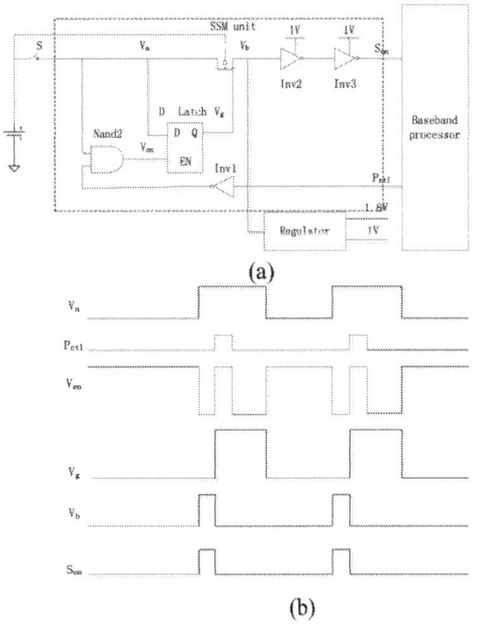


**Figure 2**.Switch status monitoring circuit

(a) switch status monitoring circuit schematic; (b) main signal timing diagram

## 4. Power Management Unit

The PMU consists of rectification circuit, voltage regulator circuit, and wake-up circuit; among which the former two have the same functions as those of the passive chip. The wake-up circuit is a specific module for passive chip that can change the power supply mode based on the strength of received RF energy; which can increase the working distance of chip on the one hand and decrease the consumption to extend the service life [8], [9] of battery on the other hand.

The wake-up circuit consists of two independent sub-circuits that control the two output terminals of rectification circuit respectively. Fig. 3 shows the schematic diagram of one circuit in which $V_{REC}$ connects with the output terminal of rectification circuit, $V_{bat}$ with the positive electrode of battery, and $V_O$ with the input terminal of voltage regulator circuit. The serial resistance divider (which can be diode-connected MOS instead) connecting with $V_{REC}$ divides the output voltage of rectification circuit proportionately that generates two voltage signals i.e. $V_{sel1}$ and $V_{sel2}$. $V_{sel1}$ equaling to $0.75V_{REC}$ connects with inverter INV1, while $V_{sel2}$ equaling to 0.25 $V_{REC}$ connects with inverter INV2. Threshold voltages of the two inverters are represented by $V_{m1}$ and $V_{m2}$ respectively. Only when $V_{sel1} > V_{m1}$ and $V_{sel2} < V_{m2}$ (namely $4 V_{m1}/ 3 < V_{REC} < 4V_{m2}$), the grid potential of PMOS MP is low, and MP is turned on, then the battery can feed the voltage regulator circuit that will then supply power to chip with 1V and 1.8 V. In other cases, the battery will be turned off, namely only the output voltage $V_{REC}$ is within the specified range said above can the battery supply power. The minor value ($V_{min}$) is determined by the sensitivity of ASK demodulator. For this design, the minimum RF signal amplitude that the demodulator can correctly demodulate is 300mV; to leave a certain margin, the output voltage of the rectification circuit when the input RF signal amplitude is 160mV is set as $V_{min}$; the major value ($V_{max}$) is able to meet the minimum required output voltage for normal operation [10]. With this design, the battery will be turned on only when the RF energy received reaches the sensitivity of ASK demodulator but not enough to activate the chip; which can increase the working distance of the chip with no waste of battery energy. In this design, the battery will be turned on when $0.3V < V_{RECL} < 1.1V$ or $0.4V < V_{RECH} < 2V$.

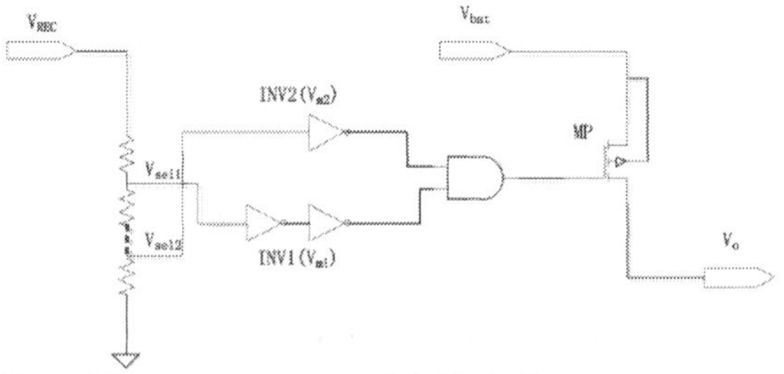


**Figure 3**.Power management module block diagram

## 5. Baseband Processor

The special tag baseband processor herein shown in Fig. 4, consists of decoder, encoder, CRC module, timing synchronization module, memory read-write control module, random-number generator, collision arbitration module and finite state machine, etc.; among which, the finite state machine is the core that consists of de-jitter circuit, switching frequency recording module, state-machine-enabled module, synchronization circuit, command processing module and Mealy-type finite state machine. In addition to performing a series of commands and operations specified by protocol, the processor is also required to record the switching frequency. As there is dither signal that may cause error count when a

DOI: 10.1051/matecconf/20167510002

mechanical switch is switching, this paper designs a de-jitter circuit to eliminate the effects.

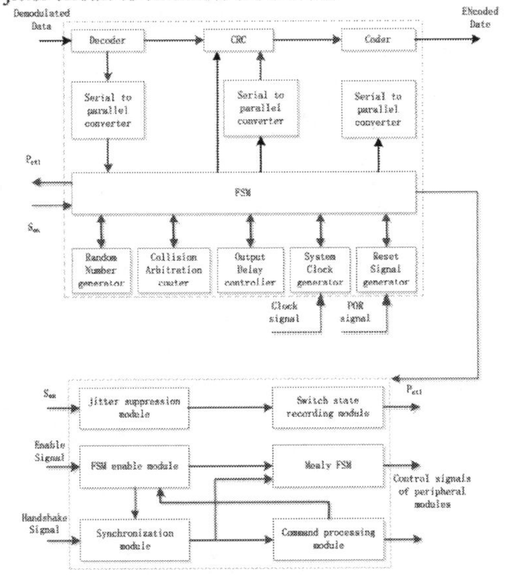


**Figure 4**.Block diagram of baseband processor system

The fundamental principle for the de-jitter circuit is to use a bidirectional counter, as shown in Figure 5, that can sample signals "$S_{on}$", and conduct addition or subtraction to the counter values respectively based on the high level or low level of "$S_{on}$" [11]-[13]. If the signal "$S_{on}$" is under high level, the counter will count the clock pulses and pull the "$S_{ons}$" high when the count value reaches the set standard. Conversely, if the signal "$S_{on}$" is under low level, the counter will do the subtraction to the count value till zero and pull the "$S_{on}$" low. As shown in Fig. 5, this method effectively eliminates the effects by dither signals and improves the system reliability.

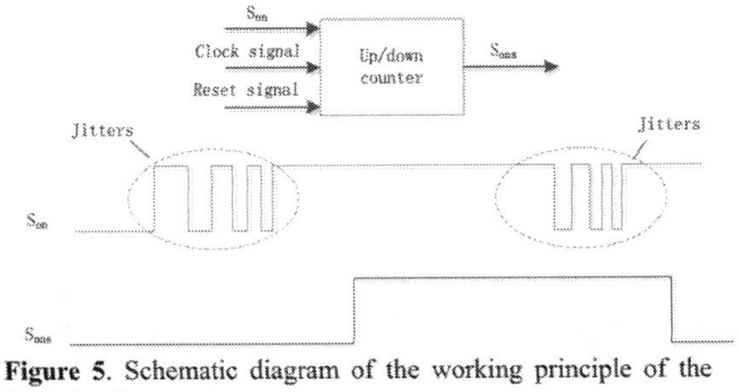


**Figure 5**. Schematic diagram of the working principle of the jitter circuit

## 6. Simulation Analysis

The chip adopts TSMC 0.18μm CMOS design, and Figure 6 shows the complete chip layout. The analog circuit uses Spectre for simulation analysis, and the chip consumes 9.56μW while working.

Simulation results of SSM are shown in Fig. 7. The delay time of the unit is 92μs and will not have negative effects on the system due to the low switching frequency. Part of the control logic of SSM and PMU is directly fed by the battery, so there is still very low power consumption while the tag is in sleep mode, and the quiescent current under this mode is only 645 pA according to the simulation results

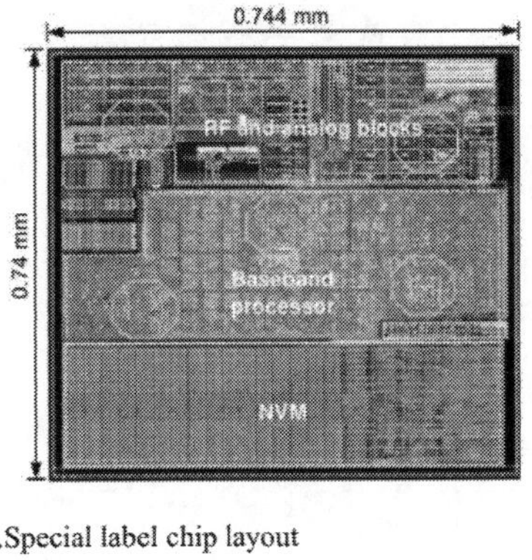


**Figure 6** .Special label chip layout

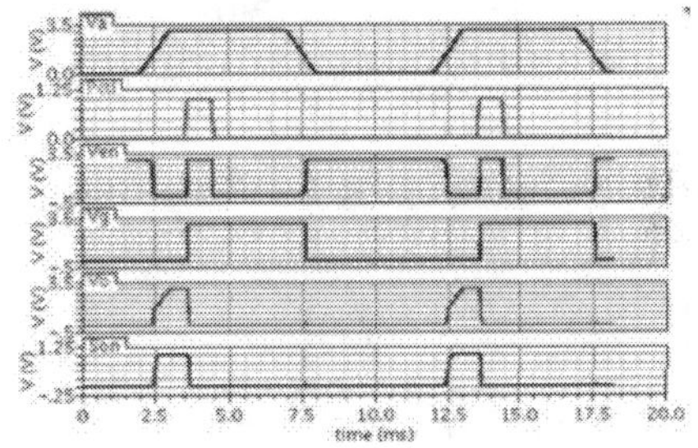

**Figure 7**.The simulation waveform of the key signal in the switch state monitoring circuit.

Fig. 8 shows the co-simulation waveform of PMU, rectification circuit and voltage regulator circuit, wherein $V_{RECL}$ is the low output voltage of the rectification circuit, Reg_out_low is the low output voltage of the voltage regulator circuit (1V), and Reg_in_low is the low input voltage of the regulator circuit. It is obvious that Reg_in_low is zero when the rectification output voltage is very low initially, and the battery is turned off; then, as the rectification output voltage rises, when 0.3V $<V_{RECL}$ <1.2V, namely the tag is in the RF field but receives insufficient RF energy, the battery will be turned on to feed the Reg_in_low (Reg_in_low is the battery voltage of about 3V); Finally, when the rectification output voltage rises and stabilizes at about 1.4V, as $V_{RECL}>$ 1.2V and the battery turns off, the voltage regulator circuit will still be activated by the rectification circuit (Reg_in_low = $V_{RECL}$).

As the diode-connected MOS plays as the serial resistance divider of wake-up circuit as well as inverter, the deviation between the divided voltage proportion and the inverter threshold is small due to the same effects on MOSs from TT, FF and SS process corners and different temperatures. However, since the fluctuations of supply voltage will affect the values of the divided voltages $V_{sel1}$ and $V_{sel2}$ of rectification output voltage, and the threshold voltage of inverter, the circuit will be affected by changes of the operating voltage (battery voltage) to some extent

[14-16].It shows the simulation results of the turning on/off threshold voltages of the wake-up circuit under different battery voltages. Wherein, bat_low_on and bat_low_off are the turning on threshold voltage and turning off threshold voltage of the wake-up circuit of regulated low input terminal respectively; while bat_high_on and bat_high_off are the turning on threshold voltage and turning off threshold voltage of the wake-up circuit of the regulated high input terminal respectively. Visibly, as the battery voltage increases, the threshold voltage of the wake-up circuit will increase to some extent, but the fluctuations of the turning on/off thresholds will not affect the normal operation of the chip [17].

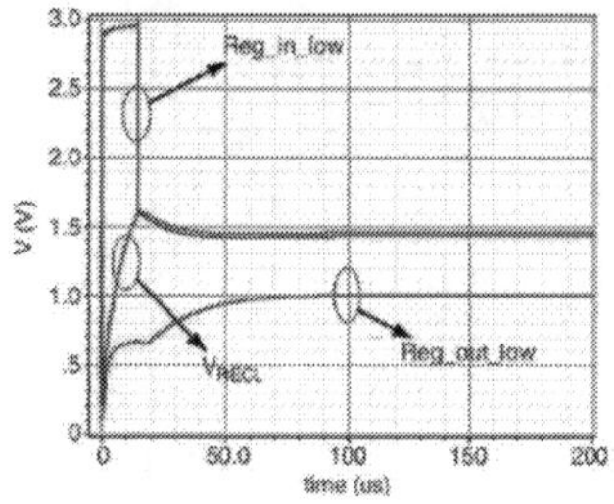


**Figure 8**. Power management module simulation waveform

## 7. Conclusion

With the wide application of RFID technology, people put forward higher demands for the functions of RFID chips, such as monitoring the environment of objects through embedding the sensors in the chip, which is significant to logistics and supply chain management and other fields. This paper introduces a design of special tag chip for container management. It can record the opening frequency of container during transportation and comply with requirements by ISO/IEC18000-6C. The major difference between this chip and the common passive tag chip lies in the system framework, switch-state monitoring unit, power management unit and anti-shake design in baseband processor. Thus, this study focuses on those abovementioned aspects, and the corresponding simulation results are discussed.

## References

1. K. Souri, K. A. A. Makinwa. A 0.12 mm2 7.4μW micropower temperature sensor with an inaccuracy of ±0.2℃ (3σ) From -30℃ to 125℃[J]. IEEE J. Solid-State Circuits, 2011, 46(7): 1693–1700
2. D. Yeager, F. Zhang, A. Zarrasvand, et al.. A 9.2μA gen 2 compatible UHF RFID sensing tag with -12dBm ensitivity and 1.25μVrms input-referred noise floor[C]. IEEE International Solid-State Circuits Conference Proceedings, San Francisco, 2010, 52–53
3. Kong Jin'ou. Maxwell Equation. Beiing: Higher Education Press. 2004
4. Y. Wang, G. Wen, W. Mao, et al.. Design of a passive UHF RFID tag for the ISO/IEC18000-6C protocol[J]. 2011, J. Semicond., 32(5): 055009
5. Impinj,Inc.. Gen 2 tag clock rate—what you need to know[EB/OL]. .im in . n/a i ati ns/ Ta C atc 200 1001. d , Oct 1, 2005
6. Wang Zheng. Studies on Air Interface Protocol and System Design for UHF RFID [D]. Tianjin: Tianjin University, 2011, 102-108
7. G. De Vita, F. Marraccini, G. Iannaccone. Low-voltage low-power CMOS oscillator with low temperature and process sensitivity[C]. IEEE International Symposium on Circuits and Systems, New Orleans, 2007, 2152–2155
8. R. Barnett, J. Liu. A 0.8V 1.52MHz MSVC relaxation oscillator with inverted mirror feedback reference for UHF RFID[C]. IEEE Custom Integrated Circuits Conference, San Jose, 2006, 769–772
9. Y.Wang,J. Liu, L. Xie, et al.. An ultra-low-power oscillator with temperature and process compensation for UHF RFID transponder[J]. Radioengineering, 2013, 22(2): 505-510
10. B. Razavi. Design of analog CMOS integrated circuits[M]. New York: McGraw-Hill, 2000.
11. K. Ueno, T. Hirose, T. Asai, et al.. A 300 nW, 15 ppm/℃, 20 ppm/V CMOS voltage reference circuit consisting of subthreshold MOSFETs[J]. IEEE J. Solid-State Circuits, 2009, 44(7): 2047–2054
12. F. Aghlmand, M. Atarodi, S. Saeedi. Low phase noise on-chip oscillator for implantable biomedical applications [C]. IEEE International Symposium on Circuits and Systems Proceedings, Rio De Janeiro, 2011, 213-216
13. S. Ayazian, V. A. Akhavan, E. Soenen. A photovoltaic-driven and energy-autonomous CMOS implantable sensor[J]. IEEE Transactions on Biomedical Circuits and Systems, 2012, 6(4): 336-343
14. K.Choe, O. D. Bernal, D. Nuttman, et al.. A precision relaxation oscillator with a self-clocked offset-cancellation scheme for implantable biomedical SoCs[C]. IEEE International Solid-State Circuits Conference Proceedings, San Francisco, 2009, 402-403
15. H. Okada, T. Itoh, T. Masuda. Development of custom CMOS LSI for ultra-low power wireless sensor node in health monitoring systems[C]. IEEE Sensors Proceedings, Limerick, 2011, 1197-1200
16. Wangyao, Wangguangjun. Demodulation circuit for ultra high frequency radio frequency identification tag chip[P]. China, invention patent, ZL.201010568305, April 30, 2012 Wang Yao, Wen Guangjun. Demodulation Circuit for Ultral High Frequency RFID Tag Chips [P]. China, Patent of Invention ZL.201010568305, April 30, 2012.
17. R.E.Barnett. High efficiency RF to DC conversion and ultra-low-power analog front end circuits for low-cost field-powered UHF RFID[D]. Dallas: University of Texas at Dallas, 2007, 30-36

# New Broadband Predistortion Linearizer Using Balanced Circuits Structure for Satellite TWTAs

Changwei Chen, Yong Cao and Kaiyu Qin
*Institute of Astronautics and aeronautics Science and Technology, the University of Electronic Science and Technology of China*

**Abstract.** The power amplifies is a key component influencing the power efficiency of communication system. The travelling wave tube is widely used in satellite system because of its high efficiency, apposite weight and broadband characteristic, but its strong nonlinear characteristic generates intermodulation distortion, which degrades the data quality of transmitted signal and causes interference in adjacent channels. The analogue predistortion structure using Schottky diodes could improve the linearity well, but parasite element in the Linearizer limits predistortion broadband, reluctant to satisfy the growing high volume data transmission requirement. This paper presents a new balanced structure of Linearizer for the travelling wave tube amplifier. We have analyzed the model of linearizer and the TWTA and used to calculate the intermodulation distortion and optimize the Linearizer nonlinear expansion characteristic. A frequency bandwidth of 1.2 GHz Linearizer has been demonstrated with a K-band TWTA between 19.8 and 21.0 GHz. The measurement result shows that the linearizer delivers about 2~6dB of gain expansion and 20~60 degrees phase shift expansion, and the cascaded LTWTA shows good linearity improvement of 17.8dB at 6dB input power back off point.

## 1 Introduction

With the rapid growth of high volume data in satellite communication systems, the modern millimeter wave satellite communication use non-constant envelope digital modulation techniques and demand for the wideband, high efficiency, and linear high-power amplifiers. The traveling wave tube amplifier is widely used in space application because of its high efficiency [1], but it has strong nonlinear characteristic. The nonlinearity generates intermodulation distortion (IMD), which causes interference in adjacent channels. To compensate these nonlinear distortions, many linearization techniques have been proposed, such as feedback, analog predistortion, feedforward, baseband digital predistortion. Among these technique such as baseband prediction, Practical DSP-based linearization systems of baseband predistortion can give a satisfactory IMD correction, but it is hard to reach bandwidth over 100MHz because of its complexity, cost, power consumption, and availability of required DSP components [2], [3]. The feedforward technique is too complex to implement on satellite.

Analogue predistortion technique has been the only choice in space applications because of their low power consumption, compactness, high reliability, and wide absolute bandwidth and easy to use.

There has been lots of analog predistortion structure proposed using Schottky diodes [4], and [5] or FETs [6] as the nonlinear device to produce the Inverse nonlinear distortion. It is relatively difficult to design a broad band Linearizer with parasite element in the Linearizer. The Linearizer bandwidth in [7, 8] is only 400MHz.In this paper, a broadband millimeter wave predistortion TWT linearizer is presented. The linearizer use Symmetrical structure, which expands the bandwidth and improves the linearizer input/output VSWR, and any additional isolator isn't required for matching.

## 2 Linearizer characterize analysis

### 2.1 The analysis and compensation of the TWTA nonlinearity

The TWTA is a memoryless and strong nonlinear amplifier, and the signal is treated as narrow band-limited signal, Let the input signal be

$$x(t) = r(t)\cos[\omega_0 t + \phi(t)] \tag{1}$$

where $\omega_0$ is the carrier frequency, and r(t) and $\psi$(t) are the modulated envelope and phase, respectively. The corresponding output is written as

$$y(t) = A[r(t)]\cos\{\omega_0 t + \phi(t) + \Phi[r(t)]\} \tag{2}$$

The output voltage can be expressed function of input signal with Saleh model [9] as eq. (2). The model presents nonlinear models of TWT amplifiers that are based on a simple two-parameter formula for each of the four aforementioned functions. For each of several cases examined these formulas fit TWT measurements accurately.

$$\begin{cases} A[r(t)] = \dfrac{\alpha_a r(t)}{1+\beta_a r^2(t)} \\ \Phi[r(t)] = \dfrac{\alpha_\varphi r^2(t)}{1+\beta_\varphi r^2(t)} \end{cases} \quad (3)$$

The Predistortion Linearizers are widely used in TWT amplifier linearization at microwave and higher frequencies. The Linearizer is cascaded before the TWT amplifier which provides the opposite nonlinear transfer response to TWT amplifier both in amplitude and phase characteristic as shown in Fig. 1. The gain expansion of the Linearizer compensates the gain expression of the TWT amplifier, and the phase increase of the Linearizer cancel the phase change of the TWT amplifier.

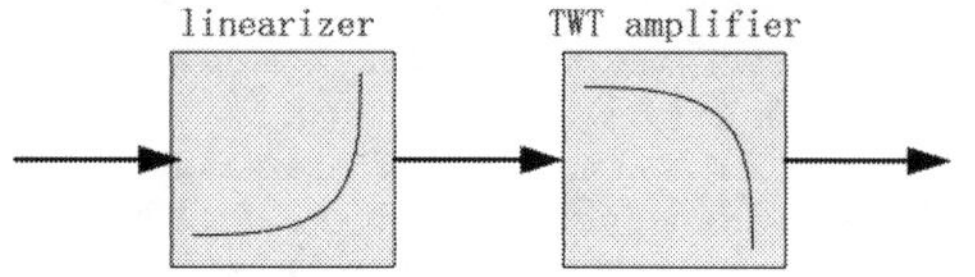


**Figure 1.** Predistortion Linearizer transfer characteristic schematic diagram

Using cascade net theory, the ideal AM-AM and AM-PM characteristic of predistortion linearizer is deduced, the ideal predistortion linearizer model is as (4) and (5)

$$r(t) = \begin{cases} \dfrac{\alpha_a - \sqrt{\alpha_a^2 - 4\beta_a \rho^2(t)}}{2\beta_a \rho(t)}, 0 < \rho(t) \le G_s \\ G_s, G_s \le \rho(t) < \infty \end{cases} \quad (4)$$

$$\theta(t) = \frac{-\alpha_\phi \left(\alpha_a - \sqrt{\alpha_a^2 - 4\beta_a \rho^2(t)}\right)^2}{4\rho^2(t)\beta_a^2 + \beta_\phi \left(\alpha_a - \sqrt{\alpha_a^2 - 4\beta_a \rho^2(t)}\right)^2}$$

$$0 \le \rho(t) \le \infty \quad (5)$$

where, $G_s = \dfrac{1}{\sqrt{\beta_a}}$ , $\rho(t)$ is the input voltage of the linearizer, $r(t)$ is the output voltage of the linearizer, $\theta(t)$ is the phase shift when the signal pass through the linearizer, $\alpha_a, \beta_a, \alpha_\varphi, \beta_\varphi$ is the parameter of the Saleh model.

For simplifying the IMD calculation, the nonlinear characteristics could be expressed as complex power series:

$$v_{in} = g_1 A_{in} + g_3 A_{in}^3 + g_5 A_{in}^5 \quad (6)$$

Employing standard minimum mean-square-error curvefitting procedure, one obtains the required optimum values of g1, g2 and g3 of the complex power series. Using the quadrature model of TWTA [6] calculate the IMD of the cascaded linearized TWTA, optimized the IMD and get the optmum inverse characteristic of TWTA as the design goal of the linearization design. The optimum value of Linearizer gain expansion and phase expansion is 3.8dB and 37°. Accurate gain and phase expansion is required for satisfactory IMD correction in band. The IMD curve of the linearized TWTA showed as the Fig. 2 after optimum.

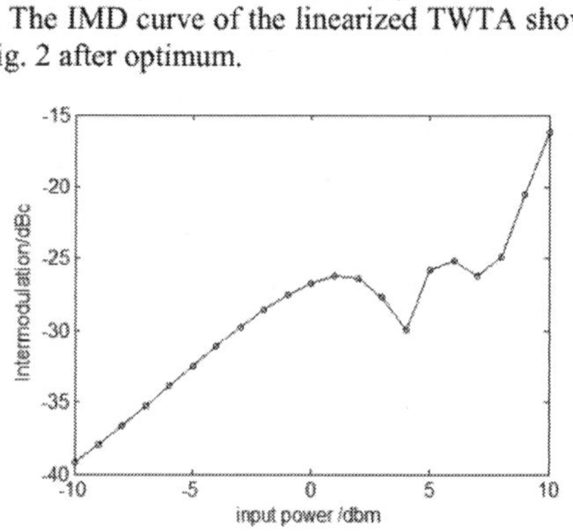


**Figure 2.**optimum IMD curve of the linearized TWTA.

## 2.2 Reflective schottky diode predistorter using balanced structure

The schematic diagram of the proposed predistortion Linearizer is shown in Fig. 3.

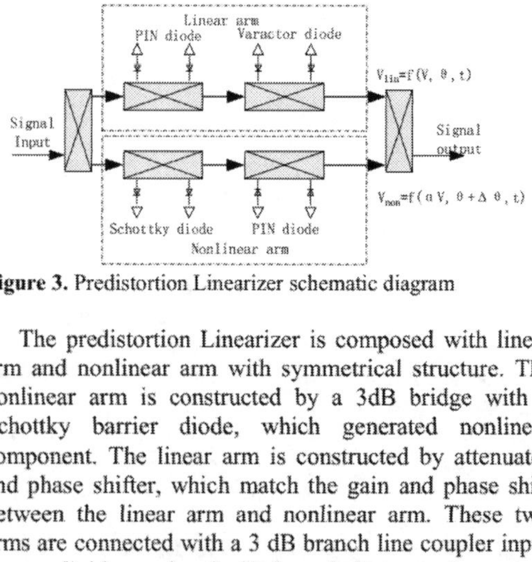


**Figure 3.** Predistortion Linearizer schematic diagram

The predistortion Linearizer is composed with linear arm and nonlinear arm with symmetrical structure. The nonlinear arm is constructed by a 3dB bridge with a Schottky barrier diode, which generated nonlinear component. The linear arm is constructed by attenuator and phase shifter, which match the gain and phase shift between the linear arm and nonlinear arm. These two arms are connected with a 3 dB branch line coupler input power divider and a 3 dB branch line coupler output power divider. This Symmetrical structure improves the bandwidth and VSWR of the Linearizer.

In the predistortion Linearizer schematic diagram, the $v_{LIN}$ is the output signal from the linear arm, it is a function of the input amplitude, phase and the group delay. The linear arm output signal could be written

$$V_{LIN} = f(v, \theta, t) = V[\cos(\omega t + \theta)] \quad (7)$$

The signal $v_{NON}$ is the output signal from the nonlinear arm. After the adjust of the amplitude and phase, it could be written as

$$V_{NON} = f(\alpha v, \theta + \Delta\theta, t + \Delta t)$$

$$= \alpha V[\cos(\omega t + \theta + \Delta\theta)] \quad (8)$$

where, the variable $\alpha$ is amplitude ratio between the linear arm and nonlinear arm. It is decided by the nonlinear generator which is composed of schottky diode, and attenuator composed of PIN diode. It is a function of input signal amplitude.

The variable $\Delta\theta$ is phase difference between the linear arm and nonlinear arm. It is decided by the nonlinear generator and phase shifter. It is a function of input signal amplitude.

When the two signals are combined, the result is

$$V_{out} = V[\cos(\omega t + \theta) + \alpha \cos(\omega t + \theta + \Delta\theta)] \quad (9)$$

Using the trigonometric function transformation, the formulas (8) could be converted as

$$V_{out} = gV[\cos(\omega t + \theta + \phi)] \quad (10)$$

where,

$$g = \sqrt{1 + \alpha^2 + 2\alpha \cos\Delta\theta} \quad (11)$$

$$\phi = \arctan \frac{\alpha \sin\Delta\theta}{1 - \alpha \cos\Delta\theta} \quad (12)$$

The vector diagram of Fig. 4 illustrates the subtracting a linear arm signal from a nonlinear arm signal at high and low amplitude level. The gain of the nonlinear arm $V_{NON}$ decreases as the input power increases. Appropriate adjustment of the amplitude ratio $\alpha$ and angle difference $\Delta\theta$ between the linear and nonlinear arm allows the output signal level change in positive direction same as required expansion transfer function.

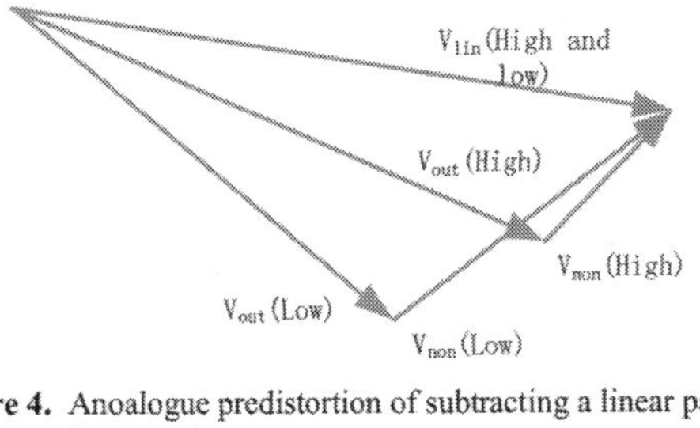


**Figure 4.** Anoalogue predistortion of subtracting a linear path from a nonlinear path

In the section 2.1, satisfactory broad bandwidth IMD correction needs consistent gain and phase expansion in band. From the formula (10) to (12), the output signal gain and phase is associated with the amplitude ratio $\alpha$ and angle difference $\Delta\theta$ between the linear and nonlinear arm, regardless of the flatness of the linear and nonlinear arm.

Parasite element in the Linearizer deteriorates the gain and phase flatness of Linearizer linear or nonlinear arm. The symmetrical structure of the Linearizer could be used to keep the amplitude ratio $\alpha$ and angle difference $\Delta\theta$ consistent in the band under influence of the parasite element in the Linearizer.

The photograph of the manufactured Linearizer module is shown in Fig. 5. The Linearizer is realized in a very compact size of 18×6mmm and its substrate used is alumina with the dielectric constant of 9.9 and thickness of 0.254mm. The used schottky diode is HSCH-5332 of Agilent. The PIN diode used in attenuator is HPND4005 of skyworks. The Silicon hyperabrupt junction varactors diode used in phase shifter, which has about 60°adjust range, is SMV2019 of skyworks.

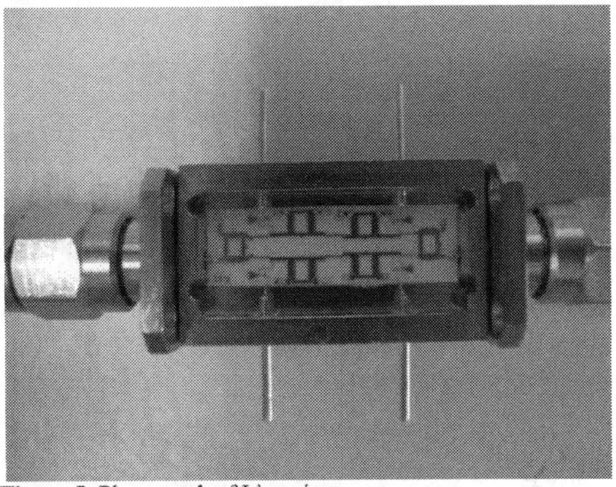

**Figure 5.** Photograph of Linearizer

## 3 Experimental results

The AM-to-AM and AM-to-PM of the calculated above is realized, and could be tuned with adjusting the bias the control voltage of the attenuator and the phase shifter. The shape tuning capability [8] could be gotten through the bias with the control voltage of the nonlinear generator. The AM-to-AM and AM-to-PM expanding nonlinear characteristic matched with TWTA nonlinear curve is a key to the linearization.

The AM-to-AM and AM-to-PM expanding nonlinear characteristic has been tested using Agilent Technologies' PNX series large signal network analyzer N5244, the AM-to-AM and AM-to-PM curve result figure is shown in Fig. 6. The measurement result shows that the Linearizer delivers about 2-6dB of gain expansion and 20-60 degrees phase shift expansion characteristic and could be adjustable and be consistent in band.

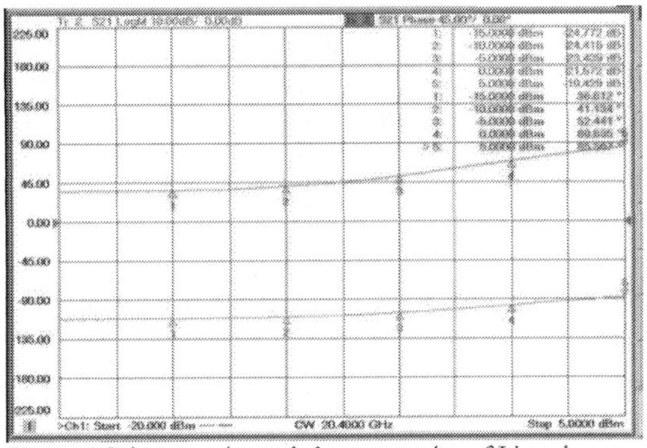

**Figure 6.**Gain expansion and phase expansion of Linearizer

The output power transfer characteristic of the TWTA with and without lineariztion has been tested using Agilent Technologies' signal generator and power meter the curve result figure is shown in Fig. 7. The power sweep from 0dB IBO (input power back off) to 20 dB IBO. The output power transfer characteristics figure shows the good performance of the Linearizer.

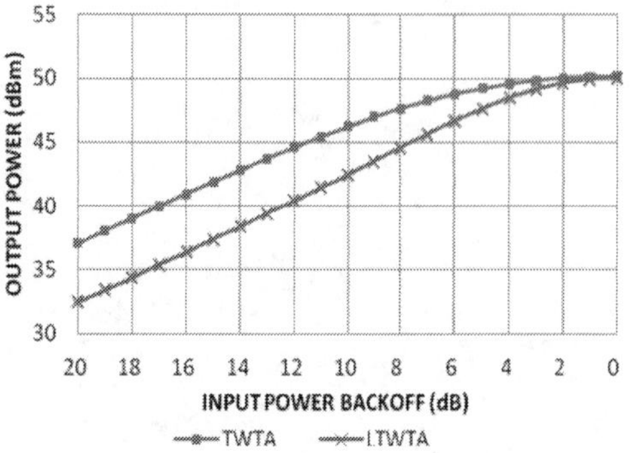


**Figure 7.** Output power transfer characteristics of the TWTA with and without lineariztion

The IMD of the linearised TWT amplifier has been tested using Agilent Technologies' PNX series large signal network analyzer N5244, Two-tone IMD performance of TWT amplifier with TWTA and TWTA with predistorter and feedforward amplifier was tested separately. The tone spacing is selecting to be 5MHz. The IMD result figure is shown in Fig. 8.

The performance of the Linearizer is evaluated with a 100 W K-band TWTA between 19.8 and 21.0GHz. An amplifier CHA3689 and an attenuator CHA3091 of UMS are used between the Linearizer and the TWTA is used to compensated the loss of Linearizer and adjust the TWT required input power.

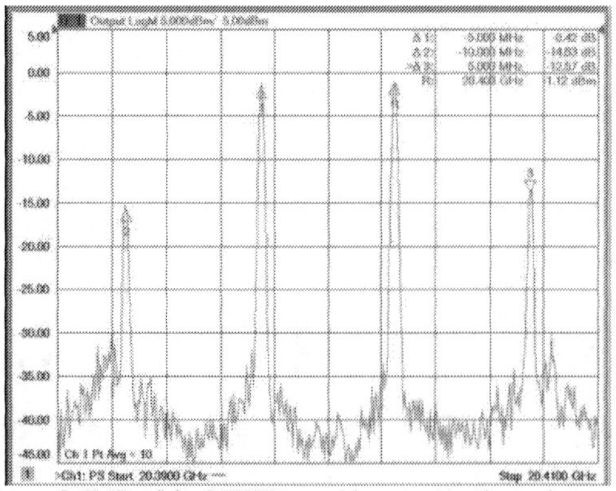

**Figure 8.** IMD of the LTWTA working at saturation point

The cascade linear TWTA gain compression is compensated to 0.8dB. The phase shift compression is compensated to -5.8°at the input saturation point. The IMD performance of the LTWTA is improved, which 2.1dB at 0dB IBO, and 17.8dB improvement at 6.5dB IBO which is improvement optimum point as shown in Fig. 9.

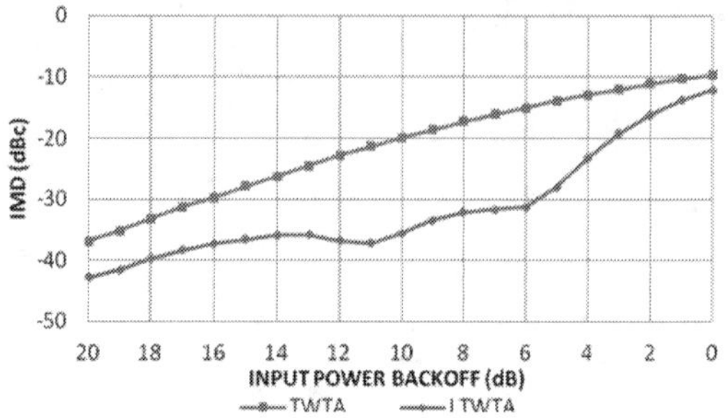


**Figure 9.** The IMD performance of the TWTA and linearised TWTA

## 4 Conclusions

The model of the nonlinear TWT and Linearizer is established and predict the optimum value of Linearizer gain expansion and phase expansion with quadrature model of TWTA. With the optimum value of Linearizer, A two-way linearizer signal subtraction is analyzed and gives the condition of accurate nonlinear gain and phase expansion. Using balance structure of linearizer, the parasite element influence is cancelled and extends the bandwidth of the linearizer. Good input and output VSWR characteristic make it connect to amplifier without isolator due to the symmetrical structure. The cascaded LTWTA shows good nonlinearity improvement of 2.1dB at 0dB IBO and 17.8dB at 6.5dB IBO.

## References

1. E. Cuignet, E. Tonello, J. Maynard, P. Boone, *IVEC2013*, 14th ,Pag: 1 – 2(2013).
2. A. Katz, R. Gray, R. Dorval, IEEE micro. Mag., J.**10**, 7, pp. 20~27 (2009).
3. X. Feng, B. Feuvrie, A.S. Descamps, Y. Wang, Electr. Lett., J., 50, (24), pp. 1882–1884(2014).
4. A. Katz, R. Gray, R. Dorval, IEEE Trans. Electr. Dev., 56, 5, pp. 959 - 964 (2009).
5. A. Katz, R. Sudarsanam, C. Aubert, MTTS Int. Microw. Symp. Dig., pp. 661–664 (1985).
6. A. Katz, R. Domal, IEEE MTT-S Digest, pp. 437–440 (1994)
7. H.Y. Jeong, S.K. Park, N.S. Ryu, Y.C. Jeong, *EGAAS*, pp. 597-600(2005).
8. S.C .Bera, V. Kumar, S. Singh, D.K. Das, **23**, 4, pp. 211 – 213(2013).
9. A. A. M. Saleh, Tans. on Comm., **29**, 11, pp. 1715–1720. (1981).

MATEC Web of Conferences 75, 10004 (2016) DOI: 10.1051/matecconf/20167510004
*ICMIE 2016*

# A Study on High Density Gate-Oxide Anti-Fuse PROM Memory Cell Program Features

Keqin Yang, Suge Yue and Shijin Lu

*Center of Radiation Hardening, Beijing Microelectronic Institute of Technology, China*

**Abstract.** The program consistency of high density gate-oxide anti-fuse PROM (programmable read only memory) memory cell is considered in this paper. To solve this problem, we do research on the mechanism and models of gate-oxide break down. A test chip based on 2T memory cell is also designed for the experiment. During the test, we have found that the program consistency of 2T cell is really pessimistic. What's more, the program voltage has effect on the consistency. Through research and test, we have got the optimal program voltage-6.5V for 2T memory cell in 180nm technology. To further improve the consistency, we modify the 2T memory cell and propose a 3T memory cell. It consumes 18% more area, but the standard deviation of equivalent resistance decreases 15.3% in the worst situation. The deviation can decrease 80.3% at most. The program consistency is greatly improved.

## 1 Introduction

Anti fuse is a very significant kind of programmable interconnected cell. The semiconductor device based on anti fuse has very advanced performance in the following aspects(1)Non volatile cell(2)radiation hardened cell (3)high reliability (4) confidentiality(5)100% testability (6)small area, high velocity and low power consumption [1]-[3].

At present, the most typical and mature anti fuses are ONO (Oxide-Nitride-Oxide) type, α-Si (Amorphous Silicon) type and gate oxide type. Compared with ONO and α-Si anti fuse, gate oxide anti fuse has its particular advantage: it is entirely compatible with the standard CMOS technology. This advantage makes the cost decreased greatly. The design procedure also becomes more convenient.

PROM is a kind of high reliable device which is applied in many important fields. It has cruel requirements on the memory cell's performance and reliability. Conventional PROM uses e-fuse memory cell. E-fuse memory cell may get together again after the cell is has been fused. The cell which is not programmed may also be fused due to high current during the read motion. These problems make poor reliability of e-fuse PROM.

Gate-oxide anti fuse is a good choice for PROM memory cell. Compared with e-fuse memory cell, gate-oxide anti fuse memory cell improves a lot in reliability, anti-radiation performance and confidentiality. Gate-oxide anti fuse's performance corresponds to the ONO type's and α-Si type's and it can cut the cost greatly.

Gate-oxide anti-fuse memory cell also has a very typical defect. That is the poor consistency of the programmed cell. The equivalent resistance of memory cell distribute from several hundred ohms to several hundred kilo ohms. The mega ohms situation exits, too. This defect increase the difficulties in designing the sense amplifier. The read speed of PROM is also facing a big challenge. If the equivalent resistance is too large, the circuit may even fail. The program consistency of gate-oxide anti fuse memory cell is awaiting for improve right now.

In this paper, we study on the mechanism and model of gate oxide break down and design test chips for experiment. The 2T memory cell's program consistency is carefully studied. We have also built a model for the programmed memory cell equivalent resistance. With this model, we assess the program consistency, read speed and subthreshold current of 2T memory cell. We perform our research on the basis of the whole memory cell, not only the anti-fuse device. The results can be directly used in PROM memory cell design application. We have also proposed a modified 3T memory cell. This structure consumes 18% more areas but it can improve the program consistency greatly.

## 2 Gate-oxide anti fuse PROM memory cell

### 2.1 Classic 3T memory cell

The most classic gate-oxide anti fuse memory cell is 3T cell. Most of the present mature gate-oxide anti-fuse PROM products are based on this structure. 3T cell consists of a N-type or P-type gate-oxide anti fuse, a high voltage block transistor and an address access transistor as showed in Fig. 1. High voltage transistor is used to protect the address access transistor during the program mode. In the design, all cell transistors are N-type, in this way higher density has been got. With the standard CMOS 180nm technology, the area of 3T cell is 4.2 $\mu m^2$. [4]

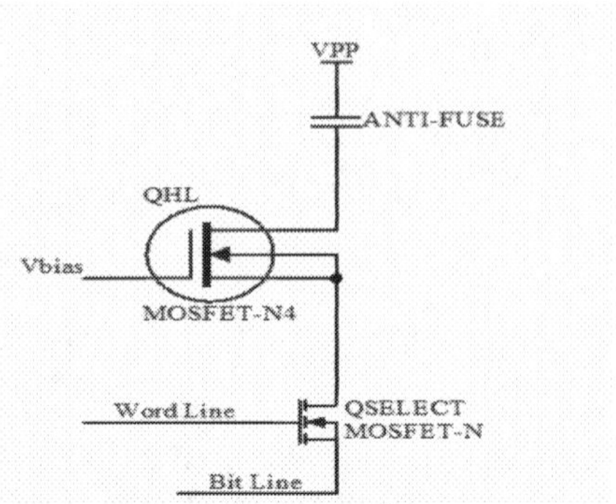


**Figure 1.** The structure of classic 3T cell

This structure has high reliability, the address access memory is well protected. But it has very cruel requirements on the size design of high voltage block transistor and address access transistor. What's worse, different technology requires different sizes. The design procedure consumes tremendous time and energy. The high voltage block is a device with thick oxide. This makes the cell consume many more areas. But very regretfully, the program consistency defect of gate-oxide anti-fuse cell has not been dealt with yet.

### 2.2 2T structure memory cell

Kilopass proposed the following 2T structure in Fig. 2. It consists of two identical N-type transistors. One is used as memory body, the other is the address access transistor. This structure reduces the design complexity substantially. At the same time, the cell's area and power consumption are also substantially reduced. In standard CMOS 180nm technology, the 2T cell we have designed consumes only 1.28 $\mu m^2$.

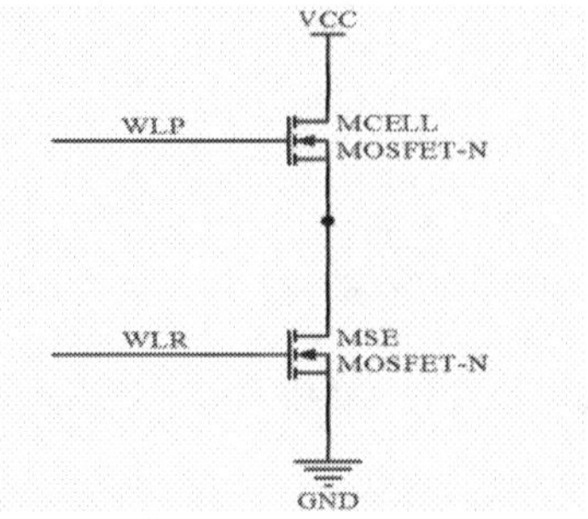


**Figure 2.** 2T memory cell

Compared to the classic 3T cell, the 2T cell has no high voltage block. During the program operation, high program voltage will break down the access transistor or have a negative effect on the access transistor's lifetime. So, we must avoid too high program voltage. The 2T cell also has the poor program consistency, which is the primary problem.

### 2.3 The choice of high density cell and research target

According to the results above, we have found that if the threat of high voltage can be resolved, the 2T cell is a very suitable choice for high density PROM. Therefore, we choose the 2T cell to do research on and consider the following two targets: (1) The failure situation of address access transistor. (2) The program consistency of the cell. We will obtain the best program of 2T cell after the study. On the basis of the 2T structure, we will trade off on the area consumption and modify the 2T cell structure to achieve better program consistency.

## 3 High density memory cell program performance study

### 3.1. Mechanism of gate-oxide anti fuse break down

When there are no defects in the $S_iO_2$ layer, the breakdown of the gate oxide is an avalanche breakdown. It corresponds to the impact ionization model. The original carriers accumulate enough energy due to the high electric field intensity. These carriers have impact on the lattice of $S_iO_2$ ,generating secondary carriers-pairs of hole and electron. The secondary carriers generate next generation carriers. Carriers increase greatly and the current magnify in exponent. When the voltage come to the critical voltage, there is a sudden change in current. The gate oxide is irreversibly broken down. [4]

But there are no $S_iO_2$ layers which have no defects. So the intrinsic breakdown of gate oxide is an impact ionization-thermal transmission process. In the recent experiment, the test current has deviated from the normal F-N current and direct tunneling current under the added electric field intensity. With the increase of the added intensity, the test current increases obviously, it proves that the impact ionization has already occurred. But the heat generated is not enough for the medium fusing. Hence, after removing the added electric field, the medium can come back to the normal shape. With the increase of the added electric field intensity, the conduct currents mainly come from the filiform tubes formed by all kinds of defects. The heat generated can make the filiform tubes fusing and form the permanent conduct tunnel. The fusing state is testified by experiment. [5]

According to the impact ionization-thermal transmission model, the gate-oxide breakdown is a localized breakdown. Different breakdown spot positions correspond to different models.

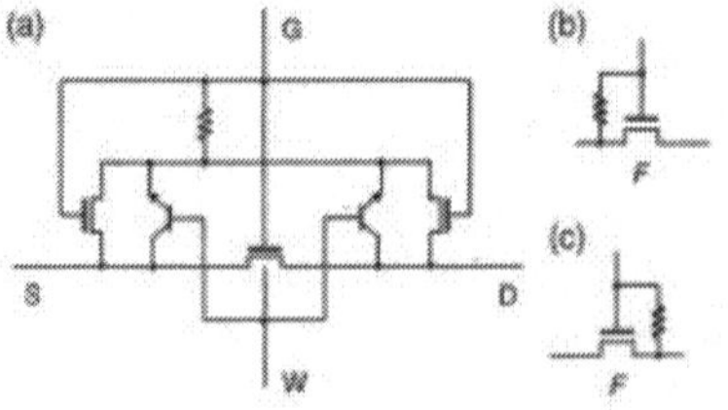


**Figure 3.** Three models of different breakdown spot position

From Fig. 3 [6], three mode are given:(a)the spot is in the channel(b)the spot is in the drain(c)the spot is in the source. From the cross-section of the programmed gate-oxide anti fuse, we can also find that the breakdown is a localized process.

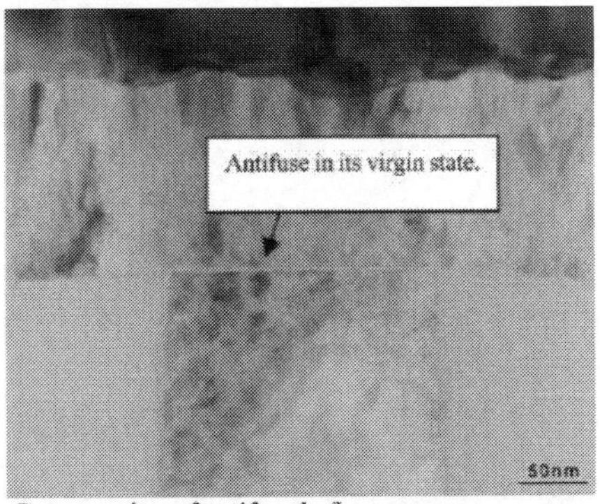


**Figure 4.** Cross-section of antifuse before program

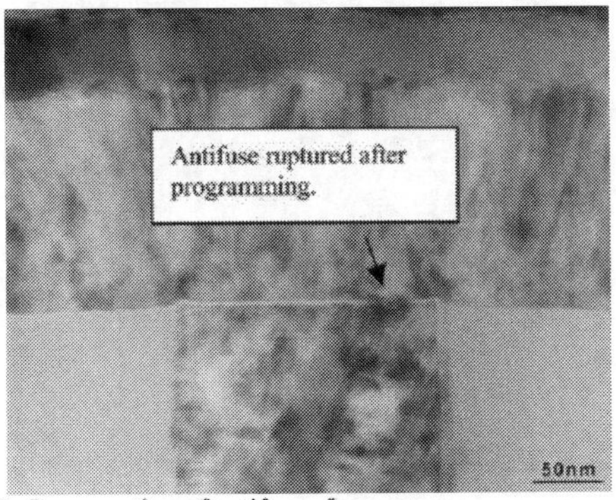


**Figure 5.** Cross-section of antifuse after program

The cross-sections in Fig. 4 and Fig. 5 certificate localized breakdown model and impact ionization-thermal transmission model [7]. The defects of the $S_iO_2$ layers are generated during the manufacture procedure. We can't orientate the defects' position right now. Therefore, we choose the program voltage as the main value and try the best to find the best program voltage. Further, we modify the 2T structure and consume more area. In this way, we have got a 3T cell structure, which improve the program consistency greatly.

### 3.2 The choice of program voltage

The program voltage is neither too high nor too low. The low voltage cannot break down the anti fuse or cost too much time. The high voltage may break down the address access transistor. So the program voltage has a feasible region: $[V_{min}, V_{max}]$.

Assume the programmed cell equivalent resistance is $R$, $R = \frac{V_{cc}}{I_{BL}}$, $V_{cc}$ is the supply voltage, and $I_{BL}$ is the current in the chosen Bit Line. Assume $\mu$ and $\sigma$ the mean and standard deviation of the programmed cell equivalent resistance. The mean value is used to describe the power consumption and read speed. The standard deviation value is to measure the program consistency. Obviously, the smaller $\sigma$ is, the better consistency is. Considering the read speed and power consumption, The value of $\mu$ is acceptable in $[10k\Omega, 30k\Omega]$. [8] We give the model to be solved as follow:

Known that:

$$\sigma = f(V_{pp}), \mu = g(V_{pp}), V_{pp} \in [V_{min}, V_{max}]$$

To get $V_{pp,opt}$, which makes

$$\sigma_{min} = f(V_{pp,opt}), \mu = g(V_{pp,opt}) \in [10k\Omega, 30k\Omega].$$

The voltage effect on the cell program feature is derivate from the defects of gate oxide layers. The number and position of the defect, the heat needed for fusing contribute to the different electric features of the breakdown gate oxide. To a typical technology, the distribution of the defects has statistic laws. Thus, we can find the best program voltage for the corresponding technology. In this way, we can get the best program features of memory cell in the typical technology.

### 3.3 Modified 3T structure

Based on the 2T structure, we have proposed a 3T cell structure showed in Fig. 6. This structure adds a identical parallel NMOS as the memory body. This change makes the anti fuse break down more easily. The program consistency has improved a lot. It also improve the reliability of the address access transistor. The area of the 3T cell is 1.51 $\mu m^2$ in standard CMOS 180nm technology. The 3T cell is still acceptable for high density design.

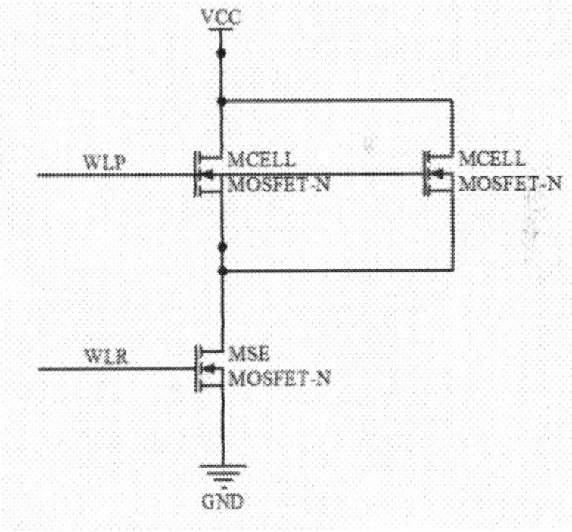


**Figure 6.** Modified 3T memory cell

## 4 Experiment design and results analysis

### 4.1 Preparation of test chips

The standard CMOS 180nm technology design rules are applied in the circuit design. To get better and more pure results of the cell program features, we have designed a 1x19 memory cell array showed in Fig. 7. It consists of nine 2T cells,6 modified 3T cells and 4 other cells for further study. The schematic is below. The chip is packaged in DIP28.

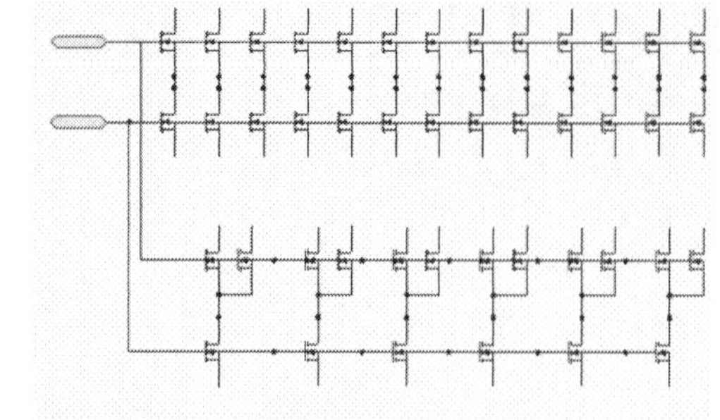

**Figure 7.** Schematic of test chip

### 4.2 Experiment scheme

(1) The set of the voltage value

The intrinsic gate oxide breakdown voltage is 5.8V [4] in 180nm standard CMOS technology. Considering the voltage drop in the test chip and the deviation, we choose 6V, 6.5V, 7V, 7.5V, 8V for study.

(2)Experiment procedure

1) 25 test chips in total. Read the cell memory before program, record the bit line current.

2) 5 test chips per voltage, do program operation on each chip and read the cell after program. Record the bit line current after program.

3) Analyze the test data and get the result.

### 4.3 Experiment results

#### *4.3.1 8V circumstance*

The 8V program voltage made the address access transistor break down in each chip.Thus,8V>$V_{max}$ .Due to the $\pm 0.25V$ deviation, we assumed that $7.75V \geq V_{max}$ .It shows that if the program voltage is too high, the address access transistor may be broken down and the circuits fail.

#### *4.3.2 6V circumstance*

Under the 6V program voltage, only 3 memory cell broke down in the total 5 chips. It was the early breakdown phenomenon which can be eliminated through screening. The result tells that $6V < V_{min}$ .Because of the $\pm 0.25V$ deviation, we assume that $6.25V \leq V_{min}$ .

#### *4.3.3 6.5V,7V,7.5V circumstances*

The three voltages all give the correct program results. Therefore, we got:

$$V_{min} < 6.5V < 7.5V < V_{max} \tag{1}$$

Due to $\pm 0.25V$ deviation, we assume that:

$$V_{min} \leq 6.25V < 7.75V \leq V_{max} \tag{2}$$

According to the results of 6V and 8V results, we can get the feasible region of program voltage: $[6.25,7.75]V$ .

After dealing with the data, the results are as follows: the distributions of equivalent resistances are showed in Fig. 8, 9 and 10.

(1)6.5V circumstance

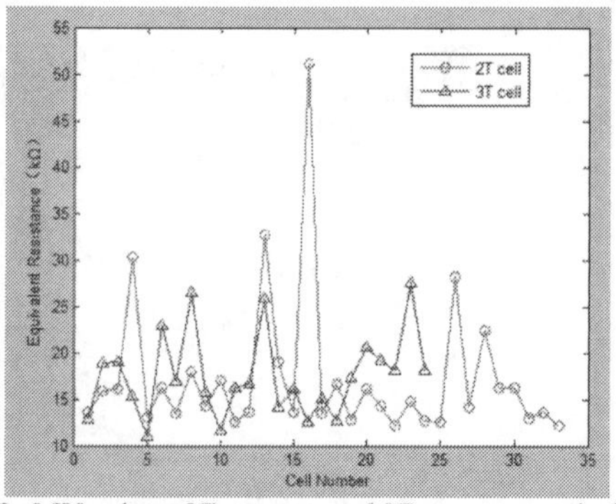


**Figure 8.** 6.5V voltage 2T structure and 3T structure equivalent resistances distribution

2T structure: $\sigma = 7.9079\text{k}\Omega, \mu = 17.3479\text{k}\Omega$

3T structure: $\sigma = 4.5402\text{k}\Omega, \mu = 17.5330\text{k}\Omega$

(2)7V circumstance

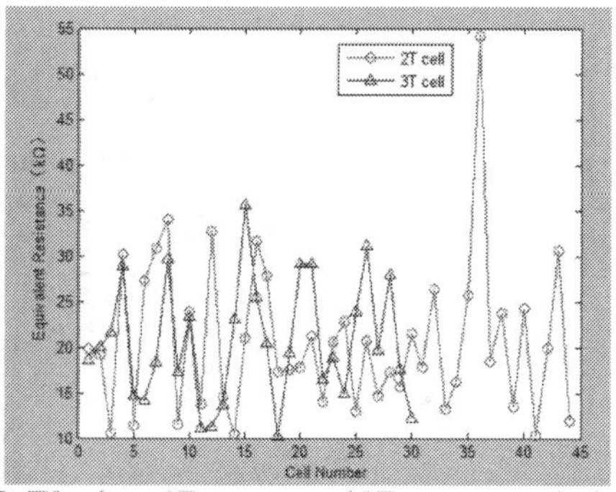


**Figure 9.** 7V voltage 2T structure and 3T structure equivalent resistances distribution

2T structure: $\sigma = 8.4118\text{k}\Omega, \mu = 20.7058\text{k}\Omega$

3T structure: $\sigma = 6.7020\text{k}\Omega, \mu = 20.6003\text{k}\Omega$

(3)7.5V circumstance

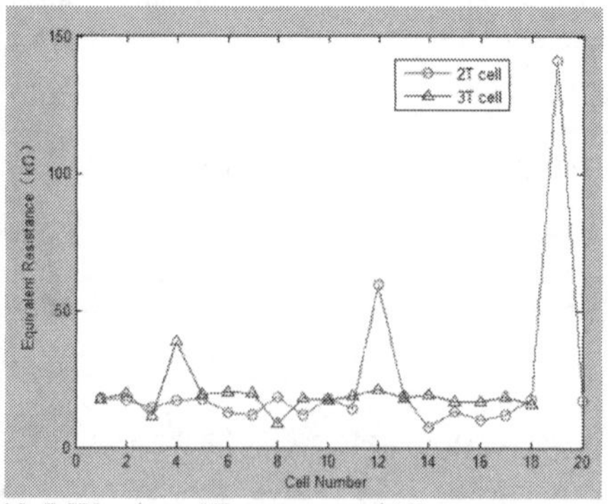


**Figure 10.** 7.5V voltage 2T structure and 3T structure equivalent resistances distribution

2T structure: $\sigma = 29.5480\text{k}\Omega, \mu = 23.5483\text{k}\Omega$

3T structure: $\sigma = 5.8086\text{k}\Omega, \mu = 18.8656\text{k}\Omega$

DOI: 10.1051/matecconf/20167510004

### 4.4 Results analysis

First of all, the feasible region of program voltage is $[6.25,7.75]V$.

Due to the $\pm 0.25V$ deviation, we can ensure $V_{pp,opt}$ from 6.5V,7V and 7.5V.Under the three program voltages, the mean values of 2T cell equivalent resistances are all in $[10k\Omega,30k\Omega]$, fulfilling the requirements. The parameter $\sigma$ which measures the program consistency is 7.9079 $k\Omega$, 8.4118 $k\Omega$, 29.5480 $k\Omega$ under 6.5V, 7V and 7.5V.It accords with the exponent model approximately: With the increase of program voltage, the cell program consistency performance decrease in exponent. The mean values of modified 3T cells are also in $[10k\Omega,30k\Omega]$.The largest value of 3T structure's $\sigma$ is 6.7020 $k\Omega$.The value is 15.3% less than the 2T's best situation. The 3T cell consumes 18% more area.15.3% is the worst case. The consistency can be improved 80.3% in 7.5V circumstance. Finally, we ensure $V_{pp,opt} = 6.5V$.

## 5 Conclusion

The program features of the 2T high density PROM gate-oxide anti-fuse memory cell have been studied in this paper. We have built a model for the programmed memory cell equivalent resistances. With the model, we have assessed the memory cell program consistency. The effect of the program voltage has also been researched. The result showed that lower program voltage made better cell program consistency in 2T cell. The best program voltage was 6.5V for 2T cell. After trading off between power consumption and read speed, the programmed cell equivalent resistance was between $[10k\Omega,30k\Omega]$. A modified 3T structure was proposed to improve the cell program consistency. The proposed cell consumed 18% more area. The standard deviation of the 3T cell equivalent resistances after program decreased 15.3% in the worst case. It could decrease 80.3% at most. The cell program consistency was improved greatly.

## References

1. R. Sam, J. J. Wang, Y.-M. Sun, et ak SET characterization and mitigation in RTAX-S antifuse FPGAs .Aero Conf IEEE, **1** (2009)
2. McCollum JohrL ASIC versus antifuse F] PGA reliability. Aero Conf IEEE, **1** (2009)
3. G. Wang, P. Li, W. li, G.-J. Zhang, X.-D. Xie, J. Jiang. Research and Application of Antifuse. The Introduction of Material A: Abstract part. **25** 6, (2011). 6:30-33.
4. H.-K. Cha, I. Yun, J.-B. Kim, B.-C. So, K.H. Chun, et al. A 32-KB Standard CMOS Antifuse One-Time Programmable ROM Embedded in a 16-bit Microcontroller. IEEE Journal of Solid-State Circuits, **41** 9, September (2006): 2115-2124.
5. X.-L. Hao, S.-Y. Yao, Y.-Q. Zhao, Impact ionization and breakdown in MOSFET gate oxide. Research & Progress of SSE. **24** 1, (2004) 2:15-19.
6. B. Kaczer, R. Degraeve, A. D. Keersgieter, K. Van deMieroop, V. Simons, and G. Groeseneken, Consistent modelfor short-channel nMOSFET after hard gate-oxide breakdown, IEEE Trans. Electron Devices, **49** 3, pp. 507–513, Mar. 2002.
7. F. Li, X.-Y. Yang, A. T. Meeks, J. T. Shearer, K. Y. Le, Evaluation of Antifuse in a 3D-OTP Memory, IEEE Transactions on Device and Materials Reliability, **4** 3, September (2004) 416-421.
8. H.-G. Qiu, Y.-L. Deng, G.-X. Pei, M. Du, X.-H. Li, J.-J. Peng, Simulation of the gate oxide thickness's effect on PROM performance, Bulletin of Science and Technology, **29** 8, Aug. (2013) 61-65.

# A Novel 3-Input AND/XOR Gate Circuit for Reed-Muller Logic Applications

Yuan Yang, Yinshui Xia and Libo Qian

*Faculty of Electrics Engineering and Computer Science, Ningbo University, 315211 Ningbo, Zhejiang, China*

**Abstract.** 3-input AND/XOR is the basic complex gate of Reed-Muller logic. Low energy consumption is important for Reed-Muller logic circuit implementation. Against the drawbacks of the published gate-level and transistor-level 3-input AND/XOR gate design in power and power delay product (PDP), a low energy consumption 3-input AND/XOR gate is proposed by employing multi-rails and hybrid-CMOS techniques to improve its speed and short the signal transimission path. Under 55nm CMOS process, post-simulations in different process corners are carried out by using HSPICE and compared with the published circuits. Simulation results show that the proposed circuit has advantages over published designs. For typical process corners, the improvement of the proposed circuit can be up to 27.21%, 19.23% and 35.39%, respectively, in terms of power, delay and power delay product.

## 1 Introduction

As the rapid development of design technology and CMOS process shrinking, the design concerns arise to comprehensive requirements of performance, area and power. Currently, almost all of the digital circuits are implemented by the traditional Boolean (TB) logic based on AND, OR and NOT. Actually, digital circuits can also be implemented by Reed-Muller (RM) logic based on AND/XOR. Studies show that statistically half of the circuits implemented by the RM logic can achieve better performance than those by TB logic. Moreover, compared to the TB logic, RM logic has the following advantages: firstly, RM logic circuit is easy to be mapped to the field programmable gate array (FPGA) and has good testability properties [1]-[2], which provide an effective way to solve the problem of "verification" in integrated circuit design; secondly, RM logic is much simpler in the implementation of some logic functions such as arithmetic components and parity function [3], which not only brings about smaller area, but also has the potential advantages of power and speed. However, RM logic was not widely used because of the absence of corresponding complex gate circuits and EDA tools. In recent years, some 3-input AND/XOR gates based on gate level and transistor level are published, but far from satisfaction in terms of power consumption, delay and Power Delay Product (PDP).

In this paper, a novel transistor level based 3-input AND/XOR gate design is proposed. Under 55 nm CMOS process, the post-simulations of the circuit under different process corners are carried out by HSPICE and compared with the published circuits. The simulation results show that the proposed circuit has advantages in power consumption and delay.

## 2 Published 3-Input AND/XOR gates

3-input AND/XOR is the basic complex gate of RM logic that performs the following equation:

$$AB \oplus C = \overline{AB}C + AB\overline{C} = \overline{AB} \oplus \overline{C} \quad (1)$$

XOR and OR, denoted by ⊕ and + respectively, are binary operations. According to the structure of 3-input AND/XOR, its design method can be divided into two paths: gate level and transistor level.

Gate level-based 3-input AND/XOR gate is implemented by cascading one AND (or NAND) gate and one XOR (or XNOR) gate. In recent years, researchers already place a high value on XOR-XNOR circuits because they are basic building blocks in various circuits especially in Arithmetic circuits, Parity Checks, Error-detecting.

Depending on the difference in XOR gates, various kinds of AND/XOR gate circuits are proposed in Fig. 1 (a)-(c). In Fig. 1 (a), a 3-input AND/XOR gate is implemented by cascading one NAND gate and one XOR gate in complementary CMOS structure [4-5], which consists of a pull-up PMOS network and a pull-down NMOS network. As a result, the gate circuit has a symmetrical structure and operates with full output voltage swing, but it requires significant transistors and high power consumption. In Fig. 1 (b), a 3-input AND/XOR gate [6] is implemented by cascading one AND gate and one XOR gate [7] in pass-transistor logic (PTL) structure. PTL allows the source terminal of the MOS transistor is connected to input line. Consequently, the number of transistors and the related parasitic capacitance are reduced. However, PMOS and NMOS have poor performance on transmitting the signal "LOW"

 DOI: 10.1051/matecconf/20167510005

and "HIGH", respectively, leading to threshold voltage loss at some nodes with certain input Combinations in PTL circuits. To make the XOR gate operate with full voltage swing at all nodes, a transmission gate (TG) is added in the XOR gate [8]. In addition, a static CMOS inverter is cascaded to improve the driving capability of PTL circuits. Based on TG logic, a 3-input AND/XOR gate is implemented by cascading one NAND gate and one XOR gate [9], as shown in Fig. 1 (c). The XOR gate is composed of two transmission gates and three static CMOS inverters. Thus, it has a full voltage swing at all nodes for all input patterns, so that it can be operated at lower supply voltage. Similarly, the inverters enhance the driving capability.

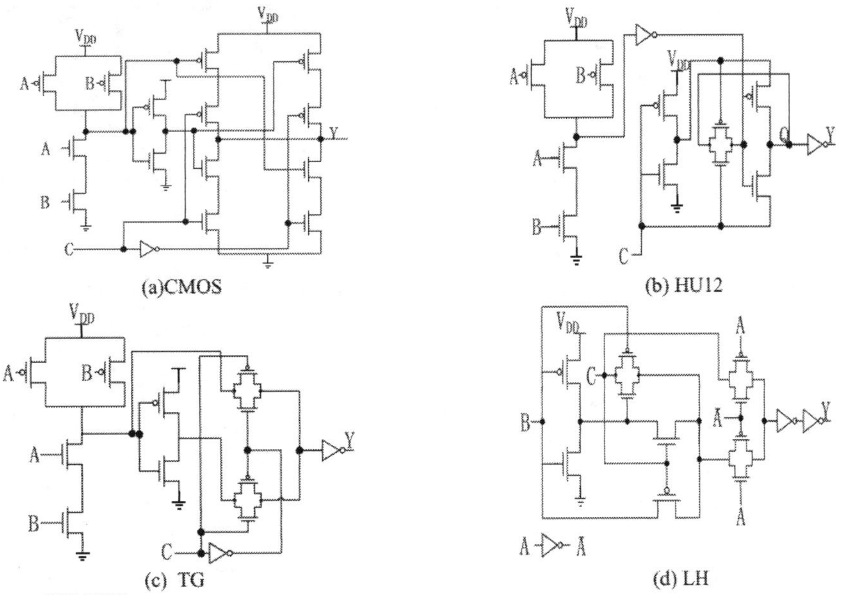

**Figure 1.** Published 3-Input AND/XOR gates

Based on transistor level, a 3-input AND/XOR gate [10] is designed as shown in Fig. 1 (d). It is composed of one XOR gate, two transmission gates and two static CMOS inverters. The input signal A and its complementary input signal control two transmission gates so that only one TG can work at the same time. This design has a full voltage swing at all nodes for all input patterns and negligible short-circuit power dissipation, which lead to lower power consumption. Furthermore, two static CMOS inverters are cascaded after the circuit to improve the driving capability.

Based on the discussion above, the published AND/XOR gates have different disadvantages, the complementary CMOS logic based 3-input AND/XOR gate circuit requires most numbers of MOS transistors and has a high power consumption; the PTL or TG based 3-input AND/XOR gate circuits have long critical paths and the more numbers of internal nodes consume extra power consumption; the transistor level based 3-input AND/XOR gate circuits has a long critical paths. Therefore, the performance of AND/XOR has yet to be further improved.

## 3 Proposed 3-Input AND/XOR gate

Based on hybrid-CMOS techniques, a 16T 3-input AND/XOR gate circuit is proposed, as shown in Fig. 2 (a). The circuit contains two substructures and has a full voltage swing at all nodes for all input combinations.

The basic function of the gate circuit is as follows. When AC=00, P1 and P3 turn on and pass a strong "HIGH" signal level to node Q and then a strong "LOW" to the output. When BC=00, P2 and P3 turn on and pass a strong "HIGH" signal level to node Q and then a strong "LOW" to the output. When AC=01, N1 and N2 turn on and pass a strong "LOW" to node Q and then a strong "HIGH" to the output. Under five input combinations above, only substructure 1 works. Due to the node Q is connected to power supply or ground, is has full voltage swing under lower supply voltage, and there is no direct path between power supply and ground, which reduce the power consumption.

When ABC=111, N3, N4 and N5 turn on and pass two weak "HIGH" to the node Q. In order to eliminate threshold voltage loss, the third transmission path, which include P4 and P5, is added, meanwhile, corresponding performance is also improved. When ABC=101, P4, P5, N3 and N5 turn on and N3 and N5 pass a strong "LOW" to the node Q and a strong "HIGH" to the output. When ABC=110, N3 and N4 turn on and pass a strong "LOW" to the node Q and a strong "HIGH" to the output. Under above three input combinations, only substructure 2 works, and all internal nodes have full voltage swing. Due to multi-rails of structure 2 when ABC=111 or 101, the equivalent resistance of transmission path is smaller, which improve the charge and discharge speed of output, then enhance the working efficiency of the circuit. The cascade inverter is used to strengthen the circuit driving capability. The proposed design is modular and simple, which simplify the layout and optimize chip area potentially. The layout of the proposed design is shown in Fig. 2 (b).

DOI: 10.1051/matecconf/20167510005

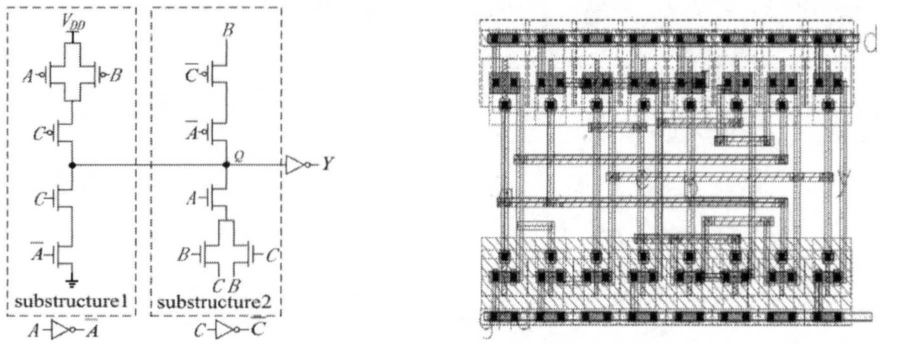


(a) The structure of the proposed AND/XOR gate

(b) The layout of the proposed AND/XOR gate

**Figure 2.** Proposed circuit

## 4 Simulation and comparison

The circuit simulation is carried out using HSPICE under 55nm CMOS process at 1.20V supply voltage. The operating frequency is 100MHZ. The gate area ratio of PMOS and NMOS are about 2:1 while the physical W/L sizes of PMOS and NMOS are 240nm/60nm and 120nm/60nm, respectively. To model the realistic environment, the input terminals are cascaded with two-stage inverters and the output drives are four parallel inverters. The simulation test bench used is shown in Fig. 3. All possible input combinations at the gate inputs are simulated. The transient waveforms are shown in Fig. 4. It can be seen that the proposed circuit has correct function for all of the input cases.

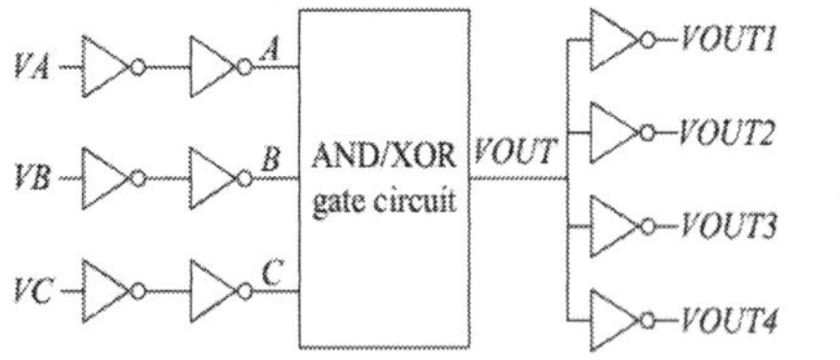


**Figure 3.** Test bench for three-input AND/XOR gates

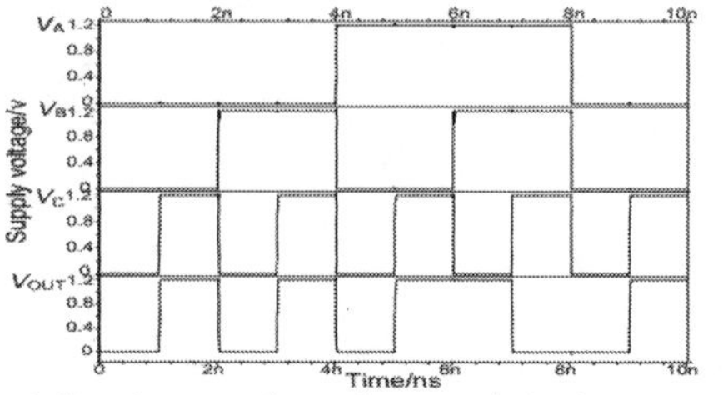


**Figure 4.** Transient waveform of proposed circuit

To compare the performance difference among the proposed gate and the published ones in Fig. 1, all of circuits are post-simulated under the same conditions. The average power consumption is measured during 100 operating cycles. The transient performances are characterized by 50% delay, which is defined as the time when the output voltage reaches the 50% of its steady state value. The PDP is used to evaluate the performance of the circuit. The improvements of power consumption, Delay and PDP are expressed by Power%, Delay% and PDP%, respectively. The comparison for circuits in Fig. 1 and Fig. 2 is shown in Table 1.

**Table 1.** The electrical performance comparison for different gate circuits

| Circuits | Power/μw | Delay/ns | PDP/μw•ns | Power% | Delay% | PDP% |
|---|---|---|---|---|---|---|
| CMOS | 2.46 | 0.22 | 0.54 | 19.51% | 4.54% | 22.22% |
| TG | 2.72 | 0.23 | 0.63 | 27.21% | 8.70% | 33.33% |
| HU | 2.50 | 0.26 | 0.65 | 20.08% | 19.23% | 35.39% |
| LH | 1.99 | 0.25 | 0.50 | 0.50% | 16.00% | 16.00% |
| Proposed | 1.98 | 0.21 | 0.42 | — | — | — |

As shown in Table 1, the proposed circuit has the lowest PDP and the improvement ratio of PDP can be up to 35.39%. The reason behind this is that the circuits like TG, HU and CMOS have cascaded structure, which increase the capacitance of internal nodes and power consumption and capacitance. Moreover, due to longer transmission path of cascaded structures, these circuits also have longer time delays. As the output of structure based on complementary CMOS is directly connected to power supply or ground, it has a good driving capability without the additional tailing inverter. For circuit LH, it has a larger delay because it needs pass the signal from the XOR to the transmission gate for some input patterns. In addition, the proposed circuit is the better and more competitive one.

In order to evaluate the capability of these circuits at different frequencies, the circuits are post-simulated at frequencies in the range from 100MHZ to 1400MHZ at 1.20V supply voltage. The curves of power and PDP versus operation frequency in the range from 100MHz to

1400MHZ are shown in Fig. 5(a) and (b). It can be seen that the proposed circuit can work reliably and has advantage in power and PDP.

The simulations are also carried out to check scaling capabilities of power supply voltages. Fig. 6 shows the curve of PDP versus supply voltage varying from 1.6V down to 0.8V under 100MHz. It can be seen that the proposed circuit has the best voltage scaling capability in terms of PDP.

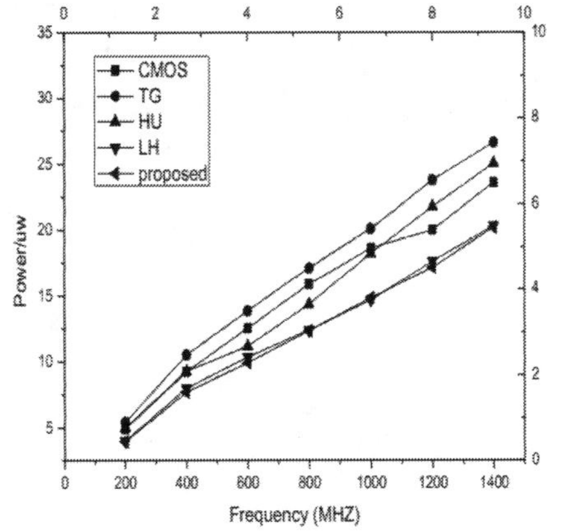


(a) Power consumption under TT process corner

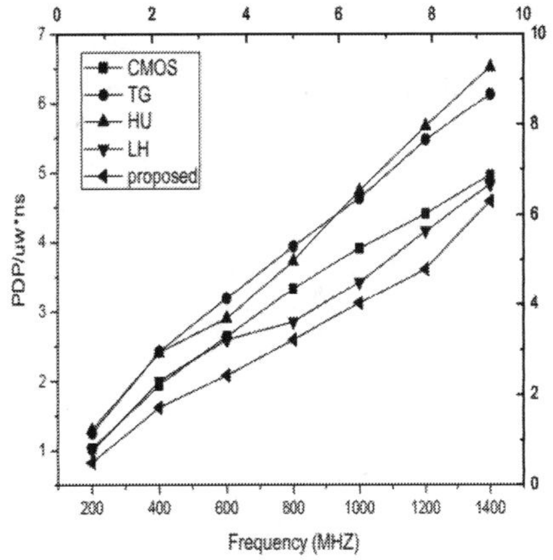


(b) PDP under TT process corner

**Figure 5.** Performance under TT process corner

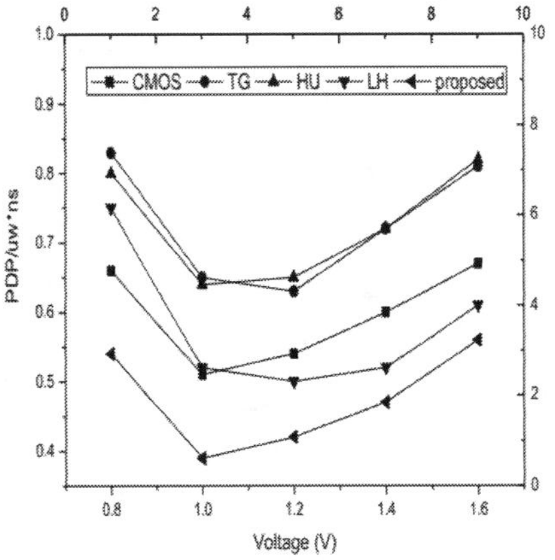


**Figure 6.** PDP versus supply voltage variatio

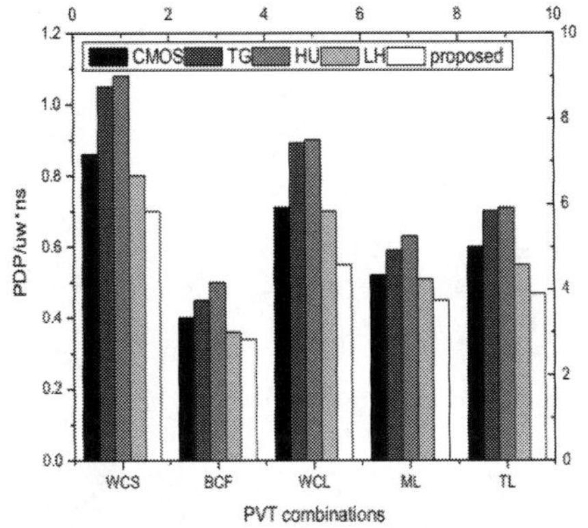


**Figure 7.** PDP under different PVT combinations

To further evaluate the effects of processes corners, voltage and temperature variations, post-simulations under different process corners are also carried out. Five standard STA conditions are chosen: (1) Worst Case Slow (WCS): SS process corner,125℃ and 1.08V; (2) Best Case Fast (BCF): FF process corner,-40℃ and 1.32V; (3) Worst Case Cold (WCL): SS process corner, -40℃ and 1.08V; (4) Maximal Leakage (ML): FF process corner, 125℃ and 1.32V; (5) TL (Typical Leakage): TT process corner, 125℃ and 1.20V. Simulation results are shown in Fig. 7. It is observed that the proposed gate circuit has the lowest PDP in various cases, demonstrating the advantage and feasibility in hybrid-CMOS logic.

## 5 Conclusion

In this paper, a transistor level 3-input AND/XOR gate is proposed. Multiple transimission paths are employed to elimate the threshold voltage loss and improve its performance. Moreover, in order to decrease its delay, the hybrid-CMOS techniques are used to short the signal transimission path. The 16T 3-input AND/XOR gate is post-simulated under 55nm CMOS process, and the results are compared with four published circuits in terms of power consumption, delay and PDP. Comparison shows the proposed AND/XOR gate has the lowest PDP compared with those published AND/XOR gates and the improvement ratio of PDP can be up to 35.39%. Moreover, the post-simulations under five different PVT combinations are also carried out and show that the proposed circuit has the lowest PDP.

## References

1. Y. S. Xia, K.Y. Mao, X.E. Ye, Journal of Computer-Aided Design & Computer Graphics, **19**, 1522-1527(2007)
2. L.Y. Wang, Y.S. Xia, X.X. Cheng, Journal of Computer-Aided Design & Computer Graphics, **24**,961-967 (2012)

3. M.H. Moaiyeri, R. F. Mirzaee, K. Navi, et al, International Journal of Electronics, **97**, 647-662(2010)
4. S. Purohit, M. Margala, IEEE Transactions on Very Large Scale Integration (VLSI) Systems, **20** ,1327-1331(2012)
5. S. Wairya, G. Singh, Vishant, et al, Proceedings of IEEE International Conference on Engineering ,1-7(2011)
6. J.P. Hu, Z.L. Li, Y.S. Xia. Energy Education Science and Technology Part A: Energy Science and Research, **30**, 85-92(2012)
7. N. Zhuang, H.M. Wu. IEEE Journal of Solid-state Circuits, **27**,840-844(1992)
8. S.S. Mishra, A.K. Agrawal, R.K. Nagaria. International Journal on Emerging Technologies, **1**,1-10(2010)
9. S. Nishizawa, T. Ishihara, H. Onodera. Proceedings of IEEE 14th International Symposium on Quality Electronic Design,703-708 (2013)
10. H. Liang, Y.S. Xia, L.B. Qian, et al, Journal of computer-Aided Design & Computer Graphics, **27**, 940-945 (2015)

# AUTHOR INDEX

# AUTHOR INDEX